037599

Feeding in Domestic Vertebrates

From Structure to Behaviour

Feeding in Domestic Vertebrates

From Structure to Behaviour

Edited by

V. Bels

Professor
Muséum National d'Histoire Naturelle
Paris, France

www.cabi.org

CABI Publishing is a division of CAB International

CABI Publishing
CAB International
Wallingford
Oxfordshire OX10 8DE
UK

CABI Publishing
875 Massachusetts Avenue
7th Floor
Cambridge, MA 02139
USA

Tel: +44 (0)1491 832111
Fax: +44 (0)1491 833508
E-mail: cabi@cabi.org
Website: www.cabi-publishing.org

Tel: +1 617 395 4056
Fax: +1 617 354 6875
E-mail: cabi-nao@cabi.org

A catalogue record for this book is available from the British Library,
London, UK.

A catalogue record for this book is available from the Library of Congress,
Washington, DC.

ISBN-10:   1-84593-063-0
ISBN-13: 978-1-84593-063-9

Typeset by SPi, Pondicherry, India.
Printed and bound in the UK by Biddles Ltd, King's Lynn.

Contents

Contributors

C. Agreil, *INRA – Unité d'Ecodéveloppement, Site Agroparc – Domaine Saint Paul, 84914 Avignon Cedex 9, France.*

R. Baumont, *INRA – Unité de Recherches sur les Herbivores – Centre de Clermont/Ferrand-Theix, 63122 St-Genès-Champanelle, France.*

S. Baussart, *Muséum Nationale d'Histoire Naturelle rue Paul Pastur 11, B – 7800 – Ath, France.*

V. Bels, *FRE 2696/USM 302 – Adaptation et Évolution des Systèmes Ostéo-Musculaires, Département Écologie et Gestion de la Biodiversité, Muséum National d'Histoire Naturelle, 57, Pavillon d'Anatomie Comparée, BP 55, rue Buffon, 75005 Paris, France.*

M. Benvenutti, *INTA EEA Cerro Azul, C.C. 6 (3313) Cerro Azul, Misiones, Argentina.*

T.S. Brand, *Animal Production Institute: Elsenburg, Department of Agriculture: Western Cape, Private Bag X1, Elsenburg 7607, South Africa.*

A.W. Crompton, *Museum of Comparative Zoology, Harvard University, 26 Oxford St, Cambridge, MA 02138, USA.*

I.C. de Jong, *Animal Sciences Group of Wageningen UR, Division Animal Resources Development, Research Group Animal Welfare, PO Box 65, 8200 AB Lelystad, The Netherlands.*

M. Doreau, *INRA – Unité de Recherches sur les Herbivores – Centre de Clermont/Ferrand-Theix, 63122 St-Genès-Champanelle, France.*

L. Dybkjær, *Danish Institute of Agricultural Sciences, Department of Animal Health, Welfare and Nutrition Research Centre Foulum, PO Box 50, DK-8830 Tjele, Denmark.*

J.M. Forbes, *School of Biology, University of Leeds, Leeds LS2 9JT, UK.*

H. Fritz, *CNRS – Centre d'Etudes Biologiques de Chizé UPR 1934 du CNRS, 79360 Beauvoir-sur-Niort, France.*

R.Z. German, *Department of Physical Medicine and Rehabilitation, Johns Hopkins Hospital, Phipps 174, 600 N Wolfe St, Baltimore, MD 21205, USA.*

T. Gidenne, *INRA, Centre de Toulouse, Station de Recherches Cunicoles, BP 27, 31326 Castanet-Tolosan, France.*

P.C. Glatz, *Livestock Systems, South Australian Research and Development Institute, Roseworthy Campus, Roseworthy, South Australia 5371, Australia.*

I.J. Gordon, *CSIRO – Davies Laboratory, PMB PO Aitkenvale, Qld 4814, Australia.*

R.M. Gous, *Animal and Poultry Science, University of KwaZulu-Natal, Pietermaritzburg 3200, South Africa.*

S.W.S. Gussekloo, *Division of Anatomy and Physiology, Department of Pathobiology, Faculty of Veterinary Sciences, Utrecht University, Yalelaan 1, 3584 CL Utrecht, The Netherlands.*

S.W. Herring, *Department of Orthodontics, Box 357446, University of Washington, Seattle, WA 98195, USA.*

K.A. Houpt, *Cornell University, College of Veterinary Medicine, Ithaca, New York 14853-6401, USA.*

S. Ingrand, *INRA – Transformation des systèmes d'élevage – Centre de Clermont/Ferrand-Theix, 63122 St-Genès-Champanelle, France.*

B. Jones, *Roslin Institute, Roslin BioCentre, Midlothian EH25 9PS, Scotland, UK.*

P. Koene, *Ethology Group, Department of Animal Science, Wageningen University, PO Box 338, 6700 AH Wageningen, The Netherlands.*

F. Lebas, *Cuniculture, 87a Chemin de Lassère, 31450 Corronsac, France.*

M. Meuret, *INRA – Unité d'Ecodéveloppement, Site Agroparc – Domaine Saint Paul, 84914 Avignon Cedex 9, France.*

Z.H. Miao, *Livestock Systems, South Australian Research and Development Institute, Roseworthy Campus, Roseworthy, South Australia 5371, Australia.*

B.L. Nielsen, *Danish Institute of Agricultural Sciences, Department of Animal Health, Welfare and Nutrition Research Centre Foulum, PO Box 50, DK-8830 Tjele, Denmark.*

W. Pittroff, *Department of Animal Science, University of California, 1 Shields Ave, Davis, CA 95616, USA.*

T.E. Popowics, *Department of Oral Biology, Box 357132, University of Washington, Seattle, WA 98195, USA.*

F.D. Provenza, *Department of Forest, Range and Wildlife Sciences, Utah State University, Logan, UT 84318-5230, USA.*

Y.J. Ru, *Danisco Animal Nutrition, 61 Science Park Road, The Galen #06-16 East Wing, Singapore Science Park III, Singapore 117525.*

P. Soca, *Facultad de Agronomía, Universidad de la Republica, Ruta 3, km 363, CP 60000, Paysandú, Uruguay.*

A.J. Thexton, *Division of Physiology, Biomedical Sciences, King's College, Strand, London, UK.*

K. Thodberg, *Danish Institute of Agricultural Sciences, Department of Animal Health, Welfare and Nutrition Research Centre Foulum, PO Box 50, DK-8830 Tjele, Denmark.*

I. Veissier, *INRA – Unité de Recherches sur les Herbivores – Centre de Clermont/Ferrand-Theix, 63122 St-Genès-Champanelle, France.*

E.-M. Vestergaard, *Danish Institute of Agricultural Sciences, Department of Animal Health, Welfare and Nutrition Research Centre Foulum, PO Box 50, DK-8830 Tjele, Denmark.*

J.J. Villalba, *Department of Forest, Range and Wildlife Sciences, Utah State University, Logan, UT 84318-5230, USA.*

Preface

Domestication of vertebrates has been based on the understanding of needs of animals in their natural environment and the success of this domestication throughout human history was mainly and is largely dependent on the knowledge of their feeding behaviour. The captive environment can be highly different from the natural habitat, and genetic selection has produced morphological, physiological and behavioural traits for improving all nutrition aspects playing a key role in animal maintenance, welfare, growth and reproduction. Since the early time of domestication, the control of animal nutrition is probably critical in the success of animal production, which can be described in terms of welfare of the animals, their speed to produce significant quantity of meat and their ability to have more offspring. This success cannot be effective without a complete understanding of the feeding behaviour that permits an animal to select and intake the food material providing responses to complex physiological demands. For feeding, animals must always control their movements and their postures. They have to respond to availability or distribution of food in an artificial environment. This environment is often characterized by absence of predators, but also often by the presence of a large number of conspecifics. Food is provided at predictable, but relatively limited, locations, which can drastically increase the stress that must be managed by each individual for obtaining nutrients and water.

This volume has been constructed around the complex relationships between structures permitting to produce a series of performances that shape a determined feeding behaviour. When I suggested compiling this volume with the primary help of a series of contributors, I emphasized that rather different approaches provide a lot of data that could be reviewed to open interesting perspectives in our knowledge of feeding behaviour in two main types of domestic vertebrates, birds and mammals. On the one hand, the feeding behaviour is deeply studied for quantifying, determining and simulating the feeding mechanisms involving not only the trophic structures such as jaws and tongue but also all other body structures such as limbs and neck. These mechanisms permit to understand the functional relationships between the properties of the structures and their performances. But we always have in mind that properties of these structures can be affected by other unknown selective forces. On the other hand, the feeding behaviour itself is deeply observed and quantified for understanding and modelling the complex relationships between the food and the behavioural and physiological responses of animals. These responses are strongly dependent on the sensory processes and their integration into the neural system. They also depend on environmental conditions and are strongly influenced by 'internal' factors that could be defined as the 'psychophysiological' state of each animal. Finally, it is clear that foraging and feeding behaviour directly influence a lot of characteristics of the environment exploited by determined species.

I am grateful to all the contributors of this volume. All of them provided highly valuable chapters to improve our knowledge of feeding function in birds and mammals. They presented different conceptual and methodological approaches that help to integrate all aspects of feeding in domestic vertebrates, from structures to their behaviour and ecological impact in the ecosystems. Probably this volume is one of the few compilations where highly different biological disciplines, from functional morphology to behavioural ecology, have been put together.

It is now time to thank colleagues and friends for their help and support. First, I would like to thank all the contributors for their participation in this volume. Their contributions will provide highly valuable ideas that can inspire all readers, from students to professionals, in animal production. Also, a majority of the contributors have also participated with a lot of other referees to add a scientific value to this volume by their help and precious advice to the editor. I would particularly thank H. Berkhoudt, E. Decuypere, A.W. Illius, K. Kardong and I. Maertens for their help. Second, I particularly wish to express my gratitude to J.M. Forbes who not only supported the idea of this volume but also accepted to write the conclusion. He opens clear perspectives that highlight the need of such integrated framework to understand the feeding behaviour in domestic vertebrates. Third, I am very grateful to many colleagues for support during the production of this volume. I would particularly thank S. Baussart who provided much help at different steps of the production of this volume.

Finally, I strongly wish to dedicate this volume to my family for their encouragement and patience along the long way to creating this volume.

V. Bels

1 Introduction

V. Bels

FRE 2696/USM 302 – Adaptation et Évolution des Systèmes Ostéo-Musculaires, Département Écologie et Gestion de la Biodiversité, Muséum National d'Histoire Naturelle, 57, Pavillon d'Anatomie Comparée, BP 55, rue Buffon, 75005 Paris, France

Introduction

The primary concern of all animals is to find and ingest food to cover their dietary needs. Feeding is defined as 'a continuous activity which, in many higher animals, is interrupted by period of non-feeding' (Forbes, 2000). Evolutionary pressures have constructed efficient, rapid and adjustable series of movements of morphological structures that permit the gain of nutriments and energy necessary for fitness of the animals. In vertebrates, the diversification of this activity underlying the success of nutritive processes has played a key role in animal ecological diversity, although the feeding system represents only modifications of the same basic set of homologous skeletal structures either directly connected by articulations or by contact with soft tissues. Therefore, feeding efficiency in the two main lineages of domestic animals, birds and mammals, results from the strong relationship between structures, performances, behaviour and fitness (Fig. 1.1).

During the last 150 million years, birds have developed a highly specialized jaw apparatus for food capture. The majority of domestic birds use their jaws to gather, strip and manipulate pieces of food of different sizes that are then thrown down the throat. From a functional point of view, birds have effectively lost the chewing component of the amniote jaw mechanism probably present in their ancestors

(Zusi, 1984, 1993). The trophic system, from braincase to jaw apparatus, shows specialized features described by Gussekloo (Chapter 2, this volume). Accordingly, drastic changes in the morphology of this system produce a highly kinetic skull, characterized by a great versatility of the jaws, to be a strong manipulative tool not only in feeding but also in a series of behaviours such as grooming and nest building (Bels and Baussart, Chapter 3, this volume). In addition, the kinetic skull can produce large gapes with a relatively small jaw depression (Bock, 1964; Zusi, 1993). Therefore, the jaw apparatus can be seen as an optimal system for food catching and delivering this food material into the oesophagus, which stores swallowed food until it can enter the stomach for processing. In domestic birds, this apparatus does not play any large role in food oral transformation. Birds 'chew' with the gizzard, which is a specialized portion of the stomach.

Approximately 120 million years ago in the reptilian lineage that became mammals, a series of unique properties arose in the feeding structures (Hiiemae, 2000). First, the shapes and joints of the jaws changed so that they could be used to break down food unilaterally and bilaterally in relationship with food position in the mouth. Second, the novel jaw joint and organization in adductor muscles allowed transverse latero-medial movements of the lower jaw against the upper jaw. Third, the

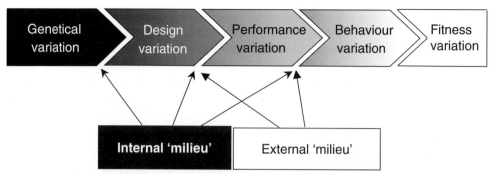

Fig. 1.1. The relationship between structures, performances, behaviour and fitness follows a conservative pattern within each lineage of vertebrates. Measure performances are strongly influenced by the internal and external milieux. (See Provenza and Villalba, Chapter 13, this volume.)

teeth of mammals were subjected to radical changes along their evolutionary history. Briefly, the upper and lower molars changed to tribosphenic molars, with complex surfaces and cusps that fit together in a dynamic way during occlusion occurring as soon as the animal closes the jaws, and therefore produce a chewing stroke (Popowics and Herring, Chapter 5, this volume). All these morphological novelties can be associated to novel feeding performances, and specialization in these performances for a particular type of food is empirically reflected by strong morphological and functional adaptations in the trophic structures involved in the selection, processing and digestion of the food. However, recent studies in ungulates demonstrated that the ability to exploit various food resources primarily results from differences in body mass between species (Pérez-Barbería and Gordon, 2001).

The set of structures (designs) used in the feeding behaviour can be roughly divided into trophic structures and postcranial structures including the axial and appendicular skeletons. The postcranial contribution to the efficiency of feeding behaviour is often neglected. Feeding postures probably play a key role in animal well-being during feeding and drinking (Bels and Baussart, Chapter 3; Koene, Chapter 6; Nielsen *et al.*, Chapter 10, this volume). Specialized structures of the trophic system perform a rather complex series of movements for exploiting different available resources of the environment (Gussekloo, Chapter 2; German *et al.*, Chapter 4; Popowics and Herring, Chapter 5, this volume). Neuromotor control

of the association between trophic, axial and limb movements depends on multiple factors, which must act simultaneously at different levels of the neural system. For example, grazing in herbivores such as cattle and sheep includes an integration of movements of locomotor, neck and trophic systems (Baumont *et al.*, Chapter 14; Agreil *et al.*, Chapter 17, this volume). Feeding postures described in birds also result from integrated movements of all these body structures (Bels and Baussart, Chapter 3, this volume). It is clear that this integration will play a key role in efficiency of food prehension.

Movements of structures are described in terms of performances that together constitute a complex feeding behaviour strongly constrained by the 'internal' and 'external' milieux (Fig. 1.1; Provenza and Villalba, Chapter 13, this volume). The external milieu is of course characterized by its complex physical properties (i.e. palatability, but see Forbes, Chapter 19, this volume). Feeding in water and feeding in air are subject to completely different constraints. There are also strong differences between birds and mammals as they show completely different cranial and postcranial structures (Gussekloo, Chapter 2; Popowics and Herring, Chapter 5, this volume). Various types of feeding have been described in aquatic vertebrates. For example, filter feeding permits the animal to obtain food suspended in the water column, and suction and ram feeding permit the animal to obtain food of different sizes in the water column (Lauder, 1985, 2000). In contrast, food-mass feeding is the only feeding behaviour used by vertebrates in terrestrial environments. It permits the animal to obtain food by exploiting either other whole organisms or pieces of these organisms.

Basically, feeding behaviour can be divided into a series of motor actions, called performances, of the hard and soft tissues determining structures that result in satiation of a large number of physiological demands. Satiation obviously refers to a decline in the consumption of an ingestive stimulus (McSweeney and Murphy, 2000). Three main constraints limit the ability of animals to meet these physiological demands (Provenza and Villalba, Chapter 13, this volume): metabolic, digestion and ingestion constraints.

The behaviour is often divided into four phases based on the functional relationship between performances of the structures and the food material (Schwenk, 2000). Each phase is regulated by different 'internal' (physiological) and 'external' (environmental) demands. The first phase called ingestion or prehension (intake) permits the animal to catch the food from the environment. Food prehension has been the subject of a lot of theories (e.g. glucostatic theory, thermostatic theory, energostatic theory). Based on the observation that animals tend to minimize the discomfort of over- or undereating some nutriments, a 'Minimal Total Discomfort' theory, strongly related to animal experience integrating food preferences and aversions, has been more recently proposed (Forbes, 1999, 2000; Forbes and Provenza, 2000). All these theories are based on our actual knowledge of the complex processes in the brain. They need to integrate a large number of parameters of the food (i.e. taste, odour, texture), physiological and cognitive factors that not only differ between species but also are largely variable from one individual to another. They also must integrate the physiological and physical constraints of the trophic system used for food intake. Two main types of food prehension have been reported in terrestrial vertebrates: jaw prehension and lingual prehension (Schwenk, 2000). In domestic birds, jaw prehension is probably the method of food intake (Gussekloo, Chapter 2; Bels and Baussart, Chapter 3, this volume). In domestic mammals, jaw prehension described as *bite* (Baumont *et al.*, Chapter 14; Agreil *et al.*, Chapter

17, this volume) is the basic food prehension motor pattern, but the tongue can largely increase the success of food intake (Fig. 1.2).

The second phase is called intraoral transport, corresponding to displacement of the food within the oral cavity (Hiiemae, 2000). In mammals, transport is divided into stage I and stage II (Fig. 1.3). Schwenk (2000) notes that this phase involves transport jaw and tongue cycles, which can be interlaced with deglutition (swallowing) cycles in mammals. The third phase is called processing, corresponding to food intraoral transformation. In mammals, this phase is called mastication. Food transport and mastication probably occur simultaneously, as shown by the analysis of the masticatory process of hay and sugarbeet by cattle (Fig. 1.4). The last phase is called pharyngeal emptying, and corresponds to swallowing or deglutition, which is, for Schwenk (2000), restricted to mammals. The swallowing cycle can be intercalated between the masticatory and transport cycles (Hiiemae, 2000).

Conservative Motor Pattern

In all these feeding phases, the structures undergo a set of cycles corresponding to coordinated motor actions of the trophic elements. These actions determine a series of performances that together form the complex behaviour. Behaviour can be viewed as a 'filter' between performances and fitness. Performances are the measured level of a given property of one or several motor actions produced by a complex set of muscular contractions (Domenici and Blake, 2000). For example, displacements and speed of mandibles in the vertical and lateral directions during the masticatory process of plant material by ruminants are typical performances explaining the intrabuccal transformation of the food before swallowing (i.e. Baumont *et al.*, Chapter 14, this volume).

Functional analyses of performances during feeding cycles have been used for providing a generalized pattern of neuromotor control of feeding behaviour in tetrapods. Two main models of the feeding cycle describing the motor control of the basic element of the

Fig. 1.2. Comparative jaw and tongue actions during prehension of sugarbeet in cattle.

food process have been suggested (Bramble and Wake, 1985; Reilly and Lauder, 1990; Lauder and Gillis, 1997). These models attempted to identify the pleomorphic characteristics of the feeding cycle in tetrapods (and in amniotes). One key characteristic is that 'tongue intraoral food processing' is one innovation of the feeding cycle in vertebrates invading the terrestrial environment (Lauder and Gillis, 1997). Basically, these models are constructed on the idea of their neural control. Two kinds of controls are now suggested in the literature: 'central pattern generator' (CPG) and 'swallowing center' (Jean, 1984), located in the pontine-medullary region of the hindbrain. However, recent studies on neural mechanism controlling these movements in infant mammals remain to be deeply analysed to propose a general neuromotor control of the feeding cycles in mammals (German and Crompton, 1996; German *et al.*, Chapter 4, this volume). Furthermore, feeding cycles cannot be studied without checking the coordina-

tion between swallowing and respiratory functions. In both models, the movements and motor control of two main trophic structures are integrated: hyobranchium (mainly tongue) and jaw apparatus. Respective movements of lower and upper jaws of this last apparatus are rather different in mammals (Schwenk, 2000), and kinesis in birds modulates movement of the upper bill (Gussekloo and Bout, 2005). Relative displacements of upper and lower jaws produce the gape angle, which is one basic element in both models (Fig. 1.5). Gape involves the same pattern divided into four stages: slow opening (SO), fast opening (FO), fast closing (FC) and slow closing (SC)/power stroke (PS). But splitting jaw opening into slow and fast stages is not always clear in all tetrapods (Hiiemae, 2000; McBrayer and Reilly, 2002). The tongue follows a forward/backward cycle in the mouth with beginning of retraction occurring at the change of the slope of jaw movements from SO to FO or during the early time of the FO stage

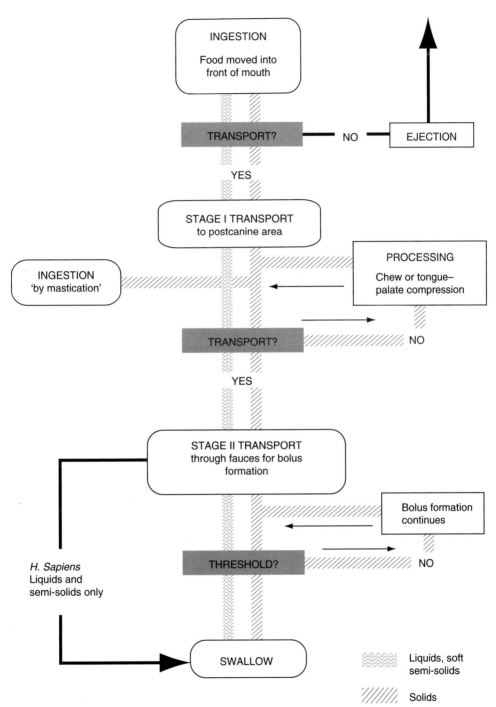

Fig. 1.3. Process model in mammals showing the complexity of the feeding sensory motor pattern (from Hiiemae, 2000). This model includes the process of two kinds of food. The shaded boxes indicate the distinction between mechanical elements of the model and postulated sensorimotor 'gates'.

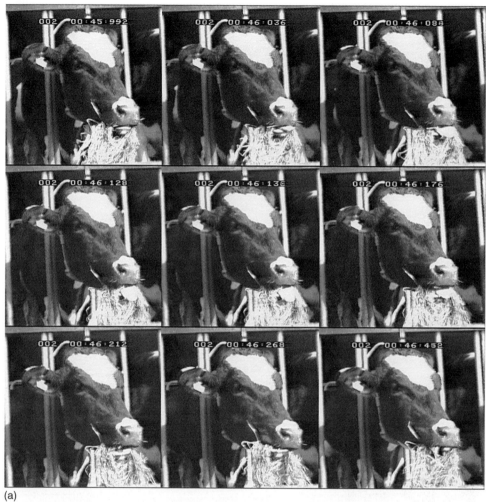

(a)

Fig. 1.4. During mastication in cattle, the food is transported from the front of the mouth to the pharyngeal cavity and reduced between the teeth simultaneously. This complexity in the processing of the food is controlled by movements of the tongue within the oral cavity. Tongue movements assure in the mean time positioning of the food between upper and lower row of teeth and its posterior displacement within the oropharyngeal cavity. (a) Hay.

(Hiiemae, 2000; Bels and Baussart, Chapter 3; German *et al.*, Chapter 4, this volume; see also Baumont *et al.*, Chapter 14, this volume). Some differences between both models have been emphasized by several authors working on the feeding cycle of various vertebrates, from food intake to swallowing (McBrayer and Reilly, 2002). One difference concerning the presence of the SO stage II in the model of Bramble and Wake (1985) is shown in

Fig. 1.5. During this stage the tongue is 'fitted' to the food item producing the 'plateau' recorded in the gape profile. In the model of Reilly and Lauder (1990), tongue adhesion to the food can simply occur during the regular increase of the gape called SO stage. Although some differences occur, these models can be used for predicting properties of jaw and tongue movements during food intake, transport and mastication, and swallowing in all domestic

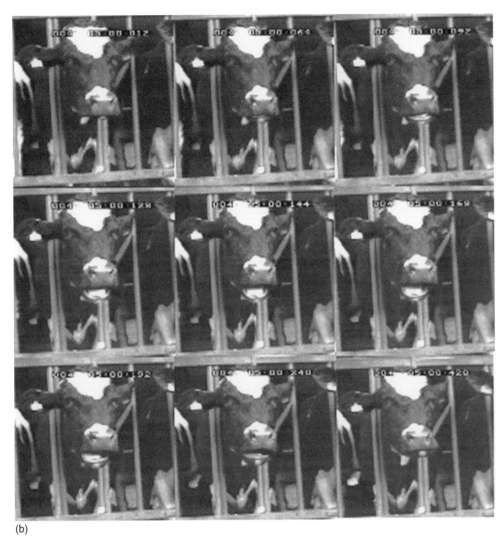

(b)

Fig. 1.4. *Continued* (b) Sugarbeet.

vertebrates, including birds and mammals. It is clear that both models can be used for determining the variability of individual responses to various foods with different properties (Fig. 1.6).

Each stage of the feeding cycle can be regulated as demonstrated by two examples in this volume. First, the majority of studies of feeding strategies and mastication of herbivores are based on the study of characteristics of 'bites'. In cattle, a difference between bites related to food prehension with low amplitude and high frequency and mastication with high amplitude and low frequency has been demonstrated (Baumont *et al.*, Chapter 14, this volume). Several categories of bites have been used for understanding feeding strategies in various environments by herbivores (Agreil *et al.*, Chapter 17, this volume). It would be pertinent to analyse the modulation of the motor control of each stage from these categories of bites to show how animals are able to regulate their feeding and efficiency in relationship with environmental constraints.

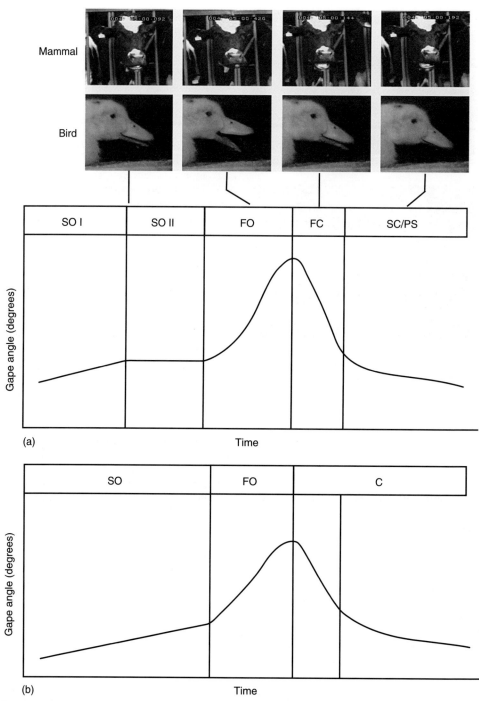

Fig. 1.5. Classical models to explain the neuromotor control of feeding behaviour in tetrapods. (a) Model of Bramble and Wake (1985). (b) Model of Reilly and Lauder (1990).

(a)

(b)

Fig. 1.6. Ruminants are able to manipulate food within the buccal cavity by tongue action. The food material can be reduced on one side of the jaw apparatus or on both sides. (a) Typical frame of crushing a piece of sugarbeet by adult cattle. (b) Typical frame of hay reduction on both sides of the jaws by adult cattle.

An Integrative Approach

One key feature of feeding in farm animals is that they meet different environmental conditions. First, they do not have to spend a lot of their time and energy to find food because food availability is controlled by humans and provided *ad libitum* or regulated for various purposes (de Jong and Jones, Chapter 8, this volume). Second, they have to regulate time and energy to intake food (i.e. grazing in herbivores and feeding in free-range systems). Of course, keeping conditions strongly influence the feeding responses of these animals to their demand. In all cases, food accessibility is organized to reduce competition between individuals for food and water and to favour their welfare and growth. Consequently, feeding strategies are not only regulated by animals themselves along their intake behaviour but

are always manipulated by humans in charge of animal production. Since the early time of modern animal production, feeding controls have been proposed not only for each type of farm animal but also for each age and physiological state of these animals. In these controls, three main factors emerge: animal physiology, animal welfare and environmental constraints created by producers. The literature is full of references proposing specific feeding controls or regimes for all kinds of domestic herbivores and omnivores. Several aspects of the relationship between feeding behaviour and human control have been deeply treated in a majority of chapters of this volume to show the strong interaction between the 'internal' (physiological and welfare demands) and the 'external' (environmental constraints) milieux (see Provenza and Villalba, Chapter 13, this volume). Herbivores such as cattle and sheep must gather various foods with different flavours at the bunk or on the pasture (Baumont *et al.*, Chapter 14; Gordon and Benvenutti, Chapter 15; Pittroff and Soca, Chapter 16; Agreil *et al.*, Chapter 17, this volume). In horses, preferences in available food in relationship with keeping conditions play a key role in their welfare and growth (Houpt, Chapter 12, this volume). For pigs, Nielsen *et al.* (Chapter 10, this volume) clearly demonstrate the relationship between feeding behaviour throughout their life not only with the change of age but also with keeping conditions (see also Glatz *et al.*, Chapter 18, this volume). In rabbits, Gidenne and Lebas (Chapter 11, this volume) emphasize the effects of food properties and environmental factors of feeding behaviour. In birds, a large number of different aspects of feeding behaviour are also regulated by keeping conditions (Koene, Chapter 6; de Jong and Jones, Chapter 8; Brand and Gous, Chapter 9; Glatz *et al.*, Chapter 18, this volume). Even with rather different methodological and conceptual approaches, all these chapters in this volume tend to integrate the complexity of factors affecting the feeding performances and behaviour. Furthermore, although changes of food properties can influence feeding along the growth in all domestic birds, these changes are dramatically more important for mammals. Lactation is a ubiquitous feature of all domestic mammals and affects the feeding behaviour not only in the young animal but also in the mother. For example, by measuring various parameters of lactating dairy cows in early to peak lactation, DeVries *et al.* (2003) demonstrated that feeding behaviour is highly repeatable within cows, but variable between cows and across stages of lactation. Thus, significant morphological, biomechanical, neurological and behavioural transformations must occur to assure efficient food intake at different growth stages of these vertebrates (German *et al.*, Chapter 4, this volume).

A rapid survey of the large number of researches on the feeding behaviour in domestic (farm) animals under various human conditions shows a great variety in scientific approaches. But in the last two decades strong links of several disciplines appeared to create more integrative approaches that provided empirical data to support various models of feeding behaviour and control (Pittroff and Soca, Chapter 16; Agreil *et al.*, Chapter 18, this volume). The main aim of this volume is to focus on the integration of behavioural, functional, physiological and biomechanical approaches by covering three main areas. Although there are still many differences in the chapters as stated by Forbes (Chapter 7, this volume), this volume tends to present one of the major holistic approaches of feeding behaviour by collecting together contributions in three main, often separated, areas: (i) biomechanics and functional morphology including relevant analysis of the structures (Gussekloo, Chapter 2; Bels and Baussart, Chapter 3; German *et al.*, Chapter 4; Popowics and Herring, Chapter 5; Baumont *et al.*, Chapter 14, this volume); (ii) ethology by behavioural and functional analyses of the feeding behaviour in various environmental, more or less controlled, conditions to propose models relating behaviour and physiology (Koene, Chapter 6; Forbes, Chapter 7; de Jong and Jones, Chapter 8; Brand and Gous, Chapter 9; Nielsen *et al.*, Chapter 10; Gidenne and Lebas, Chapter 11; Houpt, Chapter 12; Glatz *et al.*, Chapter 18, this volume); and (iii) ecology by analyses of environmental implications of the feeding behaviour and underlying mechanisms (Provenza and Villalba, Chapter 13; Gordon and Benvenutti,

Chapter 15; Pittroff and Soca, Chapter 16; Agreil *et al.*, Chapter 17; Glatz *et al.*, Chapter 18, this volume).

The first area relates to the feeding mechanisms per se and their potential evolution in relationship with changes of structures and performances. Since the early time of functional studies of behaviour, a large number of authors have used domestic vertebrates for determining mechanisms and suggesting possible evolutionary trends. Biomechanical studies on chicken, mallard and pigeon for birds, and pig (and other domestic vertebrates such as guinea pigs and rodents) for mammals are exemplary in such an approach. These animals are studied to determine the basic characteristics of feeding behaviour and cycles (Bels *et al.*, 1994; Schwenk, 2000). Based on such data, the functional morphologists have moved to wild species for testing these evolutionary trends explaining the biological diversity in various aquatic and terrestrial environments (Dullemeijer, 1974).

The second area relates to the understanding of the feeding behaviour and its implication in functioning and well-being (e.g. nutrition, reproduction and welfare) that play a key role in animal production. This understanding needs a complete knowledge of the relationship between food properties and feeding mechanisms. For example, this knowledge permits to explain the particle reduction of the food during the masticatory process and the moulding of a bolus for efficient swallowing. Based on these data, 'optimal functioning' of animals that results in welfare can be integrated to their fitness. This fitness can be measured by a series of quantitative data depicting health, growth and reproductive success that contribute to economically relevant animal production. A large number of studies cover a lot of physiological, functional and behavioural topics to attempt to provide the empirical data and subsequent models for improving animal production (Koene, Chapter 6; Forbes, Chapter 7; de Jong and Jones, Chapter 8; Brand and Gous, Chapter 9; Nielsen *et al.*, Chapter 10; Gidenne and Lebas, Chapter 11; Houpt, Chapter 12; Provenza and Villalba, Chapter 13; Baumont *et al.*, Chapter 14; Gordon and Benvenutti,

Chapter 15; Pittroff and Soca, Chapter 16, this volume). A survey of all these chapters clearly shows that they interact with each other, but strong interactions also occur with functional analyses presented in the first part of this volume. The chapters on feeding in mammals are exemplary in showing this interaction. Popowics and Herring (Chapter 5, this volume) provide the bases to explain the mechanism of mastication in mammals by extracting the major features of jaw apparatus and teeth. German *et al.* (Chapter 4, this volume) describe the complexity of mechanism and neuromotor control during the ontogenetic changes of the feeding apparatus in the same mammals. All these data are basic for explaining the complexity of the feeding response in mammals such as rabbits (Gidenne and Lebas, Chapter 11, this volume), ruminants (Baumont *et al.*, Chapter 14; Gordon and Benvenutti, Chapter 15; Agreil *et al.*, Chapter 17, this volume) and horses (Houpt, Chapter 12, this volume).

One key question for all of these analyses is that of optimization of the feeding behaviour. The studied structures and underlying mechanisms used for food intake and mastication (when it occurs) can be determined as optimized either for one kind (i.e. abrasive plant material in ruminants) or several food types. For example, even terrestrial birds can select a variety of food, from grains to small invertebrates, and the jaw apparatus assures the efficiency of food intake. But, the complexity of physiological control for each individual can largely regulate the performances of these structures. In their model, Pittrof and Soca (Chapter 16, this volume) state the complexity of regulating physiological control in maximizing energy intake and minimizing time and energy spent for feeding. The third area probably emerging more recently covers the effect of various traits of the complex feeding behaviour on the agri-environments. This area is even more complex because it does not only request an understanding of the feeding mechanism and performances of animals facing controlled food sources (i.e. texture, size, mode of distribution). Remarkably, studies of grazing by ruminants that received more attention for 20 years provide strong integration of ecological approaches with behavioural and physiological studies (Provenza and Villalba, Chapter 13; Agreil *et al.*, Chapter 17, this volume).

One interesting key point emerges from all the chapters in this volume that are related to feeding in mammals. Our understanding of the feeding mechanism and its implication needs two contradictory approaches. This point has been emphasized by Provenza (2003) for many years, but needs to be re-emphasized in such a volume. On the one hand, we need to develop statistical analyses to clarify the functional relationships between the structures, the performances and the behaviour. On the other hand, we also need to understand the individual variability, which plays a major role in animal production. Provenza and Villalba (Chapter 13, this volume) indicate that marked variability is common 'even in closely related animals', not only for nutrients because each individual is different in terms of metabolism but also in the ability to cope with toxins that can, for example, be influenced by previous experience and learning. Several chapters highlight this key point of individual difference in feeding behaviour (Koene, Chapter 6; Provenza and Villalba, Chapter 13; Baumont *et al.*, Chapter 14; Agreil *et al.*, Chapter 17, this volume). Agreil *et al.* suggest a validation of consistency of inter-individual difference in their recording of grazing behaviour in ruminants. Finally, both types of quantitative observations provide a lot of empirical data used for modelling feeding behaviour to improve the efficiency of nutrition (Pittroff and Soca, Chapter 16, this volume). These models must integrate highly different data from behavioural recording to neurohormonal and physiological information. It is rather interesting to show that a rough division separates scientists interested in monogastric and digastric mammals. Often, studies of feeding behaviour in pigs are associated with analyses of feeding behaviour in domestic birds (Koene, Chapter 6; Glatz *et al.*, Chapter 18, this volume). In contrast, analyses of feeding in ruminants form a different corpus of literature. In terms of anatomical, biomechanical and functional analysis, the literature separates studies in mammals from studies in birds.

In birds, it is clear that questions related to the environmental conditions of feeding are of primary interest. As emphasized by Forbes (Chapter 7, this volume), the neuromotor control of food choice and intake is a complex question. This control plays the key role in the feeding strategies that can be highly variable following the environmental constraints (Koene, Chapter 6; Forbes, Chapter 7; de Jong and Jones, Chapter 8; Glatz et al., Chapter 18, this volume). Empirical bases provided on mechanisms and postures (Gussekloo, Chapter 2; Bels and Baussart, Chapter 3, this volume) can greatly help to interpret the feeding strategies selected by birds, which not only respond to nutritional demands (de Jong and Jones, Chapter 8; Brand and Gous, Chapter 9, this volume) but also to minimization of discomfort (Forbes, Chapter 7, this volume).

Conclusion

This volume is an attempt to provide a synthesis of several aspects of feeding, which is 'one of the primary concerns of all animals', by collecting together various approaches of the complex function of feeding in birds and mammals and the 'internal' and 'external' factors regulating this function. It is probably a key opportunity for specialized readers to find results collected from completely different disciplines of biology related to questions on feeding function. It should be a tool for gaining in-depth understanding of feeding in domestic vertebrates and should open clear perspectives for future researchers, as emphasized by Forbes (Chapter 19, this volume) in the conclusion.

References

Bels, V.L., Chardon, M. and Vandewalle, P. (1994) *Biomechanics of Feeding Vertebrates*. Advances in Comparative and Environmental Physiology 18. Springer-Verlag, Berlin, Germany.

Bock, W.J. (1964) Kinetics of the avian skull. *Journal of Morphology* 114, 1–42.

Bramble, D.M. and Wake, D.B. (1985) Feeding mechanisms of lower tetrapods. In: Hildebrand, M., Bramble, D.M., Liem, K.F. and Wake, D.B. (eds) *Functional Vertebrate Morphology*. Harvard University Press, Cambridge, Massachusetts.

DeVries, T.J., von Keyserlingk, M.A.G. and Beauchemin, K.A. (2003) Diurnal feeding pattern of lactating dairy cows. *Journal of Dairy Science* 86, 4079–4082.

Domenici, P. and Blake, R.W. (2000) *Biomechanics in Animal Behaviour*. BIOS Scientific Publishers, Oxford, UK.

Dullemeijer, P. (1974) *Concepts and Approaches in Animal Morphology*. Van Gorkum, Assen, The Netherlands.

Forbes, J.M. (1999) Minimal total discomfort as a concept for the control of food intake and selection. *Appetite* 33, 371.

Forbes, J.M. (2000) Consequences of feeding for future feeding. *Comparative Biochemistry and Physiology – Part A: Molecular & Integrative Physiology* 128(3), 461–468.

Forbes, J.M. and Provenza, F.D. (2000) Integration of learning and metabolic signals into a theory of dietary choice and food intake. In: *Proceedings of the IX International Symposium on Ruminant Physiology*. CAB International, Wallingford, UK, pp. 3–19.

German, R.Z. and Crompton, A.W. (1996) Ontogeny of suckling mechanisms in opossums (*Didelphis virginiana*). *Brain, Behaviour and Evolution* 48(3), 157–164.

Gussekloo, S.W.S. and Bout, R.G. (2005) Cranial kinesis in palaeognathous birds. *Journal of Experimental Biology* 208, 3409–3419.

Hiiemae, K. (2000) Feeding in mammals. In: Schwenk, K. (ed.) *Feeding: Form, Function and Evolution in Tetrapod Vertebrates*. Academic Press, San Diego, California, pp. 411–448.

Jean, A. (1984) Brainstem organization of the swallowing network. *Brain, Behaviour and Evolution* 25, 109–116.

Lauder, G.V. (1985) Aquatic feeding in lower vertebrates. In: Hildebrand, M., Bramble, D.M., Liem, K.F. and Wake, D.B. (eds) *Functional Vertebrate Morphology*. Gustav Fischer Verlag, New York, pp. 179–188.

Lauder, G.V. (2000) Biomechanics and behaviour: analysing the mechanistic basis of movement from an evolutionary point of view. In: Domenici, P. and Blake, R.W. (eds) *Biomechanics in Animal Behaviour*. BIOS Scientific Publishers, Oxford, UK, pp. 19–32.

Lauder, G.V. and Gillis, G.B. (1997) Origin of the amniote feeding mechanism: experimental analysis of outgroup clades. In: Sumida, S.S. and Martin, K.L.M. (eds) *Amniote Origins: Completing the Transition to Land*. Academic Press, San Diego, California.

McBrayer, L.D. and Reilly, S.M. (2002) Testing amniote models of prey transport kinematics: a quantitative analysis of mouth opening patterns in lizards. *Zoology* 105, 71–81.

McSweeney, F.K. and Murphy, E.S. (2000) Criticisms of the satiety hypothesis as an explanation for within-session changes in responding. *Journal of the Experimental Analysis of Behavior* 74, 347–361.

Pérez-Barbería, F.J. and Gordon, I.J. (2001) Relationships between oral morphology and feeding style in the Ungulata: a phylogenetically controlled evaluation. *Proceedings of the Royal Society of London Series B Biological Sciences* 268, 1023–1032.

Provenza, F.D. (2003) Twenty-five years of paradox in plant–herbivore interactions and 'sustainable' grazing management. *Rangelands* 25, 4–15.

Reilly, S.M. and Lauder, G.V. (1990) The evolution of tetrapod feeding behavior: kinematic homologies in prey transport. *Evolution* 44, 1542–1557.

Schwenk, K. (ed.) (2000) *Feeding: Form, Function and Evolution in Tetrapod Vertebrates*. Academic Press, San Diego, California.

Zusi, R.L. (1984) A functional and evolutionary analysis of rhynchokinesis in birds. *Smithsonian Contributions to Zoology* 395, 1–40.

Zusi, R.L. (1993) Patterns of diversity in the avian skull. In: Hanken, J. and Hall, B.K. (eds) *The Skull: Patterns of Structural and Systematic Diversity*, Vol. 2. University of Chicago Press, Chicago, Illinois, pp. 391–437.

2 Feeding Structures in Birds

S.W.S. Gussekloo

Division of Anatomy and Physiology, Department of Pathobiology, Faculty of Veterinary Sciences, Utrecht University, Yalelaan 1, 3584 CL Utrecht, The Netherlands

Introduction

In domestic vertebrates two large phylogenetic groups can be distinguished: mammals and birds. Although the general anatomy of the two groups shows many characteristics that can be linked back to the vertebrate body plan, large differences can be found between the two groups. The main reason for the differences between the two groups is the large phylogenetic distance, but the different natural history and the ability of flight also have an effect on the anatomy of birds. The digestive tract of birds differs from that of mammals in many aspects, such as the absence of teeth, which are functionally replaced by a muscular stomach where mechanical food reduction takes place. The relative overall length of the digestive tract of birds is shorter than that of mammals (Denbow, 2000) and it shows in general less differentiation between the different parts of the intestinal tract.

In this chapter an overview is given of the functional morphology of the structures that play a role in the food uptake and digestion in domestic birds. Previous data are used for the general descriptions and also for reference of non-domestic species (Humphrey and Clarck, 1964; Ziswiler and Farner, 1972; Nickel *et al.*, 1973; McLelland, 1975, 1979, 1993). A number of atlases are also published with good illustrations of the anatomy of the

domestic species (e.g. Ghetie *et al.*, 1976; Yasuda, 2002). What will be presented here is a general description with some attention to specific morphological structures. It must be remembered, however, that the digestive system is very variable and its morphology can change under influence of fluctuations in food supply (see Starck, 1999, 2003 for a review). Although a large number of birds are held in captivity for a plethora of reasons, the focus of this chapter will be on the following groups: (i) the galliforms, which include chicken (*Gallus gallus*), turkey (*Meleagris gallopavo*) and guinea-fowl (*Numida meleagris*); (ii) anseriforms, which include ducks and geese; and (iii) ratites, of which the ostrich (*Struthio camelus*), the emu (*Dromaius novaehollandiae*) and rhea (*Rhea americana*) are the species that are kept most in captivity. Finally, the columbiforms will be included in the description, of which the pigeon (*Columba livia*) is the most well-known representative.

The terminology used in the chapter when related to birds will follow the official avian anatomical terminology as defined in the *Nomina Anatomica Avium* (Baumel *et al.*, 1993). When referring to non-avian species, the veterinarian anatomical terminology will be used as described in the illustrated veterinary anatomical nomenclature (Schaller, 1992).

©CAB International 2006. *Feeding in Domestic Vertebrates: From Structure to Behaviour* (ed. V. Bels)

Anatomical Structures of Food Acquisition and Intraoral Transport

On the outside, the clearest anatomical feature of the digestive tract of birds is the bill. The base of the lower bill (mandibulae) is formed by two bony elements that are fused at the bill tip (symphysis). The base of the upper bill (maxillae) consists of two ventral bars, a single dorsal and in most domestic species two lateral bars that connect the dorsal and ventral bars. This construction results in a sturdy configuration, which enables the birds to move the upper bill as a whole. In ratites, however, the lateral bar is not rigid but consists of a flexible ligament, and therefore the upper bill loses some of the stiffness, which affects the feeding behaviour (Gussekloo and Bout, 2000 a,b). In galliforms, anseriforms and columbiforms the upper bill can be elevated or depressed around a hinge between the upper bill and the cranium (os frontale, frontal-nasal hinge). These movements of the upper bill are induced by a number of muscles, which are situated behind and below the eye, and transferred onto the upper bill by a complex of bony elements. The full mechanics of the avian skull have been thoroughly described previously (Bock, 1964) and movement patterns were later experimentally confirmed (van Gennip and Berkhoudt, 1992; Gussekloo et al., 2001). In ratites the hinge between the cranium and upper bill is absent and movement of the upper bill as a whole is not observed (Gussekloo, 2000). The exact function or advantage of the movable upper bill is unknown but possible functions might be to increase gape, or increase closing speed (see Herrel et al., 2000; Bout and Zweers, 2001).

The muscles in the avian cranium that contribute to the movement of the bills can be divided into four different functional groups (Bühler, 1981): (i) openers of the lower bill; (ii) openers of the upper bill; (iii) closers of the lower bill; and (iv) closers of both upper and lower bill (Fig. 2.1). Each of these groups will be discussed shortly without thorough description of variation in morphology between species.

The lower bill is depressed by the musculus depressor mandibulae, which originates on the caudolateral side of the skull and inserts on

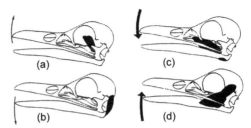

Fig. 2.1. Diagram of the four groups of jaw muscles. (a) Openers of the upper bill (m. protractor pterygoidei et quadrati). (b) Opener of the lower bill (m. depressor mandibulae). (c) Closers of both upper and lower bill (m. pterygoideus). (d) Closers of the lower bill (m. adductor complex). (Adapted from Bühler, 1981.)

the caudoventral side of the mandibulae. This muscle is comparable in function and orientation to the musculus digastricus in mammals. In pigeons, chicken, turkeys, ducks and geese multiple parts of the muscle can be distinguished (Ghetie et al., 1976; Bhattacharyya, 1980; van Gennip, 1986), while in ratites (Struthio, Rhea, Dromaius) such a subdivision is not present (Webb, 1957; Gussekloo, unpublished data, 1998).

The main opener of the upper bill (m. protractor pterygoidei et quadrati) has its insertion on the quadrate and parts of the pterygoid bone and its origin on the mediocaudal wall of the orbit. The main function of the muscle is to retract and elevate the quadrate. Rotation of the quadrate results in cranial movement of the bony elements in the palate, which will then push against the ventral part of the upper bill. These forces on the ventral part of the upper bill result in an upward rotation and therefore in opening of the upper bill. Although for several species multiple parts of the m. protractor pterygoidei et quadrati have been described, it is found that all these parts function as a single unit and distinction between parts seems highly subjective.

The closing of the lower bill is achieved by a large muscle complex situated on the lateral side of the head and behind the eye. This adductor complex (m. adductor mandibulae externus) has its origin on the lateral side of the cranium and dorsolateral side of the orbit and has its insertion on the dorsal part of the mandibulae. The muscle is often subdivided in rostral, ventral and caudal parts, but this subdivision is not clear in ratites (Webb, 1957; Gussekloo, unpublished data, 1998). Three smaller muscles also contribute to the closing of the lower bill. The first muscle (m. pseudotemporalis superficialis) is relatively small and runs alongside the adductor muscle. It has its origin medial from the adductor muscle and on the inside of the mandibulae. The other two muscles (m. pseudotemporalis profundus and m. adductor mandibulae ossis quadrati) also insert on to the mandibulae but have their origins on the quadrate and can therefore theoretically also have an effect on the closing of the upper bill.

A large muscle complex (m. pterygoideus) acts both as a closer of the upper bill and as a closer of the lower bill. The origin is on the pterygoid bones and the insertion is on the caudomedial part of the lower jaw. A contraction of the muscle results in the retraction of the pterygoid bones, and through their coupling with the upper bill, in a closing of the upper bill. At the same time the contraction results in an upward rotation of the lower bill. In general four parts of the pterygoid muscle can be distinguished, but in ratites these parts are strongly connected, which makes distinction difficult. A final muscle known as the m. retractor palatini has also been described in a number of avian species, but in ratites this muscle seems to be continuous with, or even part of, the pterygoid muscles.

A strong keratin layer known as the rhamphotheca covers both the upper and lower bill. This keratin layer is constantly produced to compensate for the constant wear on the rims of the rhamphotheca as a result of feeding and preening. Between species there are large variations in the anatomy of the rhamphotheca. In chicken, turkey and pigeon, very few distinctive features can be found in the rhamphotheca. Both the upper and lower beak appear to be a single entity. In ducks and ratites different parts can be distinguished in the upper part of the rhamphotheca. In ducks the rostromedial part of the rhamphotheca forms a nail or dertrum, which consist of very hard keratin while the rest of the bill is much softer. In ratites very distinct grooves are present that separate the medial nail-like section

from the lateral parts. Although the duck and ratite morphologies share some characteristics, Parkes and Clark (1966) showed that the ratite morphology is a specific characteristic for that taxon.

When all avian taxa are considered it is clear that the overall anatomy of the rhamphotheca is very flexible and that many adaptations are possible. Most of these adaptations can be directly related to the feeding behaviour of a specific species. Galliforms and pigeons are mainly peck feeders and the rims of the bill are therefore relatively slender and sharp, without any further characteristics. The originally aquatic anseriforms all have lamellae on the rims of their bills that are used for filter feeding, but which might be adapted for more terrestrial feeding as well. In ratites the ramphotheca shows very little adaptation and the rims are relatively blunt and rounded. This can be correlated to the fact that ratites use their bills mainly for holding a food item without any other specialized behavioural actions (Gussekloo and Bout, 2005a).

The rest of the description of the anatomical structures of the feeding apparatus will be based on their function during feeding behaviour. The basal pattern of food uptake is considered to be similar to the pecking behaviour of pigeons and chicken (Zweers, 1985; Zweers et al., 1994, 1997). Based on specific movement patterns, the complete pecking sequence can be divided into a number of behavioural phases (Zweers, 1982). A pecking sequence starts with 'head fixation', in which the bird keeps the head still and focuses on the food item, and probably gathers visual information about the food item itself and its position in space. Although this gathering of visual information is part of the feeding sequence, the anatomy of the eye and the visual pathways in birds will not be discussed (see Güntürkün, 2000 for a recent review). The fixation phase is followed by the (final) approach phase in which the head is moved in a straight line towards the food item and the bill is opened. Subsequently, the food item is picked up between the bill tips.

In anseriforms (and other non-domestic species) that use their bill to locate food, a highly sensitive organ is present at the bill tip, while this organ is less developed in species that show mainly pecking behaviour (Gottschaldt,

1985). In chicken, however, a concentration of mechanoreceptors is found at the bill tip that might have a similar function in feeding (Gentle and Breward, 1986). In pigeons the bill tip organ is well developed with similar receptors also situated along the rims of the beak and palate (Malinovsky and Pac, 1980). The bill tip organ consists of papillae that protrude from the dermis through the outer hard keratinized layer of the rhamphotheca, but each papilla is covered by soft papillary horn. Different mechanoreceptors are present in each individual papilla. Within the papilla there is a topographical distribution of Grandry corpuscles and other free nerve endings near the tip of the papilla and Herbst corpuscles near the base. A more detailed description of the bill tip organ is given by Gottschaldt (1985) and Gentle and Breward (1986). A more recent description of the individual mechanoreceptor is given by Necker (2000).

The Oropharynx

The oropharynx is the combined space of the oral cavity and the pharynx (Fig. 2.2). The two cavities are combined in terminology in birds because clear landmarks such as the soft palate or demarcating ridges such as the plica glossopharyngeus of mammals are absent. Some authors have attempted to describe the boundary between the oral cavity and the pharynx with variable results. Lucas and Stettenheim (1972) used the embryology of the visceral arches and placed the boundary between the choanae and the rima infundibuli, the entrance to the tuba auditiva or Eustachian tube. In the mallard the boundary between the oral cavity and the pharynx has been described at the level of the caudal lingual papillae (Zweers et al., 1977). The structures found in the oropharynx are highly variable between species, depending on several factors including diet and ecology. The main structures found in the roof of the oropharynx are the opening of the nasal cavities (choanae) and the opening of the tuba auditiva. The main structures on the bottom of the oropharynx are the tongue and the glottis.

A number of papillae can be found dispersed over different locations in the oropharynx. In almost all species papillae are present at the edges of the choana (papillae choanales), at the base of the tongue (papillae linguae caudales), caudal to the larynx and caudal to the rima infundibuli (papillae pharyngis caudodorsales). These papillae can be found in most of the domestic species, except in ratites, in which only the papillae behind the larynx (papillae pharyngis caudoventrales) are clearly recognizable. No other features are present in the oropharynx of the ratites although a small medial ridge is present on the palate (ruga palatina mediana), which can also be found in anseriforms. In the galliform species, however, several rows of caudal pointing papillae are present on the palate (papillae palatinae). These papillae play an important role in the intraoral transport of food items. Geese also possess papillae on the palate (Greschik, 1935), but these are very different in shape than the papillae of the galliforms and might have a different function. A remarkable feature of the waterbound anseriforms are the rows of lamellae (maxillary lamellae) on the lateral sides of the oropharynx, which are used for filtering immersed food particles from water (Kooloos, 1986).

The tongue plays an important role in intraoral transport in many avian species. The complexity of the role is directly reflected in the complexity of the anatomy of the tongue. In this description we focus on the external characteristics of the tongue and their function during feeding. A general description of intrinsic and extrinsic musculature and the skeleton of the tongue is given by McLelland (1975); a comparative study on external features is provided by Marvin and Těšik (1970).

In ratites the main mode of intraoral transport is by means of catch-and-throw behaviour in which the food is transported by its own inertia (Tomlinson, 2000; Gusseklloo and Bout, 2005a). The tongue is only used to push a food bolus from the pharynx into the oesophagus (Tomlinson, 2000). The tongue is therefore very short and relatively broad without clear adaptations other than the papillae linguae caudales that stabilize the food bolus during the final transport into the oesophagus. In the galliforms and pigeons the main feeding type is pecking behaviour, combined with intraoral transport using catch-and-throw, slide-and-glue or a combination of the two

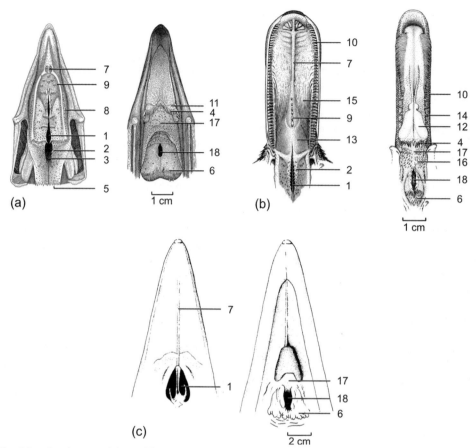

Fig. 2.2. Oropharynx of three different species. (a) Chicken (*Gallus gallus*: Heidweiller *et al.*, 1992). (b) Mallard (*Anas platyrhynchos*: Kooloos *et al.*, 1989). (c) Greater rhea (*Rhea americana*: Gussekloo, 2000). The left image shows the roof of the oropharynx and the right image, the floor of the oropharynx. (Figures of the chicken and the mallard are used with kind permission of Springer Science and Business Media and the author.) (1) choanae, (2) papilla choanales, (3) rima infundibuli, (4) papillae linguae caudales, (5) papilla pharyngis caudodorsales, (6) papillae pharyngis caudoventrales, (7) ruga palatina mediana, (8) ruga palatinae laterales, (9) papillae palatinae, (10) lamellae rostri (maxillary/mandibulary lamellae), (11) alae linguae, (12) torus linguae, (13) torus palatina, (14) papillae linguae lateralis, (15) sulci palatini, (16) papillae pharynges, (17) radix linguae, (18) glottis.

(Zweers, 1982; Van der Heuvel and Berkhoudt, 1998). In slide-and-glue feeding behaviour small food items are stuck to the tongue by saliva and subsequently transported caudally by retraction of the tongue. This can be repeated several times alternating with securing the food item between the bills to complete the intraoral transport. Slide-and-glue transport requires a tongue that has an even surface and that can be used throughout the oral cavity. The tongues of chickens and doves are therefore relatively long and slender. The tongue of the chicken has lateral wings at the base (alae linguae) that act in combination with the papillae linguae caudales as scrapers when the tongue is retracted during final intrapharyngeal food transport. The retraction of the tongue is followed by retraction of the larynx, in which the papillae pharyngis caudoventrales stabilize the food items during transport.

In the anseriforms, such as the mallard, feeding is more complex, which has had a large effect on the anatomy of the tongue. A clear character of the anseriform tongue is the large swelling near the base of the tongue. This torus linguae plays a role in specific filter feeding and drinking behaviour of the anseriforms in which the torus acts as a piston (Zweers et al., 1977; Kooloos et al., 1989). The piston function is improved by a swelling of the palate (torus palatinus) opposite to the torus on the tongue. The final intraoral transport of food items in the mallard is also different from that in pigeons and chicken. Just rostral to the torus of the tongue food items are displaced laterally, where they are pushed caudally by the large papillae that are present on the lateral border of the tongue (papillae linguae laterales) and guided by the grooves in the roof of the oropharynx (sulci palatini). This action is repeated until the food item is in the back of the oropharynx. This method of food transport has been described as a conveyor belt by Kooloos (1986). Final transport of the food into the oesophagus is achieved by a retraction of the tongue and larynx.

Very little is known about the surface of the avian tongue, probably due to the large variety in morphologies throughout the taxon. In peck-feeding species the tongue may be partly keratinized as in chicken, while a fully fleshy tongue is found in filter feeders such as many anseriforms. Also the papillae on the tongue are not well described, in contrast to the mammalian papillae. The only distinction that has been made is between lenticular and major and minor conical papillae in domestic fowl (Komarek et al., 1982).

Inside the oropharynx a number of salivary glands are present, but a general description is impossible because of interspecific variation (McLelland, 1993). The salivary glands are not aggregated like mammalian glands, but often distributed over the whole oropharynx, and therefore hard to recognize. In the roof of the oropharynx salivary glands are situated in the front (glandula maxillaris), in the palate (gl. palatinae) and around the choanae and rima infundibuli (glandulae sphenopterygoideae). On the floor of the oropharynx salivary glands are present near the bill tip (gll. mandibulares rostrales), near the tongue (gll. mandibulares caudales) and

around the glottis (gll. cricoarytenoideae). In several species each individual gland can be organized in patches, which results in subdivision and nomenclature based on the location of these patches in the oropharynx. The majority of the salivary glands have multiple openings into the oropharynx, which are clearly visible with the naked eye. Besides the salivary glands in the walls of the oropharynx, some glands can be present on the tongue as well. In the chicken two types of lingual glands are described: one on the tip of the bill (gl. lingualis rostralis) and one on the back of the tongue (gl. lingualis caudalis; Saito, 1966; Homberger and Meyers, 1989). In the pigeon three different salivary glands are described: two on the dorsal surface of the tongue (gl. lingualis rostralis dorsalis, gl. lingualis caudalis dorsalis) and one on the ventral surface of the tongue (gl. lingualis ventralis; Zweers, 1982).

The products of the salivary glands have functions in the intraoral transport (e.g. in slide-and-glue transport), but mainly in the lubrication of the food. In many domestic avian species (turkeys, geese, domestic fowl, pigeons) amylolytic activity has been described (Ziswiler and Farner, 1972), although chickens and turkeys produce very little amylase compared to songbirds (Jerrett and Goodge, 1973).

During intraoral transport the food items are repeatedly tested for taste and edibility, often in positions related to the stationary phases of pecking behaviour (Zweers et al., 1977; Berkhoudt, 1985; Zweers, 1985). In all domestic birds taste buds are present directly behind the bill tips, where the first contact with a food item is made. Taste buds are also present in relation to the gl. palatina that surround the choanae, and on the base of the tongue (near the torus linguae in the mallard). This correlates with the position of the food item just at the beginning of intrapharyngeal transport. Finally, a taste area is present on the bottom of the pharynx, which can be used for final testing of the food items before the food is transported into the oesophagus. In the pigeon and the chicken additional taste areas are located in the floor of the oropharynx lateral to the tongue.

Taste buds in birds can be divided into two major groups based on their position in the oropharynx. The taste buds are either distributed solitarily in the mucosa (solitary buds) or in association with the opening of a salivary gland (glandular buds), but this distinction is not always clear. The reason for the close relation between taste buds and salivary glands is that the saliva allows for the dissolving of taste substances and for transport to the taste buds (Belman and Kare, 1961). In birds this is of great importance because the taste buds do not open directly into the oral cavity (Berkhoudt, 1985). Three types of taste buds have been described (type I, II and III), but it seems that the domestic birds possess only one of them. In galliforms and pigeons type I is found, an ovoid taste bud with a core of sensory and supporting cells, surrounded by a sheath of follicular cells. In anseriforms type II is found, which is a long and narrow taste bud (Berkhoudt, 1985).

The Oesophagus

Retraction of the larynx pushes the food items into the oesophagus and into the remaining digestive tract. An overview of the digestive tract of the chicken is given in Fig. 2.3. Comparable data of the different parts of the digestive tract of some domestic species are given in Table 2.1.

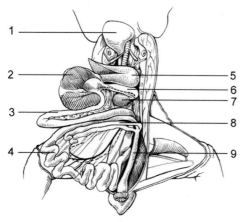

Fig. 2.3. Complete digestive tractus of the chicken (*Gallus gallus*). (1) crop, (2) gizzard, (3) duodenum, (4) jejunum, (5) liver, (6) proventriculus, (7) spleen, (8) caecum, (9) rectum. (Adapted from Dyce *et al.*, 2002.)

The oesophagus transports food from the oropharynx into the stomach. Just caudal of the pharynx the oesophagus lies in the midline (medial plane) of the neck and is attached to the larynx and trachea by connective tissue. Closer to the body the oesophagus lies very loose in the surrounding tissue and can move relative to other structures. The length of the cervical part of the oesophagus (pars cervicalis) in its rest position is shorter than the cervical column. This means that the position of the caudal part of the oesophagus is variable and depends on the posture of the cervical column. When the neck is unextended and in the natural S-shape, the oesophagus lies ventral to the cervical column in the cranial part of the neck, and dorsal and to the right of the cervical column in the caudal part of the neck. When the neck is extended the oesophagus is stretched and forms a single tube on the ventral side of the cervical column.

In cross-section, from the inside out, four layers can be distinguished in the oesophagus. The most internal layer is the mucous membrane (tunica mucosa). This is covered by a weakly developed tela submucosa. Around the tela submucosa is an external muscle layer (tunica muscularis) consisting of a longitudinal muscle layer (stratum longitudinale) and a circular muscle layer (stratum circulare). The outside of the oesophagus is covered by a fibrous layer, the tunica adventitia. The mucous layer consists of an internal longitudinal muscle layer (lamina muscularis mucosae) and a thick epithelial layer (lamina epithelialis mucosae). In pigeons and ducks this last layer can be cornified into a stratum corneum (Zietzschmann, 1911; Kaden, 1936), although Kudo (1971) has shown that this cornification in the chicken and pigeon is incomplete. Within the tunica mucosa, oesophageal glands (gl. esophageae) are present that produce mucus. This mucus is of great importance for additional lubrication of the food, since birds produce relatively little saliva (Denbow, 2000). Although the oesophageal glands can be of great importance, the structure and density of these glands can differ between species (Landolt, 1987).

Just cranial to the thoracic entrance, the oesophagus is expanded to form the crop (ingluvies), which will be discussed later. The oesophagus continues in the thorax (pars thoracica) dorsal to the trachea, in between the

Table 2.1. Comparative data on the lengths and weights of the digestive tracts in domestic birds. (From Denbow, 2000.)

Species	Body weight (kg)	Oesophagus Length (mm)	Oesophagus Total %	Oesophagus Weight (g)	Proventriculus and gizzard Length (mm)	Proventriculus and gizzard Total %	Proventriculus and gizzard Weight (g)	Small intestine Length (mm)	Small intestine Total %	Small intestine Weight (g)	Caecum Length (mm)	Caecum Weight (g)	Rectum Length (mm)	Rectum Total %	Rectum Weight (g)	Total Length (mm)	Total Length/body weight
Chicken																	
Leghorn[a]	1.2	136	9.9	8.2	86	6.3	26.7	1082	78.9	29.5	127	5.2	68	5.0	2.3	1372	1.1
Broiler[b]	3.0	140	6.4	16.8	101	4.7	43.5	1796	82.7	73.6	188	10.7	134	6.2	5.1	2171	0.7
Turkey	3.0	123	5.7	8.5	110	5.1	52.9	1853	85.7	85.3	278	20.1	75	3.5	4.4	2161	0.7
Japanese quail[a]		75	11.5		38	5.8		510	78.1		100		30	4.6		653	
Domestic duck[b]	2.2				130	4.9		2110	79.9		140		90	3.4		2640	1.2
Emu[b]	53.0	310	11.7		260	4.0		5200	79.4		120		300	4.6		6550	0.1
Rhea[c]	25.0	790	12.1		310			1400			480		400				
Ostrich[c]	122.0				480			6400			940		8000				

[a]Data from Fitzgerald (1969).

[b]Data from Herd (1985).

[c]Data from Fowler (1991).

extrapulmonary primary bronchi, and connects to the stomach on the left side of the body. In mammals the oesophagus can be closed both cranially and caudally by a sphincter in the oesophageal wall, but such sphincters are not present in most birds (Mule, 1991).

Since most domestic birds are unable to reduce the size of their food items in the first part of the digestive tract due to the lack of teeth, relatively large items have to pass the oesophagus. This is reflected in the relatively large diameter of the oesophagus of birds in comparison to other vertebrates. The presence of longitudinal folds on the inside of the oesophagus (plicae esophagi) makes it possible for the oesophagus to extend its diameter even further.

In most avian species an enlargement of the oesophagus is present just cranial to the thoracic inlet. This evagination of the oesophagus is known as the crop (ingluvies), and its size and shape can vary among species. The simplest form of crop is found in species from the taxa anseriforms and ratites, in which the crop is not more than a spindle-shaped thickening in the oesophagus. In some cases the organ is so small that it can hardly be recognized within the oesophagus. The most complex form is found in galliforms and pigeons (Columbidae). In the galliforms a clear saclike structure is present, which in some cases can become very large. In pigeons the crop is even more elaborate, and its structure is more complex with two clearly recognizable diverticula on either side of the oesophagus. Within the crop, glands (gl. ingluviei) can be present, although the distribution varies per species. In the chicken the glands are completely absent from the body (fundus) of the crop, and can only be found near the opening to the oesophagus (ostium ingluviei; Hodges, 1974).

The function of the crop is storage of food, but is directly linked to the filling of the stomach, which means that if the stomach is empty, food is transported directly to the stomach, and when the stomach is filled, the entrance to the crop is opened and food is stored there. There have been many disputes about the role of the crop in digestion because digestive enzymes are found within the crop fluid. It is, however, unclear whether the origin of these enzymes is either from the salivary glands, the food or bacterial sources (Denbow,

2000). Although amylase activity (Philips and Fuller, 1983; Pinchasov and Noy, 1994) as well as the hydrolysis of starch (Bolton, 1965) have been shown in the crop, it is generally accepted that the importance of digestion in the crop must be minimal (Ziswiler and Farner, 1972; Denbow, 2000).

A remarkable adaptation can be found in the crop of pigeons. During the breeding season the crop produces a yellowish-white fat-rich secretion known as crop milk that is used to feed the nestlings. The crop milk is a holocrine secretion of the epithelium, which is mainly produced in the lateral diverticula. During the productive period of the crop the walls of the diverticula thicken and have an increased vascularization to provide for the necessary nutrients. The crop milk resembles strongly the milk produced by mammals, except for the fact that carbohydrates and calcium are missing in crop milk.

The Stomach

The stomach (Fig. 2.4) of birds can be divided into three parts, which are not always clearly distinct. The most proximal part of the stomach (proventriculus gastris) produces the gastric acid and pepsinogen; the second compartment, the gizzard (ventriculus gastris), has a function in mechanical breakdown of food. An intermediate zone (zona intermedia gastris) between the two parts can often be distinguished, although not in all species, such as some anseriforms (Swenander, 1902). Although the morphology of the stomach in general is highly adaptive to the diet of the species, all domestic birds described here are herbivorous or granivorous and therefore the variation in the morphology of the stomach is limited. The stomach is situated on the left side of the body and is, just as in mammals, curved to the right. This means that also a curvatura minor and major can be identified, although these are not as clear as in mammals due to the differentiation of the avian stomach. Because of the curvature the entrance to the duodenum is located more cranially than the most caudal part of the gizzard, and the actual entrance to the duodenum lies very cranially in the lumen of the gizzard.

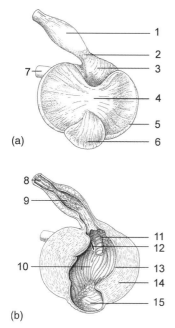

(a)

(b)

Fig. 2.4. Stomach of the chicken (*Gallus gallus*). (a) Exterior from the lateral left side. (b) Interior from the lateral left side. (1) proventriculus, (2) intermediate zone, (3) m. tenuis craniodorsalis, (4) facies tendinea, (5) facies anularis, (6) m. tenuis caudoventralis, (7) duodenum, (8) plica esophagi, (9) papillae proventriculi, (10) corpus ventriculi, (11) saccus cranialis, (12) ostium ventriculopyloricum, (13) cuticula gastris, (14) m. crassus caudodorsalis, (15) m. crassus cranioventralis. (Adapted from Dyce *et al.*, 2002.)

The proventriculus

The proventriculus is a spindle-shaped organ that produces the gastric acids and enzymes and has very little function in storage. A remarkable deviation from this general model is found in the ostrich, in which the proventriculus is enlarged and displaces the ventriculus to the right side of the body (Pernkopf and Lehner, 1937). The large size of the proventriculus indicates that it is used for storage, especially since storage of food cannot take place in the rather small crop of the ostrich.

The border between the oesophagus and the proventriculus can easily be distinguished internally, based on the absence of the internal ridges in the wall of the proventriculus that are

present in the wall of the oesophagus. In the greater rhea (*Rhea americana*) the boundary between the oesophagus and proventriculus is even clearer due to a centimetre-high muscular ridge (Feder, 1972).

The wall of the proventriculus consists of four layers: the mucous membrane (tunica mucosa proventriculi), the submucosa (tela submucosa), the muscular tunic (tunica muscularis) and the serosa (tunica serosa). The mucous membrane produces mucus, which protects the epithelium. In the chicken the amount of mucus produced is directly related to the functional state of the stomach, with higher production after feeding (Chodnik, 1947). A pattern of small sulci proventriculi and rugae proventriculi is present in the tunica mucosa around the openings of the deep proventricular glands (gll. proventriculi profundus). The deep proventricular glands produce the gastric fluids and are also situated in the tunica mucosa. In almost all domestic species these glands consist of multiple lobules with a single duct for excretion. The openings of the draining ducts in the proventricular wall are often clearly visible with the naked eye. In chicken the openings of the deep proventricular glands are located on top of papillae (papillae proventriculi), which are abundant throughout the mucosa. Beside the deep proventricular glands in some species superficial proventricular glands are also present, although these are absent in chicken and columbiforms. The superficial glands open into the lumen of the stomach at the base of the small sulci of the tunica mucosa. Below the glands a small muscular lamina is present (lamina muscularis mucosae) that might have a function in the emptying of the glands.

The intermediate zone

The intermediate zone of the stomach or gastric isthmus (zona intermedia gastris) is, in most cases, slightly lighter in colour, and in the case of most domestic species clearly smaller in diameter than the other parts of the stomach.

The wall of the intermediate zone is in general thinner than that of the proventriculus, mainly because of the absence of gastric glands, and shows, at least in chicken, high concentration of elastic fibres (Calhoun, 1954). The size and number of sulci, rugae and papillae in the tunica mucosa are strongly reduced, which results in a very smooth mucosa. In the chicken an internal lining is present, which resembles the cuticula of the gizzard, although not as thick. This layer seems to be a combination of mucus produced by the intermediate part combined with the product of the gizzard glands (Hodges, 1974). It has been hypothesized that the intermediate zone acts as an active sphincter to separate the proventriculus from the gizzard. Very little evidence for this has been found, although increased muscle mass has been described in geese (Schelpelmann, 1906) and rheas (Feder, 1972).

The ventriculus (gizzard)

The gizzard acts as the organ for grinding and mechanical breakdown of food, which is reflected by the very thick muscular walls. Externally, the gizzard looks like a flattened sphere, with a very clear aponeurosis (facies tendinea) on the flattened side, while the edges of the gizzard are clearly muscular (facies anularis). The gizzard is not completely cylindrical, and a number of characteristics can be distinguished. On the cranial and caudal sides of the gizzard the rumen of the gizzard is smaller than in the central part. The smaller parts are known as the saccus cranialis and saccus caudalis and the middle part is called the body (corpus ventriculi). The muscles of the gizzard can be divided into four semi-autonomous parts, all originating from the thick aponeuroses. Two thick muscles are located on the caudodorsal and cranioventral side (m. crassus caudodorsalis and m. crassus cranioventralis), and two thin muscular parts are found on the craniodorsal and caudoventral side (m. tenuis craniodorsalis and m. tenuis caudoventralis). The thick muscles consist of an inner and outer layer and interconnect the two aponeuroses directly, and form therefore the major part of the facies anularis. The wall of the cranial and caudal sacs is mainly formed by the thin muscles and seems continuous with the inner layer of the thick muscles. The muscle of the cranial sac (craniodorsal thin muscle) is united with the caudodorsal thick muscle, and

the caudal sac muscle (caudodorsal thin muscle) with the cranioventral thick muscle. On the outer surface of the gizzard the connections between the dorsal and ventral muscles can be seen as a cranial and caudal groove (sulcus cranialis and sulcus caudalis).

On the inside of the gizzard a layer of secretion is present, which protects the muscles of the gizzard from the degrading effects of the gastric acids and mechanical action of food particles. This layer, known as the cuticula gastris, consists of a carbohydrate/protein complex (Luppa, 1959; Eglitis and Knouff, 1962; Webb and Colvin, 1964; Glerean and Katchbarian, 1965; Michel, 1971) and is secreted by the ventricular glands (gll. ventriculares) and pyloric glands (gll. pyloricae). The yellowish-green colour of the cuticula, however, is not produced by these glands, but is probably the effect of regurgitation of bile (Groebbels, 1929; Norris, 1961). The cuticula is continuously produced to compensate for the continuous wear due to the mechanical action of the gizzard. In some anseriforms, however, the whole cuticula is replaced at once (McAtee, 1906, 1917). The cuticula is variable in thickness with the thicker part in the body of the gizzard and the more slender part in the cranial and caudal sacs. In some domestic species, such as chicken and columbiforms, the cuticula becomes extremely thickened in the dorsal and ventral parts and forms grinding plates (Flower, 1860; Schelpelmann, 1906; Cornselius, 1925). The grinding plates are elliptical in form, with an elevated wall and a flat middle section. The thickness of the plates is asymmetrical so that the thickest part of the cuticula is opposite the thickest part of the muscles on the opposing site of the lumen. On the most central surface of the cuticula a pattern of grooves and ridges is present, which seems to correlate with the grooves and ridges in the mucosa (rugae ventriculi, sulci ventriculi). In the turkey at least, the arrangement of these grooves and ridges is related to the gastric motility and seems optimized for mechanical effect (Dzuik and Duke, 1972). In the corpus of the gizzard the ridges run longitudinally, while the ridges in the cranial and caudal sacs run in several directions. The direction of the grooves is the result of the unidirectional compression of the corpus part, while the cranial and caudal sacs compress in various directions. A more detailed description of the contraction pattern of the gizzard of the turkey is given by Dzuik and Duke (1972).

The wall of the gizzard consists of four layers: the mucous membrane (tunica mucosae ventriculi), the submucosa (tela submucosa), the muscle tunic (tunica muscularis) and the serosa (tunica serosa). The ventricular and pyloric glands are located in the mucous membrane and can have different distributions depending on the species. In the pigeon the glands are distributed in rows throughout the gizzard (Swenander, 1902), while in chicken the glands are clustered in groups of 10–30 glands of which 5–8 glands open into a single crypt (Eglitis and Knouff, 1962). Since the glands produce the cuticula, there is a direct relation between the thickness of the cuticula and the number of glands present.

The submucosa of the gizzard consists of connective tissue with connective bands going into the muscular tunic. The function of the submucosa is to assure a strong connection between the mucosa and underlying muscles, which allows for effective grinding action. This is supported by the fact that the submucosa is best developed under the grinding plates (Schelpelmann, 1906). The main muscle masses are found in the muscular tunic and have been described above.

To increase the functionality of the ventriculus many birds ingest (small) stones known as grit. These grit particles enter the ventriculus and act as grindstones to enhance the grinding action of the muscles. Although the grit can improve digestion it was found not to be essential for digestion (McLelland, 1979). The grit itself is dissolved by the gastric acids and can therefore also be an additional source of calcium (Lienhart, 1953). The exact constitution of the grit differs per species and a thorough description of grit can be found elsewhere (Meinertzhagen, 1954).

The connection between the gizzard and the duodenum is known as the pylorus (pars pylorica gastris). Within most domestic birds this part is relatively small and the actual border between the pylorus and the duodenum is not more than a mucosal fold. In some ratites, however, the pylorus is clearly recognizable as a widened part of the intestine.

The Intestines

Microstructure

From the pylorus the digestive tract continues as the intestines, in which the different parts do not have distinct morphological differentiation as in mammals. The morphology of the wall of the intestine is almost equal over the full length, and consists, as the whole digestive tract, of four layers. From the lumen to the external lining the four layers of the intestinal wall are the mucosa (tunica mucosa), the submucosa (tela submucosa), the muscle tunic (tunica muscularis) and the serosa (tunica serosa). The surface of the intestinal wall is enlarged by a combination of projections (villi intestinales) and evaginations (glandulae (cryptae) intestinales) of the mucosa. The epithelial cells of the mucosa originate within the glandulae and migrate up to the tip of the villi. Three types of cells can be distinguished within the mucosal epithelium: chief cells, goblet cells and endocrine cells. The chief cells are the absorptive cells of the epithelium, which possess microvilli in the cell membrane for further enlargement of the functional surface area. The goblet cells produce the mucus that covers the internal surface of the intestine. The centre of the villi is composed of connective tissue, with large numbers of capillaries. Smooth muscles from the lamina muscularis mucosae enter the villi and are used for the movement of the villi. Movement of the villi results in the change of the total surface area available for absorption. Inside the mucosa lymphoid structures are also present. In most cases these are very small, but in some cases they aggregate and become large and visible to the naked eye. One of the most clear examples of these structures is the caecal tonsils found in the medial wall of the caeca of chicken (King, 1975), and proximal and distal bands of lymphoid tissue in the digestive tract of ducks (Leibovitz, 1968).

The muscle layer of the intestine consists of a relatively thin longitudinal layer (stratum longitudinale) and a thicker circular layer (stratum circulare). The thickness of the muscular tunic can vary along the length of the digestive tract, especially the longitudinal layer. In the chicken the muscles become more profound more distally in the intestines.

Macrostructure

The avian intestines consist of two parts: the small intestine and the large intestine. The small intestine is subdivided into three parts, which are not always clearly recognizable. These parts are the duodenum, the jejunum and the ileum.

The first part of the small intestine is the duodenum, which forms a loop (ansa duodeni) with a descending part (pars descendens) and an ascending part (pars ascendens). The loop of the duodenum points caudally and is located at the ventral side of the body cavity. The duodenal loop encloses the pancreas, which will be discussed later. Both the pancreas and the liver have ducts that end in the duodenum. Depending on the species the pancreas has one, two or three ducts, while in most cases the liver has two separate ducts that end on the duodenum. In general, the ducts enter the duodenum at the distal part of the ascending part of the duodenum, although in some species the ducts end on the proximal part of the descending part. The end of the duodenum is defined to be located where the ascending part of the duodenal loop crosses cranial to the radix mesenterica. The duodenum arches here (flexura duodenojejunalis) and continues as the jejunum.

The jejunum lies on the right side of the body, and is the longest part of the intestine. In most domestic species, including ratites and galliforms, the jejunum is suspended in the mesenteries without the presence of clear organization, which resembles strongly the anatomy of both mammals and reptiles. The jejunum is separated from the ileum per definition by the diverticulum vitellinum, the remains of the embryonic yolk sac. The diverticulum is variable in size between species, but is often small or absent in columbiforms. In most domestic species the transition from ileum to rectum is marked by the papilla ilealis. This is a protrusion of the terminal end of the ileum into the lumen of the rectum, which can be closed by a sphincter muscle (musculus sphincter ilei; Mahdi and McLelland, 1988). The final part of the intestine consists of the rectum and the caeca. Because of the special function of the caeca these will be discussed

separately. The rectum is relatively short and is located at the dorsal wall of the body cavity. It starts just caudal of the gizzard and runs in a straight line to the coprodeal part of the cloaca. It has been reported that some folds might be present in the rectum of some ratites, although these might not be permanent (Mitchell, 1896; Beddard, 1911). In general, the rectum has small villi and relatively few glands, with very little goblet cells (Clauss et al., 1991). A sphincter muscle (musculus sphincter rectocoprodei) separates the rectum from the coprodeum. The coprodeal part of the cloaca opens into the urodeum, on which the ureters and the genital tract end. The most distal part of the cloaca is the proctodeum, which opens externally through the ventus.

The caeca

In the avian body plan two caeca can be found at the beginning of the rectum. In almost all species the caeca connect to the intestine separately in the lateral walls of the intestine. In the ostrich, however, the two caeca connect to each other and have a single combined opening into the intestine. In both the duck and the chicken, sphincter muscles (musculus sphincter ceci) have been described around the opening to caeca (ostium ceci). This sphincter in combination with the caecal villi assures a good closure for even small food particles, while fluids can always pass the sphincter (Ferrando et al., 1987). As in mammals, the caeca are connected to the ileum by mesenteric connections known as the plica ilieocecalis; as a result of this connection the caeca follow the orientation of the ileum. This means that from the entrance of the caeca inward, the orientation is first cranial, followed, in the case of long caeca, by a looping back to caudal.

The size of the caeca can vary between species, with some species having very large caeca with a function in fermentation, while in other species the caeca are completely absent. In species with highly reduced caeca, such as the columbiforms, the caeca consist only of lymphoid tissue and have no function in digestion. In anseriforms and galliforms the caeca are relatively long and slender and resemble strongly the morphology of the intestines. The external morphology does not show many distinctive characters although three different regions are described based on the diameter of the caecum and the morphology of the internal wall. The base of the caecum (basis ceci) near the opening into the intestine has a relatively small diameter and a thick wall with large villi, the body of the caecum (corpus ceci) is wider with a thin wall, small villi and longitudinal folds, while the end of the caecum (apex ceci) is somewhat expanded, has small villi and both longitudinal and transverse folds (Ferrer et al., 1991). This subdivision is very clear in galliforms but in anseriforms the differences between the different parts are much smaller.

In many herbivorous bird species such as ostriches and rheas, the caeca can become extremely large, both in length and diameter. Besides an enlargement of the caecum as a whole, the caeca of ratites appear to be subdivided into small compartments. This is the result of submucosal internal spiral folds over the full length of the caecum, and not, as in other birds, due to teniae (Jacobshagen, 1937).

Pancreas

The pancreas produces the majority of digestive enzymes and also has an endocrine function. As in all vertebrates, the pancreas is located in the first intestinal loop of the duodenum, but is not restricted to this area. It can extend more cranially in the direction of the stomach and can be found in the mesenteries up to the level of the spleen. The pancreas is a thin lobulated organ without a clear morphology and can vary in colour between pale yellow and bright red. According to the modern nomenclature four different lobes can be distinguished in the pancreas (Mikami and Ono, 1962; Mikami et al., 1985): the dorsal lobe (lobus pancreatis dorsalis), the ventral lobe (lobus pancreatis ventralis), the third lobe (lobus pancreatis tertius) and the splenic lobe (lobus pancreatis splenalis). The dorsal lobe runs alongside the descending part of the duodenum and is divided into two parts (pars dorsalis and pars ventralis), which are separated by a clear groove (Paik et al., 1974). The ventral lobe runs along the ascending part of the duodenum. The third lobe, which was formerly considered a part of the ventral lobe, is also situated on that side. The splenic lobe is considerably smaller than the

other three and extends cranially in the mesentery and ends near the spleen. The different lobes are interconnected by parenchymous tissue, although the level of connection is variable between species and individuals. Species with little connections between the lobes are ducks and pigeons, in the latter species so much that the splenic lobe is sometimes completely isolated from all others.

The digestive enzymes are transported to the duodenum through the pancreatic ducts, which end on the distal ascending part of the duodenum on the inside of the duodenal loop. The draining ducts of the liver (bile ducts) end on the duodenum in this same position and the order in which pancreatic and bile ducts end on the duodenum may vary (Paik et al., 1974). The common outflows of the pancreatic and bile ducts form a papilla (papilla duodeni), which in chicken can be 5 mm long (Fehér and Fáncsi, 1971). Within the most distal part of the pancreatic ducts there are sphincter muscles (musculus sphincter ductus pancreatici) present that can control the outflow of pancreatic juice. The number of pancreatic ducts varies between species, with a maximum of three (ductus pancreaticus dorsalis, ventralis and tertius). Two ducts are present in the turkey and most anseriforms, and three ducts can be found in the pigeon and the chicken. The different ducts can have different origins within the lobes and might have a variable contribution in the transport of enzymes. In the chicken most of the transport is through the ventral duct and only small cranial parts of the pancreas are drained by the other two ducts.

The Liver

The liver is a very large gland that is located around the heart and lies in the ventral part of the body cavity. The area on the ventral side of the liver, which folds around the heart, is known as the impressio cardiaca. On the dorsal side the liver folds around a large number of organs, which results in an irregular surface with a number of possible impressions (impressio proventricularis, i. ventricularis, i. duodenalis, i. jejunalis, i. splenalis), including an impressio testicularis in male specimens. The colour of the liver in adult individuals of most domestic species is red-brown. In hatchlings, however, the colour is yellow due to the pigments that are present in the yolk, which are absorbed along with the lipids (McLelland, 1979). The liver is divided into a right (lobus dexter hepatis) and a left lobe (lobus sinister hepatis). The latter is often subdivided into a lateral (pars lateralis) and a medial part (pars medialis) in domestic species. In the chicken and turkey this subdivision is very clear due to a very deep incision (incisura lobaris). On the different lobes several papillae can be distinguished, although the presence of these varies highly between species.

On the visceral surface of the liver a transverse groove is present that stretches over both the liver lobes. In this area, known as the hilus of the liver, both the hepatic arteries and the vena portae hepatis enter the liver and the bile ducts leave the liver. The hepatic arteries emerge from the coeliac artery, which is a direct branch of the aorta. The vena portae hepatis contains the blood leaving the other organs of the digestive tract that is first transported through the liver before it is transported to the caudal vena cava by the very short vena hepatica.

The main function of the liver in relation to digestion is the production of bile. The bile is produced by the hepatocytes of the liver and transported first by small capillary bile canaliculi (canaliculi biliferes). The bile canaliculi unite to form the lobar bile ducts (ductus biliferes), which then unite to form the left and right hepatic ducts (ductus hepaticus dexter/sinister) named after the lobe they originate from. In species possessing a gall bladder (vesica biliaris) the left and right hepatic ducts connect and form the common hepatoenteric duct (ductus hepatoentericus communis), which transports the bile to the intestine and can be compared to the ductus choledochus in mammals. A second branch of the right hepatic duct continues into the gall bladder. This duct is called the ductus hepatocysticus and can be compared to the ductus cysticus in mammals. From the gall bladder the bile is transported to the duodenum in a separate duct known as the cysticoenteric duct (ductus cysticoentericus). In species without a gall bladder, such as the columbiforms and some ratites, the second branch of the right hepatic duct ends directly into the duodenum as the right hepatoenteric duct (ductus hepatoentericus dexter), although in the ostrich this duct is completely degenerated (Newton and Gadow, 1896). In the goose the common hepatoenteric duct splits to form an additional accessory hepatic duct (ductus hepatoentericus accessorius) with a separate opening into the duodenum (Simić and Janković, 1959).

As mentioned before, the bile draining ducts enter the duodenum in general in close relation to the pancreatic ducts in the distal part of the ascending duodenum. In ostriches and columba, however, the common hepatic duct enters the duodenum in the proximal part of the descending limb.

Acknowledgements

I would like to thank Dr Herman Berkhoudt for his useful comments on the chapter and providing me with additional literature. I would like to thank Martin Britijn and Professor (Dr) G.A. Zweers for providing the original drawings of the oropharynx of the mallard and the duck, and Springer Science and Business Media for granting me permission to use these. I am also grateful to Professor (Dr) C.J.G. Wensing for granting me permission to use the original drawings used in Figs 2.3 and 2.4.

References

Baumel, J.J., King, A.S., Breazile, J.E., Evans, H.E. and Vanden Berge, J.C. (1993) *Handbook of Avian Anatomy: Nomina Anatomica Avium*, 2nd edn. Nuttal Ornithological Club, Cambridge, Massachusetts.

Beddard, F.E. (1911) On the alimentary tract of certain birds and on the mesenteric relations of the intestinal loops. *Proceedings of the Zoological Society London* 1, 47–93.

Belman, A.L. and Kare, M.R. (1961) Character of salivary flow in the chicken. *Poultry Science* 40, 1377.

Berkhoudt, H. (1985) Structure and function of avian taste receptors. In: King, A.S. and McLelland, J. (eds) *Form and Function in Birds*, Part III. Academic Press, San Diego, California, pp. 463–496.

Bhattacharyya, B.N. (1980) The morphology of the jaw and tongue musculature of the common pigeon, *Columba livia*, in relation to its feeding habit. *Proceedings of the Zoological Society Calcutta* 31, 95–127.

Bock, W. (1964) Kinetics of the avian skull. *Journal of Morphology* 114, 1–42.

Bolton, W. (1965) Digestion in the crop. *British Poultry Science* 6, 97–102.

Bout, R.G. and Zweers, G.A. (2001) The role of cranial kinesis in birds. *Comparative Biochemistry and Physiology Part A* 131, 197–205.

Bühler, P. (1981) Functional anatomy of the avian jaw apparatus. In: King, A.S. and McLelland, J. (eds) *Form and Function in Birds*, Vol. 2. Academic Press, London.

Calhoun, M.H. (1954) *Microscopic Anatomy of the Digestive System*. Iowa State College Press, Ames, Iowa.

Chodnik, K.S. (1947) A cytological study of the alimentary tract of the domestic fowl (*Gallus domesticus*). *Quarterly Journal of Microscopic Science* 89, 75–87.

Clauss, W., Dantzer, V. and Skadhauge, E. (1991) Aldosterone modulates Cl secretion in the colon of the hen (*Gallus domesticus*). *American Journal of Physiology* 261, R1533–R1541.

Cornselius, C. (1925) Morphologie, Histologie, und Embryologiedes Muskelmagen der Vögel. *Gegenbauers Morphologischer Jahrbücher* 54, 507–559.

Denbow, D.M. (2000) Gastrointestinal anatomy and physiology. In: Whittow, G.C. (ed.) *Avian Physiology*, 5th edn. Academic Press, San Diego, California.

Dyce, K.M., Sack, W.O. and Wensing, C.J.G. (2002) *Textbook of Veterinary Anatomy*. Saunders, St Louis, Missouri.

Dzuik, H.E. and Duke, G.E. (1972) Cineradiographic studies of gastric motility in turkeys. *American Journal of Physiology* 222, 159–166.

Eglitis, I. and Knouff, R.A. (1962) An histological and histochemical analysis of the inner lining and glandular epithelium of the chicken gizzard. *American Journal of Anatomy* 111, 49–66.

Feder, F.H. (1972) Zur mikroskopischen Anatomie des Verdauungsapparates beim Nandu (*Rhea americana*). *Anatomischer Anzeiger* 132, 250–265.

Fehér, G. and Fáncsi, T. (1971) Vergleichende Mophologie der Bauchspeicheldrüse von Hausvögeln. *Acta Veterinaria Hungarica* 21, 141–164.

Ferrando, C., Vergara, P., Jiménez, M. and Goñalons, E. (1987) Study of the rate of passage of food with chromium-mordanted plant cells in chickens (*Gallus gallus*). *Quarterly Journal of Experimental Physiology* 72, 251–259.

Ferrer, R., Planas, J.M., Durfort, M. and Moreto, M. (1991) Morphological study of the caecal epithelium of the chicken (*Gallus gallus domesticus* L.). *British Poultry Science* 32, 679–691.

Fitzgerald, T.C. (1969) *The Coturnix Quail Anatomy and Histology*. Iowa State University Press, Ames, Iowa.

Flower, W.H. (1860) On the structure of the gizzard of the Nicobar Pigeon, and other granivorous birds. *Proceedings of the Zoological Society London* 1860, 330–334.

Fowler, M.E. (1991) Comparative clinical anatomy of ratites. *Journal of Zoo and Wildlife Medicine* 22, 204.

Gentle, M.J. and Breward, J. (1986) The bill tip organ of the chicken (*Gallus gallus* var. *domesticus*). *Journal of Anatomy* 145, 79–85.

Ghetie, V., Chitescu, St., Cotofan, V. and Hildebrand, A. (1976) *Anatomical Atlas of Domestic Birds*. Editura academiei republicii socialiste Romania, Bucharest, Romania.

Glerean, A. and Katchbarian, E. (1965) Estudo histologico e histoquimico da moela de *Gallus* (*gallus*) *domesticus*. *Revista de Farmacia e Bioquimica da Universidade de Sao Paolo* 2, 73–84.

Gottschaldt, K.M. (1985) Structure and function of avian somatosensory receptors. In: King, A.S. and McLelland, J. (eds) *Form and Function in Birds,* Part III. Academic Press, San Diego, California.

Greschik, E. (1935) Die Zunge von *Anser albifrons* und *Anser erythropus* L. *Kócsag Budapest* 8, 7–17.

Groebbels, F. (1929) Über die farbe der cuticula im muskelmagen der vögel. *Zeitschrift fuer Vergleichende Physiologie* 10, 20–25.

Güntürkün, O. (2000) Sensory physiology: vision. In: Whittow, G.C. (ed.) *Avian Physiology*, 5th edn. Academic Press, San Diego, California.

Gussekloo, S.W.S. (2000) The evolution of the palaeognathous birds. Thesis, Leiden University, The Netherlands.

Gussekloo, S.W.S., Vosselman, M.G. and Bout, R.G. (2001) Three-dimensional kinematics of skeletal elements in avian prokinetic and rhynchokinetic skulls determined by roentgen stereophotogrammetry. *Journal of Experimental Biology* 204, 1735–1744.

Gussekloo, S.W. and Bout, R.G. (2005a) The kinematics of feeding and drinking in palaeognathous birds in relation to cranial morphology. *Journal of Experimental Biology* 208 (17), 3395–3407.

Gussekloo, S.W. and Bout, R.G. (2005b) Carnial Kinesis in palaeognathous birds. *Journal of Experimental Biology* 208 (17), 3409–3419.

Heidweiller, J., van Loon, J.A. and Zweers, G.A. (1992) Flexibility of the drinking mechanism in adult chickens (*Gallus gallus*) (Aves). *Zoomorphology* 111, 141–159.

Herd, R.M. (1985) Anatomy and histology of the gut of the emu (*Dromaius novaehollandiae*). *Emu* 85, 43–46.

Herrel, A., Aerts, P. and de Vree, F. (2000). Cranial kinesis in geckoes: functional implications. *Journal of Experimental Biology* 203, 1415–1423.

Hodges, R.D. (1974) *The Histology of the Fowl*. Academic Press, London.

Homberger, D. and Meyers, R.A. (1989) Morphology of the lingual apparatus of the domestic chickens, *Gallus gallus*, with special attention to the structure of the fasciae. *American Journal of Anatomy* 186, 217–257.

Humphrey, P.S. and Clarck, G.A. (1964) The anatomy of waterfowl. In: Delacour, J. (ed.) *The Waterfowl of the World*, Vol. IV. Country Life Ltd, London.

Jacobshagen, E. (1937) Mittel- und Enddarm. In: Bolk, L., Göppert, E., Kallius, E. and Lubosch, W. (eds) *Handbuch der Vergleichende Anatomie der Wirbeltiere*, Vol. 3. Urban-Schwarzenberg, Berlin and Vienna.

Jerrett, S.A. and Goodge, W.R. (1973) Evidence for amylase in avian salivary glands. *Journal of Morphology* 139, 27–46.

Kaden, L. (1936) Über epithel und Drüsen des Vogelschlundes. *Zoologischer Jahrbücher, Abteilung Anatomie und Ontogenie des Tiere* 61, 421–466.

King, A.S. (1975) Aves lymphatic system. In: Getty, R. (ed.) *Sisson and Grossman's the Anatomy of the Domestic Animals*, Vol. 2. Saunders, Philadelphia, Pennsylvania.

Komarek, V., Malinovsky, L. and Lemez, L. (eds) (1982) *Anatomia Avium Domesticarum et Embryologia Galli*, Vols I, II. Priroda, Bratislava.

Kooloos, J.M. (1986) A conveyer-belt model for pecking in the mallard (*Anas platyrhynchos* L.). *Netherlands Journal of Zoology* 36, 47–87.

Kooloos, J.G.M., Kraaijeveld, A.R., Langenbach, G.E.J. and Zweers, G.A. (1989) Comparative mechanics of filter feeding in *Anas platyrhynchos*, *Anas clypeata* and *Aythya fuligula* (Aves, Anseriformes). *Zoomorphology* 108, 269–290.

Kudo, S. (1971) Electron microscopic observations on avian esophageal epithelium. *Archeologie Histologie Japan* 33, 1–30.

Landolt, R. (1987) Vergleichend funktionelle Morphologie des Verdaaungstrakte der Taube (Columbidae) mit besonderes Berücksichtigung der adaptiven Radiation der Fruchttauben (Treroninae). Teil I. *Zoologischer Jahrbücher Anatomie* 116, 285–316.

Leibovitz, L. (1968) *Wenyonella philiplevinei*, N.SP., a coccidial organism of the white pekin duck. *Avian Diseases* 12, 670–681.

Lienhart, R. (1953) Recherches sur le role des cailloux contenus dans le gésier des oiseaux granivores. *Bulletin de Societe Scientific de Nancy* 12, 5–9.

Lucas, A.M. and Stettenheim, P.R. (1972) *Avian Anatomy. Integument*. Part I. *Agricultural Handbook*, No. 362, United States Department of Agriculture, Washington, DC.

Luppa, H. (1959) Histogenetische und histochemische Untersuchungen am epithel des embryonalen Hühnermagens. *Acta Anatomica* 39, 51–81.

Mahdi, A.H. and McLelland, J. (1988) The arrangement of the muscle at the ileo-caeco-rectal junction of the domestic duck (*Anas platyrhynchos*) and the presence of anatomical sphincters. *Journal of Anatomy* 161, 133–142.

Malinovsky, L. and Pac, L. (1980) Ultrastructure of the herbst corpuscle from beak skin of the pigeon. *Zeitschrift fur Mikroskopisch Anatomische Forschung* 94, 292–304.

Marvin, F. and Těšik I. (1970) Comparative anatomical study of the tongue of the fowl, turkey and guineafowl. *Acta Veterinaria Brunensis* 39, 235–243.

McAtee, W.L. (1906) The shedding of the stomach lining by birds. *Auk* 23, 346.

McAtee, W.L. (1917) The shedding of the stomach lining by birds, particularly as exemplified by the anatidae. *Auk* 34, 415–421.

McLelland, J. (1975) Aves digestive system. In: Getty, R. (ed.) *Sisson and Grossman's the Anatomy of the Domestic Animals*, Vol. 2. Saunders, Philadelphia, Pennsylvania.

McLelland, J. (1979) Digestive system. In: King, A.S. and McLelland, J. (eds) *Form and Function in Birds*. Academic Press, London, pp. 69–181.

McLelland, J. (1993) Apparatus digestorius [systema alimentarium]. In: Baumel, J.J., King, A.S., Breazile, J.E., Evans, H.E. and Van den Berge, J.C. (eds) *Handbook of Avian Anatomy: Nomina Anatomica Avium*, 2nd edn. Nuttal Ornithological Club, Cambridge, Massachusetts.

Meinertzhagen, R. (1954) Grit. *Bulletin of the British Ornithological Club* 74, 97–102.

Michel, G. (1971) Zur Histologie und Histochemie der Schleimhaut des Drüsen- und Muskelmagens von Huhn und Ente. *Maryland Veterinary Medicine* 23, 907–911.

Mikami, S.I. and Ono, K. (1962) Glucagon deficiency induced by extirpation of alpha islets of the fowl pancreas. *Endocrinology* 71, 464–473.

Mikami, S.I., Taniguchi, K. and Ishikawa, T. (1985) Immunocytochemical localisation of the pancreatic islet cells in the Japanese quail, *Coturnix coturnix japonica. Japanese Journal of Veterinary Science* 47, 357–369.

Mitchell, P.C. (1896) On the intestinal tract of birds. *Proceedings of the Zoological Society London* 1896, 136–159.

Mule, F. (1991) The avian oesophageal motor function and its nervous control: some physiological, pharmacological and comparative aspects. *Comparative Biochemistry and Physiology A* 99, 491–498.

Necker, R. (2000) The somatosensory system. In: Whittow, G.C. (ed.) *Avian Physiology*, 5th edn. Academic Press, San Diego, California, pp. 57–69.

Newton, A. and Gadow, H. (1896) *A Dictionary of Birds*. Black, London.

Nickel, R., Schummer, A. and Seiferle, E. (1973) *Lehrbuch der Anatomie der Haustiere, band V: Anatomie der Hausvögel*. Paul Parey's Verlag, Berlin, Hamburg, Germany.

Norris, R.A. (1961) Colors of stomach linings of certain passerines. *Wilson Bulletin* 73, 380–383.

Paik, Y.K., Fujioka, T. and Yasuda, M. (1974) Comparative and topographical anatomy of the fowl. LXXVIII. Division of pancreatic lobes and distribution of pancreatic ducts. *Japanese Journal of Veterinary Science* 36, 213–229.

Parkes, K.S. and Clark, G.A. (1966) An additional character linking ratites and tinamous, and an interpretation of their monophyly. *Condor* 68, 459–471.

Pernkopf, E. and Lehner, J. (1937) Vorderdarm. Vergleichende beschreibung des vorderdarm bei den einzelen klassen der kranioten. In: Bolk, L., Göppert, E., Kallius, E. and Lubosch, W. (eds) *Handbuch der Vergleichende Anatomie der Wirbeltiere*, Vol. III. Urban-Schwarzenberg, Berlin and Vienna.

Philips, S.M. and Fuller, R. (1983) The activity of amylase and a trypsin-like protease in the gut contents of germ free and conventional chicken. *British Poultry Science* 24, 115–121.

Pinchasov, Y. and Noy, Y. (1994) Early postnatal amylolysis in the gastrointestinal tract of turkey poults *Meleagris gallopavo. Comparative Biochemistry and Physiology A* 107, 221–226.

Saito, I. (1966) Comparative anatomical studies of the oral organs of the poultry. IV. Macroscopical observations of the salivary glands. *Bulletin of Faculty of Agriculture Univresity Miyazaki* 12, 110–120.

Schaller, O. (1992) *Illustrated Veterinary Anatomical Nomenclature*. Ferdinant Enke Verlag, Stuttgart.

Schelpelmann, E. (1906) Über die gestalende wirkun verschiedene Ernärung auf die organe der gans, inbesondere über die funktionelle anpassung an sie nahrung. *Arch. Entw. Mech. Org* 21, 500–595.

Simić, V. and Janković, N. (1959) Ein Beitrag zur Kenntnis der Morphologie und Topographie der Leber beim Hausgeflügel und der Taube. *Acta Veterinaria Beograd* 9, 7–34.

Starck, J.M. (1999) Phenotypic flexibility of the avian gizzard: rapid, reversible and repeated changes of organ size in response to changes in dietary fibre content. *Journal of Experimental Biology* 202 (22), 3171–3179.

Starck, J.M. (2003) Shaping up: how vertebrates adjust their digestive system to changing environmental conditions. *Animal Biology* 53(3), 245–257.

Swenander, G. (1902) Studien über den bau des Schlundes und Magens der Vögel. *K. Norske Videnks Selk. Skr.* 6, 1–240.

Tomlinson, C.A. (2000) Feeding in palaeognathous birds. In: Schwenk, K. (ed.) *Feeding: Form, Function and Evolution in Tetrapod Vertebrates*. Academic Press, San Diego, California.

Van der Heuvel, W. and Berkhoudt, H. (1998) Pecking in the chicken (*Gallus gallus domesticus*): motion analysis and stereotypy. *Netherlands Journal of Zoology* 48, 273–303.

van Gennip, E.M.S.J. (1986) The osteology, arthrology and mycology of the jaw apparatus of the pigeon (*Colomba livia* L.). *Netherlands Journal of Zoology* 36(1), 1–46.

van Gennip, E.M.S.J. and Berkhoudt, H. (1992) Skull mechanics in the pigeon, *Colomba livia*. A three-dimensional kinematic model. *Journal of Morphology* 213, 197–224.

Webb, M. (1957) The ontogeny of the cranial bones, cranial peripheral and cranial parasympathetic nerves, together with a study of the visceral muscles of *struthio*. *Acta Zoologica* 38, 1–123.

Webb, T.E. and Colvin, J.R. (1964) The composition, structure, and mechanism of formation of the lining of the gizzard of the chicken. *Canadian Journal of Biochemistry and Physiology* 42, 59–70.

Yasuda, M. (2002) *The anatomical atlas of guallus.* University of Tokyo Press, Tokyo.

Zietzschmann, O. (1911) Der Verdauungsapparat der Vögel. In: *Ellenbergers Handbuch der Vergleichende Mikroskopische Anatomie der Haustiere*, 18th edn. Springer-Verlag, Berlin, Germany, pp. 1023–1031.

Ziswiler, V. and Farner, D.S. (1972) Digestion and the digestive system. In: Farner, D.S. and King, J.R. (eds) *Avian Biology*, Vol. 2. Academic Press, London, pp. 343–430.

Zweers, G.A. (1982) Pecking of the pigeon (*Columba livia* L.). *Behaviour* 81(1/2), 173–230.

Zweers, G.A. (1985) Generalism and specialism in avian mouth and pharynx. *Fortschritte der Zoologie* 30, 189–201.

Zweers, G.A., Gerritsen, A.F.Ch. and Kranenburg-Voogd, P.J. (1977) Mechanics of feeding of the mallard (*Anas platyrhynchos* L.; Aves Anseriformes). In: Hecht, M.K. and Szalay, F.S. (eds) *Contibutions to Vertebrate Evolution*, Vol. 3. Karger, Basel, Switzerland.

Zweers, G.A., Berkhoudt, H. and Van den Berge, J.C. (1994) Behavioral mechanisms of avian feeding. *Advances in Comparative and Environmental Physiology* 18, 243–279.

Zweers, G.A., Van den Berge, J.C. and Berkhoudt, H. (1997) Evolutionary patterns of avian trophic diversification. *Zoology* 100, 25–57.

3 Feeding Behaviour and Mechanisms in Domestic Birds

V. Bels[1] and S. Baussart[2]

[1]FRE 2696/USM 302 – Adaptation et Évolution des Systèmes Ostéo-Musculaires, Département Écologie et Gestion de la Biodiversité, Muséum National d'Histoire Naturelle, 57, Pavillon d'Anatomie Comparée, BP 55, rue Buffon, 75005 Paris, France; [2]Muséum Nationale d'Histoire Naturelle rue Paul Pastur 11, B – 7800 – Ath, France

Introduction

All domestic animals differ from wild ancestors because a series of morphological, physiological and behavioural traits are manipulated under the process of artificial selection. Price (1984) defines domestication as 'a process by which a population of animals becomes adapted to man and the captive environment, by some combination of genetic changes occurring over generations and environmentally induced developmental events recurring during each generation'. Domesticated birds belonging to different phylogenetic lineages have undergone one or multiple domestication episodes that changed their genetic diversity and modify a majority of biological traits. These biological changes can be observed through a series of modifications of their behavioural activities (Appleby et al., 2004).

The first domestication of birds was probably associated with producing meat in a short period of time. For meat production, about 12 species of wild birds have been selected in different areas of the world and at different times in human history: (i) terrestrial birds (i.e. ratites, chicken, pigeon, turkey, fowl, quail, goose and pheasant); and (ii) aquatic birds (duck). Rapidly, a larger number of other species, from the same lineages (i.e. anseriforms) or not, have

attracted the attention of humans; a large number of species from almost all bird lineages are bred in captivity for different purposes, from leisure to meat production.

In domestic birds, two major traits have been selected: a rapid growth rate and a greater body size. This selection results in accompanying large modifications of physiological demands regarding food and liquid ingestion. Several behavioural aspects in various species with specialized structures and mechanisms for food selecting, grasping and finally ingesting have been influenced by these new demands, and one key response to selection of increased meat production is the decrease in frequency of the behaviours with higher energy costs (Kear, 2005).

The success of selection of new behavioural traits is strongly influenced by the interaction between the adaptive plasticity, the individual differences within each species and the selective procedure (Price, 2002). These new traits resulting from a transition from the wild to domestication over generations are selected by considerable changes of the environmental constraints, such as the space available and food provisioning that can decrease time and energy spent for foraging and grasping food. Probably, one critical point for such successful selection is the knowledge of all characteristics of food that affect the feeding behaviour (i.e. nutritional requirements, food preference, feeding pattern). The relationship between food properties and the success in selected traits for meat production cannot be solved by a complete understanding of the feeding behaviour and mechanism of domestic birds.

The aim of this chapter is to provide a review with some examples showing the interrelationship between behavioural, physiological and mechanical properties that play a key role for understanding the feeding behaviour in domestic birds. We then show how feeding behaviour in birds with various trophic systems share some basic functional characteristics of the feeding cycle suggested for tetrapods (Bramble and Wake, 1985; Reilly and Lauder, 1990). The ability of modulating these characteristics provides the functional basis for improving the nutrition of domestic birds.

Feeding Postures

Feeding is a complex behaviour involving neuromotor control of a network of muscles and skeletal elements under the influence of sensory inputs and motivational control (Fig. 3.1). Dubbeldam (1989, 1991) stated that, per definition, the motor feeding systems in birds involve movements of the head and vertebral column, beak and tongue. He determined that positions of head and beak depend upon movements of the cervical column controlled by an 'orientating system'. One emergent phenomenon of feeding in all vertebrates is that food (and liquid) ingestion is the result of coordinated movements not only of two (neck and head) but of all the components of the musculoskeleton: the trophic system (including beak), the axial system (including cervical area) and the appendicular systems including limbs and their attachment to the vertebral column (Bels et al., 2002). In their extensive analysis, Van den Heuvel and Berkhoudt (1998) report that chickens take up, by locomotion, an 'adequate position' for pecking, and as soon as this position is reached, they execute a series of head/jaw movements determining the complete feeding sequence. This adequate position corresponds to the final stage of a modal-action-pattern (MAP) (Barlow, 1968, 1977) controlled by simultaneous motor actions of the axial and locomotor systems before the neck and head begin to move for food ingestion. Thus, three motor systems need to be integrated into a complex network under the control of the neural system and modulated by sensory information during feeding. We suggest qualifying 'adequate position' by posture

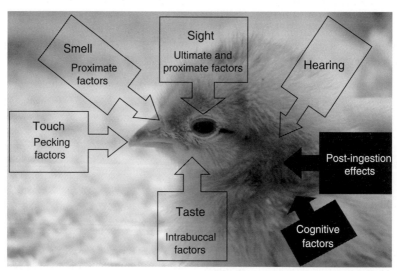

Fig. 3.1. Sensory informations regulate the feeding behaviour of all birds under the control of motivational effect.

because the posture can be defined as a rather fixed body position resulting from a complex MAP of the appendicular and axial muscular system. Postures are constrained by the construction and the shape of the skeletal elements and the musculature allowing their movements. A variety of motivational and environmental inputs affect the neuromotor network under the MAP that produces the postures. These inputs can act to adjust the frequency of the postures within or between the feeding sequences for improving efficient gain of food each time. These adjustments can be produced by motivational effects under hormonal control (i.e. leptin) or strongly related to environmental factors (i.e. density of animals, absence of predators). But the postures completing a final stage of one MAP are stereotyped. These postures provide the 'mechanical environment' for efficient neck and head movements during all behaviours such as the phases of feeding, from food catching to transport towards the oesophagus. Domestic ducks are particularly interesting models to investigate the integration of the motor patterns of axial, trophic and locomotor systems, and their modulations to reach these postures during feeding, because these birds are able to grasp food on land and in water. Consequently they must adopt determined postures for efficient food ingestion in relationship with the environmental constraints.

The 'posture' can be easily determined in the two main lineages of representative birds used for meat production: galliforms and anseriforms. Among their 'everyday behaviours' (see McFarland, 1981) adult chickens show a repertoire of body postures in relationship with the behavioural contexts. Two postures can be easily determined by the positions of the limbs and the axial and cervical skeleton: 'exploratory' and 'feeding'. The movements within the chain of vertebrae determine the position of the head in both postures. Chickens can take a large series of postures (i.e. foraging postures) between the 'exploratory' posture with head in high position and the 'feeding posture' with head oriented towards the substratum for grasping food (Fig. 3.2). The exploratory posture occurs during exploration of the environment. Adult animals stand or move with head held in a high position. This posture allows us to observe the proximate and

ultimate environmental factors and can become exaggerated in some circumstances (i.e. stress). The foraging postures are usual postures used by the animal while searching for food. Adult chickens move at a rather constant slow speed with the head held at various distances above the substrate. One typical foraging posture is taken by almost horizontal position of the cervical column during locomotion. The feeding posture occurs as soon as the animal fixes on one source of food. In galliforms, the whole body rotates towards the food and the head is moved to the food by movements within the cervical chain (Fig. 3.2). In their highly descriptive analysis of pecking in chicken, Van den Heuvel and Berkhoudt (1998) describe a fixation phase occurring before any food approach and prehension by the head. During this phase, the head of the bird is held above the food with the eyes fully open. This phase is the result of the feeding posture taken by the animal just prior to food prehension. In this chapter, we do not quantify all postures, but simply provide a basis for demonstrating the complexity of the integrative network of motor control between the appendicular and axial skeleton before any action of the trophic system for food grasping.

In anseriforms, animals also take a fixed 'feeding posture' before grasping food on substratum. This posture can be observed during feeding on artificial food and during grazing on grass. These postures occur between locomotor movements. In these postures, the whole body is held horizontally and both hind limbs are often placed next to each other or fixed in stance phases of the locomotor cycle. Then the head is moved towards the food by cyclic neck movements (Fig. 3.3). In domestic ducks able to feed on land and in water, different postures must be taken for terrestrial and aquatic feeding. Different ecological strategies of feeding behaviour of wild ducks have been identified (Green, 1998). Ducks feed closest to the surface by bill dipping and gleaning, at a greater depth by bill dipping and neck dipping, or at the greatest depths mainly by upending and neck dipping. Some species are able to dive for food deep in the water column. Several studies have shown that ecological niches are not overlapping. Different niche needs have to be exploited to produce a different

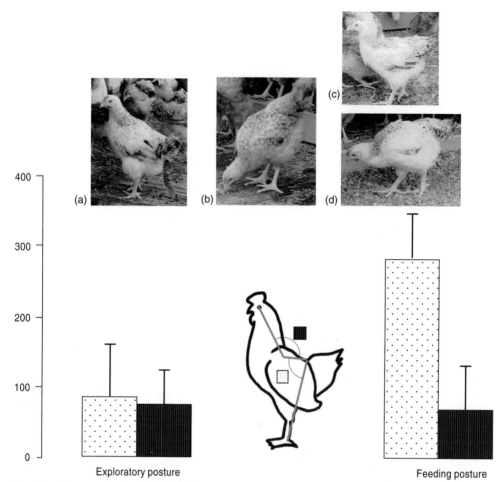

Fig. 3.2. Mean angles (± standard deviation) describing body postures recorded during a series of behavioural activities in chicken. Two angles were simply measured on the body of the animal to show the relative position of the appendicular (limb), axial (body) and cervical (neck) skeletons in two highly different behavioural activities: 'exploratory' and 'feeding' postures. This last posture corresponds to the 'adequate' position of the body previously reported for describing the mechanism of feeding in chicken (Van den Heuvel, 1992). Between both extremely different postures, chicken can take various postures under environmental factors (i.e. stress) and motivational status (i.e. satiety): ▨ angle between a line through the distal segment of the hind limb and a line on the axial skeleton; ▪ angle between a line on the axial skeleton and a line on the cervical chain.

feeding MAP and consequently posture for ingesting food. In the majority of domestic ducks, two modes of feeding postures can be observed. In a first type of aquatic feeding, the animal swims on the surface of the water and must control movement of the body for allowing the beak to be introduced into the water at a very shallow depth. The head always remains at the surface of the water. Zweers and Berkhoudt (1991) suggest that this feeding mechanism called 'filter feeding' is simply an extension of pecking (Zeigler *et al.*, 1971; Zweers *et al.*, 1977, 1994a,b,c). This suggests that motor control of cervical movements should be similar in pecking and filter feeding. In a second type of aquatic feeding, the body is moved towards the deeper source of food by movements of the hind limbs. The body is

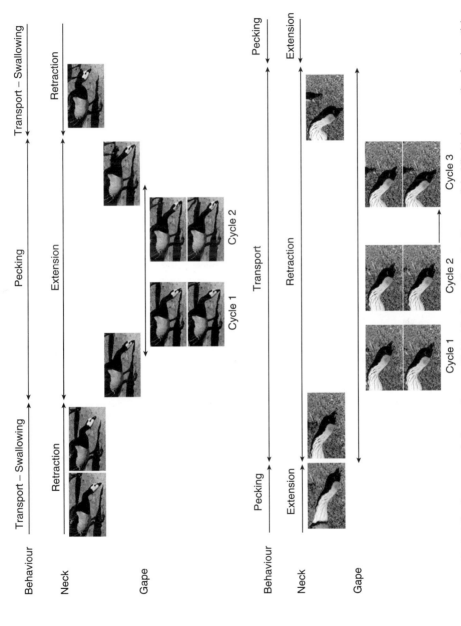

Fig. 3.3. Typical 'feeding posture' of anseriforms with hind limbs positioned rather parallel while birds are eating food particles on the substratum. These postures permit to ingest food on the substratum and transport the food through the buccal cavity.

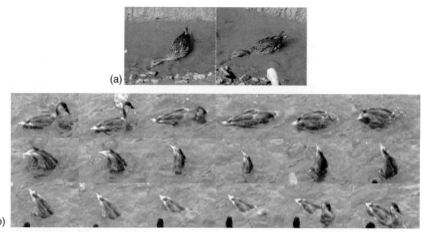

Fig. 3.4. Two typical postures of duck feeding on food in water. (a) The animal controls the body position to be able to feed on food particles at shallow depths. (b) Upending feeding posture. The animal rotates the body to be able to position the head near the food source at greater depths. This posture results from an MAP performed by sequential action of the limbs and the body musculature.

partially maintained at the surface and serves to bring the head into the water at the more efficient position for catching the food (Fig. 3.4). The cervical movements needed to position the head near the food source should also be similar to those observed in pecking. Only motor control of the whole body should be modulated in relationship with the position of the food source. This ability of modulating 'feeding posture' in terrestrial and aquatic environments probably results from behavioural plasticity in birds that are able to exploit different sources of food at the interface of two completely different environments.

In domestic birds, 'novel' feeding postures are also strongly triggered in animals facing particular environmental constraints, mainly related to the artificial devices used for delivering food and liquid. Such behavioural novelty needs to build a new MAP corresponding to a motor response under particular sensory inputs. Faced with artificial devices, each individual must rapidly adjust at the early stages. Probably this response depends on the behavioural plasticity showing a rather important interindividual difference (Fig. 3.5). When they are placed into a classical building for poultry production, all chickens must find the drinking nipple (and cup) placed in a relatively higher position, compared to the 'natural condition' where birds always

drink from water sources on the substratum. Animals bred in captivity need a complete reorganization of the integrated neuromotor control of the posture for gaining water and food in these first hours after hatching. No kinematic studies or neuromotor analyses have shown how this drinking posture and mechanism is initiated. It should also be interesting to compare the response of domestic and wild species faced with these kinds of artificial devices for liquid and food delivery. This response could also be associated with the 'cerebrotypes' described recently in birds (Iwaniuk and Hurd, 2005), showing species sharing similarities in locomotor behaviour, mode of capture or learning ability are housed together. Either artificial selection has modified the brain/behaviour relationship along with domestication or, for any reason, particular 'cerebrotypes' of birds have been selected because they respond to the demands of domestication: animals with 'adapted' behaviour for producing meat in a short period of time.

Head/Neck Systems in Feeding Behaviour

Birds are toothless vertebrates with jaws covered by keratinized sheaths forming the beak.

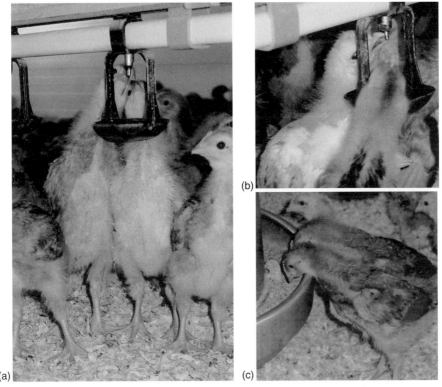

Fig. 3.5. Drinking and feeding postures of chicken at different ages when they face the classical device for water distribution in the production building. These postures are a completely new motor response under the neural control and the modulatory effects of sensory inputs.

In domestic birds, the specialized beak participates in a wide variety of behaviours including food and liquid ingestion, feather pecking and grooming (Fig. 3.6), aggressive behaviour and parental care.

As soon as the feeding posture is assumed, a series of gape cycles (Fig. 3.3) is performed in phases: pecking, grasping, mandibulation and swallowing. In all of these phases, the beak is the unique prehensile organ for food grasping and then plays a major role in association with the hyolingual apparatus for food transport through the oral cavity towards the oesophagus. Both elements of the trophic system used for food catching and transporting towards the oesophagus are moved by a complex cervical system in terrestrial and aquatic feeding. All the movements of the cranio-cervical morphological complex phases are viewed as a stimulus–response chain

(Zeigler, 1976; Zeigler et al., 1980, 1994; Bermejo and Zeigler, 1989; Bermejo et al., 1989) forming a rhythmic stereotyped behaviour. The rhythmic feeding cycles are under the control of a neural network called central-pattern-generator (CPG), which remains to be largely investigated. But some motor parameters of the actions of the neck, jaws, tongue and hyobranchium can be modulated by sensory information, mainly on food properties.

Three main conceptual approaches are used to explore the question of why animals are built as they are and how they work for exploiting resources of their environment. A series of detailed analyses in various domestic birds permits an understanding of the design of organs (e.g. beak, tongue and skull), tissue systems and their performances in feeding on various types of food (Zweers, 1985; Zweers and Berkhoudt, 1991; Zweers et al.,

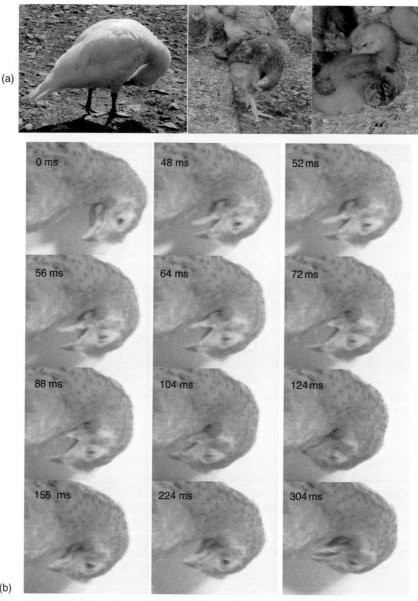

Fig. 3.6. (a) Three examples of behaviours involving beak action and neuromotor control of the cranio-cervical system. (b) Series of pictures illustrating the grooming behaviour in chicken, showing the main role played by the upper and lower jaws.

1994a,b,c). First, the mechanism of food prehension in different kinds of domestic birds (i.e. pigeon, chicken, duck) has been investigated as a classical biological model for studying the neuromotor control mechanisms and discussing the relationship between behav-

ioural and motor control of precise mechanisms such as pecking, straining, probing and filter feeding (Zeigler and Witkovsky, 1968; Zeigler et al., 1980; Kooloos, 1986; Bermejo and Zeigler, 1989; Bermejo et al., 1989; Zweers and Berkhoudt, 1991; Rubega, 2000).

All these studies help to define a 'mechano-space' to suggest an evolutionary hypothesis for deriving the specialized feeding behaviour from ancestral pecking mechanisms (Zweers and Berkhoudt, 1991; Zweers et al., 1994b,c). Second, the neural mechanism underlying the central control of feeding, drinking and sensory feedback has been deeply studied in several birds including chicken and duck (Dubbeldam, 1984, 1989). Third, the basic mechanism of the relationship between the structures of the trophic and neck systems has been largely studied in a wide diversity of birds, mainly for understanding the motor control in a phylogenetic context (Zweers and Berkhoudt, 1991).

As illustrated in Figs 3.3 and 3.4, when the animal is 'fixed' in a 'feeding posture', the position of the head relative to the food is controlled by complex movements of the cervical column. This column is a complex multi-articulated chain establishing head postures and motions (Zweers et al., 1994a,b,c). A large number of detailed analyses demonstrate differences in movement pattern of the neck as the result of anatomical difference. Van der Leeuw (1992) determined a 'General Anatid' pattern in the movement of the cervical column with strong modulation in relationship with scaling effects and probably drinking mechanisms. Differences in neuromotor patterns were also demonstrated between two highly different domestic birds – the mallard and the chicken – during drinking and pecking. Functionally, movements of the vertebrae in feeding chickens follow simple geometric principles for optimizing angular rotation efficiency with movement pattern showing simultaneous rotations in some joints and not in the others. In contrast, neck movements in mallard show a typical pattern of successive, rather than simultaneous, rotations in the rostral part of the neck based on a rather different geometric principle. These different mechanisms were associated with different patterns in neck motor control between chicken and mallard, showing that both anatomy and motor control of the cervical system are highly specialized for performing the same behaviour. But these authors discovered a similar mechanism in feeding in water and on land for ansiform birds, suggesting that neck function in both

environments is controlled by the same neuro-motor pattern. Based on a large series of studies, Zweers et al. (1994b) developed a 'comparator model' to describe the static and dynamic phases of pecking behaviour. In this model, three units control the pecking success: a 'sensor-selector' unit that receives inputs from the food item; a 'comparator' unit comparing information between successive motor commands in one feeding sequence; and a 'highest value passage' unit initiating the motor command of muscles acting for food acceptance (or rejection).

Trophic System in Feeding Mechanism

A large number of experimental studies have presented quantitative data on the feeding mechanism in domestic birds (Kuenzel, 1983; Zweers et al., 1994a–c). In the majority of these studies, successive phases have been determined along the feeding process from food grasping to swallowing. All these phases correspond to jaw and hyolingual apparatus cycles forming a typical eating-response sequence and repeated in a feeding bout or sequence (Zweers, 1982). The feeding mechanism, from grasping to swallowing, is based on the intraoral lingual transport as suggested for all tetrapods. Figure 3.7 presents successive movements of jaws and tongue in chicken pecking very small food items. In the grasping cycle, the upper and lower bills are used alone for catching food particles, whereas the tongue plays a key role in food movement within the buccal cavity during all other feeding cycles, from positioning to swallowing food particles. In food positioning, gape increases only slightly and the tongue moves forward and backward within the buccal cavity. One key point in domestic birds is the effect of particle size on the feeding behaviour. A series of investigations have been used to test the effect of this size on behavioural parameters of pecking (Wauters et al., 1997; Picard et al., 2000). Figure 3.8 summarizes some general features of particle selection by chickens. Particle size has long been an area of interest for improving feeding efficiency in domestic birds. This size must

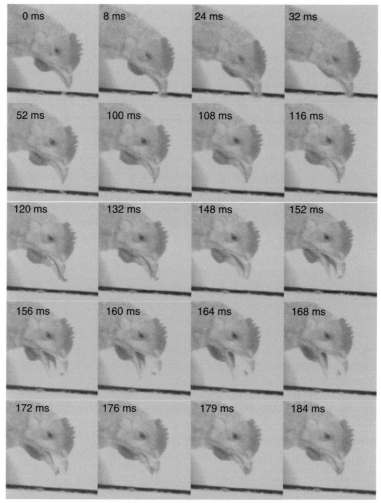

Fig. 3.7. Typical sequence of three phases of feeding behaviour in chicken. The first cycle corresponds to a typical pecking phase (0–116 ms) and the second cycle to a transport cycle (120–184 ms).

encompass both the real size of the food and the texture of the particle. It affects directly the feeding mechanism itself by providing various sensory inputs and therefore modulating the motor control of the movements of the trophic structures from grasping to transport. In a combined functional and behavioural study on the effect of particle size, it is of primary importance to define exactly what is particle size. Particle size can be easily calculated by the geometric mean diameter (GMD). But this size must also include a measure of dispersion (i.e. size range and frequency) provided to the eating animals because the grasping process and

efficiency can be affected by relatively low variation. If determining the mechanism of feeding birds must be studied under standard conditions (i.e. peas of similar diameters), geometric standard deviation providing the range of variation among the different particle sizes distributed to the animals must be known in behavioural studies closer to conditions met in animal production (Nir *et al.*, 1990, 1994, 1995). Birds can respond variably to different particle sizes, and clear preference in particle size of different foods has been demonstrated, probably for one main reason. The sensory perceptions, including visual, tactile, olfactory

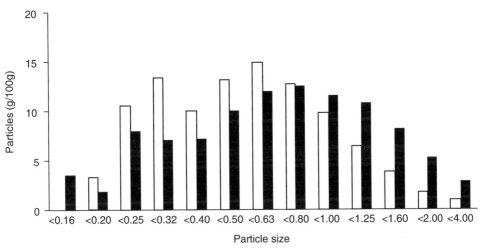

Fig. 3.8. The diameter of particles plays a primary role in food selection by chicken. Particles with diameters greater than 0.8 mm are preferred for soya–maize mixture. (Modified from Wauters *et al.*, 1997; Picard *et al.*, 2000.)

stimuli and even taste of the food, are a crucial link between behavioural responses and the success of food ingestion.

One key point in jaw movements in birds is the ability of the upper jaw to move relative to the braincase. Cranial kinesis has been the subject of a large number of studies in various birds eating various kinds of food. However, in their exhaustive analysis, Bout and Zweers (2001) state that the functional advantages of a kinetic cranium seem to be relatively modest, and actual studies seem to support the hypothesis of no correlation between the type of kinesis and the feeding behaviour of birds (Zusi, 1984; Gussekloo *et al.*, 2001). In all feeding cycles, gape is the result of movements of upper and lower jaws. Gape angle follows the same pattern divided into four stages: slow opening (SO), fast opening (FO), fast closing (FC) and slow closing (SC), as demonstrated in the series of frames corresponding to a typical pecking cycle. During the SO stage (= fixation phase, Van den Heuvel and Berkhoudt, 1998), the tongue moves forward along the floor of the lower jaw. During the FO stage (= approach phase, Van den Heuvel and Berkhoudt, 1998), the beak moves towards the food while the eyes partially close (Fig. 3.9). Peak gape angle is directly correlated with the size of the food particles. At maximal opening, only the tips of the beak are in contact with the food and the

tongue is never visible in the buccal cavity for large food items (pea, Van den Heuvel and Berkhoudt, 1998) or small particles (Figs 3.7 and 3.8). The eyes are not fully closed, suggesting that the animal is potentially able to receive visual information as grasping occurs (Fig. 3.9). At peak gape angle, the bill tips close suddenly for grasping the food during the FC stage. During the FC stage, the head is withdrawn upwards and the eyes fully open again. Then begins a series of different gape cycles. In some cases, the bird executes one gape cycle at low amplitude. These cycles observed for any kinds of food serve to reposition the food within the buccal cavity. With small particles, the animal performs a series of pecking and repositioning cycles before transporting the food through the buccal cavity.

During the pecking cycle, the tongue remains in the buccal cavity. Although the tongue does not touch the food at contact between the food item and the beak, the hyolingual system follows in a cyclic movement within the buccal cavity. The upper and lower jaws are placed around the food particles, and then the beak is suddenly closed on the food item and the bird moves the head upward. Transport of the food up to the pharynx occurs during a variable number of cycles involving jaw and hyolingual movements. Zweers (1982) suggests that the first transport cycle in the

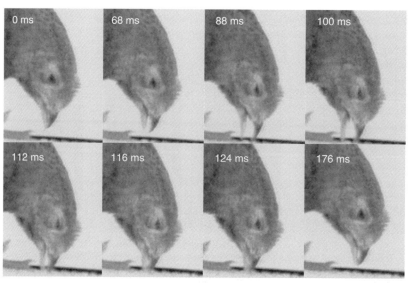

Fig. 3.9. Series of images from high-speed film (250 fields/s) showing the closing/opening movements of the eye during pecking in chicken.

chicken combines two mechanisms: slide-and-glue and throw-and-catch. In the 'slide-and-glue' mechanism, the food item comes in contact with the mucus on the protracting tongue in the buccal cavity. This occurs during the SO stage. The second 'throw-and-catch' mechanism occurs during the upward motion of the withdrawing head, while the beak opens rapidly and wide (FO stage) and the tongue retracts, moving the food posteriorly. At this moment, the food is either still glued to the tongue or is lost by the retracting lingual surface. All transport cycles are either a combination of the two mechanisms or just a 'slide-and-glue' (Zweers, 1982). Closing of the bill results again in a contact called 'catch' for large food items (Van den Heuvel, 1992). Van den Heuvel (1992) reports a modulation in gape cycle along the transport phases of the food: at the 'final transporting cycle', the FO and FC stages are slower in comparison with pecking cycles. The FC stage is followed by slow closing of the jaw apparatus (SC). All the stages of gape cycles are either a by-product of tongue movement in the buccal cavity (SO stage) or the result of sequential contraction of jaw muscles as demonstrated by the extensive electromyographic (EMG) analysis proposed by

Van den Heuvel (1992). Opening of the jaws in pecking cycles is initiated by contraction of the m. protractor quadrati et pterygoidei that occurs before EMG activity in the m. depressor mandibulae. Closing of the jaws begins before or starts at the moment of the first high amplitude of contraction of the palate retractors – m. pterygoideus dorsalis lateralis. Then, a few milliseconds later, dorsal adductors of the mandible and the m. pseudotemporalis profundus, retractor of the mandible, contract (Fig. 3.10).

Similar detailed analyses have shown the complex neuromotor relationship between gape and hyolingual movement to explore the pecking and straining mechanisms in anseriforms. Ducks perform a series of feeding behaviours with jaws and hyolingual apparatus either in the terrestrial or aquatic environment. Using different techniques, including high-speed filming, cinefluoroscopy and computer simulation, Kooloos (1986, 1989) has extensively studied transport of food (and water) from outside, through the oral cavity, and to the oesophagus. He documented pecking, straining and drinking to propose the mechanism used by mallard in these behavioural activities. Figure 3.11 (a,b,c) illustrates the differences

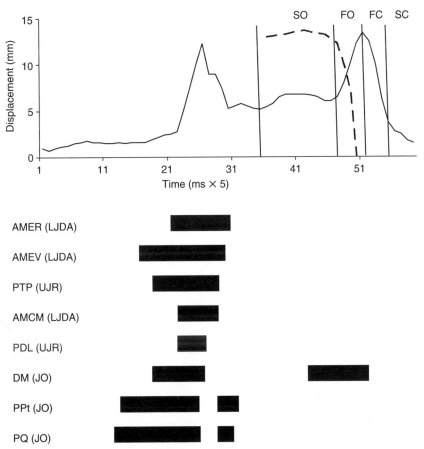

Fig. 3.10. Gape and tongue movements during a typical pecking behaviour in chicken. The first cycle corresponds to a pecking cycle with jaw prehension and the second cycle to a transport cycle involving jaw and tongue movements (Modified from Van den Heuvel, 1992). Jaw movements are presented by the black line and tongue movements by the dotted line. The successive stages of the gape cycle are indicated. SO: slow opening; FO: fast opening; FC: fast closing; SC: slow closing. Contractions of muscles involved in jaw and tongue movements are presented by black blocks. AMER: m. adductor mandibulae externus (rostral part); AMEV: m. adductor mandibulae externus (ventral part); PTP: m. pseudotemporalis profundus; AMCM: m. adductor mandibulae caudalis (medial part); PDL: m. pterygoideus dorsalis lateralis; DM: m. depressor mandibulae; PPt: protractor pterygoidei; PQ: m. protractor quadrati. The action of each muscle is given in parentheses. LJDA: lower jaw dorsal adduction; UJR: upper jaw retraction; JO: jaw opening.

in pecking and straining mechanisms in mallard (Kooloos, 1989). In pecking, the food is grasped by beak tips and moved backward through the buccal cavity by lingual protraction/retraction cycles. In our high-speed recording (Fig. 3.12) of feeding behaviour on small particles, we observed a complex behaviour involving three main types of feeding cycles that can be functionally associated. First,

the bird grasps the food by jaw ingestion and, in the mean time, moves the already present food into the buccal cavity posteriorly towards the oesophagus. Second, the bird moves the food posteriorly by a large tongue protraction/retraction cycle. In such a feeding cycle, the tongue may be protracted out of the buccal cavity and even touch the substratum. In both cycles, gape angle remains rather small. Third,

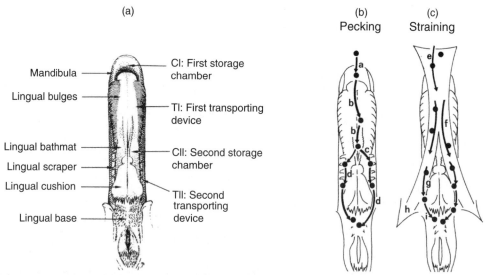

Fig. 3.11. Mechanisms of pecking and straining in mallard (From Kooloos, 1989).

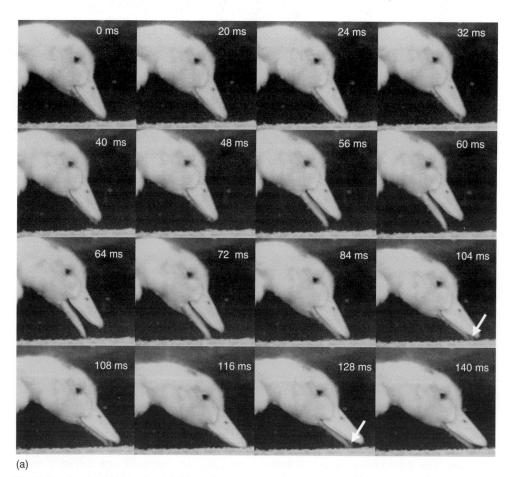

(a)

Fig. 3.12. (a) Successive images of food transport in duckling at the beginning of a feeding sequence filmed by high-speed video NAC-250 (250 fields/s). The tongue plays a key role for moving the food towards the oesophagus as demonstrated by Kooloos (1986, 1989). In several cycles, we observe that the tongue moves outside the buccal cavity.

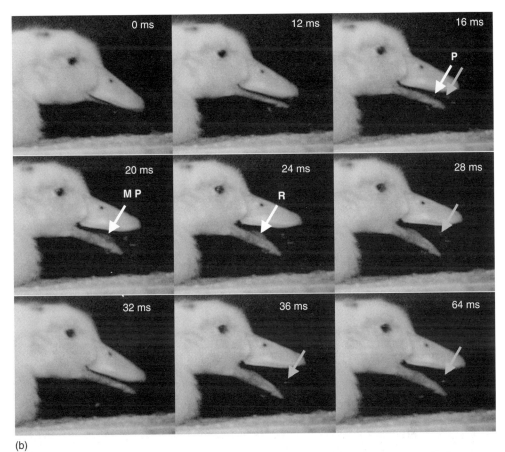

(b)

Fig. 3.12. *Continued* (b) Successive images of food transport in duckling at the end of a feeding sequence filmed by high-speed video NAC-250 (250 fields/s). The tongue plays a key role for moving the food towards the oesophagus. Relative movements of the jaws and the tongue produce posterior displacement of the food towards the oesophagus. The white arrows indicate tongue movements and the grey arrows indicate the displacement of the food particles. MP: maximal protraction; P: protraction; R: retraction.

the bird executes a cycle with larger gape and the food is moved backward to the pharynx by major tongue movement (Fig. 3.12). The tongue movement induces some food to move along the side of the tongue and some food moves backward by projection between the upper and lower jaws as in chickens. For food in water, two mechanisms occur, as experimentally well-determined by several studies: a water-pumping system and a filtering system (Kooloos *et al.*, 1988; Kooloos and Zweers, 1989; Zweers *et al.*, 1994a).

Conclusion

This chapter shows the complexity of neuromotor control of feeding performances in domestic birds. We use examples of galliform and anseriform birds to determine the postures needed for feeding in relationship with the structure of the cervico-cranial system and describe rapidly the feeding cycles involving jaw and hyolingual movements as described by Bramble and Wake (1985) and Reilly and Lauder (1990) for generalized tetrapods.

References

Appleby, M.C., Mench, J.A. and Hughes, B.O. (2004) *Poultry Behaviour and Welfare.* CAB International, Wallingford, UK.
Barlow, G.W. (1968) Ethological units of behavior. In: Ingle, D. (ed.) *The Central Nervous System and Fish Behavior.* Chicago University Press, Chicago, Illinois, pp. 217–237.
Barlow, G.W. (1977) Modal action patterns. In: Sebeok, T.A. (ed.) *How Animals Communicate.* Indiana University Press, Bloomington, Indiana, pp. 98–134.
Bels, V.L., Chardon, M. and Vandewalle, P. (1994) *Biomechanics of Feeding in Vertebrates: Advances in Comparative and Environmental Physiology* 18. Springer-Verlag, Berlin, Germany.
Bels, V.L., Gasc, J.P. and Casinos, A. (2002) *Vertebrate Biomechanics and Evolution.* BIOS, Oxford, UK.
Bermejo, R. and Zeigler, H.P. (1989) Prehension in the pigeon. II. Kinematic analysis. *Experimental Brain Research* 75, 577–585.
Bermejo, R., Allan, R.W., Houben, D., Deich, J.D. and Zeigler, H.P. (1989) Prehension in the pigeon. I: Descriptive analysis. *Experimental Brain Research* 75, 569–576.
Bout, R.G. and Zweers, G.A. (2001) The role of cranial kinesis in birds. *Comparative Biochemistry and Physiology Part A* 131(1), 197–205.
Bramble, D.M. and Wake, D.B. (1985) Feeding mechanisms in lower vertebrates. In: Hildebrand, M., Bramble, D.M., Liem, K.F. and Wake, D.B. (eds) *Functional Vertebrate Morphology.* Belknap Press, Cambridge, Massachusetts, pp. 230–261.
Dubbeldam, J.L. (1984) Brainstem mechanisms for feeding in birds; interaction or plasticity: a functional anatomical of the pathways. *Brain, Behavior and Evolution* 25, 85–98.
Dubbeldam, J.L. (1989) Shape and structure of the avian brain–an old problem revisited. *Acta Morphologica Neerlando-Scandinavica* 27(1–2), 33–43.
Dubbeldam, J.L. (1991) The avian and mammalian forebrain. In: Andrew, R.J. (ed.) *Neural and Behavioural Plasticity.* Oxford University Press, Oxford, UK, pp. 65–91.
Green, A.J. (1998) Comparative feeding behaviour and niche organization in a Mediterranean duck community. *Canadian Journal of Zoology – Revue Canadienne de Zoologie* 76(3), 500–507.
Gussekloo, S.W., Vosselman, M.G. and Bout, R.G. (2001) Three-dimensional kinematics of skeletal elements in avian prokinetic and rhynchokinetic skulls determined by Roentgen stereophotogrammetry. *Journal of Experimental Biology* 204, 1735–1744.
Iwaniuk, A.N. and Hurd, P.L. (2005) The evolution of cerebrotypes in birds. *Brain, Behavior and Evolution* 65, 215–230.
Kear, J. (2005) *Ducks, Geese, and Swans Anseriformes.* Oxford University Press, Oxford, UK.
Kooloos, J.G.M. (1986) A conveyer-belt model for pecking in the mallard (*Anas platyrhynchos* L.). *Netherlands Journal of Zoology* 36, 47–87.
Kooloos, J.G.M. (1989) Integration of different feeding mechanisms in anseriform birds. *Fortschritte der Zoologie* 35, 33–37.
Kooloos, J.G.M. and Zweers, G.A. (1989) Mechanics of drinking in the mallard (*Anas platyrhynchos*, Anatidae). *Journal of Morphology* 199, 327–347.
Kooloos, J.G.M., Kraaijeveld, A.R., Langenbach, G.E.J. and Zweers, G.A. (1988) Comparative mechanics of filter feeding of *Anas platyrhynchos, Anas clypeata* and *Aythya fuligula* (Aves, Anseriformes). *Zoomorphology* 108, 269–290.
Kuenzel, W.J. (1983) Behavioural sequence of food and water intake: its significance for elucidating central neural mechanisms controlling in birds. *Bird Behaviour* 5, 2–15.
McFarland, D. (1981) *The Oxford Companion to Animal Behavior.* Oxford University Press, Oxford, UK.
Nir, I., Melcion, J.P. and Picard, M. (1990) Effect of particle size of sorghum grains on feed intake and performance of young broilers. *Poultry Science* 69, 2177–2184.
Nir, I., Hillel, R., Shefet, G. and Nitsan, Z. (1994) Effect of particle size on performance. 2. Grain texture interactions. *Poultry Science* 73, 781–791.
Nir, I., Hillel, R., Ptichi, I. and Shefet, G. (1995) Effect of particle size on performance. 3. Grinding pelleting interactions. *Poultry Science* 74, 771–783.
Picard, M., Le Fur, C., Melcion, J.-P. and Bouchot, C. (2000) Caractéristiques granulométriques de l'aliment: le 'point de vue' (et de toucher) des volailles. *INRA Productions Animales* 13, 117–130.
Price, E.O. (1984) Behavioural aspects of animal domestication. *Quarterly Review of Biology* 59(1), 1–32.
Price, E.O. (2002) *Animal Domestication and Behaviour.* CAB International, Wallingford, UK.
Reilly, S.M. and Lauder, G.V. (1990) Evolution of tetrapod feeding behavior: kinematic homologies in prey transport. *Evolution* 44, 1542–1557.

Rubega, M.A. (2000) Feeding in birds: approaches and opportunities. In: Schwenk, K. (ed.) *Feeding: Form, Function and Evolution in Tetrapod Vertebrates.* Academic Press, San Diego, California, pp. 395–408.

Van den Heuvel, W.F. (1992) Kinetics of the skull in the chicken (*Gallus gallus domesticus*). *Netherlands Journal of Zoology* 42(4), 561–582.

Van den Heuvel, W.F. and Berkhoudt, H. (1998) Pecking in the chicken (*Gallus gallus domesticus*): motion analysis and stereotypy. *Netherlands Journal of Zoology* 48, 273–303.

Van der Leeuw, A.H.J. (1992) Scaling effects on cervical kinematics in drinking anatidae. *Netherlands Journal of Zoology* 42(1), 23–59.

Wauters, A.M., Guibert, G., Bourdillon, A., Richard, M.A., Melcion, J.P. and Picard, M. (1997) Choix de particules alimentaires chez le poussin: effet de la taille et de la composition. *Journées de la Recherche Avicole* 2, 201–204.

Zeigler, H.P. (1976) Feeding behavior of the pigeon. *Advances in the Study of Behavior* 7, 285–389.

Zeigler, H.P. and Witkovsky, P. (1968) The main sensory trigeminal nucleus in the pigeon: a single-unit analysis. *Journal of Comparative Neurology* 134(3), 255–264.

Zeigler, H.P., Green, H.L. and Lehrer, R. (1971) Patterns of feeding behavior in the pigeon. *Journal of Comparative and Physiological Psychology* 76, 468–477.

Zeigler, H.P., Levitt, P. and Levine, R.R. (1980) Eating in the pigeon (*Columba livia*): response topography, stereotypy and stimulus control. *Journal of Comparative and Physiological Psychology* 94, 783–794.

Zeigler, H.P., Bermejo, R. and Bout, R.G. (1994) Ingestion behaviour and the sensorimotor control of the jaw. In: Davies, M.N. and Green, P.R. (eds) *Perception and motor control in birds*, Springer-Verlag, Berlin, Germany, pp. 201–222.

Zusi, R. (1984) A functional and evolutionary analysis of rhynchokinesis in birds. *Smithsonian Contributions to Zoology* 395, 1–40.

Zweers, G.A. (1982) Pecking of the pigeon (*Columba livia* L.). *Behaviour* 81(1/2), 173–230.

Zweers, G.A. (1985) Generalism and specialism in avian mouth and pharynx. *Fortschritte der Zoologie* 30, 189–201.

Zweers, G.A. and Berkhoudt, H. (1991) Recognition of food in pecking, probing and filter-feeding birds. *Acta XX Congressis Internationalis Ornithologici*, 897–901.

Zweers, G.A., Gerritsen, A.F.Ch. and Kranenburg-Voogd, P.J. (1977) Mechanics of feeding of the mallard (*Anas platyrhynchos* L.; Aves Anseriformes). In: Hecht, M.K. and Szalay, F.S. (eds) *Contributions to Vertebrate Evolution*, Vol. 3. Karger, Basel, Switzerland, pp. 1–109.

Zweers, G.A., Berkhoudt, H. and Van den Berge, J.C. (1994a) Behavioral mechanisms of avian feeding. In: Bels, V.L., Chardon, M. and Vandewalle, P. (eds) *Biomechanics of Feeding in Vertebrates: Advances in Comparative and Environmental Physiology* 18. Springer-Verlag, Berlin, Germany, pp. 241–279.

Zweers, G.A., Bout, R.G. and Heidweiller, J. (1994b) Motor organization of the avian head–neck system. In: Davies, M.N.O. and Green, P.R. (eds) *Perception and Motor Control in Birds.* Springer-Verlag, Berlin, Germany, pp. 201–221.

Zweers, G.A., Bout, R. and Heidweiller, J. (1994c) Generalism and specialism in avian mouth and pharynx. *Fortschritte der Zoologie* 30, 189–201.

4 Ontogeny of Feeding in Mammals

R.Z. German,[1] A.W. Crompton[2] and A.J. Thexton[3]
[1]Department of Physical Medicine and Rehabilitation, Johns Hopkins Hospital,
Phipps 174, 600 N Wolfe St, Baltimore, MD 21205, USA; [2]Museum of Comparative
Zoology, Harvard University, 26 Oxford St, Cambridge, MA 02138, USA; [3]Division
of Physiology, Biomedical Sciences, King's College, Strand, London, UK

Introduction

One of the defining features of the class Mammalia is the presence of mammary glands (albeit in a much different form in monotremes such as the echidna or the platypus); another is the fact that the infants of all species obtain their nutrition, in the form of milk, by suckling from the maternal teat. It is not surprising, therefore, that infant mammals share numerous features, reflecting not only the commonalities of adult mammals but also the common feature of the infant feeding function. Feeding movements on the maternal teat differ significantly from the later, adult feeding movements, although this may be more evident in omnivores and carnivores. One of the more extreme differences is found in the opossum where only small-amplitude oral movements occur when the infant is attached to the teat but wide jaw opening and powerful closure is associated with the adult ingestion of hard food (Thexton and Crompton, 1989; German and Crompton, 1996). In contrast, during post-weaning maturation, the rabbit is reported to exhibit a reduction in jaw opening in response to soft food (Weijs et al., 1989). While the common characteristics of infant mammals are striking, the ontogenetic changes in anatomy and movement that occur during the conversion from infant to adult feeding are likely to be highly diverse. This is because different species

become specialized as adults to ingest foods with substantially different characteristics, varying from small leaves to whole animals. In this respect, explanations of weaning changes that are based purely upon anatomical and kinesiological descriptors are unlikely to generate a unifying, species-wide explanation of the diverse later stages of the ontogeny of feeding. Unfortunately, information on changes in the neural mechanisms controlling oral movement during this period is currently extremely sparse as is information on the interaction between all the foregoing factors.

One of the major concerns in feeding in any mammal is the functional separation of the paths taken by respired air and swallowed food. Because mammals are endothermic, or warm-blooded, continuous respiration is a condition of life. Interruption of that respiration by solid food blocking the airway for any extended period causes death. However, it is also important to note that aspiration of liquids into the respiratory tract is a potential problem. In both adults and infants of most terrestrial mammals, the pathways through which air and food pass are not completely physically separated in space since both air and food pass through some part of the pharynx. In adults, there is normally a separation of these activities in time, with respiration being interrupted at a specific time in the respiratory cycle to allow the swallowing to proceed, while at the same time glottal closure and positional change of the larynx may occur. The reflexes, which exist in adults to protect the airway from aspiration, may not always be as well developed in the young of some species (Bamford et al., 1992) as they will be later in life. It should be appreciated that swallowing itself is a mechanism for airway protection, even in sleep (Page et al., 1995). In infants, some anatomical protection against aspiration also exists which derives from the high larynx that allows the airway to open into the nasopharynx, so producing nearly separate physical channels.

Infant Morphology

Some of the striking anatomical features that characterize adult mammals are present in infants, but others are modified or even absent. The presence of a temporomandibular joint, tongue (a large muscular hydrostat), pharynx, larynx and the unique mammalian organization of craniomandibular muscles with their neural sup-

ply are all present in infants. However, the head, relative to the body, is larger at birth, and the pharynx, oral cavity and tongue are also proportionately larger than in adults. The larynx is intranarial, i.e. it opens into the nasopharynx, and thus the respiration must occur through the nose (Fig. 4.1). Finally, and perhaps most importantly, infants are edentulous or nearly so at birth, requiring maternal care and feeding of the very young. The existence of two sets of dentition over the juvenile and adult stages, with the development of a precision occlusion in the adult, is associated with independent feeding (Schummer et al., 1979; Osborn, 1981). However, it has been reported that newborn rat pups, isolated from the mother, can ingest softened food using ingestive movements that resemble those of adult animals, the main difference being that it is a random, non-directed activity (Hall, 1985).

The high position of the larynx and the intranarial position of the epiglottis in mammalian infants (and some adults) create a space that is contiguous with the valleculae and in which food can be temporarily accumulated. In lateral view radiographs this appears as a

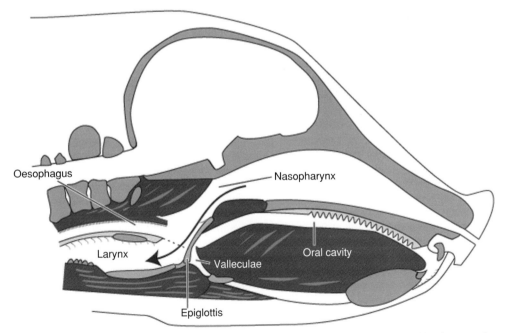

Fig. 4.1. Infant oropharyngeal anatomy. In midline sagittal section, the airway, indicated by the curved arrow, crosses the path that a swallowed bolus takes from the oral cavity through the valleculae into the oesophagus. Striped shading indicates muscle and bone, and cartilage is drawn in light grey.

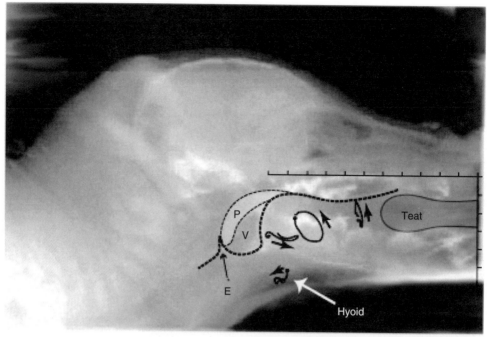

Fig. 4.2. Lateral view radiograph of infant pig while suckling, indicating movement of tongue and hyoid markers (m: mid-tongue; mp: mid-posterior; fp: far posterior) during suckling. The small arrows indicate direction of movement over time. P: outline of the soft palate; V: vallecular space; E: epiglottal cartilage. The bold dotted line is the outline of the epiglottis and of the dorsal surface of the tongue, which is continuous with the vallecular space, and anterior to the epiglottis. The orbits of movement of the tongue markers between the teat and the hyoid marker are mid tongue, mid posterior and far posterior markers.

roughly triangular space, which is delimited by the back of the tongue, the soft palate and the front of the anatomically high epiglottis (Figs 4.1 and 4.2); this will be referred to as the 'vallecular' space.

Infant Function

During the ingestion of maternal milk, the liquid has to be 'extracted' from the teat, transported through the oral cavity, through the pharynx, to the point of swallowing, and then transported through the oesophagus.

The mechanism that infant mammals use for obtaining milk from the teat is one based on tongue movement, with assistance from the myoepithelial cells of the maternal breast that help release and squirt the milk into the infant's mouth. Different authors have described different mechanisms of milk acquisition. Ardran and co-workers outline three ways in which milk can be obtained over a series of papers (e.g. Ardran

and Kemp, 1955, 1959; Ardran *et al.*, 1958). The first is based on an upward jaw movement, which carries the tongue upwards to exert a force against the teat. The second is a 'stripping motion' of a travelling wave of tongue elevation working on the teat to express milk. Finally, a third mechanism is a downward movement of the mid-posterior tongue increasing the space within either the oral cavity or pharynx and causing suction (also see German *et al.*, 1992; Thexton *et al.*, 2004). These data were first described for man, goats and sheep (Ardran and Kemp, 1955, 1959; Ardran *et al.*, 1958), but also now for opossums (German and Crompton, 1996), non-human primates (German *et al.*, 1992) and pigs (German *et al.*, 1992; Thexton *et al.*, 1998). In all of these animals, suckling consists of successive cycles during which milk is accumulated in the valleculae, even though the duration of accumulation may in some cases be very brief. While data exist for a number of domestic animals, the most complete and extensive data exist for piglets, suckling from an artificial nipple (German et al., 1992, 1996, 1998, 2004; Thexton et al., 1998, 2004). Data that compare maternal teats to artificial ones are not extensive, but are suggestive that, with respect to mechanics and physiology, the artificial teat is an acceptable model for understanding normal function (Ardran et al., 1958).

During infant pig feeding, the general kinematic pattern is a series of usually 2–4 suck cycles, during which the vallecular space is filled with milk. This set of cycles terminates in a single cycle, a suck/swallow cycle, during which the valleculae empty into the oesophagus, while at the same time fresh milk is being transported back over the anterior tongue. This serial pattern of multiple suck and single suck/swallow cycles repeats often as many as 20 times, without interruption. In the middle of such a sequence, an animal may break off the teat for a few seconds, and then immediately resume suckling. This pattern of behaviour lasts until satiation is reached (German et al., 1997).

The movement of the pig tongue, viewed in lateral projection videoradiography, consists of a series of wavelike surface elevations and depressions that progress in a pharyngeal or posterior direction (Fig. 4.2). The tongue elevations rise progressively and make contact with the palate, the milk being pushed posteriorly by the tongue elevations (Thexton et al., 2004). A radio-opaque marker in the posterior one-third of the tongue moves in essentially a circular orbit, i.e. moving almost equally in dorsoventral and antero-posterior directions. This movement is suggestive of one produced by the contraction of two sets of muscle fibres, arranged approximately at right angles to each other but contracting out of phase with each other. In support of this hypothesis, the more anterior radio-opaque markers move in a nearly vertical (dorso-ventral) path, while more posteriorly, adjacent to the valleculae, markers move only antero-posteriorly (Fig. 4.2). The movements of the tongue are nevertheless synchronized with the smaller movements of the hyoid and jaw although, as the volume of fluid in the valleculae increases, the hyoid becomes progressively displaced in a ventro-caudal direction.

During this activity, the edges of the anterior tongue are elevated so that they wrap approximately 60% of the way around the teat (Gordon and Herring, 1987; Thexton et al., 2004). Radiographically, in infant pigs, the most anterior tongue maintains a nearly stationary contact with the teat although some compression of the teat is radiographically evident in about 30% of the cycles. There is, however, no obvious travelling wave of compression acting sequentially along the length of the teat, i.e. no 'stripping'. The travelling wave of depression on the tongue appears to originate under the teat tip and moves in a pharyngeal direction. When this depression in the surface of the tongue is below and caudal to the teat orifice, radio-opaque milk can be seen issuing from the teat, consistent with the production of a pressure gradient. A microtransducer positioned intraorally during radiographic visualization to record pressures in these tongue depressions confirms that such cyclical depressions are associated with sharp cyclical falls of pressure (Thexton et al., 2004; Fig. 4.3).

The movements observed during suckling are consistent with a relatively fixed pattern of motor activation of a compliant structure (the tongue) so that different local movements may arise, depending upon the compliance of the teat and volume of intraoral material against which the tongue is acting. The level of pressure developed in the anterior mouth is also dependent upon the effectiveness of the 'tongue to palate' and 'tongue to teat' seals and also upon the volume of fluid entering the mouth, i.e. suckling on a teat with a restricted orifice will result in greater level of negative pressure intraorally than when fixed volumes of milk are delivered automatically by pump. The implication is that intraoral pressure is not a directly controlled parameter. The level of pressure measured in the vallecular space by a microtransducer is independent of that developed more anteriorly in the mouth because of the width and firmness of the tongue/palate seal between those two regions; in the pig vallecular pressure is positive when the vallecular space is emptied into the lower pharynx/oesophagus.

Changes during Weaning

At some point in time postnatally, the suckling animal ceases to obtain milk from the mother and obtains both liquid and solid nutriment

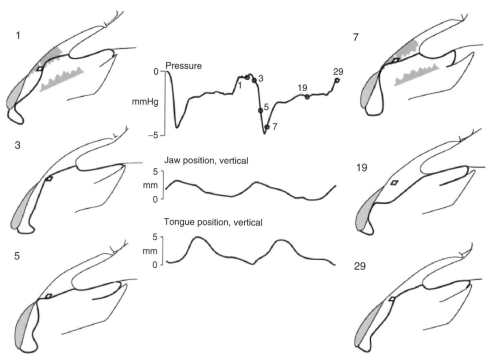

Fig. 4.3. Changes in intraoral pressure during infant pig suckling. The numbers are frame numbers, where each frame is 10 ms apart. The diamond indicates the location of the pressure transducer, and the signal from this transducer is plotted in the first graph. Vertical opening of the jaw is measured between the upper incisor tooth and the lower incisor. The vertical tongue position is measured between the hard palate and a marker near the dorsal surface of the tongue (not shown in these figures). The location of the tooth row is indicated in frames 1 and 7. The soft palate is shaded and lies above the vallecular space.

from another source. Both the time after birth at which this commences and the time taken for the completion of the transition are variable. In man, cultural influences may distort the period, extending it as far as 6 years in some societies, although in the absence of obvious cultural influences the transition period is far shorter; commercial and logistic considerations may also modify the naturally occurring tempo of change in food livestock. The ability or tendency to suckle consequently coexists with the ability to drink or lap, e.g. while the guinea pig may not cease suckling until 2 weeks of age, it can process hard food on day 1. Experimental data on the factors involved in the conversion of the motor activities and their time course are limited, complicated by the fact that the duration of the coexistence of the different modes of ingestion appears to be variable across species.

When infant opossums are forced to change from suckling to drinking, by removal from the pouch at the age of 66 days, a change back to suckling, even when the sole source of nutrient is by teat (German and Crompton, 1996), does not occur, although weaning normally does not occur until approximately 100 days (Farber, 1988). In contrast, at 17–21 days of age the pig can easily be switched from suckling to drinking and back again to suckling, simply by changing the source of available milk (Thexton *et al.*, 1998). This is consistent with the finding that pigs can be weaned at any time between 17 and 35 days (Gu *et al.*, 2002), although the traditional weaning date seems to be around the 49th day of life. It would therefore seem that there are species differences in the reversibility of weaning.

Prior to the onset of solid food intake (around day 18), rat pups spend most of their active time suckling. Over the next 8 days, feeding and drinking activities become more frequent, while suckling gradually begins to decline. By day 28, the food and water intake (expressed relative to body weight) reaches adult levels (Thiels et al., 1990) but pups may nevertheless continue to suckle until day 34. The factors involved in the gradual suppression of suckling and the switch to drinking appear to be linked to maturational changes in the level of serotonin within the brain. Evidence suggesting this comes from the finding that the administration of serotonin agonists suppresses suckling in 10-day-old rat pups. Conversely, up to 2 weeks after normal weaning, the administration of a serotonin antagonist (methysergide) returns the animal to suckling behaviour (Williams et al., 1979; Stoloff and Supinski, 1985). It is, however, not certain that the effect is solely related to central mechanisms; it has been suggested that the effects could be specific to nipple attachment, involving peripheral rather than central serotoninergic mechanisms (Bateman et al., 1990).

It should also be noted that dopaminergic mechanisms of the basal ganglia have been implicated in the maturational changes in feeding. In the pig, early weaning is associated with changes in dopamine metabolism in the caudate nucleus (Mann and Sharman, 1983). In the neonatal rat, virtually complete destruction of central dopaminergic neurons produces no major dysfunction in feeding but, in the adult rat, dopamine-depleting lesions give rise to severe impairments of ingestive behaviour (Bruno et al., 1984), and unilateral lesions produce contralateral deficits when feeding on solid foods (Whishaw et al., 1997). While this could suggest that the basal ganglia are involved in some way in the development of mastication, the neuronal plasticity of dopaminergic cells in the neonate (Sandstrom et al., 2003) makes too simplistic an interpretation of these experiments unwise.

The activities of suckling, drinking and eating represent specific types of movement that require equally specific patterns of electromyographic (EMG) activity in the oral musculature to produce those movements. The data on these activities are unfortunately currently severely limited by the species that have been adequately studied (Liu et al., 1998).

As indicated above, in the infant pig there is a short period before weaning during which the animal can be switched between drinking from a dish and suckling on a bottle simply by changing the source of available milk. In both suckling and drinking there are minimal jaw and hyoid movements but there are extensive movements of the tongue (Fig. 4.4). In drinking cycles, the movement occurs in all parts of the tongue, whereas in suckling, the movement is largely confined to the mid-posterior part of the tongue. In both activities there is a strong similarity in the timing of the bursts of EMG activity in the various muscles, but in drinking, the level of EMG activity in geniohyoid and sternohyoid muscles, among others, is significantly lower than is found in suckling. In both cases milk is transported towards, and is accumulated in, the vallecular space over a variable number of cycles. This space is then emptied in a single cycle during which, compared to the preceding cycles, there is a marked posteriorly directed movement of the back of the tongue, associated with an increased level of EMG activity in hyoglossus, styloglossus and omohyoid muscles.

With maturation of the infant and the change of the physical nature of food ingested, a major change occurs in the way food is transported intraorally. This is due to the fact that, while liquids flow and can therefore be transported pharyngeally by pumping or squeezing of the tongue as the jaw opens and closes a limited amount (Thexton and McGarrick, 1988), solid food items cannot be transported in this manner. The intraoral transport of solid food requires that it is moved as a bulk item, and in mammals this is normally performed by retracting the tongue with the food item resting on its dorsal surface. If the food is to be moved freely into or through the mouth there is a requirement that the gape be sufficiently large at the time of tongue retraction for the food item to avoid contact with teeth or rugae. The required extra clearance may be small in herbivorous animals but in opossums and in domestic cats this extra jaw opening is a large-amplitude and rapidly performed movement (Thexton and Crompton, 1989; Thexton and McGarrick, 1989). The subsequent rapid jaw closure then leaves the food item trapped between the teeth or held against palatal

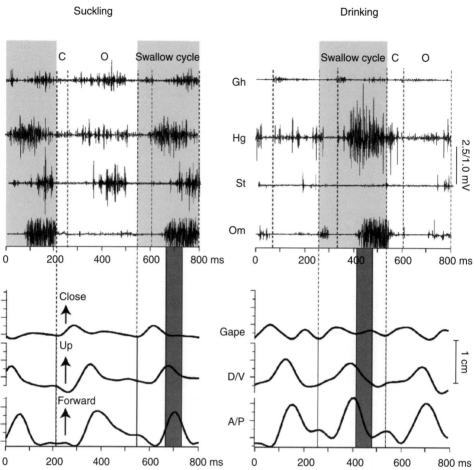

Fig. 4.4. Movement and EMG profiles for several suckling cycles, including one swallow cycle of infant pig feeding. These data are from the same animal in the same feeding session. Suckling data are on the left, and drinking on the right. Gh: geniohyoid; Hg: hyoglossus; St: sternohyoid; Om: omohyoid. The lower panels are movements of markers in the tongue. Gape is the distance between the upper and lower incisor teeth. D/V is the dorsal/ventral movement of a tongue marker lying below the hard/soft palate junction. A/P is anterior/posterior movement of a marker in the most posterior aspect of the tongue. Arrows indicate the direction of movement.

rugae; this then allows a protracting tongue to pass under it, so repositioning the food item more posteriorly on the tongue surface. A somewhat similar action has been described in the rabbit (Cortopassi and Muhl, 1990). Depending on species, this process may be repeated several times before commencement of cycles of jaw movement causing triturition of the food. In many mammals the subsequent breakdown of food and its mixing with saliva converts it into a bolus that can flow. The tongue mechanisms used to transport this

softened bolus in a pharyngeal direction may then have characteristics similar to those seen when transporting liquids. In all non-human mammals, so far studied, the processed food then accumulates in the vallecular space (over a number of cycles that vary from 1 to 20, depending on species) before the vallecular space is emptied in a single cycle and the contents passed down the oesophagus.

As indicated above, the changes in the patterns of EMG activity and of jaw/tongue movements that are required to convert from suckling to drinking are relatively minor in a bilaterally symmetrical activity. Although conversion to the well-known alternating unilateral actions of chewing (e.g. Oron and Crompton, 1985; Langenbach et al., 1992, 2001) requires greater changes, it has to be noted that the processing of hard food is bilaterally symmetrical in some adult mammals (Weijs, 1975) so that, across species, the main changes after weaning would appear to be reduced to the ability to produce intraoral transport of solid food and to exert adequate breakage force via the teeth.

In a variety of carnivores and omnivores, jaw opening in each ingestive cycle consists of two successive stages. The first stage of opening is of almost constant duration and amplitude irrespective of whether the food ingested is solid or liquid. During the second stage of opening, when ingesting liquid, little further opening occurs, but this period is associated with tongue retraction. In contrast, during the ingestion of solid and semi-solid food, large angles of jaw opening occur during the second stage of opening, but this period is still associated with tongue retraction. In the case of the ingestion of solid and semi-solid food, a characteristic pattern of EMG activity (in which all the muscles moving the hyoid are simultaneously activated) is added to the EMG pattern seen in the same period in lapping cycles (Thexton and Crompton, 1989; Thexton and McGarrick, 1994); this additional activity pulls the hyoid backwards as well as pulling the mandible down towards it. The pattern of the additional EMG activity is the same as that seen in the jaw-opening reflex, a reflex that can be elicited experimentally by mechanical stimulation of oral mucosa, teeth, etc.

If the jaw-opening reflex is in any way associated with the ability to handle solid food, the reflex might be expected to be present in the weaned animal but absent in the infant. Several studies have shown that in the newborn of several altricial species, the response to oral stimulation consists only of a low-amplitude, long-latency, long-duration burst of EMG activity in the digastric and associated muscles. The short-latency jaw-opening reflex appears abruptly some time later but before the onset of weaning (Thexton and Griffiths, 1979; Thexton and McGarrick, 1984). The relatively sudden appearance of the short-duration, short-latency reflex EMG activity is not consistent with the slow process of myelination of nerve pathways but, because it can be switched off and on pharmacologically (Thexton et al., 1988) during the critical period of its appearance, it is consistent with a change in central synapses. The implication is that at least some of the neural mechanisms appropriate for ingesting solid food are present and functional well before they are utilized in ingestive behaviour by the maturing animal. Interestingly, the guinea pig, which eats solids within hours of birth, has an adult jaw-opening reflex latency at birth (Thexton and McGarrick, 1984), and the pig, which will mouth solids at an early age, has a near adult reflex latency even before birth (Thexton et al., 1995).

The development of the ability to process solid food has a number of other important components such as eruption of teeth and the change of dentition from deciduous to permanent, the growth and change of type of jaw muscle, changes in the form and size of skull/jaw and changes in neural control. One of the most obvious changes that is temporally associated with weaning is the eruption of deciduous teeth and, in later development, the eruption of a permanent dentition. The development of the dentition is, however, very variable across species and the interested reader is referred to Butler (1983) and Widdowson (1939).

The growth of the jaw muscles and the change in their fibre type composition with weaning and maturation is again complex and species-dependent. One of the major problems is to determine whether the maturational changes in the muscles that are 'preprogrammed' within the muscle represent a response to higher levels of mechanical activity, to changed nutrition, to changed endocrine influences or to changed patterns of activation by the motor nerves. It is clear that the relative mass, fibre orientation and fibre type of jaw-closing muscles change with maturation (Herring and Wineski, 1986; Sato et al., 1998; Gojo et al., 2002). The behavioural transition from infant suckling to adult mastication is also associated with a disproportionate increase in the mass of the masseter, relative to the digastric (Anapol and Herring, 1989), and there is subsequently a relative increase in masseter bulk and activity compared to the temporal muscle (Langenbach and Weijs, 1990; Iinuma *et al.*, 1991).

Following the change of food preference, away from milk, the documented sum of the changes occurring in the dentition, in the skull and jaw, and in the growth and maturation of the jaw muscles might seem to be a reasonable basis on which to base an explanation of increasing ability to process solid food. This does, however, overlook one of the functions of the tongue involved in the process. As solid food is broken into smaller particles or otherwise formed into a bolus, food particles may be displaced into the buccal sulcus (depending upon whether or not complete cheeks are present) or displaced medially onto the tongue. The tongue not only replaces food on to the occlusal surfaces of the teeth but also, in the case of particulate food, preferentially selects the larger particles for this process. This selection process (Voon *et al.*, 1986; van der Glas *et al.*, 1992) results in a major increase of efficiency over random selection of large and small particles. While some aspects of the selection process might be the result of undirected mechanical events, there is evidence that sensory input and an active process of selection by the tongue is also involved (Kapur *et al.*, 1990). Identification of individual particles within the selection process would appear to require a significantly more complex central processing of mechanoreceptor sensory input from the tongue than would be required simply to recognize the intraoral presence of a fluid. Furthermore, the motor response of the tongue is also fundamentally different in the two cases. The process of suckling is essentially one of transport in which the liquid bolus occupies a midline position on the tongue. The placement of unbroken solid food material between the teeth does, however, require the tongue of many species to rotate around its long axis so that the dorsal surface carrying the food material faces the gap created between the upper and lower teeth as the jaw opens. This unilateral activity is characteristic of individual cycles of mastication in many animals (Oron and Crompton, 1985; Thexton and McGarrick, 1989; Huang *et al.*, 1994) where there is periodic change in the working and the balancing sides. The development of central processing of lingual sensory input and the development of asymmetrical motor control of the tongue consequently appear to be crucial to the development of adult feeding, at least in some mammals.

Conclusion

The level of information available on the processes involved in converting from the ingestion of maternal milk to the independent consumption of solid foods is currently inadequate. While in the short term this is exacerbated by the use of different animal species with different adult feeding stratagems, it is only after adequate cross-species baseline information is available that any unifying hypotheses will arise.

References

Anapol, F. and Herring, S.W. (1989) Length-tension relationships of masseter and digrastic muscles of miniature swine during ontogeny. *Journal of Experimental Biology* 143, 1–16.

Ardran, G.M. and Kemp, F.H. (1955) A radiographic study of the tongue in swallowing. *Dental Practitioner* 5, 252–263.

Ardran, G.M. and Kemp, F.H. (1959) A correlation between suckling pressures and the movement of the tongue. *Acta Paediatrica* 48, 261–272.

Ardran, G.M., Kemp, F.H. and Lind, J. (1958) A cineradiographic study of bottle feeding. *British Journal of Radiology* 31, 11–22.

Bamford, O., Taciak, V. and Gewolb, I.H. (1992) The relationship between rhythmic swallowing and breathing during suckle feeding in term neonates. *Pediatric Research* 31, 619–624.

Bateman, S.T., Lichtman, A.H. and Cramer, C.P. (1990) Peripheral serotonergic inhibition of suckling. *Pharmacology, Biochemistry and Behavior* 37(2), 219–225.

Bruno, J.P., Snyder, A.M. and Stricker, E.M. (1984) Effect of dopamine depleting brain lesions on suckling and weaning in rats. *Behavioral Neuroscience* 98(1), 156–161.

Butler, P.M. (1983) Evolution and mammalian dental morphology. *Journal de Biologie Buccale* 11(4), 285–302.

Cortopassi, D. and Muhl, Z.F. (1990) Videofluorographic analysis of tongue movement in the rabbit (*Oryctolagus cuniculus*). *Journal of Morphology* 204(2), 139–146.

Farber, J.P. (1988) Medullary inspiratory activity during opossum development. *American Journal of Physiology* 254, R578–R584.

German, R.Z. and Crompton, A.W. (1996) Ontogeny of suckling mechanisms in opossums (*Didelphis virginiana*). *Brain, Behavior and Evolution* 48(3), 157–164.

German, R.Z., Crompton, A.W., Levitch, L.C. and Thexton, A.J. (1992) The mechanism of suckling in two species of infant mammal: miniature pigs and long-tailed macaques. *Journal of Experimental Zoology* 261(3), 322–330.

German, R.Z., Crompton, A.W., McCluskey, C. and Thexton, A.J. (1996) Coordination between respiration and deglutition in a preterm infant mammal, *Sus scrofa. Archives of Oral Biology* 41(6), 619–622.

German, R.Z., Crompton, A.W., Hertweck, D.W. and Thexton, A.J. (1997) Determinants of rhythm and rate in suckling. *Journal of Experimental Zoology* 278(1), 1–8.

German, R.Z., Crompton, A.W. and Thexton, A.J. (1998) The coordination and interaction between respiration and deglutition in young pigs. *Journal of Comparative Physiology A* 182(4), 539–547.

German, R.Z., Crompton, A.W., Owerkowicz, T. and Thexton, A.J. (2004) Volume and rate of milk delivery as determinants of swallowing in an infant model animal (*Sus scrofia*). *Dysphagia* 19(3), 147–154.

Gojo, K., Abe, S. and Ide, Y. (2002) Characteristics of myofibres in the masseter muscle of mice during postnatal growth period. *Anatomy, Histology and Embryology* 31(2), 105–112.

Gordon, K.R. and Herring, S.W. (1987) Activity patterns within the genioglossus during suckling in domestic dogs and pigs: interspecific and intraspecific plasticity. *Brain, Behavior and Evolution* 30(5–6), 249–262.

Gu, X., Li, D. and She, R. (2002) Effect of weaning on small intestinal structure and function in the piglet. *Archive für Tierernahrung* 56(4), 275–286.

Hall, W.G. (1985) What we know and don't know about the development of independent ingestion in rats. *Appetite* 6(4), 333–356.

Herring, S.W. and Wineski, L.E. (1986) Development of the masseter muscle and oral behavior in the pig. *Journal of Experimental Zoology* 237(2), 191–207.

Huang, X., Zhang, G. and Herring, S.W. (1994) Age changes in mastication in the pig. *Comparative Biochemistry and Physiology A* 107(4), 647–654.

Iinuma, M., Yoshida, S. and Funakoshi, M. (1991) Development of masticatory muscles and oral behavior from suckling to chewing in dogs. *Comparative Biochemistry and Physiology A* 100(4), 789–794.

Kapur, K.K., Garrett, N.R. and Fischer, E. (1990) Effects of anaesthesia of oral structures on masticatory performance and food particle distribution. *Archives of Oral Biology* 35, 397–403.

Langenbach, G.E. and Weijs, W.A. (1990) Growth patterns of the rabbit masticatory muscles. *Journal of Dental Research* 69(1), 20–25.

Langenbach, G.E., Brugman, P. and Weijs, W.A. (1992) Preweaning feeding mechanisms in the rabbit. *Journal of Developmental Physiology* 18(6), 253–261.

Langenbach, G.E., Weijs, W.A., Brugman, P. and van Eijden, T.M. (2001) A longitudinal electromyographic study of the postnatal maturation of mastication in the rabbit. *Archives of Oral Biology* 46(9), 811–820.

Liu, Z.J., Ikeda, K., Harada, S., Kasahara, Y. and Ito, G. (1998) Functional properties of jaw and tongue muscles in rats fed a liquid diet after being weaned. *Journal of Dental Research* 77, 366–376.

Mann, S.P. and Sharman, D.F. (1983) Changes associated with early weaning in the activity of tyrosine hydroxylase in the caudate nucleus of the piglet. *Comparative Biochemistry and Physiology C: Comparative Pharmacology and Toxicology* 74(2), 267–270.

Oron, U. and Crompton, A.W. (1985) A cineradiographic and electromyographic study of mastication in *Tenrec ecaudatus. Journal of Morphology* 185(2), 155–182.

Osborn, J.W. (1981) *Dental Anatomy and Embryology.* Blackwell Scientific Publications, Oxford, UK.

Page, M., Jeffrey, H.E., Marks, V., Post, E.J. and Wood, A.K. (1995) Mechanisms of airway protection after pharyngeal fluid infusion in healthy sleeping piglets. *Journal of Applied Physiology* 78, 1942–1949.

Sandstrom, M.I., Nelson, C.L. and Bruno, J.P. (2003) Neurochemical correlates of sparing from motor deficits in rats depleted of striatal dopamine as weanlings. *Developmental Psychobiology* 43(4), 373–383.

Sato, I., Konishi, K., Kuramochi, T. and Sato, T. (1998) Developmental changes in enzyme activities and in structural features of rat masticatory muscle mitochondria. *Journal of Dental Research* 77(11), 1926–1930.

Schummer, A., Nickel, R. and Sack, W.O. (1979) *The Viscera of the Domestic Mammals*, 2nd rev. edn. Springer-Verlag, New York.

Stoloff, M.L. and Supinski, D.M. (1985) Control of suckling and feeding by methysergide in weaning albino rats: a determination of Y-maze preferences. *Developmental Psychobiology* 18(3), 273–285.

Thexton, A.J. and Crompton, A.W. (1989) Effect of sensory input from the tongue on jaw movement in normal feeding in the opossum. *Journal of Experimental Zoology* 250(3), 233–243.

Thexton, A.J. and Griffiths, C. (1979) Reflex oral activity in decerebrate rats of different age. *Brain Research* 175(1), 1–9.

Thexton, A.J. and McGarrick, J.D. (1984) Maturation of brainstem reflex mechanisms in relation to the transition from liquid to solid food ingestion. *Brain, Behavior and Evolution* 25(2–3), 138–145.

Thexton, A.J. and McGarrick, J.D. (1988) Tongue movement of the cat during lapping. *Archives of Oral Biology* 33(5), 331–339.

Thexton, A.J. and McGarrick, J.D. (1989) Tongue movement in the cat during the intake of solid food. *Archives of Oral Biology* 34(4), 239–248.

Thexton, A.J., McGarrick, J.D. and Stone, T.W. (1988) Pharmacologically induced changes in the latency of digastric reflexes in 1–7 day old rabbits. *Comparative Biochemistry and Physiology C* 89(2), 383–387.

Thexton, A.J. and McGarrick, J.D. (1994) The electromyographic activities of jaw and hyoid musculature in different ingestive behaviours in the cat. *Archives of Oral Biology* 39(7), 599–612.

Thexton, A.J., German, R.Z. and Crompton, A.W. (1995) Latency of the jaw-opening reflex in the premature, preweaning and adolescent pig (hanford strain miniature pig, *Sus scrofa*). *Archives of Oral Biology* 40(12), 1133–1135.

Thexton, A.J., Crompton, A.W. and German, R.Z. (1998) Transition from suckling to drinking at weaning: a kinematic and electromyographic study in miniature pigs. *Journal of Experimental Zoology* 280(5), 327–343.

Thexton, A.J., Crompton, A.W., Owerkowicz, T. and German, R.Z. (2004) Correlation between intraoral pressures and tongue movements in the suckling pig. *Archives of Oral Biology* 49(7), 567–575.

Thiels, E., Alberts, J.R. and Cramer, C.P. (1990) Weaning in rats: II. Pup behavior patterns. *Developmental Psychobiology* 23(6), 495–510.

van der Glas, H.W., van der Bilt, A. and Bosman, F. (1992) A selection model to estimate the interaction between food particles and the post-canine teeth in human mastication. *Journal of Theoretical Biology* 155(1), 103–120.

Voon, F.C., Lucas, P.W., Chew, K.L. and Luke, D.A. (1986) A simulation approach to understanding the masticatory process. *Journal of Theoretical Biology* 119(3), 251–262.

Weijs, W.A. (1975) Mandibular movements of the albino rat during feeding. *Journal of Morphology* 145(1), 107–124.

Weijs, W.A., Brugman, P. and Grimbergen, C.A. (1989) Jaw movements and muscle activity during mastication in growing rabbits. *Anatomical Record* 224(3), 407–416.

Whishaw, I.Q., Coles, B.L., Pellis, S.M. and Miklyaeva, E.I. (1997) Impairments and compensation in mouth and limb use in free feeding after unilateral dopamine depletions in a rat analog of human Parkinson's disease. *Behavioural Brain Research* 84(1–2), 167–177.

Widdowson, T.W. (1939) *Special or Dental Anatomy and Physiology and Dental Histology*, Vol. 2. John Bale, Sons and Curnow Ltd, London, 245pp.

Williams, C.L., Rosenblatt, J.S. and Hall, W.G. (1979) Inhibition of suckling in weaning-age rats: a possible serotonergic mechanism. *Journal of Comparative Physiology and Psychology* 93(3), 414–429.

5 Teeth, Jaws and Muscles in Mammalian Mastication

T.E. Popowics[1] and S.W. Herring[2]

[1]Department of Oral Biology, Box 357132, University of Washington, Seattle, WA 98195, USA; [2]Department of Orthodontics, Box 357446, University of Washington, Seattle, WA 98195, USA

Introduction

The mammalian masticatory system is an integrated functional unit consisting of diverse tissue types and multiple levels of tissue organization. The evolutionary history of each domestic species has established key differences in the craniofacial structures performing oral processing. Understanding how modern domestic species apply masticatory structures to domestic feeding conditions is an important factor for their health and breeding success. This chapter will address the structural features of the masticatory system and discuss their contributions to chewing in selected commercially important species, such as mink (*Mustela vison*), rabbit (*Oryctolagus cuniculus*) and ungulates; horse (*Equus caballus*); cattle (*Bos taurus*); sheep (*Ovis aries*); llama (*Lama glama*); and pig (*Sus scrofa*). Specifically, the mineralized tissues forming the tooth, the structures supporting the tooth, crown morphology and tooth position will be described, and the functional morphology of the jaw joint and masticatory muscles will be compared. An excellent review of feeding in mammals generally has been provided recently by Hiiemae (2000), and the reader is referred to this work for details on non-domestic mammals and for information on the tongue, ingestion and intraoral transport of food – topics not covered herein.

Anatomical Components of the Masticatory System

Mineralized dental tissues: enamel, dentin and cementum (Fig. 5.1)

The composition and structure of enamel define its ability to concentrate stress as well as to resist wear and fracture. Enamel is the most highly mineralized biological tissue, ~97% by weight (Waters, 1980). The mineral is consolidated into crystallites during development, and groups of crystallites are organized into rods (or prisms) and inter-rods (interprismatic enamel). The high mineral density of enamel results in a high degree of hardness and therefore resistance to abrasion (Waters, 1980). Rod orientation is also an important factor in wear rates, such that a high angle of interception with the occlusal surface resists wear. Therefore, the enamel crystallites tend to be parallel with the 'abrasive vector', the direction of force transmitted on to the tooth surface through the food (Fortelius, 1985; Koenigswald and Clemens, 1992; Rensberger, 2000). This abrasive vector pushes food into the occlusal surface regardless of orientation of the chewing stroke, and thus is potentially similar across mammals with different modes of mastication. The abrasive vector may account for the prevalence of radial enamel on the outer surface of mammalian teeth (Koenigswald and Pfretzschner, 1991; Koenigswald

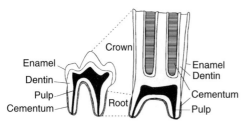

Fig. 5.1. The organization of dental tissues in sagittally sectioned lower molars of mink (left) and horse (right). In mink, the enamel layer covers the underlying dentin and pulp cavity and cementum is limited to root surfaces. The crown of young horses is taller than the roots (hypsodont) and erupts in association with wear. When the enamel covering the crown surface is worn away, dentin appears enclosed by enamel crests. Cementum occurs on root surfaces but also supports the high enamel crests on the crown surface.

and Clemens, 1992). Enamel's high mineral content also corresponds with its high compressive strength, but the resulting tissue is brittle and weak in tension (Waters, 1980; Currey, 1999). Tensile stresses that develop within the enamel tend to split rods apart rather than fracturing them transversely (Rasmussen *et al.*, 1976). Enamel is particularly prone to fracture at points of interface between crystallites of different orientation. Such a weak interface typically occurs at the boundary of inter-rod and rod enamel.

Resistance to enamel fracture occurs through complexity of the crystallite pathways both within enamel rods and between rods and inter-rods. These differences in orientation are generally referred to as 'decussation', and in the case of enamel rods, can be observed in sectioned enamel as Hunter–Schreger bands. Hunter–Schreger bands appear as groups of longitudinally sectioned enamel rods adjacent to transversely sectioned rods (Rensberger, 1993) and represent enamel rods diverging in the horizontal or tangential plane (Koenigswald *et al.*, 1987). The plane in which differences in rod orientation occur is termed the 'rod decussation plane', and is parallel to the long axes of the diverging rods (Rensberger, 2000). Because tensile stresses oriented perpendicular to enamel rods tend to generate cracks at rod boundaries, decussation inhibits crack growth

by diverting the propagating cracks in different directions (Fig. 5.2). The greatest fracture resistance is achieved when the decussation plane is oriented in parallel with the largest tensile stresses. In this case, the developing cracks are orthogonal to the decussation plane, and the advancing crack tip undergoes the greatest possible directional change and loss of energy within the decussation plane. In contrast, if the developing cracks are oriented parallel to the decussation plane, cracks propagate easily along rod boundaries and little fracture resistance is achieved (Koenigswald *et al.*, 1987; Rensberger and Pfretzschner, 1992; Rensberger, 1993, 1995). The second type of decussation between inter-rod and rod crystallites establishes another plane of decussation and resistance to tangential or horizontal cracks (Fig. 5.2).

Dentin is the living connective tissue underlying the enamel crown (Fig. 5.1). The structure includes a mineralized meshwork of collagen fibrils and a wide variety of non-collagenous proteins (Goldberg and Smith, 2004). Dentin tubules radiate from the pulp cavity through the dentin thickness, and while intertubular dentin forms the main body of the crown and root dentin, peritubular dentin forms a more mineralized tissue that lines dentin tubules. The odontoblasts that form dentin line the periphery of the pulp cavity and their elongated cell processes extend into the fluid-filled dentin tubules. Dentin sensitivity and pain have been attributed to movement of fluid within dentin tubules triggering closely associated nerve fibres (Pashley, 1996). Unlike enamel, injured dentin can be repaired, at least in part. The odontoblasts, whose cell bodies reside inside the pulp cavity, can secrete new dentin. If these cells have been impaired, new cells may be generated to secrete a less organized reparative dentin on the inner surface. Overall, dentin is less mineralized than enamel, containing ~70% hydroxyapatite by weight and ~30% organic matrix and water, and thus is softer and more flexible (Waters, 1980). In consequence, it undergoes greater wear when exposed at the crown surface, but also confers deformable support to the brittle enamel crown.

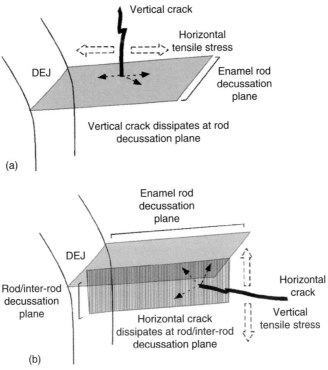

Fig. 5.2. Schematic illustration of crack-stopping attributes of (a) horizontal rod decussation and (b) inter-rod–rod decussation. (a) Changes in rod orientation within the horizontal decussation plane inhibit the propagation of vertical cracks produced by horizontal tensile stresses. (b) Changes in crystallite orientation within the vertical inter-rod–rod decussation plane inhibit the propagation of horizontal cracks generated during vertical tensile stresses. DEJ: dentino-enamel junction.

Cementum is a mineralized tissue covering the tooth root that receives the attachment of the periodontal ligament (Fig. 5.1) and supports the tooth within the jaw. In mineral and protein composition, cementum largely resembles dentin and bone (Bosshardt and Selvig, 1997). Cementoblasts secrete acellular cementum during root development and tooth eruption. Mineralized fibres of the periodontal ligament become embedded in the matrix. In the fully erupted and functioning tooth, acellular cementum lines the coronal two-thirds of the root surface. In many herbivorous species, acellular cementum also forms on the crown surface of incisors and molars, filling deep crevices in the enamel and supporting the surrounding crown structure

(Michaeli *et al.*, 1980; Janis and Fortelius, 1988; Lowder and Mueller, 1998). During tooth function, cellular cementum is deposited at root apices.

Because dentin and cementum are vital tissues that can be repaired to some extent, damage to enamel is potentially the most serious injury for a tooth.

Periodontium

During chewing, the bite stresses that are concentrated on the tooth surface are transmitted to the cranioskeleton. The joint between a tooth and its surrounding bone, known as the periodontium, is the site of application of occlusal force to the maxilla and mandible. The periodontium includes the periodontal ligament and a thin layer of cortical bone that forms the tooth socket or alveolus. The periodontal ligament consists of bundles of type I collagen fibres that connect the cementum of the tooth root to the alveolus for the transfer of tooth stresses to adjacent bone. Fibre bundles radiate in different directions from the root surface, thus some fibres are always under tension even during off-axis loads. A high concentration of glycosaminoglycans occurs in the periodontal space and is presumed to act as a shock absorber during a bite (Beertsen et al., 2000; Berkovitz, 2004).

The periodontium provides an adjustable viscoelastic support system for the teeth. The periodontal vascular network is highly elaborated and is connected to the venous spaces of the neighbouring alveolar bone marrow (Matsuo and Takahashi, 2002). Both blood pressure and the hammock arrangement of periodontal ligament fibres have been implicated in maintaining the position of the tooth within its socket and in resisting occlusal loads (Picton and Wills, 1978; Myhre et al., 1979). Occlusal forces are presumed to be transmitted to the alveolar bone by the ligament fibres. The alveoli themselves are supported by trabeculae of cancellous bone that connect them with each other and with the cortical bone of the jaws.

Morphology of teeth and arrangement in jaws

Tooth number and molar morphology vary among domestic species (Table 5.1). All domestic mammals possess heterodont dentitions, meaning that the tooth rows are divided into tooth classes, i.e. incisors, canines, premolars and molars. Incisors and canines are located anteriorly in the oral cavity and are typically involved in food prehension. The posterior cheek teeth are the sites of mastication of all ingested food. The central function of molar crown shape is to concentrate stress during chewing. Some crowns include cusps, which function in a mortar-and-pestle-like crushing action with basined surfaces on opposing molars, whereas other cusped molar

crowns form shearing blades (Fig. 5.1, left; Lucas and Luke, 1984). Molars may also show ridges of enamel instead of distinct cusps, and shearing blades also occur on the edges of these crests (Fig. 5.1, right). The details of crown morphology are highly species-specific and have been the subject of many publications (Hillson, 1986).

Modification of tooth morphology by wear

During tooth function, wear facets develop on the occlusal surface that further shape the functional morphology of the tooth. Wear facets appear as regions of polished enamel on opposing molar features, and occur largely through tooth-to-tooth contact during chewing, known as attrition. Abrasion, enamel wear occurring through tooth contact with food, appears as pits or striations within facets. A high rate of attrition or high stresses during attritial chewing generates well-defined wear facets. The formation of such facets on a blade edge will have a sharpening effect (Popowics and Fortelius, 1997). In contrast, abrasive foods such as plant material containing phytoliths (Walker et al., 1978) or grit will mark the enamel with pits or striations and have a rounding effect on cusp tips or blades. The extent of pitlike abrasion has been found to increase with abrasive particle size (Maas, 1994).

Occlusal wear occurs in all mammals, but reaches an extreme in herbivorous species. Whereas the dentitions of domestic carnivores are fully functional upon eruption, the cheek teeth of most herbivores must undergo substantial wear to obtain a functioning crown shape. In these species, wear removes the cementum and thin enamel from the tips of the cusps of the newly erupted molar crown, thus establishing a series of enamel crests on the crown surface (Fig. 5.1, right; Osborn and Lumsden, 1978; Michaeli et al., 1980). The greater hardness of enamel relative to dentin means that the enamel wears at a slower rate than the dentin, so the enamel crests eventually protrude above the neighbouring dentin. In addition, wear can result in ridges across the crown, called lophs. Such ridge–valley relief is more prominent in browsers or mixed feeders

Table 5.1. Summary of characteristics of selected domestic mammals.

	Dental formula	Molar type	Enamel characteristics	Predominant jaw-joint movement
Carnivora				
Mink[a–c]				
Mustela vison	I 3/3, C 1/1, P 3/3, M 1/2	Carnassial	Horizontal Hunter–Schreger bands	Rotational
Perissodactyla				
Horse[d,e]			Horizontal Hunter–Schreger bands	
Equus caballus	I 3/3, C 0-1/0-1, P 3-4/3-4, M 3/3	Lophodont, hypsodont	Rod/inter-rod crystallite decussation	Translational
Artiodactyla[f–h]				
Bovidae				
Cattle				
Bos Taurus				
Goat	I 0/3, C 0/1, P 3/2-3, M 3/3			
Capra hircus				
Sheep		Selenodont, hypsodont		
Ovis aries			Horizontal Hunter–Schreger bands	
Camelidae				Translational
Llama			Rod/inter-rod crystallite decussation	
Lama glama	I 1/3, C 1/1, P 2/1, M 3/3			
Suidae				
Pig		Multi-cusped, bunodont		
Sus scrofa	I 3/3, C 1/1, P 4/4, M 3/3			
Lagomorpha				
Rabbit[g,i]				
Oryctolagus cuniculus	I 2/1, C 0/0, P 3/2, M 3/3	Lophodont, hypselodont	Irregular enamel	Translational

[a]Ewer (1998); [b]Lariviére (1999); [c]Stefen (1997); [d]Bennett and Hoffman (1999); [e]Fortelius (1985); [f]Kawaii (1955); [g]Nowak (1999); [h]Pfretzschner (1992); [i]Michaeli *et al.* (1980).

than in grazing ungulates, which undergo greater wear (Fortelius and Solounias, 2000).

Increases in tooth height are called hypsodonty (tooth height exceeding its length or width) or hypselodonty (continuous apical growth, or ever-growing), and typically occur in species with molar dentitions strongly modified by wear. The elongated bases of hypsodont or hypselodont teeth are enclosed within the jaw and erupt over the course of the animal's lifetime in association with crown wear (Janis, 1988). Hypsodonty and hypselodonty are assumed to have evolved in response to selection pressures associated with herbivory (Williams and Kay, 2001). Many plant materials are rich in cellulose, which adds bulk to the diet but little nutritional value. Intake of such foods must occur in high volume in order to meet the animal's nutritional requirements, and the increased food processing amplifies tooth wear. Grasses, which are fracture-resistant and often contain phytoliths or are ingested with grit further increase tooth wear during chewing (Rensberger, 1973; Fortelius, 1985). High-crowned molars (Fig. 5.1, right) and ever-growing teeth thus compensate for the loss of dental tissue through increased wear in grazing species. Because dental wear is such a central feature in the normal molar function of grazing animals, the absence of sufficient tooth wear (e.g. from diets that are too soft for domestic species) can lead to pathologies as serious as those of too much wear.

Jaws

Mammals are unique among vertebrates in having a one-bone lower jaw, the mandible, which articulates with the squamosal bone of the skull. The upper jaw, like that of other tetrapods, consists of the premaxillary bone, which bears the incisors, and the maxillary bone, which bears the canines and cheek teeth. Paired premaxillaries and maxillaries are joined in the ventral midline to form a bony secondary palate. Tooth-bearing is not the only function of the premaxillary and maxillary bones, which also form the floor and lateral walls of the nasal cavity and support the lips and nasal cartilages. Furthermore, except for a few fibres of the masseter muscle, the muscles of mastication originate from the zygomatic arch, pterygoid region and temporal fossa, not from the upper jaw. Thus, the morphology of the upper jaw is not always closely tied to feeding behaviour.

In contrast, the lower jaw is dominated by its role in feeding. In addition to housing the dentition, the mandible is a major attachment for muscles that move the tongue and the sole insertion for the jaw-opening and jaw-closing muscles. The mandible can be considered to be divided into zones, each dependent on local functional activity (Moss, 1968). The mandibular body extends from the joint at the condyle to the symphysis that joins the two sides anteriorly, the alveolar process contains the teeth, and the ascending ramus is comprised of two muscle attachment areas: the coronoid process above the body and the angular process below the body (Fig. 5.3). The mandibular symphysis is primitively a movable joint, but in various groups of mammals, primarily herbivorous species, the symphysis has become less mobile or has fused completely. When patent, the mandibular symphysis allows each mandibular body to rotate around its own long axis (Lieberman and Crompton, 2000).

Temporomandibular joint and masticatory movements

The jaw joint is variously known as the temporomandibular joint (TMJ), squamomandibular joint or craniomandibular joint. Although the least precise, the TMJ terminology is the most often employed and will be used here. The TMJ is formed relatively late in ontogeny

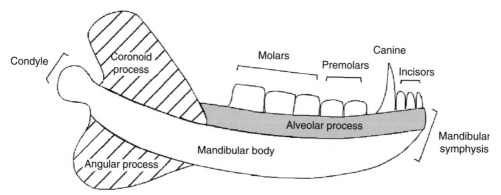

Fig. 5.3. Lateral view of a generalized mammalian mandible showing its organization into functional compartments. The coronoid and angular processes are sites of attachment for the jaw-closing muscles, and the condyle forms the mandibular portion of the jaw joint. The alveolar process includes the bone supporting the teeth. Mammalian dentitions are typically divided into tooth classes: incisors, canines, premolars and molars.

and is unusual among synovial joints in that the articular cartilage is a secondary tissue derived from the periosteum of the intramembranously ossified mandible and squamosal bone (Smith, 2001). The secondary cartilage that forms the mandibular condyle also functions as the major growth site of the mandible, and thus injuries to the TMJ area can result in serious malformations and malocclusions (Tsolakis et al., 1997). Most mammals, and all domestic species, have an intra-articular partition, the TMJ disc, which divides each joint cavity into an upper and a lower compartment. The TMJ disc is thought to allow complex movements to occur at the jaw joint. Because the disc attaches to the medial and lateral poles of the condyle, it tends to follow anterior-to-posterior movements of the mandible. These anterior-to-posterior excursions, often called translations (Fig. 5.4), therefore involve primarily the upper joint compartment, between the disc and the squamosal bone. In contrast, open–close movements of the mandible, called rotations, occur more within the lower joint compartment, between the condyle and the disc. Although the components of the TMJ are constant among mammals, there is great diversity in the relative importance of translation vs. rotation, and this is reflected in the morphology of the joint (Fig. 5.4). Where translations predominate, typically in herbivores with grinding mastication, the articular surfaces are either flat or form a track in the direction of movement. In species that emphasize rotation, exemplified by members of the order Carnivora, translation is prevented by

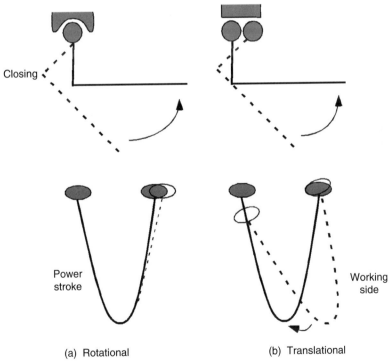

(a) Rotational   (b) Translational

Fig. 5.4. Comparison of the (a) rotational and (b) translational types of mammalian jaw joints shown during closing (upper row) and the power stroke (lower row). The articular surfaces in the (a) rotational type are highly curved and bony processes on the temporal bone limit sagittal-plane joint movement to a hingelike rotation. In some species the power stroke may involve a side-shift of one or both condyles towards the working side (Scapino, 1981). The articular surfaces in (b) translational type are either flat or form a track in the direction of movement. Opening involves an anterior movement and closing, a posterior movement of the condyle. The power stroke uses an anterior excursion of the working-side condyle and/or a posterior excursion of the balancing-side condyle.

bony processes and the TMJ takes the form of a door hinge (Smith and Savage, 1959).

As implied by the varying morphologies of the TMJ, the movements of mastication differ among mammalian species. A chewing cycle is considered to consist of: (i) an opening movement that incorporates a phase for transporting food (Thexton and McGarrick, 1994); followed by (ii) a relatively rapid (because unloaded) closing movement, which ends when the teeth engage the bolus; and finally (iii) a power stroke, during which the jaw completes its closure against the resistance of the food. The power stroke is guided in part by the inclined planes of the molar surfaces (Herring, 1993). Although there are exceptions, most mammals chew on one side at a time, which is called the working side. The rapid closing movement typically ends with the jaw deviated laterally towards the working side. During the power stroke the jaw moves anteriorly and medially back to the midline, completing the cycle. These posterior-to-anterior and lateral-to-medial excursions are minimal in species that emphasize TMJ rotational movements, primarily carnivorans, but can be very extensive in species that emphasize TMJ translations (Fig. 5.4).

Muscles of mastication and jaw mechanics

As might be expected, the muscles that close the jaw and produce the power stroke are far larger than those that open the jaw. The closing muscles are complex, subdivided and interconnected (Herring, 1992; Hiiemae, 2000), but three powerful jaw adductors are typically recognized. The temporalis, which arises from the temporal fossa on the side of the braincase, inserts on the coronoid process to pull the mandible posteriorly as it closes. The masseter arises from the zygomatic arch and inserts on the mandibular angle. There are at least two subdivisions: a superficial masseter that pulls the mandible anteriorly and a deep masseter (sometimes called zygomatico-mandibularis) that inclines more posteriorly and is often intertwined with the temporalis. The third large adductor is the medial ptery-

goid muscle, which arises from the pterygoid region of the skull and inserts on the medial side of the mandibular angle. The action of the medial pterygoid is similar to that of the superficial masseter, except that it has a strong medial component, and because the two muscles insert on opposite sides of the angle, they tend to rotate the mandibular body in opposite directions around its long axis.

In addition to these large adductor muscles, the lateral pterygoid muscle inserts on the mandibular condyle and pulls it anteriorly and medially, and muscles in the oral floor such as the digastric and geniohyoid open the jaw by pulling down and back on the mandible.

The jaw muscles serve both to move the mandible and to provide the occlusal force needed to reduce the bolus. In species with primarily rotational TMJ action, the emphasis is usually on vertical force. Vertical force is a product of muscle size (cross section) and leverage. Occlusal force increases as the teeth get closer to the TMJ (shorter jaws) and the muscles get farther away (large coronoid and angular processes). The temporalis is typically the most important muscle for species with rotational TMJs (Turnbull, 1970), a circumstance usually ascribed to the usefulness of the posterior vector of muscle pull for prey capture (Smith and Savage, 1959). Bite force is theoretically highest at the posterior end of the tooth row, but because separation of upper and lower jaws is least here (Greaves, 1985; Biknevicius and Van Valkenburgh, 1996), the largest cheek teeth are usually located in an intermediate position.

In contrast to the 'rotational TMJ' type, species with translating TMJs typically have muscles specialized to produce the anterior-to-medial power stroke rather than vertical occlusal force (Fig. 5.5). For translational movements, the most common contraction pattern is termed 'transverse' (Weijs, 1994), in which the power stroke is produced by the working side superficial masseter and medial pterygoid (sometimes lateral pterygoid as well), aided by the temporalis (and sometimes deep masseter) on the opposite (balancing) side (Fig. 5.5). This combination of muscles brings the working condyle forward and the balancing condyle backward, causing an excursion of the mandibular teeth towards the balancing side. Species with translational TMJs and the 'transverse' contraction pattern typically have large superficial masseters and medial pterygoid muscles with mechanical advantage improved by high condylar position, whereas the reduced temporalis has a short-moment arm and serves to retract the condyle rather than to provide occlusal force. Tooth rows curved in the occlusal plane in many species result in only a few teeth at a time being in occlusion so that occlusal force can be concentrated. The largest cheek teeth are situated relatively posteriorly in the tooth row, but not at its terminus. This location is thought to optimize vertical bite force while maintaining a stable mechanical system for the loosely constructed TMJs (Greaves, 1978).

Characteristics of Domestic Mammals

Mink (*Mustela vison*), Family Mustelidae, Order Carnivora (Figs 5.6 and 5.7a)

The enamel of mink conforms to that of carnivorans generally (Skobe *et al.*, 1985; Stefen, 1997, 2001). Enamel rods decussate in the horizontal plane (Table 5.1). The Hunter–Schreger bands of carnivorans form subtle waves macroscopically (Koenigswald, 1997) and microscopically that can be seen to be oriented at approximately 90° to the dentin–enamel junction (Kawaii, 1955), inclin-

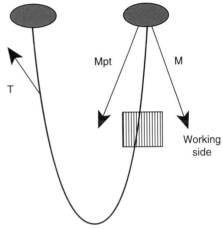

Fig. 5.5. The power stroke in translational jaw movement is produced by the working-side superficial masseter (M) and medial pterygoid (Mpt), and aided by the temporalis (T) on the opposite (balancing) side. This pattern of muscle contraction is referred to as 'transverse' (Weijs, 1994). The box on the working side represents the bolus.

ing occlusally near the cusp tip but more horizontally at the sides and in the cervical region (Hanaizumi, 1992; Hanaizumi *et al.*, 1996; 1998). The horizontal decussation of enamel rods in carnivorans indicates resistance to vertical fractures, such as might occur in response to hoop tensile strains when the tooth is loaded vertically (Fig. 5.2a). Decussation between rod

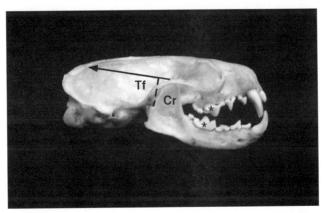

Fig. 5.6. Lateral view of the skull and mandible of the mink, *Mustela vison*. Note the upper and lower carnassial teeth (asterisks). The dominance of the temporalis muscle in this 'rotational' species is indicated by the enlarged temporal fossa (Tf) and coronoid process (Cr). The elongated moment arm of the temporalis is shown by the dashed line while the estimated action line of the muscle is solid.

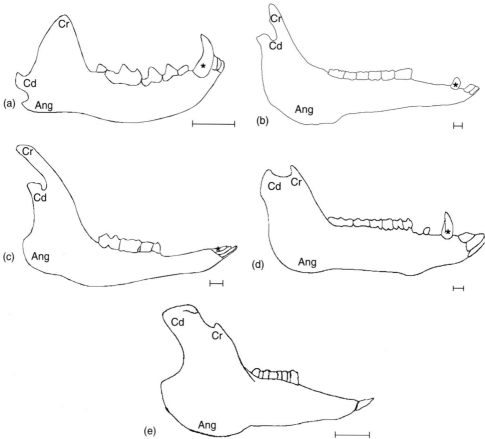

Fig. 5.7. Lateral view of the mandibles of domestic species. (a) Mink, *Mustela vison*. (b) Horse, *Equus caballus* (male). (c) Goat, *Capra hircus*. (d) Pig, *Sus scrofa*. (e) Rabbit, *Oryctolagus cuniculus*. Mandibles are shown at the same length to facilitate shape comparison. The size measure equals 1 cm. Ang: angular process, Cd: mandibular condyle, Cr: coronoid process. Note the large size of the coronoid process in mink but its reduction in the other species. In contrast, all but mink have enlarged angles and elevated condyles. An asterisk marks the lower canine tooth (absent in rabbits).

and inter-rod crystallites is absent in carnivorans; instead, the crystallites of inter-rod enamel run parallel to those of the rod (Koenigswald, 1997).

The periodontal tissues in mink are not as well studied as in its close relative, the ferret *Mustela putorius furo*; thus, the ferret's periodontium will be reviewed as a proxy species. Histological study of the developing ferret periodontium has shown that collagen fibre bundles of the periodontal ligament form attachments to cementum and bone just prior to eruption (Berkovitz and Moxham, 1990). The periodontal ligament itself is more pliable during the eruption period than when the

tooth is in function (Moxham and Berkovitz, 1989). Although there are variations between maxillary and mandibular cheek teeth, the ferret periodontal ligament resembles that of rat and man in being wider at the alveolar crest and apical root region than in the middle of the root (Fischer *et al.*, 1993). Both mink and ferrets have been used as model systems for the study of periodontal disease (Weinberg and Bral, 1999; Render *et al.*, 2001).

Generalized carnivorans have cheek teeth that function either mainly in shearing or mainly in crushing, and this duality enables them to consume a variety of foods in addition to meat. The shearing teeth are the fourth upper premolar (P^4) and the lower first molar (M_1), modified into blades by the alignment of two main cusps along the axis of the jaw (Ewer, 1998). These blade-like teeth, known as carnassials, shear pliant or tough foods by focusing stresses in foods trapped between bladed edges (Greaves, 1983; Lucas and Luke, 1984). As the jaw closes, point contacts between opposing blades progress along the length of the carnassial. The carnassial blades in mink are longer than more generalized species, extending the amount of continuous shear (Crusafont-Pairo and Truyols-Santoja, 1956; Van Valen, 1969; Popowics, 2003) and suggesting adaptation to carnivory in the wild (Fig. 5.6). Posterior to the carnassials, generalized carnivorans have molars with crushing surfaces formed between blunt cusps and opposing basined tooth surfaces. These surfaces can function to fracture brittle foods, such as nuts or seeds, in mortar-and-pestle fashion. Cracks propagate easily through such foods and do not require continuous application of occlusal stresses, in contrast to softer but tougher foods, which must be sheared (Lucas and Luke, 1984; Lucas and Teaford, 1994). The mink has a reduced crushing dentition relative to many other carnivorans, with only one small crushing posterior cheek tooth behind each carnassial (Butler, 1946; Aulerich and Swindler, 1968; Lariviére, 1999). This also points to carnivory as a mink adaptation.

The TMJs of mink and other mustelids are of the extreme rotational type, with bony processes that prevent translation, although some side-shift and movement of the mobile mandibular symphysis allow adjustments of carnassial contact (Scapino, 1981). The buttressing of the TMJs allows bite force to be maximized because muscles from the non-working side of the jaw can be recruited without danger of disarticulation (Dessem and Druzinsky, 1992). A large gape is facilitated by the very small mandibular angle (Scapino, 1974); thus large items can be positioned at the carnassials even though the jaws are short (Radinsky, 1981).

Although muscles of the tongue, hyoid and pharynx have complex programmes for transporting and swallowing food (Thexton and McGarrick, 1994; Kobara-Mates et al., 1995),

the closing muscles contract roughly in synchrony, with only a difference in magnitude of activity signalling the side of chewing (Gorniak and Gans, 1980; Dessem and Druzinsky, 1992; Thexton and McGarrick, 1994). In all carnivorans the temporalis is by far the most important jaw adductor muscle, serving to close the jaw and to resist the struggling of prey items (Smith and Savage, 1959). The digastric, an opener, is also enlarged in carnivorans to accomplish rapid wide gape (Scapino, 1976). Other muscles, especially the pterygoids, are reduced because of the unimportance of translational movements at the TMJ (Turnbull, 1970). These trends are accentuated in the hypercarnivorous mink, which shows comparatively poor development even of the masseter (He et al., 2004). The short jaws of mink and other mustelids suggest a particular specialization for large vertical closing force (Radinsky, 1981). High vertical occlusal force is also suggested by an especially large temporalis muscle in the genus Mustela. In addition to its size, the temporalis has remarkable mechanical advantage through an elongated temporal fossa and an enlarged coronoid process (Figs 5.6 and 5.7a). Type I (slow-twitch, fatigue-resistant) fibres make up 90% of the temporalis cross-sectional area in ferret (He et al., 2004), indicating that high bite forces can be sustained for extended periods. Indeed, ferret bite force is in the same range as that of the much larger dog, Canis familiaris, and ferrets easily crush chicken bones with the carnassial teeth (Dessem and Druzinsky, 1992).

In summary, all aspects of the mink feeding mechanism indicate an emphasis on vertical occlusal force, including the strictly horizontal decussation of enamel rods, the accentuated carnassials and reduced crushing teeth, the strongly rotational TMJ, and the large, fatigue-resistant temporalis muscle.

Ungulates (Orders Perissodactyla and Artiodactyla), common features (Figs 5.7b–d and 5.8–5.10)

Despite the fact that they belong to two different orders, domestic ungulates show great similarity in their adaptations of enamel structure for fracture resistance. Enamel rod decussation

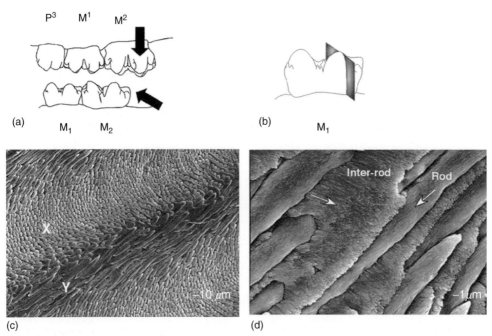

Fig. 5.8. (a) Occlusion of upper and lower dentition of the pig, *Sus scrofa*. Arrows indicate vertical and horizontal interactions of the cheek teeth. (b) First molar, showing the plane of section for scanning electron microscopy in (c) and (d). (c) Inner enamel from cusp side, showing bands of parallel-sectioned (Y) and cross-sectioned (X) rods. SEM, X500. (d) Band of parallel-sectioned rods at high magnification, showing lengthwise orientation of enamel rods and inter-rod enamel between rods. White arrows indicate orientation differences between rod and inter-rod crystallites. SEM, X4000.

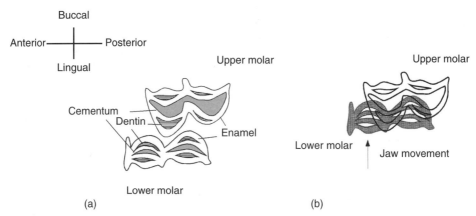

Fig. 5.9. Selenodont molar crown morphology and occlusion in the llama, *Lama glama*. (a) Right-side upper and lower molar crowns show wear-derived crescent-shaped enamel crests, and intermediary dentin and cementum pools. (b) The occlusal view of the upper molar is depicted as a transparent undersurface in order to show alignment with the lower molar's occlusal surface. Transverse jaw movement translates enamel crests of the lower molars across the crests of the upper molar. The reversed curvature of upper and lower crests generates point contacts that concentrate stress to shear food during the power stroke.

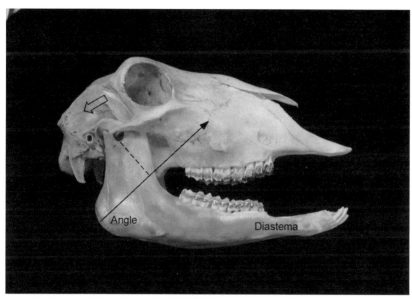

Fig. 5.10. Lateral view of the skull and mandible of the sheep, *Ovis aries*. Note absence of upper incisors and the long edentulous diastema. The mechanical advantage of the superficial masseter (solid arrow, moment arm shown as dashed line) and medial pterygoid muscle is improved by elevation of the jaw joint above the tooth row. However, the temporalis (open arrow) is reduced and pulls mainly posteriorly.

occurs in all domestic ungulates, although the Hunter–Schreger bands are not as sharply defined (Kawaii, 1955) and the decussation is somewhat less horizontal than in carnivorans (Koenigswald, 1997). As in carnivorans, this largely horizontal decussation resists the vertical cracking that occurs when occlusal compression causes high tensile stresses to develop in the horizontal circumferential direction. Unlike carnivorans, however, ungulates also show decussation between inter-rod and rod crystallites, and this decussation is in the vertical, not the horizontal, plane. Sheets of inter-rod enamel are sandwiched between radial rows of occlusally oriented enamel rods, with crystallites crossing at approximately a 90° angle (Fig. 5.8). Because the inter-rod–rod decussation is vertical, it provides resistance to horizontal cracking, such as that which might occur if an enamel crest were bent.

Periodontal studies of domestic ungulates include both anatomical descriptions (Spence, 1978) and biomechanical testing. The periodontal ligaments of pig and ox have viscoelastic properties, and their mechanical strength has been related to the quantity and position of collagen fibres (Dorow *et al.*, 2002, 2003; Pini *et al.*, 2004). All domestic ungulates have been observed to suffer from periodontal disease (Aitchison and Spence, 1984; Crabill and Schumacher, 1998; Dixon *et al.*, 1999a; Ingham, 2001; Kene and Uwagie-Ero, 2001).

With the exception of pigs, the molars of domestic ungulates are hypsodont, while the anterior dentition is de-emphasized, with maxillary incisors reduced in llamas and absent in bovid artiodactyls (cattle, sheep, goats; Table 5.1, Fig. 5.7), and canines (usually) absent in mares. A diastema, or tooth-free space, stretches between the anterior teeth and the cheek teeth. The cheek teeth vary in number among species (Table 5.1), but in all the occlusal contact area is increased. Whereas horses show enlargement of premolars to expand the cheek tooth surface, artiodactyls (including pigs) show expansion of the molars, especially elongation of the third molar (Mendoza *et al.*, 2002).

The high-crowned teeth of ungulates (except for pig) require wear to create lophs with enamel crests. The resulting enamel crests are oriented primarily anteroposteriorly, orthogonal to the medial direction of the power stroke, thus maximizing the cumulative contact surface (Koenigswald and Sander, 1997). This is analogous to the lengthening of carnassial shear facets in carnivorans. The enamel crests in the upper and lower cheek teeth often show reversed curvature such that the upper crests curve opposite to those of the lower teeth (Rensberger, 1973). In this way, occlusal stresses are concentrated on point contacts (Fig. 5.9). The reversed curvature is also likely to retain food on the crown surface for further reduction (Rensberger, 1973; Fortelius, 1985) and is analogous to the action of the two-cusp carnassial blade.

All ungulates have TMJs that permit extensive translation and have power strokes that feature lateral-to-medial movement of the lower teeth relative to the upper (Fig. 5.4). The primitive condition for mammalian chewing includes a two-phase power stroke in which the lower jaw moves medially, first closing to full intercuspation and then opening slightly to bring the cusps out of occlusion (Weijs, 1994). This primitive pattern is retained in pigs (Herring, 1976), but in the other ungulates the flattening of the molar surface has reduced the power stroke to a single closing phase.

As might be expected from the similarity of TMJs and masticatory movements, domestic ungulates resemble each other in jaw muscles and mechanics. The superficial masseter muscle is dominant (Turnbull, 1970) and its mechanical advantage is increased by enlargement of the mandibular angle and elevation of the TMJ above the tooth row (Figs 5.7b–d and 5.10; Smith and Savage, 1959). The pterygoid muscles, which aid the superficial masseter in producing the power stroke, are also relatively important (Turnbull, 1970). The grazing lifestyle of non-pig ungulates requires repetitive chewing, and it has been reported that the jaw muscles are 100% type I (slow-twitch, fatigue-resistant; reviewed by Herring, 1994). In all species that have been studied, the translational movements of the TMJs are produced by the 'transverse' pattern (Fig. 5.5), in which the superficial masseter and pterygoid muscles of one side cooperate with the

temporalis and deep masseter of the opposite side to produce movement towards the opposite side (Weijs, 1994).

In summary, domestic ungulates are distinguished by their adaptations for a translational power stroke that brings the lower molars medially across the upper molars. The molar teeth are elaborated and enlarged, and their enamel decussation resists not only vertical occlusal force but also bending stresses. TMJs and muscles are specialized for extensive movement rather than for vertical force production.

Horses (Equus caballus), Order Perissodactyla

The overall morphology of a horse molar is illustrated in Fig. 5.1, and the horse mandible is compared with those of other domestic species in Fig. 5.7. Among domestic ungulates, horses have the most robust dentition, because as hindgut-fermenters they must process high volumes of low-quality food (Janis, 1976). Horses maximize the surface area for potential food contact through increases in the complexity of enamel crests (Rensberger, 1973; Rensberger et al., 1984), which results in lophs of irregular shape but generally oriented anteroposteriorly, perpendicular to the direction of the power stroke (Fortelius, 1985). Because grazing on low-quality fodder leads to high abrasion, the crown surfaces become flattened, with low cusp relief (Osborn and Lumsden, 1978), and the edges of the enamel crests become rounded by wear. To compensate for high abrasion, the molars are very high-crowned. In horses, the sheets of inter-rod enamel are of the same thickness as the enamel rods, and thus carry an equal portion of the occlusal load (Pfretzschner, 1992). Interestingly, different horse breeds have been shown to differ in enamel hardness (Muylle et al., 1999).

Although the complexity of the crown surface is clearly an adaptation for food processing, the cementum-filled crevices that result are prone to the development of infection. The cementum wears at a faster rate than the surrounding enamel, and food becomes impacted in these areas, encouraging bacterial fermentation (Lowder and Mueller, 1998). Caries affects the first maxillary molar most frequently, and increases in incidence with age (Crabill and Schumacher, 1998).

The cheek tooth battery of horses is enlarged not only by the elaboration of the molars but also by the molarization of the premolars, except for the small and inconstant P[1] (Butler, 1952). Unlike the (non-pig) artiodactyls, horses have a complete complement of high-crowned incisors (Table 5.1), which form upper and lower cropping rows. Like the cheek teeth, the incisors require wear. If the incisors do not occlude properly (Gift et al., 1992; Dixon et al., 1999b), overgrowth occurs. Wry nose is a congenital deformity that includes a strong lateral deviation of the rostral nasal tissues and upper incisive arcade, more often producing incisal malocclusions among draft-breed horses (DeBowes and Gaughan, 1998). Fracture of the incisor teeth is common (Dixon et al., 1999a).

There is surprisingly little literature on masticatory movements in the horse, but manipulation of the skull makes it clear that they fit the ungulate pattern, with extensive translation at the TMJ (Scapino, 1972) and a power stroke that moves the lower jaw medially and somewhat anteriorly (Fig. 5.4). The mandibular symphysis is fused, and the lower cheek tooth rows are closer together than are the upper rows. Therefore, the mandible must be strongly deviated towards the working side at the beginning of the power stroke, and only one side can be in occlusion at a time. The typical ungulate adaptations of superficial masseter muscle enlargement, angular expansion and TMJ elevation (Fig. 5.7b) are exaggerated, while the temporalis is reduced to a small muscle that helps to retrude the condyle but contributes little to occlusal force (Turnbull, 1970). Preliminary electromyographic results show that horses use the transverse pattern of muscle activity to produce the translational movements of mastication, but that the superficial masseter is recruited before the other members of its group (Williams et al., 2002).

In summary, horses provide an extreme example of the typical ungulate features, with special adaptations related to consuming large quantities of a particularly resistant diet.

Cattle (Bos Taurus), *sheep* (Ovis aries), *goats* (Capra hircus), *Family Bovidae, llamas* (Lama glama) *and alpacas* (Lama pacos), *Family Camelidae, Order Artiodactyla*

Ruminant artiodactyls differ from horses in tending to feed on lower quantities of higher quality food. Even though cud-chewing results in a large number of daily chewing cycles, the overall abrasiveness is probably reduced compared to horses. The situation is probably intermediate for llamas, which like other camelids have a less complicated stomach than bovids.

These artiodactyls are characterized by a specialized molar crown morphology known as selenodonty, where the lophs are shaped as crescents, as illustrated in Fig. 5.9. In lateral view (Fig. 5.10) the lophs form cusplike peaks so that the molars appear serrated, with the height difference between cusp tips and intercusp valleys forming a high relief (Osborn and Lumsden, 1978). Enamel microstructure conforms to the general ungulate condition (Fig. 5.8). The molars in selenodonts are large and high-crowned, but less so than in horses (Sisson and Grossman, 1953), and the premolars are less molarized. Although molar malocclusions can result in insufficient wear of parts of the teeth (Aitchison and Spence, 1984; Ingham, 2001; Kene and Uwagie-Ero, 2001), the more common problem in ruminant artiodactyls is too much wear. Excessive wear of molars can lead to such poor mastication performance that cattle must be slaughtered in the prime of their reproductive life (Kene and Uwagie-Ero, 2001). Enamel erosion is accelerated by acidic diets (Rogers et al., 1997).

The selenodont artiodactyls also share a unique cropping mechanism in which the lower incisors work against a horny palatal pad. Despite not occluding with upper incisors, the lower cropping row can become pathologically worn (Aitchison and Spence, 1984; Raman, 1994). The unusual cropping apparatus is associated with a modification of the periodontal ligament, which is more strongly developed and attached on the lingual side than on the labial side of the root. The increased lingual stiffness is thought to counter the outward rotation of the incisor when herbage is pressed against the palatal pad (Spence, 1978).

The TMJ movements are of the standard ungulate type and produce extensive transverse movements of the jaw. As in the horse, the mandible is much narrower than the upper jaw, and only one side can be in occlusion at a time. The jaw muscles are also of typical ungulate proportions (Figs 5.7c and 5.10) with large superficial masseter and pterygoid muscles and small temporalis (with the exception of the temporalis in *Lama* (see below; Turnbull, 1970) and contract in the standard transverse pattern, with the medial power stroke produced by working-side superficial masseter and pterygoids and the opposite side temporalis and deep masseter (Fig. 5.5; DeVree and Gans, 1976; Williams *et al.*, 2002).

The differences between camelid and bovid domestic species relate to: (i) the robustness of the feeding apparatus; and (ii) the anterior dentition. Camelids have relatively greater molar surface area than comparably sized bovids (Herring, 1985). The lower incisors are strongly hypsodont (ever-growing in the closely related vicuña; Nowak, 1999) and the symphysis is fused. Llamas retain sharp, conical canine teeth in both upper and lower jaws, and also have a conical upper incisor lateral to the palatal pad. In cattle, sheep and goats the lower canine is spatulate and functions as an additional lower incisor in the cropping row (Fig. 5.7c). The incisors are relatively low-crowned and the mandibular symphysis is very mobile (Lieberman and Crompton, 2000). A striking difference in jaw muscle proportions is that the camelid temporalis is relatively huge (Turnbull, 1970). Although this condition may relate to canine biting, the presence of a powerful temporalis suggests that it is unusually important during mastication, in which case force could be transferred to the working-side molars by the rigidly fused symphysis. Another consequence of the large temporalis is a relatively small but well-buttressed TMJ (Herring, 1985). In contrast, the domestic bovids all have broad, loose TMJs, which are so specialized for translation that the normal curvature is reversed, with the condyle as the concave element (Bermejo *et al.*, 1993). The loose TMJ and the open symphysis of bovids permit the mandibular body to rotate around its long axis during mastication (Lieberman and Crompton, 2000).

In summary, while all the selenodont artiodactyls show typical ungulate features, llamas present an intermediate condition between the horse and domestic bovids in terms of robustness of jaw and muscle construction and dental adaptation to an abrasive diet.

Pigs (Sus scrofa), Family Suidae, Order Artiodactyla

Among living ungulates, pigs are the most omnivorous and have retained many primitive features in their feeding apparatus. While other domestic ungulates are hypsodont, the pig is an exception, possessing low-crowned (bunodont) molars with multiple rounded cusps, an adaptation that may relate to gripping small hard items (Herring, 1976). The pig also preserves a complete dentition (Table 5.1, Fig. 5.7d), bilateral occlusion and the more generalized two-phase power stroke, with an overall transverse excursion that is much less than that of other ungulates (Herring, 1976). The third molars are particularly elaborated with accessory cusps, and the canines are ever-growing.

Somewhat surprisingly, the microstructure of pig enamel conforms to the typical ungulate pattern, with both horizontal rod decussation and vertical inter-rod–rod decussation (Fig. 5.8). Thus, enamel resistance to both vertical and horizontal cracking must have evolved at an early stage of herbivory. The utility of this microstructure for pigs relates to the unpredictable modes of breakage of hard foodstuffs (Popowics *et al.*, 2001, 2004) as well as to the notable pig habit of chewing on all available materials including the bars and fencing designed to contain them, and stones found in outdoor pens (Harvey and Penny, 1985; Davies *et al.*, 2001). Another similarity with other ungulates is that despite the low-crowned morphology pig cusp tips undergo substantial wear that removes enamel to form crests (Hünermann, 1968). Mechanical testing has revealed that pig enamel is weak compared to human enamel, a property that promotes removal of the tip while preserving the remainder of the tooth (Popowics *et al.*, 2001, 2004).

The pig TMJ is of the translational type (Herring *et al.*, 2002), with a rounded condyle generally similar to that of horses and llamas, differing from the reversed curvature of the bovids. Pigs, horses and llamas also share a fused mandibular symphysis and retain functional caniniform teeth, suggesting that this suite of characteristics may be linked, possibly indicating the retention of some importance for TMJ rotation in biting behaviours.

Jaw muscle proportions are less specialized in the pig than in other ungulates. Although the superficial masseter and pterygoid muscles are emphasized (Fig. 5.7d), the temporalis is not reduced and retains the ability to contribute to occlusal force (Herring, 1985). The masticatory power stroke of pigs utilizes the standard 'transverse' pattern of working-side superficial masseter and pterygoids plus opposite-side temporalis and deep masseter (Herring and Scapino, 1973). The most interesting thing about pig mastication is the likelihood that both sides of the jaw are in simultaneous contact (Herring et al., 2001); it is not known whether the bolus is bilateral, but if so, pigs may be able to chew on both sides at the same time.

In summary, pigs have typical ungulate enamel and chewing patterns but are less specialized for herbivory than other ungulates. Their more generalized characteristics include a complete dentition with bunodont molars and a large temporalis muscle capable of generating vertical force. Unique to pigs are the ever-growing, laterally directed canine tusks and the addition of multiple accessory cusps to the molars.

Rabbits (*Oryctolagus cuniculus*), Order Lagomorpha (Fig. 5.7e)

Rabbits are grazers, like most ungulates, and have even evolved an analogue of rumination, coprophagy (Nowak, 1999). Although in many ways convergent with the ungulates (Smith and Savage, 1959), the rabbit masticatory system demonstrates some unique characteristics. Rabbits possess a distinct enamel microstructure known as irregular enamel. Bundles of enamel rods decussate in an irregular pattern thought to have evolved from Hunter–Schreger bands (Koenigswald and Clemens, 1992; Koenigswald, 1997). Addi-

tionally, the continuous eruption of both molars and incisors leads to unusual features of the periodontal ligament, which have been studied extensively for the incisors. The ligament is least stiff in the basal region where incisor development occurs and most stiff near the crown where intrusive forces are resisted (Komatsu et al., 1998). Periodontal ligament fibre bundles aligned with the orientation of stress bear the greatest load (Komatsu and Chiba, 2001).

Rabbits lack upper and lower canines and the second upper incisor is tucked behind the first (Hillson, 1986), perhaps for support. A long diastema separates the chisel-like incisors from the cheek teeth (Fig. 5.7e). The premolars are molarized and large, while the third molars are diminished (Hillson, 1986). The lophs of the cheek teeth are highly contoured and form a series of transverse ridges and valleys. The hypselodonty of the permanent incisors and cheek teeth compensates for rapid occlusal wear (Michaeli et al., 1980). In rabbits lack of wear quickly leads to overgrowth of the cheek teeth, lacerations of the tongue and cheek, incisor overgrowth, and anorexia (Harcourt-Brown, 1995).

Rabbits have an immobile, but usually unfused, mandibular symphysis. Like grazing ungulates, the lower jaw is narrower than the upper, and occlusion is unilateral only. The TMJs of rabbits are clearly of the translational type, but differ from those of ungulates in that the mandibular condyle is spherical rather than transversely elongated. This convergence with rodents is related to a power stroke that has a greater posterior-to-anterior component than that of ungulates (Weijs and Dantuma, 1981). The jaw muscles of rabbits are proportioned similarly to those of grazing ungulates (Turnbull, 1970). The superficial masseter is the most important jaw muscle, and as in ungulates its mechanical advantage is improved by elevation of the mandibular condyle and expansion of the mandibular angle (Fig. 5.7e). This muscle has some unusual fibre types and has been the subject of much research (Sciote and Kentish, 1996; Van Eijden et al., 2002; English and Widmer, 2003; Langenbach et al., 2004). Muscle contraction during mastication conforms to the 'transverse' pattern (Fig. 5.5; Weijs, 1994).

In summary, rabbits are strongly convergent with grazing ungulates. Although the enamel microstructure is different, it represents an alternative way of achieving resistance to fracture in multiple directions. Cropping involves a chisel-like mechanism not found in ungulates, however. Compared to ungulates, rabbits show greater anterior movement during the power stroke, and this is reflected in the transverse orientation of molar lophs and the shape of the TMJ.

Conclusion

Of the commercially important domestic mammals, the ungulates and the rabbit show many features in common, all related to the mastication of a resistant, abrasive and usually herbivorous diet. Enamel microstructure in these species shows crystallite decussation in at least two planes, strengthening the tooth against horizontal as well as vertical forces.

Molars are elaborated both in size and in complexity. The power stroke of mastication is a transverse lateral-to-medial sweep of the lower jaw, produced by anterior-pulling muscles (superficial masseter and pterygoids) on the working side and posterior-pulling muscles (temporalis and deep masseter) on the opposite side. At the jaw joints, the condyles can move freely anteroposteriorly. The mechanical advantage of the superficial masseter and medial pterygoid muscles is enhanced by elevation of the jaw joint and expansion of the mandibular angle. In contrast, the mink, bred domestically for fur, has a very different masticatory apparatus, featuring adaptation for vertical force to shear food with the carnassial teeth. The enamel crystallites decussate in a single plane, conferring resistance only to vertical compression. The well-buttressed jaw joint does not permit anterior-to-posterior movement, and thus the muscles are recruited to maximize bite force rather than to control condylar position.

References

Aitchison, G.U. and Spence, J.A. (1984) Dental disease in hill sheep: an abattoir survey. *Journal of Comparative Pathology* 94, 285–300.

Aulerich, R.J. and Swindler, D.R. (1968) The dentition of the mink (*Mustela vison*). *Journal of Mammalogy* 49, 488–494.

Beertsen, W., McCulloch, C.A.G. and Sodek, J. (2000) The periodontal ligament: a unique, multifunctional connective tissue. *Periodontology 2000* 13, 20–40.

Bennett, D. and Hoffmann, R.S. (1999) *Equus caballus*. *Mammalian Species* 628, 1–14.

Berkovitz, B.K.B. (2004) Periodontal ligament: structural and clinical correlates. *Dental Update* 31, 46–54.

Berkovitz, B.K.B. and Moxham, B.J. (1990) The development of the periodontal ligament with special reference to collagen fibre ontogeny. *Journal de Biologie Buccale* 18, 227–236.

Bermejo, A., González, O. and González, J.M. (1993) The pig as an animal model for experimentation on the temporomandibular articular complex. *Oral Surgery, Oral Medicine and Oral Pathology* 75, 18–23.

Biknevicius, A.R. and Van Valkenburgh, B. (1996) Design for killing: craniodental adaptations of predators. In: Gittleman, J.L. (ed.) *Carnivore Behavior, Ecology, and Evolution*, Vol. 2. Cornell University Press, Ithaca, New York, pp. 393–428.

Bosshardt, D.D. and Selvig, K.A. (1997) Dental cementum: the dynamic tissue covering of the root. *Periodontology 2000* 13, 41–75.

Butler, P.M. (1946) The evolution of the carnassial dentitions in the Mammalia. *Proceedings of the Zoological Society of London* 116, 198–220.

Butler, P.M. (1952) Molarization of the premolars in the Perissodactyla. *Proceedings of the Zoological Society of London* 121, 819–843.

Crabill, M.R. and Schumacher, J. (1998) Pathophysiology of acquired dental diseases of the horse. *Veterinary Clinics of North America: Equine Practice* 14, 291–305.

Crusafont-Pairo, M. and Truyols-Santoja, J. (1956) A biometric study of the evolution of fissiped carnivores. *Evolution* 10, 314–332.

Currey, J.D. (1999) The design of mineralised hard tissues for their mechanical functions. *Journal of Experimental Biology* 202, 3285–3294.

Davies, Z.E., Guise, H.J., Penny, R.H. and Sibly, R.M. (2001) Effects of stone chewing by outdoor sows on their teeth and stomachs. *The Veterinary Record* 149, 9–11.

DeBowes, R.M. and Gaughan, E.M. (1998) Congenital dental disease of horses. *Veterinary Clinics of North America: Equine Practice* 14, 273–289.

Dessem, D. and Druzinsky, R.E. (1992) Jaw-muscle activity in ferrets, *Mustela putorius furo. Journal of Morphology* 213, 275–286.

DeVree, F. and Gans, C. (1976) Mastication in pygmy goats *Capra hircus. Annales de la Societe Royale Zoologique de Belgique* 195, 255–306.

Dixon, P.M., Tremaine, W.H., Pickles, K., Kuhns, L., Hawe, C., McCann, J., McGorum, B., Railton, D.I. and Brammer, S. (1999a) Equine dental disease. Part 1: A long-term study of 400 cases: disorders of incisor, canine and premolar teeth. *Equine Veterinary Journal* 31, 369–377.

Dixon, P.M., Tremaine, W.H., Pickles, K., Kuhns, L., Hawe, C., McCann, J., McGorum, B.C., Railton, D.I. and Brammer, S. (1999b) Equine dental disease. Part 2: A long-term study of 400 cases: disorders of development and eruption and variations in position of the cheek teeth. *Equine Veterinary Journal* 31, 519–528.

Dorow, C., Krstin, N. and Sander, F.-G. (2002) Experiments to determine the material properties of the periodontal ligament. *Journal of Orofacial Orthopedics* 63, 94–104.

Dorow, C., Krstin, N. and Sander, F.-G. (2003) Determination of the mechanical properties of the periodontal ligament in a uniaxial tensional experiment. *Journal of Orofacial Orthopedics* 64, 100–107.

English, A.W. and Widmer, C.G. (2003) Sex differences in rabbit masseter muscle function. *Cells, Tissues, Organs* 174, 87–96.

Ewer, R.F. (1998) *The Carnivores.* Cornell University Press, Ithaca, New York.

Fischer, R.G., Klinge, B. and Attström, R. (1993) Normal histologic features of domestic ferret periodontium. *Scandinavian Journal of Dental Research* 101, 357–362.

Fortelius, M. (1985) Ungulate cheek teeth: developmental, functional and evolutionary interrelations. *Acta Zoologica Fennica* 180, 1–76.

Fortelius, M. and Solounias, N. (2000) Functional characterization of ungulate molars using the abrasion–attrition wear gradient: a new method for reconstructing paleodiets. *American Museum Novitates* 3301, 1–36.

Gift, L.J., DeBowes, R.M., Clem, M.F., Rashmir-Raven, A. and Nyrop, K.A. (1992) Brachygnathia in horses: 20 cases (1979–1989). *Journal of the American Veterinary Medical Association* 200, 715–719.

Goldberg, M. and Smith, A.J. (2004) Cells and extracellular matrices of dentin and pulp: a biological basis for repair and tissue engineering. *Critical Reviews of Oral Biology and Medicine* 15, 13–27.

Gorniak, G.C. and Gans, C. (1980) Quantitative assay of electromyograms during mastication in domestic cats (*Felis catus*). *Journal of Morphology* 163, 253–281.

Greaves, W.S. (1978) The jaw lever system in ungulates: a new model. *Journal of Zoology* 184, 271–285.

Greaves, W.S. (1983) A functional analysis of carnassial biting. *Biological Journal of the Linnaean Society* 20, 353–363.

Greaves, W.S. (1985) The generalized carnivore jaw. *Biological Journal of the Linnaean Society* 85, 267–274.

Hanaizumi, Y. (1992) Three-dimensional changes in direction and interrelationships among enamel prisms in the dog tooth. *Archives of Histology and Cytology* 55, 539–550.

Hanaizumi, Y., Maeda, T. and Takano, Y. (1996) Three-dimensional arrangement of enamel prisms and their relation to the formation of Hunter–Schreger bands in dog tooth. *Cell and Tissue Research* 286, 103–114.

Hanaizumi, Y., Kawano, Y., Ohshima, H., Hoshino, M., Takeuchi, K. and Maeda, T. (1998) Three-dimensional direction and interrelationship of prisms in cuspal and cervical enamel of dog tooth. *The Anatomical Record* 252, 355–368.

Harcourt-Brown, F.M. (1995) A review of clinical conditions in pet rabbits associated with their teeth. *The Veterinary Record* 137, 341–346.

Harvey, C.E. and Penny, R.H.C. (1985) Oral and dental disease in pigs. In: Harvey, C.E. (ed.) *Veterinary Dentistry.* W.B. Saunders, London, pp. 272–280.

He, T., Olsson, S., Daugaard, J.R. and Kiliaridis, S. (2004) Functional influence of masticatory muscles on the fibre characteristics and capillary distribution in growing ferrets (*Mustela putorius furo*) – a histochemical analysis. *Archives of Oral Biology* 49, 983–989.

Herring, S.W. (1976) The dynamics of mastication in pigs. *Archives of Oral Biology* 21, 473–480.

Herring, S.W. (1985) Morphological correlates of masticatory patterns in peccaries and pigs. *Journal of Mammalogy* 66, 603–617.

Herring, S.W. (1992) Muscles of mastication: architecture and functional organization. In: Davidovitch, Z. (ed.) *The Biological Mechanisms of Tooth Movement and Craniofacial Adaptation.* The Ohio State University College of Dentistry, Columbus, Ohio, pp. 541–548.

Herring, S.W. (1993) Functional morphology of mammalian mastication. *American Zoologist* 33, 289–299.

Herring, S.W. (1994) Functional properties of the feeding musculature. In: Bels, V.L., Chardon, M. and Vandewalle, P. (eds) *Biomechanics of Feeding in Vertebrates.* Springer-Verlag, Berlin, pp. 5–30.

Herring, S.W. and Scapino, R.P. (1973) Physiology of feeding in miniature pigs. *Journal of Morphology* 141, 427–460.

Herring, S.W., Rafferty, K.L., Liu, Z.-J. and Marshall, C.D. (2001) Jaw muscles and the skull in mammals: the biomechanics of mastication. *Comparative Biochemistry and Physiology* 131, 207–219.

Herring, S.W., Decker, J.D., Liu, Z.J. and Ma, T. (2002) The temporomandibular joint in miniature pigs: anatomy, cell replication, and relation to loading. *The Anatomical Record* 266, 152–166.

Hiiemae, K.M. (2000) Feeding in mammals. In: Schwenk, K. (ed.) *Feeding: Form, Function and Evolution in Tetrapod Vertebrates.* Academic Press, San Diego, California, pp. 411–448.

Hillson, S. (1986) *Teeth.* Cambridge University Press, Cambridge, UK.

Hünermann, K.A. (1968) Die Suidae (Mammalia, Artiodactyla) aus den Dinotheriensanden (Unterpliozän = Pont) Rheinhessens (Südwestdeutschland). *Schweizerische Palaeontologische Abhandlungen* 86, 1–96.

Ingham, B. (2001) Abattoir survey of dental defects in cull cows. *The Veterinary Record* 148, 739–742.

Janis, C. (1976) The evolutionary strategy of the Equidae and the origins of rumen and cecal digestion. *Evolution* 30, 757–774.

Janis, C.M. (1988) An estimation of tooth volume and hypsodonty indices in ungulate mammals, and the correlation of these factors with dietary preference. In: Russell, D.E., Santoro, J.-P. and Sigogneau-Russell, D. (eds) *Teeth Revisited: Proceedings of the 7th International Symposium on Dental Morphology, Paris, 1986. Mémoires du Muséum National D'Histoire Naturelle,* Vol. 53. Paris, France, pp. 367–387.

Janis, C. and Fortelius, M. (1988) On the means whereby mammals achieve increased functional durability of their dentitions, with special reference to limiting factors. *Biological Reviews* 63, 197–230.

Kawaii, N. (1955) Comparative anatomy of the bands of Schreger. *Okijimas Folia Anatomica Japonica* 27, 115–131.

Kene, R.O.C. and Uwagie-Ero, E.A. (2001) Dental abnormalities of nomadic cattle of Nigeria. *Tropical Veterinarian* 19, 191–199.

Kobara-Mates, M., Logemann, J.A., Larson, C. and Kahrilas, P.J. (1995) Physiology of oropharyngeal swallow in the cat: a videofluoroscopic and electromyographic study. *American Journal of Physiology* 268, G232–G241.

Koenigswald, W.v. (1997) Brief survey of enamel diversity at the Schmelzuster level in Cenozoic placental mammals. In: Koenigswald, W.v. and Sander, P.M. (eds) *Tooth Enamel Microstructure.* Balkema, Rotterdam, The Netherlands, pp. 137–161.

Koenigswald, W.v. and Clemens, W.A. (1992) Levels of complexity in the microstructure of mammalian enamel and their application in studies of systematics. *Scanning Microscopy* 6, 195–218.

Koenigswald, W.v. and Pfretzschner, H.U. (1991) Biomechanics in the enamel of mammalian teeth. In: Schmidt-Kittler, N. and Vogel, K. (eds) *Constructional Morphology and Evolution.* Springer-Verlag, Berlin, pp. 113–125.

Koenigswald, W.v. and Sander, P.M. (1997) Schmelzmuster differentiation in leading and trailing edges, a specific biomechanical adaptation in rodents. In: Keonigswald, W.v. and Sander, P.M. (eds) *Tooth Enamel Microstructure.* Balkema, Rotterdam, The Netherlands, pp. 259–266.

Koenigswald, W.v., Rensberger, J.M. and Pfretzschner, H.U. (1987) Changes in tooth enamel of early Paleocene mammals allowing increased diet diversity. *Nature* 328, 150–152.

Komatsu, K. and Chiba, M. (2001) Synchronous recording of load–deformation behaviour and polarized light-microscopic images of the rabbit incisor periodontal ligament during tensile loading. *Archives of Oral Biology* 46, 929–937.

Komatsu, K., Yamazaki, Y., Yamaguchi, S. and Chiba, M. (1998) Comparison of biomechanical properties of the incisor periodontal ligament among different species. *The Anatomical Record* 250, 408–417.

Langenbach, G.E.J., van Wessel, T., Brugman, P. and Van Eijden, T.M.G.J. (2004) Variation in daily masticatory muscle activity in the rabbit. *Journal of Dental Research* 83, 55–59.

Lariviére, S. (1999) *Mustela vison. Mammalian Species* 608, 1–9.

Lieberman, D.E. and Crompton, A.W. (2000) Why fuse the mandibular symphysis? A comparative analysis. *American Journal of Physical Anthropology* 112, 517–540.

Lowder, M.Q. and Mueller, P.O.E. (1998) Dental embryology, anatomy, development, and aging. *Veterinary Clinics of North America: Equine Practice* 14, 227–245.

Lucas, P.W. and Luke, D.A. (1984) Chewing it over: basic principles of food breakdown. In: Chivers, D.J., Wood, B.A. and Bilsborough, A. (eds) *Food Acquisition and Processing in Primates.* Plenum Press, New York, pp. 283–301, 576pp.

Lucas, P.W. and Teaford, M.F. (1994) Functional morphology of colobine teeth. In: Davies, A.G. and Oates, J.F. (eds) *Colobine Monkeys: Their Ecology, Behaviour and Evolution.* Cambridge University Press, Cambridge, UK, pp. 173–204, 415pp.

Maas, M. (1994) A scanning electron-microscopic study of *in vitro* abrasion of mammalian tooth enamel under compressive loads. *Archives of Oral Biology* 39, 1–11.

Matsuo, M. and Takahashi, K. (2002) Scanning electron microscopic observation of microvasculature in periodontium. *Microscopy Research and Technique* 56, 3–14.

Mendoza, M., Janis, C. and Palmqvist, P. (2002) Characterizing complex craniodental patterns related to feeding behaviour in ungulates: a multivariate approach. *Journal of Zoology (London)* 258, 223–246.

Michaeli, Y., Hirschfeld, Z. and Weinreb, M.M. (1980) The cheek teeth of the rabbit: morphology, histology and development. *Acta Anatomica* 106, 223–239.

Moss, M.L. (1968) The primacy of functional matrices in orofacial growth. *Dental Practitioner* 19, 65–73.

Moxham, B.J. and Berkovitz, B.K. (1989) A comparison of the biomechanical properties of the periodontal ligaments of erupting and erupted teeth of non-continuous growth (ferret mandibular canines). *Archives of Oral Biology* 34, 763–766.

Muylle, S., Simoens, P., Verbeeck, R., Ysebaert, M.T. and Lauwers, H. (1999) Dental wear in horses in relation to the microhardness of enamel and dentine. *The Veterinary Record* 144, 558–561.

Myhre, L., Preus, H.R. and Aars, H. (1979) Influences of axial load and blood pressure on the position of the rabbit's incisor tooth. *Acta Odontologica Scandinavica* 37, 153–159.

Nowak, R.M. (1999) *Walker's Mammals of the World.* The Johns Hopkins University Press, Baltimore, Maryland.

Osborn, J.W. and Lumsden, A.G.S. (1978) An alternative to 'thegosis' and a re-examination of the ways in which mammalian molars work. *Neues Jahrbuch für Geologie und Paläontologie, Abhandlungen* 156, 371–392.

Pashley, D.H. (1996) Dynamics of the pulpo-dentin complex. *Critical Reviews in Oral Biology and Medicine* 7, 104–133.

Pfretzschner, H.U. (1992) Enamel microstructure and hypsodonty in large mammals. In: Smith, P. and Tchernov, E. (eds) *Structure, Function and Evolution of Teeth.* Freund Publishing House, London and Tel Aviv, pp. 147–162.

Picton, D.C.A. and Wills, D.J. (1978) Viscoelastic properties of the periodontal ligament and mucous membrane. *The Journal of Prosthetic Dentistry* 40, 263–272.

Pini, M., Zysset, P., Botsis, J. and Contro, R. (2004) Tensile and compressive behaviour of the bovine periodontal ligament. *Journal of Biomechanics* 37, 111–119.

Popowics, T.E. (2003) Dental form in the Mustelidae and Viverridae (Mammalia: Carnivora). *Journal of Morphology* 256, 322–341.

Popowics, T.E. and Fortelius, M. (1997) On the cutting edge: tooth blade sharpness in herbivorous and faunivorous mammals. *Annales Zoologica Fennici* 34, 73–88.

Popowics, T.E., Rensberger, J.M. and Herring, S.W. (2001) The fracture behaviour of human and pig molar cusps. *Archives of Oral Biology* 46, 1–12.

Popowics, T.E., Rensberger, J.M. and Herring, S.W. (2004) The relationship between enamel structure and strain in the fracture behaviour of human and pig molar cusps. *Archives of Oral Biology* 49, 595–605.

Radinsky, L.B. (1981) Evolution of skull shape in carnivores 1. Representative modern carnivores. *Biological Journal of the Linnaean Society* 15, 369–388.

Raman, K.S. (1994) Abnormality of eruption and wear of incisor teeth in Nilagiri sheep. *Journal of Veterinary and Animal Sciences* 25, 166–168.

Rasmussen, S.T., Patchin, R.E., Scott, D.B. and Heuer, A.H. (1976) Fracture properties of human enamel and dentin. *Journal of Dental Research* 55, 154–164.

Render, J.A., Bursian, S.J., Rosenstein, D.S. and Aulerich, R.J. (2001) Squamous epithelial proliferation in the jaws of mink fed diets containing 3, 3′, 4, 4′, 5-pentachlorobiphenyl (PCB 126) or 2, 3, 7, 8-tetrachlorodibenzo-P-dioxin (TCDD). *Veterinary and Human Toxicology* 43, 22–26.

Rensberger, J.M. (1973) An occlusion model for mastication and dental wear in herbivorous mammals. *Journal of Paleontology* 47, 515–528.

Rensberger, J.M. (1993) Adaptation of enamel microstructure to differences in stress intensity in the Eocene perissodactyl Hyracotherium. In: Kobayashi, I., Mutvei, H. and Sahni, A. (eds) *Structure, Formation and Evolution of Fossil Hard Tissues.* Tokai University Press, Tokyo, 131–145.

Rensberger, J.M. (1995) Determination of stresses in mammalian dental enamel and their relevance to the interpretation of feeding behaviors in extinct taxa. In: Thomason, J.J. (ed.) *Functional Morphology in Vertebrate Paleontology.* Cambridge University Press, London, pp. 151–172.

Rensberger, J.M. (2000) Pathways to functional differentiation in mammalian enamel. In: Teaford, M., Smith, M.M. and Ferguson, W.J. (eds) *Development, Function and Evolution of Teeth.* Cambridge University Press, Cambridge, UK, pp. 252–268.

Rensberger, J.M. and Pfretzschner, H.U. (1992) Enamel structure in astrapotheres and its functional implications. *Scanning Microscopy* 6, 495–510.

Rensberger, J.M., Forstén, A. and Fortelius, M. (1984) Functional evolution of the cheek tooth pattern and chewing direction in Tertiary horses. *Paleobiology* 10, 439–452.

Rogers, G.M., Poore, M.H., Ferko, B.L., Kusy, R.P., Deaton, T.G. and Bawden, J.W. (1997) *In vitro* effects of an acidic by-product feed on bovine teeth. *American Journal of Veterinary Research* 58, 498–503.

Scapino, R.P. (1972) Adaptive radiation of mammalian jaws. In: Schumacher, G.-H. (ed.) *Morphology of the Maxillo-Mandibular Apparatus.* Georg Thieme, Leipzig, Germany, pp. 33–39.

Scapino, R.P. (1974) Function of the masseter–pterygoid raphe in carnivores. *Anatomischer Anzeiger* 136, 430–446.

Scapino, R.P. (1976) Function of the digastric muscle in carnivores. *Journal of Morphology* 150, 843–860.

Scapino, R.P. (1981) Morphological investigation into functions of the jaw symphysis in carnivorans. *Journal of Morphology* 167, 339–375.

Sciote, J.J. and Kentish, J.C. (1996) Unloaded shortening velocities of rabbit masseter muscle fibres expressing skeletal or α-cardiac myosin heavy chains. *Journal of Physiology* 492, 659–667.

Sisson, S. and Grossman, J.D. (1953) *The Anatomy of the Domestic Animals.* W.B. Saunders, Philadelphia, Pennsylvania.

Skobe, Z., Prostak, K.S. and Trombly, P.L. (1985) Scanning electron microscope study of cat and dog enamel structure. *Journal of Morphology* 184, 195–203.

Smith, K.K. (2001) The evolution of mammalian development. *Bulletin of the Museum of Comparative Zoology* 156, 119–135.

Smith, M.J. and Savage, R.J.G. (1959) The mechanics of mammalian jaws. *The School Science Review* 40, 289–301.

Spence, J.A. (1978) Functional morphology of the periodontal ligament in the incisor region of the sheep. *Research in Veterinary Science* 25, 144–151.

Stefen, C. (1997) Differentiations in Hunter–Schreger bands of carnivores. In: Koenigswald, W.v. and Sander, P.M. (eds) *Tooth Enamel Microstructure.* Balkema, Rotterdam, The Netherlands, pp. 123–136.

Stefen, C. (2001) Enamel structure of Arctoid Carnivora: Amphicyonidae, Ursidae, Procyonidae, and Mustelidae. *Journal of Mammalogy* 82, 450–462.

Thexton, A.J. and McGarrick, J.D. (1994) The electromyographic activities of jaw and hyoid musculature in different ingestive behaviours in the cat. *Archives of Oral Biology* 39, 599–612.

Tsolakis, A.I., Spyropoulos, M.N., Katsavrias, E. and Alexandridis, K. (1997) Effects of altered mandibular function on mandibular growth after condylectomy. *European Journal of Orthodontics* 19, 9–19.

Turnbull, W.D. (1970) Mammalian masticatory apparatus. *Fieldiana: Geology* 18, 149–356.

Van Eijden, T.M.G.J., Turkawski, S.J.J., van Ruijven, L.J. and Brugman, P. (2002) Passive force characteristics of an architecturally complex muscle. *Journal of Biomechanics* 35, 1183–1189.

Van Valen, L. (1969) Evolution of dental growth and adaptation in mammalian carnivores. *Evolution* 23, 96–117.

Walker, A.C., Hoeck, H.N. and Perez, L. (1978) Microwear of mammalian teeth as an indicator of diet. *Science* 201, 908–910.

Waters, N.E. (1980) Some mechanical and physical properties of teeth. In: Vincent, J.F.V. and Currey, J.D. (eds) *The Mechanical Properties of Biological Materials.* Society for Experimental Biology, London, pp. 99–135.

Weijs, W.A. (1994) Evolutionary approach of masticatory motor patterns in mammals. In: Bels, V.L., Chardon, M. and Vandewalle, P. (eds) *Biomechanics of Feeding in Vertebrates*, Vol. 18. Springer-Verlag, Berlin, pp. 281–320.

Weijs, W.A. and Dantuma, R. (1981) Functional anatomy of the masticatory apparatus in the rabbit (*Oryctolagus cuniculus* L.). *Netherlands Journal of Zoology* 31, 99–333.

Weinberg, M.A. and Bral, M. (1999) Laboratory animal models in periodontology. *Journal of Clinical Periodontology* 26, 335–340.

Williams, S.H. and Kay, R.F. (2001) A comparative test of adaptive explanations for hypsodonty in ungulates and rodents. *Journal of Mammalian Evolution* 8, 207–229.

Williams, S.H., Wall, C.E., Vinyard, C.J. and Hylander, W.L. (2002) Jaw-muscle motor patterns in ungulates: is there a transverse pattern? *Integrative and Comparative Biology* 42, 1336.

6 Feeding and Welfare in Domestic Animals: A Darwinistic Framework

P. Koene

*Ethology Group, Department of Animal Science, Wageningen University,
PO Box 338, 6700 AH Wageningen, The Netherlands*

Introduction

During evolution, animals adapted to their environment for survival and fitness. Their feeding adapted in terms of functional morphology, ecology and behaviour. Feeding became embedded in their time budget, space use and social life. Animals that adjusted benefited from living in the neighbourhood of man and became domesticated. Due to genetic selection and recent intensification of housing conditions, restrictions on the domestic animal became increasingly apparent as shown in their health and welfare. In particular, feeding in captivity became very different from the circumstances under which it evolved, i.e. food is provided by man and animals have less control over food intake. Problems regarding non-nutritive requirements, substitute activities and the psychological aspects of feeding developed.

To develop an understanding of the welfare in animals in terms of biological function, feeding behaviour of the domestic animal and the wild ancestors will be compared. Specific behavioural and welfare problems related to feeding will be discussed for domestic mammals and birds. Solutions will be discussed in terms of providing basic feeding requirements to alleviate suffering (or hunger) or feeding enrichment that may give pleasure. The approach is influenced by work on wild,

feral, zoo and farm animals and attempts to fit behaviour and welfare in a Darwinistic framework.

> Feeding is any action of an animal that is directed toward the procurement of nutrients. The variety of means of procuring food reflects the diversity of foods used and the myriad of animal types.
> The second meaning of feeding is the act of supplying food by humans.
> (*Encyclopaedia Britannica*, 2002)

The only definition of feeding given in the *Dictionary of Farm Animal Behavior* (Hurnik *et al.*, 1995) is 'delivery of feed to animals'. A domestic animal can be defined as one that has been bred for a long time in captivity for profit to humans, who have control over its social organization, breeding and food supply. One of the first definitions of animal welfare was: 'Animal welfare is a state of physical and psychological harmony between the organism and its surroundings' (Lorz, 1973). More and recent definitions of welfare are discussed in the last part of this chapter.

Feeding and welfare in domestic animals has been the explicit and implicit subject of a number of chapters and books. To avoid duplicating the already vast amount of existing information some are given. The link between feeding in nature and in captivity can be found in Price (2002, ch. 12). The transition from nature to captivity produces large environmental changes in the life of a domestic animal and may cause problems in different stages of its life. The dependency on humans for food is formulated as follows:

> In nature, animals spend a large share of their time and energy searching for and consuming food. Choices with regard to feeding sites, diet selection and foods consumed are often quite varied. Most captive animals, on the other hand, are dependent on humans to provide an appropriate diet, and foods offered are often relatively uniform on a daily and seasonal basis.   (Price, 2002)

A basic introduction to feeding in nature can be found in Broom (1981), Krebs and Davies (1991) and Barnard (2004). The subject of motivation has a special place in relation to feeding behaviour (Toates, 1980; Fraser and Broom, 1990). Basics of animal welfare problems are presented in Fraser and Broom (1990) and Ewbank et al. (1999). The important topic of hunger in relation to animal behaviour and welfare is perfectly covered in Lawrence et al. (1993) and Kyriazakis and Savory (1997).

In short, this chapter explores the natural feeding behaviour, domestic feeding, behavioural problems related to feeding in captivity, animal welfare and the solutions for feeding problems and poor welfare. Details of feeding and welfare problems are given in other chapters in this volume.

Natural Feeding Behaviour

Natural feeding behaviour of an animal developed in its niche in the course of evolution and was observable in its natural environment. Feeding behaviour consists of all behaviour necessary for the appetitive and consumatory phases. The appetitive phase includes the searching and localization of the food. The final consumatory phase consists of handling, manipulating, tasting and consumption.

The feeding system is the most important behaviour system for survival of an organism. All organisms prey on other organisms and most can be preyed upon. Although different names are given to organisms that eat meat (carnivores) or that eat plants (herbivores), the

questions that arise during the feeding process are much alike. Animals have to search for food, and determine the specific kind of food and the place to forage. Animals have to make many decisions during the feeding process about where to feed, what to feed on, how long to feed and how to explore potential feeding places or patches. These decisions are partly based on species-specific decision rules (Barnard, 2004), partly on what they learned during preceding feeding experiences (Forbes, 2001).

Optimal foraging

The optimal foraging theory states that foraging strategies may involve decisions that maximize the net rate of food intake, or of some other measure of foraging efficiency (Krebs, 1978). Animals optimize their foraging because they are in evolutionary competition with other animals with different feeding strategies. The ones that have the best strategy in the long term will survive and outdo other strategies and individuals with those strategies. In the course of evolution a combination of stable strategies within a species, an evolutionary stable strategy (ESS) survives. This means that the mechanisms and behaviours of feeding are shaped in a specific form in which decisions, rules and the amount of flexibility (learning) are determined. One of the decision rules for many animal species is to stay feeding in a specific place or patch as long as its intake rate is above that of the average of places (marginal value theorem) in its environment (Charnov, 1976). This means that an animal always calculates the costs and benefits of staying at a certain place, or leaves and starts foraging in another place (Bailey, 1995). The choice of the diet of foraging animals seems to be best predicted by the following rules: take the most profitable food, take less profitable prey based on the calculated profitability (Charnov, 1976) and include less profitable prey dependent on the number of food items encountered. The calculation of what food item is the most profitable is based on the energy gain, a cost benefit calculation and the time needed to find and consume the food item.

Social factors

Animals in most cases do not forage and eat alone. Added to the optimal foraging approach social factors play an important role. Animals could benefit from the presence of conspecifics or even members of other species. Animals may find the best place by following other more experienced animals, i.e. leaders in cattle (Reinhardt, 1983); use information from other animals (Galef, 1991); use food for other purposes like attraction of females in a chicken harem (Gyger and Marler, 1988). In a group, the safety may be improved by the presence of more watching and alert eyes of vigilant group-mates (many eyes hypothesis: Pulliam, 1972). Thus, animals in groups tend to allocate less time to vigilance behaviour and more time to other activities such as foraging (Quenette, 1990), and the risk of predation is lower in a group due to confusion effects (Roberts, 1996). For such beneficial purposes, mechanisms developed that control the group size of the foraging animal. Synchronization of behaviour is one of those mechanisms (Clifton, 1979; Rook and Huckle, 1995). The fusion–fission system of large grazers also has the advantage of information exchange of location and quality of foraging patches in an area to a large group of animals while avoiding too much individual competition (Prins, 1996). The disadvantages of foraging in a group are increased competition between individuals and consequently mutual interference (Case and Gilpin, 1974; Sutherland and Koene, 1982). Dependent on food distribution and availability, animals often distribute according to the food supply (i.e. ideal free distribution: Fretwell and Lucas, 1970).

Individual feeding

Fitness is the relative contribution of an individual to the gene pool of the next generations. When that contribution is large due to natural selection the animal has a high fitness. As densities increase, resources are divided into many pieces and all individuals suffer equally (scramble competition). If there is competitive hierarchy between individuals, some will not suffer but others will (contest competition). In the natural environment, those individuals that avoid competition may do as well as competitive ones. This may result in a bi-modal distribution of animals, with each mode demonstrating its own morphology (size) and behavioural characteristics. Dominant despots, who remain close to, and defend, the source of food will receive a large proportion of what is delivered. Subordinate animals, which stay out of the feeding area, will survive on what few items pass their way. Part of the population is hence in a poor and stressful condition. Depending on condition and environment, different strategies will be successful (Korte et al., 2004). Individual strategies mingle, building up complex animal societies and populations (Sutherland, 1996) often by self-organization (Camazine, 2001).

Environment of evolutionary adaptedness

The environment of evolutionary adaptedness (EEA) is the ensemble of physical and social conditions in which a particular trait was naturally selected (Tooby and Cosmides, 1990). More precisely, 'The "environment of evolutionary adaptedness" (EEA) is not a place or a habitat, or even a time period. Rather, it is a statistical composite of the adaptation-relevant properties of the ancestral environments encountered by members of ancestral populations, weighted by their frequency and fitness-consequences'. In practice, this means that animals are adapted to react on stimuli from the environment according to the rules that developed and are learned as the best strategy in their EEA.

The EEA concept is comparable to the notion of 'niche' in evolutionary biology (Case and Gilpin, 1974) and 'Umwelt' in ethology (Uexküll, 1939, 1956). Horses faced different problems than did cattle, and as a result they have different adaptations. The EEA for any specific organism is the set of reproductive problems faced by members of that species over evolutionary time. Each of these species faced reproductive problems that the others did not, and thus their EEAs are different. Within populations, different feeding strategies may be successful, being part of the same ESS, but leading to large individual differences within a species (Koolhaas et al., 1999).

Feeding behaviour examples

Mammals show a large range of feeding adaptations, such as herbivory and carnivory. Morphological, physiological and behavioural adaptations determine the efficiency of such strategies. Adaptation is visible in differences in the design of teeth. Herbivores have a design different from carnivores. Green food is normally abundant and accessible. It is, however, hard to digest. A diet of plant material must often be consumed in large quantities, because of its often low nutritive value, which becomes even lower in matured plants. On the other hand, medics (lucerne) have a high nutritive value (Koene, 1999). Breaking it down to small particles is difficult and requires strong flat teeth that can handle intensive chewing. To break down the cellulose content of plants the herbivores developed two adaptations: the foregut and the hindgut fermenters. Both rely on symbiotic microorganisms residing in their alimentary tract. Foregut fermentation is characteristic of ruminant bovids. Foregut fermenters feed on new grown parts of plants and selectively cut them with the tongue and lips (ruminants lack the upper incisors). Food is regurgitated and disintegration of food is mainly done in their different stomachs. Hindgut fermenters have their fermenting microorganisms housed in the caecum and large intestine. Food is not regurgitated and all mechanical disintegration of food must be performed during mastication. The passage of food through the hindgut fermenters is almost twice as fast as through the foregut fermenters. Hindgut fermenters survive on low-quality food, if available in large quantities. They can separate dry plant mass from tannins, both of which normally reduce efficient foregut fermenting, although tannins consumed at appropriate levels may have beneficial effects (Reed, 1995). Due to these feeding and digestion differences, both foregut and hindgut fermenters can forage without real competition next to each other in the same environment. There is ecological separation when populations are below the carrying capacity, because they do not use the same food resources, and so the principle of competitive exclusion (Hardin, 1960) is applied.

Birds that fly need highly nutritive food. Therefore, most birds eat at least some insects, which they actively capture. Some species are highly specialized; others are more varied in their diet. They possess many methods of searching food, of which most involve sight. Birds that do not fly are able to use low-value nutritional food and may be larger to digest the increased quantities of food needed. Most information about ancestral feeding behaviour in the EEA has come from wild (red jungle fowl), feral (horses) and semi-wild (cattle) animals.

Cattle feeding behaviour

The wild ancestor of cattle, the aurochs, became extinct in 1627. Therefore, we do not have a precise idea of its feeding behaviour. Probably, the favourite living area of the aurochs was sedge marshes along rivers. The last free-roaming aurochs (*Bos primigenius*) also lived in woodlands and were reported to consume grass during summer, acorns in autumn and twigs in winter (Van Vuure, 2001, 2003). It may be assumed that habitat choice and woody diets of these wild species have been affected by predators. Bovinae in general include grazers, grazer-browsers (mixed feeders) and browsers. Large species mainly graze and only browse when fresh leaves are present or when forced due to absence of grass. Cattle consume large amounts of relatively low-quality grass. They grasp the grass with their tongue and cut it with their lower incisors with a short head movement. When enough raw material is collected, rumination takes place often in the shade of trees at a safer location. During a day, cattle graze 6–9 h and ruminate 4–6 h divided in a number of graze–ruminate bouts. During hot days, the animals feed at night. Individuals tend to show the same activities in synchronization (allelomimetic behaviour). Some cattle live in environments with a large seasonal change in food supply. In the temperate zone in summer they have an abundance of green material. During winter the grass stops growing and the animals have to survive on their fat deposits. Heck cattle in The Netherlands show a large decrease in condition score during winter

(Fig. 6.1). This has of course a large impact on condition, health and mortality, and their feeding behaviour. In the Mediterranean environment the situation is comparable, but reversed, i.e. an abundance of feed in winter but shortage of feed in summer.

Horse feeding behaviour

The last Tarpan (a wild horse) died in Munich Zoo in 1887. Many studies of feral and wild horses give a good impression of the natural feeding behaviour of horses (Linklater, 2000). All equids forage primarily on fibrous foods. Although horses feed mainly on grasses and sedges, they will consume bark, leaves, buds, fruits and roots. Equids can sustain themselves in habitats that are more marginal and on diets of lower quality than can ruminants (mustangs in Nevada deserts). Equids forage during the day and night at about 60–80% of the time over 24 h. Horses mainly live in temperate zones with a strong seasonal varying environment. Wild horses (such as the Przewalski horse in Mongolia) which have hot summers and harsh winters show much less dramatic changes in condition score than cattle.

Chicken feeding behaviour

The chicken's ancestor, the red jungle fowl, spends its day foraging for food in grasslands or the understorey of bamboo forests in Southeast Asia. It prefers young succulent leaves of plants, especially of bamboo; its most appropriate name is bamboofowl (Janzen, 1976). It has a short, down-curved bill, used to peck plant material from the ground or from short vegetation. The jungle fowl is omnivorous and its diet includes seeds (rice), invertebrates and eggs. It has large, strong feet, a crucial attribute that allows it to expose seeds and roots that are inaccessible to most other animals. Its heavy build indicates a diet based on bulky vegetable matter, although the chicks depend on insects and larvae during their first few weeks (Savory et al., 1978). The chicken has a crop, which can be extended to cache food before digestion. It also has a very strong gizzard, which is used to grind seeds and the tough fibres of the vegetation consumed. To aid digestion, birds regularly swallow small stones to assist the grinding action in the gizzard. In (semi-)natural conditions, the red jungle fowl spends 60–90% of its time budget on foraging and feeding (Dawkins, 1989).

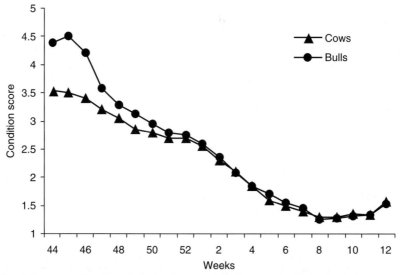

Fig. 6.1. Condition score of Heck cattle in the Oostvaardersplassen, The Netherlands, between October 2004 and April 2005. (From Koene and Munsters, in press.)

Domestic Feeding

Domestication of cattle, horse and chicken occurred during the last 10,000 years (Clutton-Brock, 1999). Domestication is the process by which a population of animals become socialized to humans and adapted to the captive environment by some combination of genetic changes occurring over generations and environmentally induced developmental events recurring during each generation (Price, 1984). Some effects of domestication can be assessed by comparison of wild and domestic animals (Schütz et al., 2001).

Environment of domestic adaptedness

We may call the domestic environment – in analogy with EEA – the environment of domestic adaptedness (EDA). In the same way, the EDA is the statistical composite of the adaptation-relevant properties of the original environment of domestication. In the dynamic process of domestication, adaptive and non-adaptive processes are discernible. In the original domestication, roughly until halfway through the last century, docility in combination with attractive characteristics determined the behavioural phenotype of domestic animal (Price, 2002). However, in the last part of that century, intensification by genetic selection and intensive housing and keeping conditions restricted the domestic animal more and more, and an increasing number of health and welfare problems emerged (Rauw et al., 1998). Selection for high egg production made laying hens more fearful and less docile (Kjaer and Sorensen, 1997). This means also that the current environment in which domestic animals are kept is rather different from the original environment in which they were domesticated (EDA). This present and current environment is called here the environment of current adaptedness (ECA). The ECA, EDA and EEA are combined in the group environments of adaptedness (EAs). Especially feeding in captivity became very different from the circumstances under which it evolved, i.e. food is provided by man. Animals work hard to obtain food. By increasing the workload the quantity consumed remains constant (Ladewig and Matthews, 1996), showing that food is an inelastic demand and the most important need for a domestic animal.

Changes in domestication

Domestication changed many characteristics like morphology and behaviour (Belyaev and Trut, 1975). It changed the skull: snout jaws and the number of teeth were reduced (cattle, pig). It resulted in a smaller brain volume (Kruska, 2005). Fat storage also changed, as shown in the modern sow, which became in the recent past 50% leaner and 50% heavier. In addition, behaviour and behavioural strategies changed, for instance 'contrafreeloading' (Inglis and Ferguson, 1986). Contrafreeloading refers to animals that choose to work for food in the presence of free available food (Duncan and Hughes, 1972). Red jungle fowl showed contrafreeloading, but white laying hens just eat from the free food in a choice situation, and thus did the opposite (Schütz et al., 2001). Probably animals work for food to obtain information about alternative food sources (exploration). Domestication induced changes in foraging, especially in the appetitive behaviours, in which the (natural) selection is relaxed (Inglis, 1983).

Potential wildness

Although animals changed in morphology and behaviour, reports show that many domestic animals are still able to become feral and survive (Stricklin, 2001), such as laying hens and roosters (Wood-Gush and Duncan, 1976; Duncan et al., 1978; Savory et al., 1978), pigs (Stolba and Woodgush, 1984), cattle (Daycard, 1990) and horses (Linklater, 2000). 'Feral horse social and spatial organization and behaviour are similar to that of other closely related equids and therefore have been largely unaltered by domestication and artificial selection.' The docility of animals as the most pronounced characteristics of domestication is lost also very quickly in feral animals. This implies that the essential characteristics of natural feeding behaviour are still present and operational in the natural environment. The process of domestication appears to be mainly a quantitative process, in which behaviour in its limited form reflects the limited environment of the domestic animal. Many domestic animals possess a high 'potential wildness' (Koene and Gremmen, 2000). Animals that are changed extremely by genetic selection (broilers and sow) will probably not be able to successfully survive under natural, wild, feral conditions. They possess a low 'potential wildness'. The degree of domestication and the degree of 'potential wildness' differ per species and breed, but are somehow related (Fig. 6.2). Docility is an important aspect of the behaviour of the domesticated animal. The degree of docility or fear of humans may be related to the degree of potential wildness. The domestic behaviour (Lankin and Bouissou, 2001) as defined by the conflict between the approach to food and the avoidance of a human may give an indicator of potential wildness. Strong species and breed differences are found in domestic behaviour and depend on the amount of food deprivation (Lankin and Bouissou, 2001). The human factor is very important in this respect. Human–animal interactions influence the behaviour and the physiology of farm animals (Hemsworth et al., 1993). Positive contacts may be pleasant for animals and even increase production. Negative contacts may be

stressful and lead to illness and lower productivity. Another method to discover the potential wildness in animals is to feralize animals experimentally (feralization model) and measure what characteristics of feeding behaviour are still present in the domestic animal (McBride, 1984) to learn the difference between ECA, EDA and EEA in feeding behaviour.

Examples of domestic feeding

Domestic mammals, especially farm animals such as cattle, horse, pig, sheep, goat, camel, yak and llama, are kept under a wide variety of conditions. In those conditions the methods of feeding and provision of food differs as well as the quantity, quality, spatial distribution, diet composition and schedule of presentation. In general, there are advantages and disadvantages for animals being domesticated. Domestic animals may benefit from being fed, resulting in better body condition and health than under natural conditions. On the other hand, in most cases the appetitive phase of the feeding is reduced – almost no foraging behaviour is necessary or performed in captivity. The food is often presented under crowded conditions, which increases competition between animals. Often also, food is provided in low quantities, which leads to chronic hunger. In addition, there is little variation in the diet. Domestication of birds (chicken, turkey, duck, goose, ostrich) has occurred in several human cultures, typically involving species kept for meat or eggs, but at times birds were domesticated for other reasons first (Gallus gallus). Examples of a number of species concerning domestication and feeding are given in the next section.

Cattle domestic feeding

Cattle were domesticated around 7000 BC. Their importance lies in providing meat, milk and working power. The different types of cattle are kept under a large variety of conditions ranging from very extensive (often found in Australia) to very intensive (The Netherlands, USA). Dairy cattle in a free-stall barn get a ration of roughage and concentrates. The amount of roughage is normally not limited and

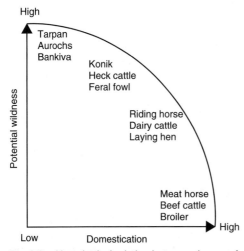

Fig. 6.2. Hypothetical relation between degree of domestication and degree of wildness (or potential wildness). The form of the curve is because many domestic animals are able to feralize to a certain degree. (From Koene and Gremmen, 2002a.)

is provided in several meals. Concentrates are mostly delivered in separate feeders. Veal calves are provided with milk in troughs or buckets (Bokkers and Koene, 2001). Often some solid food is given consisting roughage (straw, maize) or concentrated feed in quantities up to 500 g dry matter per animal per day. Most systems have small groups of about five veal calves. In a system with large groups of calves (40–80 animals) the milk is provided by an automatic milk feeder (Bokkers and Koene, 2001). Individual calves go freely to the milk feeder and may obtain their pre-programmed ration through a drinking teat. The number of calves per automatic feeding station is approximately 40.

Horse domestic feeding

The horse was the last of the most common livestock animals to be domesticated in 4000–3000 BC. Initially it was used only as a source of meat, but soon became established as a perfect means of transport until the recent past. Horses were bred doubtless not only for meat but also for riding. They are sometimes stabled and sometimes allowed to graze in pasture. They are mostly supplied with one or two meals a day often with concentrated contents and limited forage and roughage (Davidson and Harris, 2002). Concentrates or pelleted feed is eaten very quickly (1 kg/10 min) compared to raw feed in 40–50 min (Ellis, 2002).

Chicken domestic feeding

The ancestor of the modern laying hen and broiler chicken was the red jungle fowl (*Gallus gallus*). This bird was domesticated 6000–8000 years ago (Yamada, 1988; West and Zhou, 1989). Feeding of layers in commercial battery cages is mainly done in long troughs using a chain feeder. Commercial barn and free-range systems also use automatic feeders. In most systems, every hen has between 10 and 12 cm of space available at the feeder, and receives meal or pelleted feed. Feeding in broilers is similar to the system used in laying hen houses. For separating the feeding of male and female broiler breeders different types of feeding pans are set at different

heights, allowing only one sex to eat in the provided feeders.

Feeding and Mismatches in EAs

In order to perform feeding behaviour adapted to the housing and feeding condition, the traits of the animal have to match the requirements and opportunities of the environment. In this sense, it is necessary to distinguish natural and unnatural, and functional and non-functional behaviour for the animal in its environment (Table 6.1). Characteristics of the behaviour range from normal, unwanted stereotypies to self-destructive behaviour. Mismatches can be understood with the help of the EEA concept.

Environment of current adaptedness

The idea of the EEA is valuable to emphasize the fact that contemporary organisms may exist in an environment very different from that in which many of their traits were first selected. Clearly, any differences between the EEA and the present environment (ECA) could be a source of stress for which contemporary animals would try to compensate (Table 6.1). EEA adapted traits are thus frequently 'mismatched' to the current environment (ECA), and thus organisms find themselves doing the best they can to deal with contemporary stimuli using the traits they possess. In other words, because an organism's social and physical environment may evolve at a rate faster than its ability to adapt, it will find itself increasingly 'mismatched' to its environment (from EEA domesticated to EDA and intensively selected or kept in ECA). Mismatches between environments are mainly due to mismatches in feeding behaviour, based on genotype–environment interactions and limits in resource allocation (Van der Waaij, 2004).

Depending on the mismatch between ECA and EDA all sorts of behavioural and feeding problems may develop. Mismatches in appetitive and consumatory behaviour could be discerned. Appetitive mismatches are for instance: no opportunity for walking in the pasture for grazers, rooting in pigs or ground-pecking and scratching in chickens.

Table 6.1. Behaviour label, environment, animal welfare in relation to the dichotomies of natural vs. non-natural and functional vs. non-functional behaviour.

	Functional in its environment	Non-functional
Behaviour		
Natural	Natural behaviour	Unwanted behaviour (flee)
Non-natural	Adaptive behaviour	Ethopathy (self-mutilation)
Environment		
Natural	EEA and good EDA, ECA	Bad ECA
Non-natural	EDA and good ECA	Bad ECA
Welfare		
Natural	Good	Bad
Non-natural	Good	Bad

Consumatory mismatches are for example: the lack of control over the quality, quantity and availability of food, i.e. silage for grazers or pellets for chickens. Direct effects of a mismatch are especially apparent in the appetitive phase, as it often lacks in captive conditions. In a choice test, chickens prefer to peck in combination with the consumatory phase of eating (Sterrit and Smith, 1965); the chickens prefer consumatory behaviour, preceded by appetitive behaviour. The absence of the appetitive phase provides spare time or waiting time.

Spare time

When animals do not need to search for feed, a large percentage of their time budget is spare time (Koene, 1999). For some animals, the spare time may be less of a problem; others, such as carnivores and feed specialists, may develop abnormal behaviours, i.e. stereotypies or excessive laziness (Herbers, 1981). In addition, specific qualities of the food have an impact on the time budget, i.e. selective feeders are adapted to their feed and have more requirements concerning their feeding. For instance, giraffes show different amounts of tongue-playing depending on the quality of their food; lucerne increases the amount of tongue-playing probably because less rumination is needed (Koene and Visser, 1997; Koene, 1999, 2001). In addition, when animals are fed easy concentrates, the appetitive phase of feeding (foraging) is not much shorter. In order to understand what domestic animals do with their spare time, attention

must be paid to time spent feeding, time spent in passive behaviour and time spent in replacement activities (Koene, 1996b). In the case where animals have more spare time, more 'abnormal' activities are performed (Koene, 1999). These activities could be unnatural behaviour or behaviour occurring with an unnatural frequency. Whether behaviour in nature should be used as a reference is debatable (Veasey et al., 1996; Dawkins, 1998).

Behavioural problems

Many mismatches between the EAs concerning feeding are often expressed in oral behavioural problems. In horses, cattle and chickens these behaviours are long known (Fraser, 1988) and still are the most prevalent welfare problems. Stereotypies in pigs are related to feeding problems (Rushen et al., 1993). Low energy intake is one of the main causes (Lawrence et al., 1993), and use of appropriate diets can almost eliminate stereotypic behaviour (Robert et al., 1997). The occurrence of stereotypies is an indication of continuing high levels of feeding motivation after a meal. Tongue-playing in cattle seems to improve individual welfare (less ulceration) in milk-fed calves (Wiepkema et al., 1987; Seo et al., 1998). The link found between oral stereotypic tongue-playing in calves and abomasal damage seems to be reversed in horses, i.e. oral stereotypies seem to be related with high ulceration (McGreevy and Nicol, 1998). Results concerning oral stereotypies are contradictory and the subject needs more research effort.

Examples of feeding problems

Cattle mismatches

Cattle may show excessive grooming, inter-sucking, tongue-playing, tongue-rolling and pica behaviour (Fraser, 1988). Tongue-playing is shown by adult cattle in captivity (Sato *et al.*, 1994). Especially veal calves show many abnormal oral behaviours (Bokkers and Koene, 2001), like excessive licking, and biting or licking objects directed to it or to other calves. The most obvious one is tongue-playing, which occurs mostly after the calves have reached 6 weeks of age, and can take up to 6% of the time budget. Causes of tongue-playing in veal calves include the way feed is administered (large amounts of milk, no roughage), not allowing for sufficient activity during food intake (drinking out of a bucket instead of drinking via a teat) and social isolation.

Horse mismatches

Feeding problems are underlying causes of stereotypical behaviour in horses (Henderson and Waran, 2001). Horses show behaviours like weaving, crib-biting, polidipsia, aerophagia, coprophagia and wood-chewing. Weaving may be partly caused by limits for forage (Cooper *et al.*, 2005) and the supply of feeding concentrates (McBride and Cuddeford, 2001), and is probably strongly related to food-searching behaviour, i.e. problems in the appetitive phase of feeding. Oral stereotypies like crib-biting and wind-sucking occur mainly after a meal and are probably related to problems in the consumatory phase of feeding (McGreevy and Nicol, 1998): feeding motivation remains at a high level.

Chicken mismatches

In modern egg-producing systems the most obvious problem behaviours are feather-pecking (Rodenburg and Koene, 2004) and cannibalism (Newberry, 2004). These behaviours cause loss of feather cover and mortality. Laying hens may develop feather-pecking in battery cages as well as in alternative systems. In alternative systems, however, feather-pecking causes bigger problems, because it may spread throughout the entire flock. Feather-pecking seems to be a multifactorial problem with many causes and cures (Blokhuis and Wiepkema, 1998). One of the main theories on the causation of feather-pecking is the hypothesis that it is a form of redirected ground-pecking (Blokhuis and Arkes, 1984). They showed that birds housed on slatted floors showed more feather-pecking and less ground-pecking than birds housed on litter. Moreover, when birds housed on litter were transferred to slatted floors, feather-pecking increased. Feather-pecking was inversely related to foraging activity (Huber-Eicher and Wechsler, 1997). Animals that do not have adequate floor substrate may redirect their foraging behaviour to the plumage of other birds.

During their ontogeny broilers show increasing inactivity and leg problems, mainly caused by their increasing weight, which is nine times as high as that of red jungle fowl at the same age (Bokkers and Koene, 2003). It has been suggested that broiler chickens have a disturbed satiety and hunger mechanism. The satiety mechanism for eating can be expressed as the positive correlation between meal length and the length of the preceding (preprandial) interval. The hunger mechanism for eating shows a positive correlation between meal length and the length of the succeeding (postprandial) interval. Significant preprandial correlations but no postprandial correlations were found in broilers (Bokkers and Koene, 2003). In the laying chickens, both significant preprandial and postprandial correlations were found. This indicates that for regulating eating behaviour, the satiety mechanism dominates the hunger mechanism in broilers, and satiety and hunger mechanisms are equally involved in feeding decisions of laying chickens. Broilers eat to their maximal physical capacity. Probably this change in feeding regulation causes rapid growth and the indirect effect on inactivity and leg problems. Inherent and implicit to problems of broilers are the problems, especially hunger, in broiler breeders (de Jong and Jones, Chapter 8, this volume).

Feeding behaviour in chickens takes normally a large part of the daytime budget. Due to intensive housing conditions, large mismatches between ECA and EDA or EEA are apparent (Table 6.2).

Table 6.2. Estimated 'Daylight' time budgets of *Gallus gallus* under some different housing conditions and EAs.

Gallus	Environment	% Foraging	% Eating	% Abnormal	% Resting	Reference
Bankiva	Nature	<50	50	0	<50	Collias *et al.* (1966)
Bankiva	Zoo/feral	<60	60	0	10	Dawkins (1989)
Bantam	Feral	<48	48	0	10	Savory *et al.* (1978)
Layers	Pens	>0	>18	0	3	Black and Hughes (1974)
Layers	Cage	0	22	0	8	Black and Hughes (1974)
Broiler breeders	*Ad libitum* fed	>8	8	0	50	Savory *et al.* (1992)
Broiler breeders	Feed-restricted	0	0	32	6	Savory *et al.* (1992)
Broilers	Straw	2	10	0	70	Bokkers and Koene (2003)

Note: Percentage feeding consists of percentage foraging plus eating. Other behaviour is not shown, so totals are not 100%.

Welfare and Feeding

Definitions of animal welfare

There are many opinions about what animal welfare is and should include related to feeding. In some welfare regulations it seems to be clear how welfare and feeding go along as defined, for instance, in the following citation:

> Animals shall be fed a wholesome diet which is appropriate to their age and species and which is fed to them in sufficient quantity to maintain them in good health, to satisfy their nutritional needs and to promote a positive state of well-being.
> (Schedule 1, paragraphs 22–27 of the Welfare of Farmed Animals (England) Regulations 2000 (S.I. 2000 No. 1870)).

However, scientifically this relation is not at all clear, as may be shown by the next citations from Lawrence *et al.* (2004):

> Protecting farm animals from hunger and thirst is central to good welfare.

and

> However, our understanding of the impact of nutritional environments on psychological welfare and specifically hunger is much less well advanced.

This may be partly due to contradictions between welfare and health related to feeding. *Ad lib* feeding may lead to obese animals, which feel well and experience good welfare, while hunger may lead to lean healthy animals, which have strong unfulfilled behavioural needs and show signs of bad welfare. This is probably the case in broilers (Bokkers and Koene, 2003). To complicate the relation between feeding and welfare further there are many definitions of animal welfare, partly dependent on research goal or on societal actions (Broom and Johnson, 1993; Stafleu *et al.* 1996; Appleby and Hughes, 1997; Appleby, 1999). Consequently, definitions and measures of animal welfare are hard to give. Animal welfare definitions may be grouped in threes, i.e. definitions relating to: (i) animal natures, natural conditions and natural behaviour; (ii) animal bodies biological functioning, health, disease, growth, production; and (iii) animal minds, feelings pleasure, suffering, subjective experiences (Appleby, 1999). The concepts of animal nature, body and mind overlap.

The suggested relation between animal nature and welfare is debatable (Dawkins, 1998), because '[n]atural selection maximizes fitness but not necessarily the well-being of organisms' (Hofer and East, 1998). Only a few attempts have been made to define or assess animal welfare in natural circumstances (Kirkwood *et al.*, 1994). On the other hand, one of the methods of comparing species and linking natural behaviour with behaviour in captive conditions comes of age (Clubb and Mason, 2004). The research on problem behaviour still concentrates on stereotypies (Mason and Latham, 2004). In carnivores, the home range size and the daily travelled distance predicted the amount of stereotypies under captive conditions of the 35 species tested (Clubb and Mason, 2003). Stereotypic pacing and oral stereotypies may be the result of mismatches between natural mobility (Clubb and Mason, 2003) and feeding under natural conditions (Lawrence *et al.*, 1993; Koene and Visser, 1997; Mason and Mendl, 1997). The 'animal nature' approach contrasts largely the approaches of biological functioning (Broom, 1996) and the feelings approach (Duncan, 1993) or the two-stage definition (Dawkins, 2004) based on two questions: Is the animal healthy? Does it suffer?

Pluralism

The approach taken towards defining animal welfare is to use the best of three worlds and have a pluralistic standpoint, to explore and discuss the borders and overlaps given by the definitions. A definition of welfare combining the definitions of animal natures, bodies and feelings (Appleby, 1999) is Webster's definition:

> The welfare of an animal must be defined not only by how it feels within a spectrum that ranges from suffering to pleasure but also by its ability to sustain physical and mental fitness and so preserve not only its future quality of life but also the survival of its genes.

In short: 'The welfare of an animal is determined by its capacity to (i) avoid suffering and (ii) sustain fitness' (Webster, 1995). This definition promises good opportunities to compare wild as well as captive animal traits between the concepts of EAs. However, its relationship with an evolutionary approach and a way to measure it in captivity and in wild animals is not yet clear. One quite specific implication of this definition is that killing animals – e.g. in nature reserves to control the population size – reduces the individual fitness of the animal and thus reduces their welfare. Fitting in an evolutionary pluralistic approach to animal welfare – as related to the EEA – are definitions referring to positive and negative experiences (Simonsen, 1996), and their integration, especially perceiving, learning and fixing those experiences in a Darwinistic way in the brain for later use (Edelman, 1992).

Suffering and pleasure

'The hypothesis is that animals suffer when they are not able to feed as they are designed to do' (Thorpe, 1969; McFarland, 1989). McFarland's proposal is that an animal may suffer if it is unable to perform an activity that it is motivated to perform. 'In cases that are likely to be commonplace in nature, we might expect the animal not to suffer because it has been designed to cope with this type of situation' (McFarland, 1989). Suffering in animals can also be seen from a more precisely defined evolutionary standpoint (Barnard and Hurst, 1996). Animals are designed and shaped during their evolution and possess certain design limits and their species-specific decision rules. When animals are pushed over these limits, for instance by environmental or feeding demands, suffering may occur. In order to use this approach to estimate welfare, the situations the animal is designed for and not designed for must be distinguished. In the management of large herbivores in dedomestication in a nature reserve in The Netherlands a comparable distinction is made between natural suffering (the animal is designed for) and unnecessary suffering, i.e. human-caused (by fencing in the animals) (Koene and Gremmen, 2002).

Suffering is evolved in an organism to help it avoid conditions that are unfavourable for its reproduction. When the current state of an animal is different from the environment it evolved in (EEA), the mismatch between both can have a large impact on its functioning and suffering. Based on the concept of fitness, it is a characteristic of alleles/genes, making the individual often expendable. 'Organisms are designed for self-expenditure and the relative importance of self-preservation and survival, and the concomitant investment of time and resources in different activities varies with life history strategy' (Barnard and Hurst, 1996). Consequently, proximate subjective states related with ultimate functional states are naturally selected. Domestication – changing the environment from wild to captivity – makes some specific decision rules that were functional in the EEA, not functional in the EDA and ECA and may cause suffering. Suffering is thus dependent on the discrepancy between its current environment and its EDA and EEA. Furthermore, two forms of suffering must be distinguished: natural suffering (negative emotional states related to adaptive self-expenditure) and cultural or unnecessary suffering (negative emotional states related to non-adaptive self-expenditure).

This parallels the distinction between natural and unnecessary suffering, mentioned above; only the latter causes real suffering (Barnard and Hurst, 1996). Although their approach is based on sound biology, it remains difficult to find out the nature of decision rules and priorities of an animal species and its exact EEA, especially in the case where the ancestor of the domesticated animal is extinct. To corroborate this distinction further, the causes of unnecessary suffering in wild and domestic animals should be investigated (Kirkwood et al., 1994).

Mismatches between EAs are the probable cause of problems with feeding behaviour, feeding period, and synchronization and diet characteristics. On a smaller scale, minimal total discomfort (MTD, Forbes, 1999) as a measure of the 'mismatch' between inputs of nutrients and the requirements may help in finding a good measure of animal welfare related to feeding. Along this line, a large MTD is related to suffering, and a positive interpretation in terms of pleasure should be investigated (Berridge and Robinson, 1998; Roitman et al., 2004).

The study of animal pleasure has attracted less attention, because behavioural problems and suffering needed urgent solutions. Cabanac (1979, 1992) was the first to draw attention to the importance of pleasure as motivation for behaviour. More recently, feelings of pleasure and suffering and biological functioning are coupled in the concept of motivational affectional states (Fraser and Duncan, 1998). Like other characteristics of the animals, motivational states are subject to natural selection. States of suffering evolved to motivate immediate actions in emergency situations, like feeding behaviour, while states of pleasure evolved to motivate behaviour in opportunity situations, which benefit the animal in the long-term, giving an evolutionary basis of animal welfare, like sexual behaviour.

This approach is extremely important to reduce mismatches between EAs, especially by environmental changes (Duncan and Olsson, 2001).

The Future of Feeding: Matching EAs

Changing the environment

The anthropocentric view approaches changes in the environment from a human viewpoint, i.e. research and technology, properly applied, will meet all animals' needs. The zoocentric (bio- or eco-) view holds that animals' needs are fulfilled only from the animals' perspective and their social and environmental surrounding. Both approaches can be found in zoo environments, i.e. environmental engineering (Markowitsch, 1982) and the naturalistic approach (Young, 2003). In the past and present, the anthropocentric approach dominated the housing, feeding and care of domestic animals. The data show that animals demonstrate less signs of abnormal or problem behaviour in a more natural environment with a more natural feeding programme (Grandia et al., 2001). The zoocentric approach provides animals with a naturalistic environment, more natural food and feeding schedules. Largely, both approaches parallel a bottom-up and top-down method of designing the environment for animals. In the case of farm animals, the bottom-up approach is usually taken (Koene and Duncan, 2001), i.e. the animals are kept under circumstances that are as simple as possible. If animals in such environments show behaviour indicative of suffering, modification of the environment is necessary to decrease animal suffering. The second, top-down, approach starts from an environment that fulfils all natural needs of the wild or captive animal, and reduces natural complexity until an environment remains in which at least the basic needs of the animal are fulfilled. This second approach is not often taken, but is recognizable in some designs of naturalistic zoo enclosures (Koene, 2004). The bottom-up approach of designing environment provides the animals with a history of suffering. For both types of animal environments, the first objective should be to provide the animal with all its basic needs and so eliminate or reduce states of suffering such as pain, discomfort, frustration and fear, e.g. providing space to animals with high motility, such as ungulates. It is suggested that this should be described as providing environmental requirements (Duncan and Olsson, 2001). When basic needs are fulfilled, it may be possible to enhance quality of life further by environmental manipulations that lead to states of pleasure (Duncan and Olsson, 2001). An example would be providing new exploration ground, fresh pastures, unexpected new partners or even live prey. They suggest that the term 'environmental enrichment' be reserved for these latter manipulations.

As already indicated, the term environmental enrichment is often misused in situations of changing environments. The biological end point is crucial for evaluating environmental enrichment (Newberry, 1995). The term environmental enrichment is used to describe changes to the environment that have a beneficial effect on the animal in that environment, i.e. they result in improved welfare and probably pleasure (Duncan and Olsson, 2001).

Providing basic feeding requirements is thus focused on alleviating poverty of the feeding environment and reducing suffering. One way to do that is to restore appetitive behaviour absent in the intensive housing system by providing food that better fits the amount of searching, foraging, handling or diet selection in the EEA. This means that a thorough knowledge is necessary of the animal's behaviour, motivational systems and evolutionary history and – in general – adaptations of the animal both in the EEA and EDA. Another way to adapt the feeding and housing conditions to the needs of animals is to completely change the feeding or housing, for instance by expanding and making conditions more like EDA or even dedomestication, i.e. reversing the process of domestication (Koene and Gremmen, 2002) to mimic the EEA to a certain degree. The Stolba stable for pigs is an example of designing a new living environment based on knowledge of behavioural problems, evolutionary history and potential wildness, provided by research in the Pig Park in Edinburgh (Stolba and Woodgush, 1984). In horses and cattle, environmental changes focus on giving time on pasture (Hrohn, 1994)

and increasing the proportion of forage and roughage in the feed to increase forage time. In chickens many environmental changes (Jones, 2004) are proposed to reduce feather-pecking (Rodenburg and Koene, 2004), cannibalism (Newberry, 2004) in layers, and leg problems and inactivity in broilers (Bizeray et al., 2002). Projects in The Netherlands concerning laying hens are partly based on the mismatch approach (Blokhuis and Metz, 1995) or expert opinion approach (Bracke, 2001). New designs based on 'natural behaviour' lack a biological framework that makes explicit the requirements in terms of animal welfare. Comparing the ECA and the performance of the animal in that environment, based on its former adaptations, and its potential wildness produces points in space that show matches and mismatches in terms of neutral states and suffering and pleasure (Fig. 6.3). This might be the start of a possible framework that can be applied to all domestic animals. However, the problem is to find the right parameters, variables and indicators and

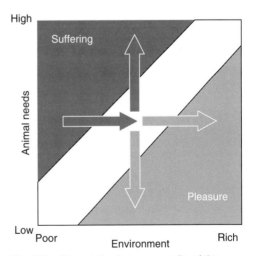

Fig. 6.3. The relation between quality of the environment and quality of animal life (animal welfare). The arrows show changes in environmental conditions. The neutral (white) area on the diagonal depicts a match between animal needs and its environment. On the x-axis the ECA (often poor), EDA (intermediate) and EEA (mostly rich) have a different position per species. Note that the environment is depicted here as the total environment; it may also be used more specifically, i.e. the feeding environment.

the right solutions using firm biological, psychological and statistical methods.

Measuring adaptedness

In addition to data already available on feeding behaviour, feeding and behavioural problems, there is a need for methods of comparing performance and behaviour of animals in different housing systems. Some will be mentioned shortly here.

Behavioural indicators

Normally, an animal tries to reduce suffering, mainly by species-specific avoidance behaviour. If it is not able or successful to reduce the suffering under the actual circumstances, suffering may become chronic. On the other hand, the animal tries to perform behaviours that increase pleasure (Berridge and Robinson, 1998). Stereotypies are often used as behavioural indicators of welfare (Mason and Latham, 2004). They tend to call stereotypy a 'do-it-yourself enrichment', but the exceptions make stereotypies unreliable as welfare indicators: 'It is clearly vital not to overlook the potential problems of animals with low or absent stereotypies: in stereotypy-eliciting circumstances, these individuals quite possibly have the worst welfare.' Dawkins states that the most practically useful and evolutionary measures of welfare are primary measures as direct threats to fitness and secondary measures that are perceived as threats to fitness (Dawkins, 1998). Positive indicators of welfare have not drawn that much attention, because they are less obvious and probably less present in intensive housing conditions. Play behaviour is the most obvious indicator (Fagen, 1981; Koene, 1998; Špinka et al., 2001). Cows on their first steps in the pasture in spring show playful behaviour, like lambs playing with each other.

Vocal indicators

In intensive and extensive animal husbandry, animals are often restricted in the execution of behaviour (i.e. consequently 'frustrated'), for instance, as a reaction to feed in troughs that restrict eating together or food that is expected but not delivered. Zimmerman and Koene showed that thwarting the consumatory behaviour in laying hens increased the number of so-called Gakel-calls (Koene and Wiepkema, 1991; Zimmerman et al., 2000a,b). Gakel-calls increased in number after thwarting of feeding, drinking, dust bathing or prelaying behaviour. Gakel-calls may be a general indicator of frustration in laying hens. In addition, in other domestic species vocalization may indicate frustration (Manteuffel, 2004). A more general effect related to feeding is that the number and the temporal structure of crowing in rooster changes after food deprivation (Koene, 1996a). Probably more than related to emotions, the crowing is related to the condition and health of the animal. The ku-call in chickens and the soft grunt in piglets may be used as positive indicators of welfare (Huber and Fölsch, 1978; Jensen and Algers, 1983/84; Crowell Comuzzie, 1993). Social contact calls that may have a positive loading are also easy to measure as indicators of positive welfare. However, in this area we still lack information.

The four whys

There is an urgent need to ask the 'why' questions (Tinbergen, 1963) about the problem behaviours in applied ethology and use the different answers explicitly (Duncan, 1995; Mason and Mendl, 1997). For example, consider the question: 'Why do animals (not) show their species-specific feeding?' and analyse the four possible answers to find cues and commands relating to the persistence for the oral behaviour, preferably according to the formula AB = C + D + E + F, i.e. Animal Behaviour is the integration of Causal, Developmental, Evolutionary and Functional forces. What causes feeding behaviour? Focus on motivation of feeding behaviour. How does the feeding behaviour develop during the individual's lifetime? Focus on learning and feeding. How has the feeding behaviour evolved in the species? Is it shown by closely related species? Focus on the EEA of feeding. Why is feeding behaviour performed in a particular way? Focus on fitness of feeding. All questions might be equally relevant in relation to estimate the nature and degree of mismatch between animal needs, shown in feeding behaviour and after feed supply, as shown in domestic feeding.

Consumer demand method

Measuring needs, motivations, frustrations and preferences for different feedstuffs and feeding contexts are essential for understanding differences in feeding behaviour between species, sexes, types and individuals. One way is to measure the price animals want to pay (consumer demand theory) to acquire a certain commodity (Ladewig and Matthews, 1996). This method is easily applicable in chickens in Skinner boxes (Bokkers et al., 2004; Rodenburg et al., 2005) to reveal differences in the need for food in different chicken groups. Currently, the consumer demand method can be used ex situ (Jensen et al., 2004) and in situ (Cooper, 2004).

Preference test

Since the early 1970s, scientists have used preference tests. These tests require animals to choose between two or more different options or environments as a means of answering questions about animal welfare. However, preference tests have been criticized (Duncan, 1978, 1992). Therefore, if preference research is used to answer questions about animal welfare, the issues of validity, the nature of the choices made by the animal and the Umwelt of the animal need to be addressed.

Animal welfare index

The multiplicity of perceptions and criteria of animal welfare has led to the design of 'animal welfare indexes' (Bartussek, 1999). They are mainly based on scientific evidence about features of housing systems and management features related to animal behaviour and welfare. A good example of the use of such an index is illustrated by the work of Bock (1990), who showed the mismatch between housing attrib-

utes and behaviour attributes of dairy cows. The behaviour and management actions of the farmer appeared to make a large contribution to the discrepancy found. Such indexes may contribute to basic ethological and veterinary methods. Assessment of animal welfare at the farm level has shown much progress (Webster and Main, 2003).

Single-case analysis and meta-analysis

Evaluation of the effects of environmental changes, i.e. providing environmental requirements and/or enrichment should often be done in situ, i.e. on the farm and in the zoo. If animal numbers are large enough, some statistical methods for evaluation are available and recommended (Young, 2003). However, in many circumstances the effects have to be measured at the individual level (zoo animals, sport animals, pets), using the method of repeatedly asking different 'questions'. For instance, by providing feeds differing in quantity, shape, nutritional contents or taste in random order, the answers of the animal in terms of feeding behaviour parameters or problem behaviours can be analysed (Fisher, 1966). Individual animals must be asked the right questions by using randomization and single-case analysis year (Barlow and Hersen, 1984; Onghena, 1992), and answers of different individuals can be combined in overall conclusions using meta-analysis by combining p-values (Sokal and Rohlf, 1969). An example of single-case and meta-analysis related to feeding is the tongue-playing and ruminating in giraffes in relation to different feeds (Koene and Visser, 1997; Koene, 1999).

Matching EAs: a synthesis and Darwinistic framework

In order to solve mismatches between the environment the animal is actually living in and the behaviour and decision and learning rules it adapted to in an earlier environment (i.e. EDA and/or EEA), a simple model shows actions to be taken to acquire knowledge and minimize the mismatch. Firstly, on individual level the match/mismatch between the animal's requirements (needs) and its environment depicts feelings, areas of suffering and pleasure, next to a neutral area (Fig. 6.3). Changing the environment may have different effects based on the level of needs of the animal. It may change (shaded arrow) suffering (mismatch) in the neutral area (match), or it may change (unshaded arrow) an animal from the neutral state to a pleasure state. The shaded arrow change provides basic requirements; the unshaded arrow enriches the environment. Changing the animal by genetic selection or changing the ontogenetic development may change the feelings of the animal (arrows in the vertical plane), but that is not the issue here (Faure, 1980). The dynamics of actions of animals, and of humans, to change the animal–environment interaction is depicted in a cybernetic model as an analogy of the feedback models of behaviour (Wiepkema, 1985). Such a Darwinistic framework helps analyse and take actions when animals are not optimally adapted to their actual environment. Within limits, humans should take actions in the same way an animal may take action. The model is a synthesis of the animal in its ECA (Istwert), from which welfare parameters are perceived that label the environment and compare it with the presumed or measured parameters of its EDA and/or EEA. The comparison motivates actions to be undertaken to change the ECA (Fig. 6.4). In more detail, the following steps are taken:

1. Animal lives in its ECA.
2. Appreciation of its ECA: expressions in feelings – neutral, pleasure, natural suffering and unnecessary suffering.
3. Perception of its ECA: the monitoring of the animal in its ECA with its feelings and behaviours. Ask the 'why' questions. Monitoring can be done with many methods (observations, indexes, preferences, 'asking' questions by changes in the environment) and analyses (multivariate, single-case and meta-analysis). Whether animals themselves are able to monitor in a comparable way is debatable.
4. Labelling of its ECA: animal and human label the ECA ranging from bad to sufficient to good. Assessment of environmental changes in terms of belonging to basic requirements or to environmental enrichment is done here.

5. Comparison of its ECA with EDA and EEA: comparison of steps 2, 3 and 4 for the three environments. Note: the same steps taken for the ECA have to be taken for the two other environments. If EEA and/or EDA and the animals living in them are not present any more and cannot be monitored (ancestors extinct), appreciation, perception and labelling of the EDA and EEC have to be estimated. Matches and mismatches are discovered. Making Venn diagrams of the ECA, EDA and EEA that show agreements, differences and challenges is helpful (Fraser *et al.*, 1997).
6. Motivation to change ECA: based on matches and mismatches of ECA and EDA and/or EEA positive or negative action is motivated (animal action as well as human action for captive animals).
7. Action to change ECA: actions of animal or human are taken to reduce the difference between ECA and EDA and/or EDA.
8. Method to change ECA: many methods to change the ECA are possible. Here the top-down (removing features from EDA or EEA) and bottom-up (adding features from the EDA or EEA) approaches are given to change the environment.

The main goal of this model is to evaluate the match between the animal and its current environment, based on the animal's former matches with the environments it evolved in and in which it was domesticated. This general approach is especially relevant for optimization of feeding the animals, and supplying food in a way the animal is able to cope with and is adapted for.

Conclusion

Feeding of domestic animals is discussed in relation to evolutionary approaches towards animal welfare. Three environments are distinguished that are important for the feeding behaviour of the animal: the environment of current adaptedness (ECA), of domestic adaptedness (EDA) and of evolutionary adaptedness (EEA). Feeding behaviour in the different environments (EAs) is shortly described. Mismatches in the appetitive (food-searching) and consummatory (diet selection and consumption)

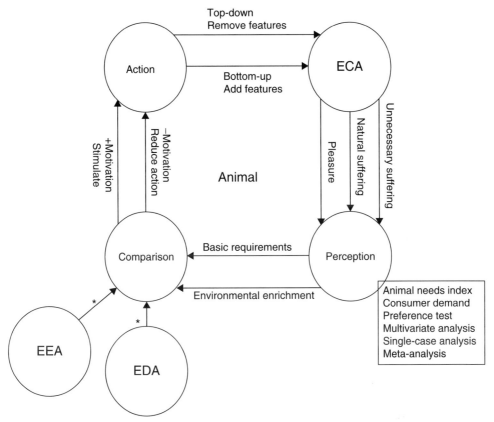

Fig. 6.4. Model of human and/or animal actions to match the current environment (ECA) with the original domestic environment (EDA) and the environment the animal evolved in (EEA). The arrows with asterisks resemble the same pathway and steps as necessary for the ECA (steps 2, 3 and 4; see text).

phase are described. Mismatches between EAs are discussed in terms of animal welfare by using evolutionary definitions of animal welfare that include suffering and pleasure. Methods for measuring mismatch and eventually a Darwinistic framework are given that naturally select the most optimal adaptation of the animal to its environment (ECA), independent of the starting point (bottom-up or top-down).

Many animal welfare problems related to feeding are multifactorial and it is impossible to give the solution. A pluralistic, multifactorial, multivariate approach in finding solutions is therefore crucial. The framework given will help show the way forward.

References

Appleby, M.C. (1999) *What Should We do about Animal Welfare?* Blackwell Scientific Publications, Oxford, UK, 192 pp.

Appleby, M.C. and Hughes, B.O. (1997) *Animal Welfare.* CAB International, Wallingford, UK, 316 pp.

Bailey, D.W. (1995) Daily selection of feeding areas by cattle in homogeneous and heterogeneous environments. *Applied Animal Behaviour Science* 45, 183–200.

Barlow, D.H. and Hersen, M. (1984) *Single Case Experimental Designs. Strategies for Studying Behavior Change.* Allyn & Bacon, Needham Heights, Massachusetts, 419 pp.

Barnard, C.J. (2004) *Animal Behaviour: Mechanism, Development, Function and Evolution.* Prentice-Hall, Harlow, UK, 726 pp.

Barnard, C.J. and Hurst, J.L. (1996) Welfare by design: the natural selection of welfare criteria. *Animal Welfare* 5, 405–433.

Bartussek, H. (1999) A review of the animal needs index (ANI) for the assessment of animals' well-being in the housing systems for Austrian proprietary products and legislation. *Livestock Production Sciences* 61, 179–192.

Belyaev, D.K. and Trut, L.N. (1975) Some genetic and endocrine effect on the selection on domestication of silver foxes. In: Fox, M.W. (ed.) *The Wild Canids.* Van Nostrand Reinhold, New York, pp. 416–426.

Berridge, K.C. and Robinson, T.E. (1998) What is the role of dopamine in reward: hedonic impact, reward learning, or incentive salience? *Brain Research Reviews* 28, 309–369.

Bizeray, D., Estevez, I., Leterrier, C. and Faure, J.M. (2002) Effects of increasing environmental complexity on the physical activity of broiler chickens. *Applied Animal Behaviour Science* 79, 27–41.

Black, A.J. and Hughes, B.O. (1974) Patterns of comfort behaviour and activity in domestic fowls: a comparison between cages and pens. *British Veterinary Journal* 130, 23–33.

Blokhuis, H.J. and Arkes, J.G. (1984) Some observations on the development of featherpecking in poultry. *Applied Animal Behaviour Science* 12, 145–157.

Blokhuis, H.J. and Metz, J.H.M. (1995) *Aviary Housing for Laying Hens.* ID-DLO Lelystad, Lelystad, The Netherlands, 196 pp.

Blokhuis, H.J. and Wiepkema, P.R. (1998) Studies of feather pecking in poultry. *Veterinary Quarterly* 20, 6–9.

Bock, C. (1990) *Zur Beurteilung Tiergerechter Laufställe für Milchvieh.* KTBL, Darmstadt, Germany, 82 pp.

Bokkers, E.A.M. and Koene, P. (2001) Activity, oral behaviour and slaughter data as welfare indicators in veal calves: a comparison of three housing systems. *Applied Animal Behaviour Science* 75, 1–15.

Bokkers, E.A.M. and Koene, P. (2003) Behaviour of fast and slow growing broilers to 12 weeks of age and the physical consequences. *Applied Animal Behaviour Science* 81, 59–72.

Bokkers, E.A.M. and Koene, P. (2003) Eating behaviour, and preprandial and postprandial correlations in male. *British Poultry Science* 44, 538–544.

Bokkers, E.A.M., Koene, P., Rodenburg, T.B., Zimmerman, P.H. and Spruijt, B.M. (2004) Working for food under conditions of varying motivation in broilers. *Animal Behaviour* 68, 105–113.

Bracke, M.B.M. (2001) Modelling of animal welfare: the development of a decision support system to assess the welfare status of pregnant sows. PhD thesis, Wageningen University, Wageningen, The Netherlands,150 pp.

Broom, D.M. (1981) *Biology of Behaviour.* Cambridge University Press, Cambridge, UK, 320 pp.

Broom, D.M. (1996) Animal welfare defined in terms of attempts to cope with the environment. *Acta Agriculturae Scandinavica Section A, Animal Science* 27(suppl.), 22–28.

Broom, D.M. and Johnson, K.G. (1993) *Stress and Animal Welfare.* Chapman & Hall, London, 211 pp.

Cabanac, M. (1979) Sensory pleasure. *The Quarterly Review of Biology* 54, 1–29.

Cabanac, M. (1992) Pleasure: the common currency. *Journal of Theoretical Biology* 155, 173–200.

Camazine, S. (2001) *Self-organization in Biological Systems.* Princeton University Press, Princeton, New Jersey, 538 pp.

Case, T.J. and Gilpin, E.G. (1974) Interference competition and Niche theory. *Proceedings of the National Academy of Sciences USA* 71, 307.

Charnov, E.L. (1976) Optimal foraging: the marginal value theorem. *Theoretical Population Biology* 9, 129–136.

Clifton, P.G. (1979) The synchronization of feeding in domestic chicks by sound alone. *Animal Behaviour* 27, 829–832.

Clubb, R. and Mason, G. (2003) Captivity effects on wide-ranging carnivores. Animals that roam over a large territory in the wild do not take kindly to being confined. *Nature* 425, 473–474.

Clubb, R. and Mason, G. (2004) Pacing polar bears and stoical sheep: testing ecological and evolutionary hypotheses about animal welfare. *Animal Welfare* 13, 533–540.

Clutton-Brock, J. (1999) *A Natural History of Domesticated Mammals.* Cambridge University Press, Cambridge, UK, 238 pp.

Collias, N.E. and Collias, E.C. (1967) A field study of the red jungle fowl in north-central India. *Condor* 69, 360–386.

Cooper, J.J. (2004) Consumer demand under commercial husbandry conditions: practical advice on measuring behavioural priorities in captive animals. *Animal Welfare* 13, 547–556.

Cooper, J.J., McAll, N., Johnson, S. and Davidson, H.P.B. (2005) The short-term effects of increasing meal frequency on stereotypic behaviour of stabled horses. *Applied Animal Behaviour Science* 90, 351.

Crowell Comuzzie, D.K. (1993) Baboon vocalizations as measures of psychological well-being. *Laboratory Primate Newsletter* 32, 5–6.

Davidson, N. and Harris, P. (2002) Nutrition and welfare. In: Waran, N.K. (ed.) *The Welfare of Horses*. Kluwer Academic Publishers, Dordrecht, The Netherlands.

Dawkins, M.S. (1989) Time budgets in red junglefowl as a baseline for the assessment of welfare in domestic fowl. *Applied Animal Behaviour Science* 24, 77–80.

Dawkins, M.S. (1998) Evolution and animal welfare. *The Quarterly Review of Biology* 73, 305–328.

Dawkins, M.S. (2004) Using behaviour to assess animal welfare. *Animal Welfare* 14, 1–14.

Daycard, L. (1990) Social structure of the population of feral cattle on Amsterdam Island (southern Indian Ocean). *Revue D'Ecologie La Terre Ethology La Vie* 45, 35–54.

Duncan, I.J.H. (1978) The interpretation of preference tests in animal behaviour. *Applied Animal Ethology* 4, 197–200.

Duncan, I.J.H. (1992) Measuring preferences and the strength of preferences. *Poultry Science* 71, 658–663.

Duncan, I.J.H. (1993) Welfare is to do with what animals feel. *Journal of Agricultural and Environmental Ethics* 6, 8–14.

Duncan, I.J.H. (1995) D.G.M. Wood-Gush memorial lecture: an applied ethologist looks at the question 'why?' *Applied Animal Behaviour Science* 44, 205–217.

Duncan, I.J.H. and Hughes, B.O. (1972) Free and operant feeding in domestic fowls. *Animal Behaviour* 20, 775–777.

Duncan, I.J.H. and Olsson, I.A.S. (2001) Environmental enrichment: from flawed concept to pseudoscience. In: Garner, J.P., Mench, J.A. and Heekin, S.P. (eds) *Proceedings of the 35th International Congress of the ISAE*, Davis, California, 73 pp.

Duncan, I.J.H., Savory, C.J. and Wood-Gush, D.G.M. (1978) Observations on the reproductive behaviour of domestic fowl in the wild. *Applied Animal Ethology* 4, 29–42.

Edelman, G.M. (1992) *Bright Air, Brilliant Fire: on the Matter of the Mind*. Allen Lane, London, 280 pp.

Ellis, A.D. (2002) *Ingestive and Digestive Processes in Equines*. Writtle College, Essex University, Essex, UK.

Encyclopaedia Britannica (2002) *Britannica 2002 Standard Edition*, CD-ROM (Windows and Macintosh). SelectSoft Publishing, San Francisco, California.

Ewbank, R., Kim-Madslien, F. and Hart, C.B. (1999) *Management and Welfare of Farm Animals*. Universities Federation for Animal Welfare (UFAW), Herts, UK, 308 pp.

Fagen, R. (1981) *Animal Play Behavior*. Oxford University Press, New York, 684 pp.

Faure, J.M. (1980) To adapt the environment to the bird or the bird to the environment? In: Moss, R. (ed.) *The Laying Hen and Its Environment*. ECSC, EEC, EAEC, Brussels-Luxembourg, pp. 19–42.

Fisher, R.A. (1966) *The Design of Experiments*. Oliver & Boyd, Edinburgh, UK (originally published in 1935).

Forbes, J.M. (1999) Minimal total discomfort as a concept for the control of food intake and selection. *Appetite* 33, 371.

Forbes, J.M. (2001) Consequences of feeding for future feeding. *Comparative Biochemistry and Physiology Part A* 128, 463–470.

Fraser, A.F. (1988) Behavioural needs in relation to livestock maintenance. *Applied Animal Behaviour Science* 19, 368–383.

Fraser, A.F. and Broom, D.M. (1990) *Farm Animal Behaviour and Welfare*. Tindall, London, 437 pp.

Fraser, D. and Duncan, I.J.H. (1998) 'Pleasures', 'pains' and animal welfare: towards a natural history of affect. *Animal Welfare* 7, 383–396.

Fraser, D., Weary, D.M., Pajor, E.A. and Milligan, B.N. (1997) A scientific conception of animal welfare that reflects ethical concerns. *Animal Welfare* 6, 187–205.

Fretwell, S.D. and Lucas, H.L. (1970) On territorial behaviour and other factors influencing habitat distribution in birds. *Acta Biotheoretica* 19, 16–36.

Galef, B.G. (1991) Information centres of Norway rats: sites for information exchange and information parasitism. *Animal Behaviour* 41, 295–301.

Grandia, P.A., Van Dijk, J.J. and Koene, P. (2001) Stimulating natural behavior in captive European brown bears by stimulating feeding conditions. *Ursus* 12, 199–202.

Gyger, M. and Marler, P. (1988) Food calling in the domestic fowl, *Gallus gallus*: the role of external referents and deception. *Animal Behaviour* 36, 358–365.

Hardin, G. (1960) *Nature and Man's Fate*. Cape, London, 375 pp.

Hemsworth, P.H., Barnett, J.L. and Coleman, G.J.N. (1993) The human–animal relationship in agriculture and its consequences for the animal. *Animal Welfare* 2, 33–51.

Henderson, J.V. and Waran, N.K. (2001) Reducing equine stereotypies using an Equiball™. *Animal Welfare* 10, 73–80.

Herbers, J.M. (1981) Time resources and laziness in animals. *Oecologioa* (*Berlin*) 49, 252–262.

Hofer, H. and East, M.L. (1998) Biological conservation and stress. *Advances in the Study of Behavior* 27, 405–525.

Hrohn, C.C. (1994) Behaviour of dairy cows kept in extensive (loose housing/pasture) or intensive (tie stall) environments. III Grooming, exploration and abnormal behaviour. *Applied Animal Behaviour Science* 42, 73–86.

Huber-Eicher, B. and Wechsler, B. (1997) Feather pecking in domestic chicks: its relation to dustbathing and foraging. *Animal Behaviour* 54, 757–768.

Hurnik, J.F., Webster, A.B. and Siegel, P.B. (1995) *Dictionary of Farm Animal Behavior*. Iowa State University Press, Ames, Iowa, 200 pp.

Inglis, I.R. (1983) Towards a cognitive theory of exploratory behaviour. In: Archer, J. and Birke, L. (eds) *Exploration in Animals and Humans*. Van Nostrand Reinhold, Cambridge, UK, pp. 72–116.

Inglis, I.R. and Ferguson, N.J.K. (1986) Starlings search for food rather than eat freely available food. *Animal Behaviour* 34, 614–616.

Janzen, D.H. (1976) Why bamboos wait so long to flower. *Annual Review of Ecology and Systematics* 7, 347–391.

Jensen, M.B., Pedersen, L.J. and Ladewig, J. (2004) The use of demand functions to assess behavioural priorities in farm animals. *Animal Welfare* 13, 527–532.

Jensen, P. and Algers, B. (1983/84) An ethogram of piglet vocalizations during suckling. *Applied Animal Ethology* 11, 237–248.

Jones, R.B. (2004) Environmental enrichment: the need for practical strategies to improve poultry welfare. In: Perry, G.C. (ed.) *Welfare of the Laying Hen*. CAB International, Wallingford, UK, pp. 215–226.

Kirkwood, J.K., Sainsbury, A.W. and Bennett, P.M. (1994) The welfare of free-living wild animals: methods of assessment. *Animal Welfare* 3, 257–273.

Kjaer, J.B. and Sorensen, P. (1997) Feather pecking behaviour in White Leghorns, a genetic study. *British Poultry Science* 38, 333–341.

Koene, P. (1996a) Temporal structure of red jungle fowl crow sequences: single-case analysis. *Behavioural Processes* 38, 193–202.

Koene, P. (1996b) The use of time-budgets of zoo mammals for captive propagation and zoo biology. In: Ganslosser, U., Hodges, J.K. and Kaumans, W. (eds) *Captive Research and Propagation*. Filander Verlag, Fürth, Germany, pp. 272–284.

Koene, P. (1998) Adaptation of blind brown bears to a new environment and its residents: stereotypy and play as welfare indicators. *Ursus* 10, 379–386.

Koene, P. (1999) When feeding is just eating: how do farm and zoo animals use their spare time? In: Heide, D.v.d., Huisman, E.A., Kanis, E., Osse, J.W.M. and Verstegen, M.W.A. (eds) *Regulation of Feed Intake*. CAB International, Wallingford, UK, pp. 13–19.

Koene, P. (2001) Giraffe. In: Bell, C.E. (ed.) *Encyclopedia of the World's Zoos*. Fitzroy Dearborn Publishers, Chicago and London, pp. 512–517.

Koene, P. (2004) Providing animal environments based on wildness, requirements and welfare. *Advances in Ethology* 26–29 September, p. 138.

Koene, P. and Duncan, I.J.H. (2001) From environmental requirement to environmental enrichment: from animal suffering to animal pleasure. In: *5th International Conference on Environmental Enrichment*. 4–9 November, Taronga Zoo, Sydney, Australia, p. 36.

Koene, P. and Gremmen, B. (2000) Domestic animals back to nature: de-domestication, feralization, naturalization? In: Ramos, A., Pinheiro Machado Filho, L.C. and Hoetzel, M.J. (eds) *34th International Congress of the ISAE*, 17–20 October, Florianopolis, Brazil. Laboratory of Applied Ethology – UFSC, Florianopolis, Brazil, 39 pp.

Koene, P. and Gremmen, B. (2002) Wildheid gewogen: samenspel van ethologie en ethiek bij dedomesticatie van grote grazers. NWO Project No 210–11–302. Wageningen Universiteit, Wageningen, The Netherlands, 200 pp.

Koene, P. and Visser, E.K. (1997) Tongue playing behaviour in captive giraffes. *Zeitschrift für Säugetierkunde* 62, 106–111.

Koene, P. and Wiepkema, P.R. (1991) Pre-dustbath vocalization as an indicator of need in domestic hens. *Proceedings of the International Conference Alternatives Animal Husbandry*, Witzenhaussen, Germany, pp. 95–103.

Koolhaas, J.M., Korte, S.M., De Boer, S.F., Van der Vegt, B.J. and Van Reenen, C.G. (1999) Coping styles in animals: current status in behavior and stress-physiology. *Neuroscience Biobehavioral Reviews* 23, 925–935.

Korte, S.M., Koolhaas, J.M., Wingfield, J.C. and McEwen, B.S. (2005) The Darwinian concept of stress: benefits of allostasis and costs of allostatic load and the trade-offs in health and disease. *Neuroscience and Biobehavioral Reviews* 29, 3–38.

Krebs, J.R. (1978) Optimal foraging: decision rules for predators. In: Krebs, J.R. and Davies, N.B. (eds) *Behavioural Ecology: An Evolutionary Approach.* Blackwell Scientific Publications, Oxford, UK, pp. 23–63.

Krebs, J.R. and Davies, D.B. (1991) *Behavioural Ecology: An Evolutionary Approach.* Blackwell Scientific Publications, Oxford, UK.

Kruska, D.C.T. (2005) On the evolutionary significance of encephalization in some eutherian mammals: effects of adaptive radiation, domestication, and feralization. *Brain, Behavior and Evolution* 65, 73–108.

Kyriazakis, I. and Savory, C.J. (1997) Hunger and thirst. In: Appleby, M.C. and Hughes, B.O. (eds) *Animal Welfare.* CAB International, Wallingford, UK, pp. 49–62.

Ladewig, J. and Matthews, L.R. (1996) The role of operant conditioning in animal welfare research. *Acta Agriculturae Scandinavia Section A, Animal Science* 27(suppl.), 64–68.

Lankin, V.S. and Bouissou, M.-F. (2001) Factors of diversity of domestication-related behaviour in farm animals of different species. *Russian Journal of Genetics* 37, 783–795.

Lawrence, A.B., Terlouw, E.M.C. and Kyriazakis, I. (1993) The behavioural effects of undernutrition in confined animals. *Proceedings of the Nutrition Society* 52, 219–229.

Lawrence, A.B., Tolkamp, B., Cockram, M.S., Ashworth, C.J., Dwyer, C.M. and Simm, G. (2004) Food, water and malnutrition: perspectives on nutrient requirements for health and welfare in farm animals. In: *Global Conference on Animal Welfare: An OIE Initiative,* 23–25 February, Paris, France, pp. 189–204.

Linklater, W.L. (2000) Adaptive explanation in socio-ecology: lessons from the Equidae. *Biological Review* 75, 1–20.

Lorz, A. (1973) *Tierschutzgesetz – Kommentar von A. Lorz.* Verlag Beck, Munich, Germany.

Manteuffel, G., Puppe, B. and Schön, P.C. (2004) Vocalization of farm animals as a measure of welfare. *Applied Animal Behaviour Science* 88, 163–182.

Markowitsch, H.J. (1982) *Behavioral Enrichment in the Zoo.* Van Nostrand Reinhold, New York.

Mason, G.J. and Latham, N.R. (2004) Can't stop, won't stop: is stereotypy a reliable animal welfare indicator? *Animal Welfare* 13, 557–569.

Mason, G.J. and Mendl, M. (1997) Do the stereotypies of pigs, chickens and mink reflect adaptive species differences in the control of foraging? *Applied Animal Behaviour Science* 53, 45–58.

McBride, G. (1984) Feral animal studies in animal science: the uses and limitations of feral animal studies to contemporary animal science. *Journal of Animal Science* 58, 474–481.

McBride, S.D. and Cuddeford, D. (2001) The putative welfare-reducing effects of preventing equine stereotypic behaviour. *Animal Welfare* 10, 173–189.

McFarland, D. (1989) Suffering in animals. In: McFarland, D. (ed.) *Problems of Animal Behaviour.* Longman, Essex, UK, pp. 34–58.

McGreevy, P. and Nicol, C. (1998) Physiological and behavioral consequences associated with short-term prevention of crib-biting in horses. *Physiology & Behavior* 65(1), 15–23.

Newberry, R.C. (1995) Environmental enrichment: increasing the biological relevance of captive environments. *Applied Animal Behaviour Science* 44, 229–243.

Newberry, R.C. (2004) Cannibalism. In: Perry, G.C. (ed.) *Welfare of the Laying Hen.* CAB International, Wallingford, UK, pp. 239–258.

Onghena, P. (1992) Randomisation tests for extensions and variations of ABAB single-case experimental designs: a rejoinder. *Behavioral Assessment* 14, 153–171.

Price, E.O. (1984) Behavioral aspects of animal domestication. *Quarterly Review of Biology* 39, 1–30.

Price, E.O. (2002) *Animal Domestication and Behaviour.* CAB International, Wallingford, UK, 297 pp.

Prins, H.H.T. (1996) *Ecology and Behaviour of the African Buffalo.* Chapman & Hall, London, 293 pp.

Pulliam, R.H. (1972) On the advantages of flocking. *Journal of Theoretical Biology* 38, 419–422.

Quenette, P.Y. (1990) Functions of vigilance in mammals: a review. *Ecologia* 6, 801–818.

Rauw, W.M., Kanis, E., Noordhuizen-Stassen, E.N. and Grommers, F.J. (1998) Undesirable side effects of selection for high production efficiency in farm animals: a review. *Livestock Production Sciences* 56, 15–33.

Reed, J.D. (1995) Nutritional toxicology of tannins and related polyphenols in forage legumes. *Journal of Animal Science* 73, 1516–1528.

Reinhardt, V. (1983) Movement orders and leadership in a semi-wild cattle herd. *Behaviour* 83, 251–264.

Robert, S., Matte, J.J., Farmer, C., Girard, C.L. and Martineau, G.P. (1993) High-fibre diets for sows: effects on stereotypies and adjunctive drinking. *Applied Animal Behaviour Science* 37, 297–309.

Roberts, G. (1996) Why individual vigilance declines as group size increases. *Animal Behaviour* 51, 1077–1086.

Rodenburg, T.B. and Koene, P. (2004) Feather pecking and feather loss. In: Perry, G.C. (ed.) *Welfare of the Laying Hen.* CAB International, Wallingford, UK, pp. 227–238.

Rodenburg, T.B., Koene, P., Bokkers, E.A.M., Bos, M.E.H., Uitdehaag, K.A. and Spruijt, B.M. (2005) Can short-term frustration facilitate feather pecking in laying hens? *Applied Animal Behaviour Science* 91, 85–101.

Roitman, M.F., Stuber, G.D., Phillips, P.E.M., Wightman, R.M. and Carelli, R.M. (2004) Dopamine operates as a subsecond modulator of food seeking. *The Journal of Neuroscience* 24, 1265–1271.

Rook, A.J. and Huckle, C.A. (1995) Synchronization of ingestive behaviour by grazing dairy cows. *Animal Science* 60, 25–30.

Rushen, J., Lawrence, A.B. and Terlouw, E.M.C. (1993) The motivational basis of stereotypies. In: Lawrence, A.B. and Rushen, J. (eds) *Stereotypic Animal Behavior: Fundamentals and Applications to Welfare.* CAB International, Wallingford, UK, pp. 41–64.

Sato, S., Nagamine, R. and Kubo, T. (1994) Tongue-playing in tethered Japanase Black cattle: diurnal patterns, analysis of variance and behaviour sequences. *Applied Animal Behaviour Science* 39, 39–47.

Savory, C.J., Wood-Gush, D.G.M. and Duncan, I.J.H. (1978) Feeding behaviour in a population of domestic fowls in the wild. *Applied Animal Ethology* 4, 13–27.

Savory, C.J., Seawright, E. and Watson, A. (1992) Stereotyped behaviour in broiler breeders in relation to husbandry and opioid receptor blockade. *Applied Animal Behaviour Science* 32, 349–360.

Schütz, K.E., Forkman, B. and Jensen, P. (2001) Domestication effects on foraging strategy, social behaviour and different fear responses: a comparison between the red junglefowl (*Gallus gallus*) and a modern laying strain. *Applied Animal Behaviour Science* 74, 1–14.

Seo, T., Sato, S., Kosaka, K., Sakamoto, N. and Tokumoto, K. (1998) Tongue-playing and heart rate in calves. *Applied Animal Behaviour Science* 58, 179–182.

Simonsen, H.B. (1996) Assessment of animal welfare by a holistic approach: behaviour, health and measured opinion. *Acta Agriculturae Scandanavica Section A, Animal Science* 27(suppl.), 91–96.

Sokal, R.R. and Rohlf, F.J. (1969) *Biometry.* W.H. Freeman, San Francisco, California, 340 pp.

Špinka, M., Newberry, R.C. and Bekoff, M. (2001) Mammalian play: training for the unexpected. *Quarterly Review of Biology* 76, 141–168.

Stafleu, F.R., Grommers, F.J., Vorstenbosch, J. (1996) Animal welfare: a hierarchy of concepts. *Animal Welfare* 5, 225–234.

Sterrit, G.M. and Smith, M.P. (1965) Reinforcement effects of specific components of feeding in young leghorn chicks. *Journal of Comparative and Physiological Psychology* 59, 171–175.

Stolba, A. and Woodgush, D.G.M. (1984) The identification of behavioural key features and their incorporation into a housing design for pigs. *Annuel Recherche Vétérinaire* 15, 287–298.

Stricklin, W.R. (2001) Evolution and domestication of social behaviour. In: Keeling, L.J. and Gonyou, H.W. (eds) *Social Behaviour in Farm Animals.* CAB International, Wallingford, UK, pp. 147–176.

Sutherland, W.J. (1996) *From Individual Behaviour to Population Ecology.* Oxford University Press, Oxford, UK, 391 pp.

Sutherland, W.J. and Koene, P. (1982) Field estimates of the strength of interference between oyster-catchers *Haematopus ostralegus. Oecologia* (*Berlin*) 55, 108–109.

Thorpe, N.H. (1969) Welfare of domestic animals. *Nature* 224, 18–20.

Tinbergen, N. (1963) On aims and methods of ethology. *Zeitschrift für Tierpsychologie* 20, 410–433.

Toates, F.M. (1980) *Animal Behaviour: A Systems Approach.* John Wiley & Sons, Chichester, UK, 299 pp.

Tooby, J. and Cosmides, L. (1990) The past explains the present: emotional adaptations and the structure of ancestral environments. *Ethology and Sociobiology* 11, 375–424.

Uexküll, J.v. (1939) Tier und Umwelt. *Zeitschrift für Tierpsychologie* 2, 101–114.

Uexküll., J.v. (1956) *Streifzüge durch die Umwelten von Tieren und Menschen.* Rohwolt, Hamburg.

Van der Waaij, E.H. (2004) A resource allocation model describing consequences of artificial selection under metabolic stress. *Journal of Animal Science* 82, 973–981.

Van Vuure, C. (2001) Retracing the aurochs. In: Gerken, B.G.M. (ed.) *Neue Modelle zu Masznahmen der Landschaftsentwicklung mit groszen Pflanzenfressern – Praktische Erfahrungen bei der Umsetzung.* Natur-und Kulturlandschaft, Höxter/Jena, Germany, pp. 261–265.

Van Vuure, C. (2003) *De oeros. Het spoor terug.* Wetenschapswinkel Wageningen UR, Wageningen, The Netherlands, 348 pp.

Veasey, J.S., Waran, N.K. and Young, R.J. (1996) On comparing the behaviour of zoo housed animals with wild conspecifics as a welfare indicator. *Animal Welfare* 5, 139–153.

Webster, A.J.F. (1995) *Animal Welfare: A Cool Eye Towards Eden: A Constructive Approach to the Problem of Man's Dominion Over the Animals.* Blackwell Scientific Publications, Oxford, UK, 273 pp.

Webster, A.J.F. and Main, D.C.J. (2003) Proceedings of the 2nd International Workshop on the Assessment of Animal Welfare at Farm and Group Level. *Animal Welfare* 12 (special issue).

West, B. and Zhou, B.-X. (1989) Did chickens go north? New evidence for domestication. *World's Poultry Science Journal* 45, 205–218.

Wiepkema, P.R. (1985) Abnormal behaviour in farm animals: ethological implications. *The Netherlands Journal of Zoology* 35, 277–299.

Wiepkema, P.R., Van Hellemond, K.K., Roessingh, P. and Romberg, H. (1987) Behaviour and abomasal damage in individual veal calves. *Applied Animal Behaviour Science* 18, 257–268.

Wood-Gush, D.G.M. and Duncan, I.J.H. (1976) Some behavioural observations on domestic fowl in the wild. *Applied Animal Ethology* 2, 255–260.

Yamada, Y.T. (1988) The contribution of poultry science to society. *World's Poultry Science Journal* 44, 172–178.

Young, R.J. (2003) *Environmental Enrichment for Captive Animals.* Universities Federation for Animal Welfare, Blackwell Scientific Publications, Oxford, UK, 228 pp.

Zimmerman, P.H., Koene, P. and Hooff, J.A.R.A.M.v. (2000a) The vocal expression of feeding motivation and frustration in the domestic laying hen, *Gallus gallus domesticus. Applied Animal Behaviour Science* 69, 265–273.

Zimmerman, P.H., Koene, P. and Hooff, J.A.R.A.M.v. (2000b) Thwarting of behaviour in different contexts and the gakel-call in the laying hen. *Applied Animal Behaviour Science* 69, 255–264.

7 Food Choice and Intake in Chickens

J.M. Forbes

School of Biology, University of Leeds, Leeds LS2 9JT, UK

Introduction

The choices birds make when confronted with more than one food have been studied in a variety of contexts, but this chapter will mainly use the domestic fowl (both growing broilers and laying hens) as its example species, as this is where much of the relevant research has been carried out. We will consider mainly the factors that affect food choice and the controls exerted by sensory and nutritional factors in the context of learned preferences and aversions. We will also link this to the control of intake of a single food, which we will propose is also subject to learning.

Many studies on food choice have been carried out without proper consideration of the nutritional context. By nutritional we mean the metabolic consequences of eating a particular food, be they beneficial or harmful. In order to make a proper study of birds' ability to choose between foods for the optimum nutritional benefit we need to know not only the composition of the foods on offer but also the weight eaten. Observations in the wild do not usually give sufficient information for proper interpretation of observations to be made and the focus here will be on experiments carried out under controlled conditions in which the appropriate measurements can be made.

Conditions necessary to demonstrate diet selection

If a proper assessment is to be made of birds' ability to construct a balanced diet from a choice of two foods on offer, certain conditions need to be met.

One food should contain a higher concentration of the nutrient in question than the bird requires in relation to the energy content, while the other food must contain a lower concentration than required. If no mixture of the two foods will provide an optimal concentration of the nutrient, then clearly the bird cannot select a balanced diet. A pair of foods spanning the optimal content for a bird with a low requirement for that nutrient might not be adequate for a bird with high requirements for that nutrient, e.g. calcium in laying and non-laying hens.

The two foods must differ in one or more sensory properties so that birds can differentiate between them before committing themselves to swallowing. Properties that birds can use to differentiate between foods include colour, flavour and position of food container. An example of the use of colour to identify foods is the ability of chicks to make appropriate selection for ascorbic acid when the deficient and sufficient foods were different colours, but not when they were visually identical (Kutlu and Forbes, 1993b). Birds are able to use food flavours as cues to their nutritional value, but not as successfully as colour (Wilcoxon *et al.*, 1971). In the absence of other cues as to the nutritional properties of foods, birds can use the position of the food in the cage. For example, chickens used to receiving methionine-deficient food consistently on one side of the cage and methionine-sufficient food on the other learned to compose a balanced diet; when the position of the foods was reversed, it took several days to re-establish the successful selection (Steinruck *et al.*, 1990).

Another factor to be taken into account in studying diet selection is the social context and history. Animals learn from their close conspecifics, usually their mothers, and rearing in isolation or away from adults can fail to provide a young animal with its normal array of food-associated knowledge. Group housing allows birds to copy from each other and even individually caged naive chicks, which did not voluntarily consume any wheat when given a choice with a standard grower food, immediately started to eat significant amounts when put in pairs, irrespective of whether the partner was formerly a wheat-eater or not (Covasa and Forbes, 1995).

The ability of chickens to make nutritionally appropriate choices when offered two foods has been studied for a variety of nutrients. Rather than list these, however, two examples will be given in detail. Coverage of other nutrients can be found elsewhere (Forbes, 1995).

Diet Selection by Chickens

Example 1: selection for protein

After energy, protein is usually considered the most important constituent of the diet. It (or rather its constituent amino acids) is required at concentrations between about 120 and 230 g/kg of dry food, depending on physiological state, with a normal energy content of 12.5 MJ ME/kg DM. (It would be more appropriate to express the protein content of foods in relation to their ME content, but this is rarely done as most foods have energy contents of around 12.5 MJ ME/kg DM.) Given a choice between a food containing 65 g protein/kg DM and one containing 280 g protein/kg DM, growing broiler chickens chose a mixture that provided 189 g/kg, which gave them growth and carcass composition very close to that of exactly similar birds given a single food containing 225 g/kg, i.e. close to that which is known to be optimal for growth (Shariatmadari and Forbes, 1993). If both foods have protein contents below the optimal

range, birds eat predominantly from the higher one, and when both foods have protein contents higher than optimal, they eat mostly from the lower. No matter how unsuitable one of the foods on offer is, birds still taste this more extreme food from time to time to ensure that it is still unsuitable.

When birds are selecting between two foods in order to support a near-optimal growth rate, birds with different potential growth rates will select different ratios of foods. In addition to broiler chicks, the study of Shariatmadari and Forbes (1993) included chicks of a layer strain, with a much lower growth potential than the broilers. The layer chicks chose a lower proportion of the high-protein food (285 vs. 65), which gave a protein content of the diet of 167 g/kg (i.e. much lower than the broilers), and their performance was close to that of birds given a single food containing 165 g protein/kg.

Mature broilers selected for high body weight selected a higher protein diet than those of a strain selected for low body weight when given a choice of HP (467 g CP/kg) and LP (82 g CP/kg) (Brody *et al.*, 1984). There was a significant increase in energy intake by low-weight birds when given a choice of HP and LP, compared to a single, complete food with 188 g CP/kg, suggesting that the latter was limiting food intake by providing an excess of protein in relation to energy intake. (Note that mature animals have a much lower requirement for protein, in relation to energy, than they do when young and growing.) In addition, birds selected for low body weight and given the single, complete food, took considerably more glucose solution than when given a choice of HP and LP. Provision of this choice of foods allowed birds to optimize both their energy and protein intakes independently, whereas a single food very rarely provides these resources in the optimum ratio.

If, as is often observed, birds choose a protein content lower than that of a single food giving maximal growth (e.g. Siegel *et al.*, 1997; Yo *et al.*, 1998), we must not propose that the birds are 'wrong'. We must accept sometimes that the food mixture that the bird finds most comfortable is not necessarily that which gives maximum performance, in an agricultural sense.

There are many other examples of the ability of poultry to select for protein (Forbes and Shariatmadari, 1994) so that it is relatively easy to make the claim that poultry can learn to choose a diet with a protein content appropriate to the needs of the bird.

Example 2: selection for phosphorus

The case of phosphorus (P) is, however, not so clear-cut. P is an essential nutrient, required for bone and eggshell formation in addition to many metabolic processes. There are several reasons to optimize P nutrition of poultry, including the imperative to avoid excessive excretion into the environment. A complicating factor is the fact that much of the P in plant materials is in the form of phytate, which is resistant to digestion; hence the importance of the proportion of dietary P that is available for absorption (available P, AP). Phytase enzymes release P from phytate and these are used as feed additives to improve the availability of P from poultry diets. Nevertheless, it is not possible to predict with accuracy the optimal rate of inclusion of phytase in diets for birds in different physiological states. Diet selection methodology might offer a means to determine optimum provision of dietary P.

Holcombe et al. (1976) observed that laying hens chose to eat more foods with intermediate total P (TP) contents (4.6–10.0 g/kg) and less foods with low (1.9 g/kg) or high (24.3 g/kg) contents, but this was not a critical test of the ability to compose an optimal diet because the availability of P was not known. It is necessary first of all to establish that one food to be offered is deficient in the nutrient in question, in this case, P. This has proved surprisingly difficult with the type of wheat-based food usually used in northern Europe because of the relatively high activity of natural phytase in wheat (Barkley, 2001). Using maize/soya-based foods, Barkley et al. (2004) confirmed that the low-AP (LAP, 1.1 g AP/kg) food they formulated was indeed deficient, as evidenced by significantly reduced egg weight compared to eggs from hens given a similar food formulated to be adequate in AP (normal AP, NAP, 2.2 g AP/kg DM) by addition of dicalcium phosphate (DCP). Laying hens were then

offered a choice between LAP and NAP foods to determine whether they would choose proportions of the two appropriate to their needs for AP.

At the peak of egg laying (25–28 weeks of age with a production of 0.98 eggs/bird/day) P-replete or P-deficient hens were fed either NAP or LAP alone or a choice of LAP and NAP or a choice of LAP and PAP (LAP with 400 FTU/kg microbial phytase). The P-deficient hens given a choice of NAP/LAP ate a significantly greater proportion of NAP (66%) than P-replete hens fed NAP in phase 1 (28%), demonstrating that P deficiency influenced subsequent selection for AP, i.e. an appetite for P (Fig. 7.1).

When the choice was between LAP and the same deficient food with added phytase (PAP), there was no effect of previous diet on the proportions of food chosen that were not significantly different from 50%. PAP included phytase at 400 FTU/kg; therefore the concentration of exogenous phytase in their diet was diluted to approximately 200 FTU/kg, which is the recommended level of inclusion in the diet for laying hens. It may be, therefore, that the birds were choosing to select close to 50% of each food, rather than eating at random. However, the P-depleted hens would have been expected to choose a higher proportion of PAP than the P-replete birds. This illustrates the need, in diet selection experiments, to use two foods whose optimal mixture will be significantly different from 50:50 if unequivocal results are to be obtained to support or refute the ability to make nutritionally appropriate diet selection.

A possible limitation of this experiment was the lack of differentiation in the sensory properties of the two foods, to which no colours or flavours were added to make it easier for the birds to learn the association between sensory properties and nutritional value. However, they should have allowed them to learn the association as the food containers were coloured differently and their positions in the cage remained constant for each bird.

Another factor limiting the birds' ability to choose wisely for AP might be the nature of the 'discomfort' induced by P deficiency. Mild deficiencies or toxicities, whose effects on

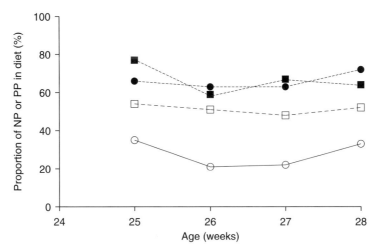

Fig. 7.1. Proportion of NP or PP eaten by laying hens. Diets: ○: NP in phase 1, choice between NP and LP foods in phase 2; □: LP in phase 1, choice between LP and PP in phase 2; ●: NP in phase 1, choice between LP and PP in phase 2; ■: LP in phase 1, choice between NP and LP foods in phase 2. (From Barkley *et al.*, 2004.)

metabolism have a delayed onset and/or which do not affect abdominal receptors, are less likely to induce sensory-paired aversions than are those inducing immediate nausea; P deficiency or excess is likely to fall into the former category and therefore not be a strong candidate for conditioned preference or aversion.

It is worth mentioning that in the experiment of Barkley *et al.* (2004) all of the birds on every treatment were offered access to ground limestone, a rich source of calcium. Those laying hens fed LAP food throughout the experiment chose to reduce their limestone intake relative to those fed the NAP food. This is likely to have been due to the mobilization of P from bone being accompanied by calcium, thereby reducing the need for dietary calcium, and by the reduced demand for calcium because of the reduction in egg production in this treatment.

In summary, therefore, although the evidence supports an ability amongst birds to select their diet according to its P content, this has not been demonstrated as clearly as has been the case for protein.

The timescale of selection

It might be expected that selection for an optimum diet would occur in response to the consequences of eating previous meals; intake of a meal of high-protein food generates high levels of absorbed amino acids and their products of metabolism that may be sensed in the liver (Rusby and Forbes, 1987) or elsewhere, creating a condition that can be alleviated by eating low-protein food. If this is true, one or more meals of HP should be followed by one or more meals of LP. In fact the situation is not as simple as this because Shariatmadari and Forbes (1992) observed many meals in which birds ate from both foods, a situation that would be anticipated as making it difficult for birds to associate the different nutritional characteristics of the two foods with their different sensory properties. Over periods of a few hours birds did not select a balanced diet, yet over several days their diet contained close to an optimum content of protein. This is similar to the observations of Yeates *et al.* (2002) with dairy cows in which short-term randomness of eating was overtaken in the longer term by organized selection. There is clearly much yet to be understood about the timescale over which eating and diet selection are controlled, but it is certain that learning is involved (see below).

However, there is no doubt that chickens are sensitive to the composition of the food they are eating and compensate for intake of one food (e.g. HP) by subsequently eating more of the other (e.g. LP) in order to make a balanced diet. For example, Forbes and Shariatmadari (1996) accustomed 4-week-old male chicks of both broiler and layer strains to selecting a diet from HP (300 g CP/kg) and LP (70g CP/kg). Then, after an overnight fast, they offered a large meal of either HP or LP and then immediately offered the choice of foods. Figure 7.2 (immediate) shows that when the initial meal was HP, the birds chose to eat significantly more of the LP food during the next hour; conversely, when a meal of LP was given first, subsequently there was significant preference for HP. When there was a delay of 45 min between the first meal and the subsequent choice of foods, the results were similar (Fig. 7.2, 1 h delay). When the initial meal was given by tube into the crop, so that it could not be seen or tasted, there was no significant preference in the hour after choice of the two foods being given 45 min after the gavage (Fig. 7.1, gavage). This demonstrates the need for the birds to identify the food by its sensory properties in order to predict the likely availability of nutrients during and after absorption of that meal; i.e. the bird has learned what to expect metabolically after eating food with a certain set of sensory properties.

On a longer timescale, when HP and LP were offered on alternate days or half days, broilers still selected a similar dietary protein content, and gained weight at the same rate as similar birds given free choice (Forbes and Shariatmadari, 1996). However, chicks prefer a single, balanced food to a choice between HP and LP as demonstrated by Rovee-Collier et al. (1982), who offered a choice between high- and low-protein foods during the 12 h of daylight but a food with an adequate protein content at night. The birds adopted a nocturnal feeding pattern (although they did eat a little food during the day), showing that if necessary the chick will eat at night in order to get a single, complete food even though it is normally a diurnal feeder. This suggests that birds are more comfortable eating a single balanced diet than mixing their own diet by choosing appropriately from two foods.

Learning

While one food might be more 'palatable' than another, and be preferred immediately after the two foods are first offered, in the longer term it is the metabolic consequences of eating the foods that will largely determine subse-

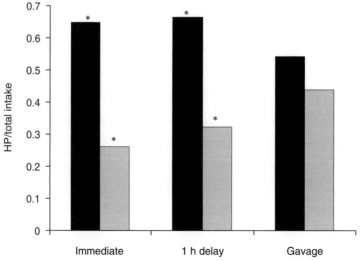

Fig. 7.2. HP intake as a proportion of total intake (HP + LP) (g/h) by broiler chickens previously given a meal of 10 g LP (solid bars) or HP (grey bars) either immediately, with 1 h delay or by gavage (see text). (From Shariatmadari and Forbes, 1997.) *: Significant difference between premeal of HP and of LP.

quent preferences. Consider the example of chicks injected with cholecystokinin (CCK) on day 1 followed by 2 h access to food of one colour, saline on day 2 followed by 2 h access to food of a different colour and, briefly, a preference test involving a choice between the two colours of food on day 3 (Covasa and Forbes, 1994). (The dose of CCK used was one that caused a mild reduction in food intake for 2 h subsequent to its injection and is likely to have caused mild nausea.) There was a trend towards preference for the saline-paired colour. This 3-day cycle was repeated twice more and the preference became stronger. Similar results were obtained subsequently in the same birds when the pairing of the colours was reversed, e.g. a preference for red food induced by pairing green food with CCK became a preference for green food when red food was paired with CCK. These results are shown in Fig. 7.3. This clearly demonstrates the rapidity with which the learned association was gained during training and lost once training was discontinued.

Similarly with ascorbic acid, a change in the 'requirement' induced by a change in environmental temperature was followed by a change in preference for ascorbic acid–supplemented food, which took about 3 days to stabilize at the new level (Kutlu and Forbes, 1993a).

It is abundantly clear from these examples that chicks learn to associate the sensory properties of foods with the metabolic consequences of eating those foods.

'Palatability'

Palatability is a much-misused word and it is probably better to avoid it altogether. However, given its prevalence in discussions on food choice and food intake, we emphasize that it is not just a function of the sensory properties of each food but also of the nutrients supplied by the food, the animal's nutrient requirements and its previous experience of foods with similar sensory properties.

'**Palatable** describes food or drink that has a pleasant taste' (http://dictionary.cambridge.org/) but it has been demonstrated that vision is more important in birds' choice of foods than taste (Wilcoxon et al., 1971). The definition should therefore be broadened to 'food that has pleasant sensory properties'. The experiment outlined above, in which chicks showed a reversible preference for the colour of food associated with saline compared to CCK (Covasa and Forbes, 1994), clearly demonstrates that the pleasantness ('palatability') an animal perceives in a food is subject to the animal's previous experience of foods

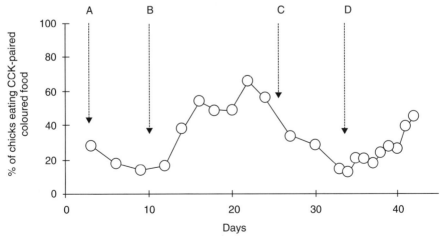

Fig. 7.3. Proportion of chicks moving towards the colour of food that they had experienced after being injected with CCK. (A) Start of training period 1. (B) End of training period 1. (C) Start of training period 2. (D) End of training period 2. (From Covasa and Forbes, 1994.)

characterized by the sensory properties of the food(s) currently on offer.

In my lectures I challenge students to bring me two chicks and a bag of nutritious coloured food and to return 1 week later by which time one bird will find the food very attractive (palatable) and the other will find it aversive (unpalatable). In the intervening week, of course, on several occasions I offer one bird the food provided for a short time after injecting with lithium chloride (or adding LiCl to the food; LiCl causes a dose-related feeling of nausea) and for the rest of the time provide food of a different colour – this bird learns to avoid the food with the colour associated with LiCl. The other bird is given the food provided for most of the time and, from time to time, food of another colour, either following an injection of LiCl or with LiCl added to the food. No student has ever taken up the challenge as they can see so clearly how the claimed result can be achieved! What price 'palatability'?

Intake of a Single Food

It has been clearly demonstrated above that learning is involved in the process of birds matching food choice to requirements, and we now speculate that the intake of a single food is subject to the learning process in which the effects of several internal and external factors are involved. We will explore what happens when a single food on offer is deficient or excessive in one or more nutrients.

Superficially, animals 'eat for calories', i.e. they regulate the amount of food they eat to meet their energy 'requirements'. However, this regulation is not exact, as shown in the comprehensive experiment of de Groot (1972) in which laying hens were offered foods ranging in energy concentration from 10.5 to 13.4 MJ ME/kg. Although the weight of food eaten per day declined with increasing energy concentration, there was a steady increase in ME intake, egg weight and body weight gain, as the ME concentration of the diet increased. Possible reasons for this will be outlined below.

The weight of food eaten is also affected by its protein content, amino acid balance, fibre content and concentrations of minerals and vitamins, as well as external factors such as environmental and social stress. While the bird might attempt to alleviate a marginal deficiency of an essential nutrient by increasing its food intake, and thus its intake of the nutrient in question, this increase in energy intake leads to increased short-term negative feedbacks from metabolites and long-term ones from increased body fat. Any serious dietary deficiency leads to reduced food intake as metabolism is slowed by the lack of the essential nutrient. Conversely, while animals can cope with moderate excesses of nutrients, ultimately these excesses can become toxic and their effects can be lessened by reducing food intake.

Timescale of intake

Requirements for nutrients vary with the animal's physiological state, environmental temperature and composition of the diet. Therefore, the optimum amount of a single food to be eaten is likely to vary. How does the animal know that the amount it is eating currently is still the optimum amount? We may gain some clues by examining sequences of daily intakes of individual birds.

Daily intakes of individual growing broiler chicks given a barley-based food were recorded by Yasar and Forbes (1999). Because daily intakes increased in an approximately linear manner as the birds grew, day-to-day variability was examined using the residuals of the regression of intake on age which were subject to autocorrelation with successive days, from the next day up to 6 days thence.

Examples of four birds are shown in Fig. 7.4; residual intake tended to be positively correlated with intakes 1 and 2 days subsequently, but negatively with the third and subsequent days. This suggests that intake is being controlled to meet requirements over a period of several days; higher than average intake might persist for 2 or 3 days but is followed by lower than average intake for the next few days.

This behaviour can be interpreted as the birds 'experimenting' by increasing or decreasing their daily intake in order to determine whether they feel more or less comfortable, and then following the line of least discomfort (see below).

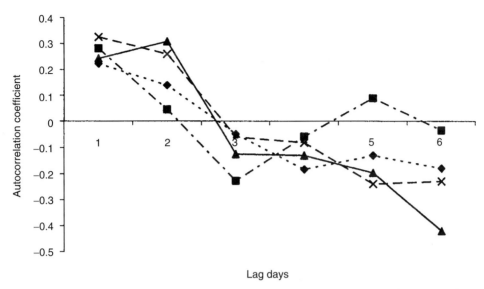

Fig. 7.4. Correlation coefficients for daily food intake of four broiler chickens against intakes 1–6 days later. (From S. Wahl and J.M. Forbes, unpublished results, 2002.)

Discomfort as a determinant of food choice and intake

The internal feelings generated by intake of an imbalanced food can be described as 'discomfort' (Forbes and Provenza, 2000), and it has been proposed that animals modify their intake and selection of foods in order to minimize this discomfort. A striking example is that in which lame and sound broilers were offered a choice of two coloured foods, one including carprofen (analgesic). The proportion of drugged food selected by lame birds was significantly greater than for sound birds, and the lamer the bird, the higher the proportion of the drugged food selected (Danbury *et al.*, 2000). Clearly, lame birds were making their choices between foods in order to minimize the discomfort caused by the lameness.

In the case given earlier, in which chicks preferred a food associated with injection of saline to one of a different colour associated with CCK injection (Covasa and Forbes, 1994), the birds were behaving in such a way as to minimize the discomfort induced by CCK. We speculate that discomforts can be generated by many things: over- or undereating of energy, protein, an essential amino acid or any other essential nutrient; by dietary bulk and distension of the digestive organs; by the social environment (behaviour of conspecifics, threat of predation) or physical environment (wind, rain, excessively cold or hot weather).

Minimization of discomfort

If an animal is to behave in such a way as to reduce discomfort (from any source), it is logical to expect that it will try to minimize that discomfort, preferably to zero. This is the reason we have adopted the concept of 'minimization of discomfort' rather than 'maximization of comfort', as discomfort has a definable minimum (of zero) while there is no attainable maximum to comfort. Consider a range of foods varying in protein/energy ratio from severely deficient to grossly excessive. Given the severely deficient food, birds are prevented from attaining their genetic target of protein deposition (or egg production in the case of laying hens); any attempt to redress the deficiency by eating more food is met by a greater load of energy-yielding nutrients, which cannot all be metabolized due to the lack of protein. Intake is adjusted to a low level in order to minimize the discomfort generated by the lack of protein and difficulty of metabolizing carbohydrates and fats. With a mildly protein-deficient food these problems are less severe and a higher daily intake is achievable, sometimes even greater than that of an optimum diet (Boorman, 1979). In the case of foods with an excess of protein over energy, discomforts are generated by the heat and toxic products of the oxidation of amino acids when they are absorbed in excess of the rate at which they are being used for protein synthesis. Food intake is therefore lower than with a balanced diet, especially in a hot environment. A graph of discomfort plotted against food protein content is, therefore, U-shaped.

The same logic can be applied to other nutrients for which it can be postulated that there is an optimum intake for a bird in any given physiological state. A current food intake that supplies more, or less, than this will generate discomfort; the greater the surplus or deficiency, the greater the discomfort.

Total discomfort

The example given above, of birds self-medicating to reduce the discomfort of lameness, is unusually simple, especially when compared to dietary discomfort in which, almost by definition, there are several factors that might potentially cause discomfort. Increasing the fibre content of a food, for example, not only provides more dietary bulk but also reduces the concentration of metabolizable energy in the food.

If it is accepted that discomfort can come from many sources, it is necessary to consider how the central nervous system integrates the various signals. A common method used in the past has been to predict the intakes of a food from each of several of its properties, for a given type of animal, and to assume that the actual intake will be the least of these predicted intakes (e.g. Poppi et al., 1994 for cattle). The problem with this approach is that it assumes that one factor controls intake until, suddenly, it is overtaken by another, which then takes total control. This seems physiologically unlikely. The effects of various feedbacks are more likely to be integrated rather than

mutually exclusive and the simplest mode of integration is summation. There are examples of additivity in physiological systems: the signals from different families of temperature receptor in controlling heat production in goats are additive (Jessen, 1981); impositions of multiple stresses on plasma corticosterone in chicks are additive (McFarlane and Curtis, 1989). Details of the likely neural pathways for additivity have been given (Forbes, 1996).

We arrive at the concept of total discomfort as the sum of many individual discomforts generated in many parts of the body by many food-related factors.

Minimal total discomfort

A single food

Combining 'minimization of discomfort' with 'additivity of discomforts' we arrive at the concept that animals eat quantities of one or more foods that lead to minimal total discomfort (MTD). As an example we use a laying hen with optimal intakes ('requirements') of 1.25 MJ ME, 3 g calcium (Ca) and 25 g protein per day, offered a single food containing 11.5 MJ ME, 17 g crude protein (CP) and 2.5 g Ca per kg. A food intake of 100 g/day is used as a starting point and the quantities of ME, Ca and CP in these weights of food are calculated. The differences between these intakes and the optimum intakes for the bird in question are calculated and these 'errors' are squared and summated, giving total discomfort (TD). Food intake is then increased by 10 g steps up to 150 g and the calculations are repeated. The results are shown in Fig. 7.5. In this example discomfort due to ME is minimal at an intake of 110 g/day, which is due to Ca at an intake of 120 g/day, while discomfort due to CP is minimal up to an intake of 150 g/day. Summation gives a TD curve whose minimum is at an intake of 120 g/day, at which point discomfort is not zero as the animal is suffering a moderate excess of ME and a deficiency of CP.

Choice between foods

The MTD approach can be used to simulate a bird with access to two or more foods. The

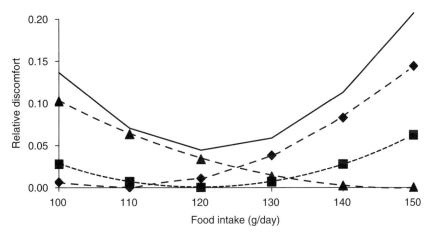

Fig. 7.5. Relative discomfort due to ME (diamonds), Ca (squares), CP (triangles) and total discomfort (continuous line) for laying hens at food intakes from 100 to 150 g/day.

model is given the bird's 'requirements' and the composition of the foods and 'experiments' with different combinations of food until it arrives at the mixture that gives MTD. Here we use the same laying hen as above, offered two foods with different protein:energy ratios, and ground limestone (LS), i.e. a choice between three foods. HP contains 11.5 MJ ME, 0.25 kg CP and 0.010 kg Ca per kg; LP contains 12.5 MJ ME, 0.15 kg CP and 0.01 kg Ca per kg; LS includes 0.35 kg Ca/kg.

It is possible to achieve optimal intakes of three nutrients when choosing between three foods as long as the composition of the foods allows this. In this case there is a mixture of the three that provides optimal intakes of ME, CP and Ca and this is shown in Table 7.1, which also includes the results of simulations with each pair of foods and each single food. The comments in the table summarize how the hen is predicted to cope in each situation. As long as LS is available Ca intake can be maintained, but in the absence of HP or LP, CP or ME, respectively, is underconsumed. The model appears to behave as hens are expected to behave but no attempt is made here to compare the predictions with real observations.

At this point the reader might be expecting some comparison of predictions of the MTD model with experimental observations. However, it is not the intention to screen numerous published sets of results searching for a few that match the model. Rather, the intention is to see that a model believed by the author to be based on physiological principles behaves in a rational manner. It is a research model, not a predictive model. There is a long way to go, especially in discovering relative weighting factors for each food resource and determining the shape of the relationships between food intake and discomfort generated by each of the nutrients of interest.

Conclusion

In order to demonstrate that birds can make nutritionally wise selection from a choice of foods it is necessary to provide cues such as different colours or flavours, whereby they can learn to associate the sensory properties of the food with the metabolic consequences of eating that food. There is clear evidence of selection for protein (and several other nutrients) in order to achieve a balanced diet, while the evidence for a P appetite is less convincing. It takes several days for birds to relearn sensory–metabolic associations after a change in either their requirement for the nutrient in question or the composition of one or both foods.

The same control mechanisms are likely to underlie the control of intake of a single food as of the intakes of more than one food given in free choice. In the absence of any ability to change the composition of its diet by changing

Table 7.1. Outcome of running the MTD model with three foods, two foods or one food available, in terms of intake (g/day) and total discomfort (TD, arbitrary units) (see text for details of hen's requirements and composition of foods).

Food(s) available	HP intake	LP intake	LS intake	Total intake	TD	Comment
HP, LP and LS	89	18	6	113	0	Perfect match
HP and LP	74	50	–	124	0.378	Overeats both ME and CP, but still very short of Ca
LP and LS	–	118	5	123	0.118	Overeats ME and undereats CP, marginally; OK for Ca
HP and LS	104	–	6	110	0.003	Undereats ME and overeats CP, marginally; OK for Ca
HP	115	–	–	115	0.41	Overeats a little to try to get Ca
LP	–	131	–	131	0.46	Overeats a lot to try to get Ca and CP; excessive ME intake
LS	–	–	9	9	2	Only eats to get exactly right Ca

the ratio of the foods eaten, it is proposed that the chicken responds to the internal discomfort generated by eating a single imbalanced food by changing its intake until a state of MTD is reached; discomforts generated by under- or oversupply of nutrients are postulated to be summated, and the bird is considered to attempt to minimize this TD. The same principle applied to the choice-feeding situation yields sensible predictions, as yet unvalidated.

It is clear that the controls of food intake and choice are multifactorial and the proposed theory is an attempt to integrate these controls into a physiologically realistic model.

References

Barkley, G.R. (2001) Phosphorus nutrition of laying hens: phytase and diet selection. PhD thesis, University of Leeds, UK.

Barkley, G.R., Forbes, J.M. and Miller, H.M. (2004) Phosphorus nutrition of laying hens: phytase and diet selection. *British Journal of Nutrition* 92, 233–240.

Boorman, K.N. (1979) Regulation of protein and amino acid intake. In: Boorman, K.N. and Freeman, B.M. (eds) *Food Intake Regulation in Poultry*. Longman, Edinburgh, UK, pp. 87–126.

Brody, T.B., Cherry, J.A. and Siegal, P.B. (1984) Responses to dietary self-selection and calories in liquid form by weight selected lines of chickens. *Poultry Science* 63, 1626–1633.

Covasa, M. and Forbes, J.M. (1994) Exogenous cholecystokinin octapeptide in broiler chickens: satiety, conditioned colour aversion and vagal mediation. *Physiology & Behavior* 56, 39–49.

Covasa, M. and Forbes, J.M. (1995) The effect of prior training of broiler chickens on diet selection using whole wheat. *Animal Production* 58, 471.

Danbury, T.C., Weeks, C.A., Chambers, J.P., Waterman-Pearson, A.E. and Kestin, S.C. (2000) Self-selection of the analgesic drug carprofen by lame broiler chickens. *Veterinary Record* 146, 307–311.

de Groot, G. (1972) A marginal income and cost analysis of the effect of nutrient density on the performance of White Leghorn hens in battery cages. *British Poultry Science* 13, 503–520.

Forbes, J.M. (1995) *Voluntary Food Intake and Diet Selection in Farm Animals*. CAB International, Wallingford, UK.

Forbes, J.M. (1996) Integration of regulatory signals controlling forage intake in ruminants. *Journal of Animal Science* 74, 3029–3035.

Forbes, J.M. and Provenza, F.D. (2000) Integration of learning and metabolic signals into a theory of dietary choice and food intake. In: Cronje, P. (ed.) *Ruminant Physiology: Digestion, Metabolism, Growth and Reproduction.* CAB International, Wallingford, UK, pp. 3–19.

Forbes, J.M. and Shariatmadari, F.S. (1994) Diet selection for protein by poultry. *World's Poultry Science Journal* 50, 7–24.

Forbes, J.M. and Shariatmadari, F.S. (1996) Short-term effects of food protein content on subsequent diet selection by chickens and the consequences of alternate feeding of and high and low protein foods. *British Poultry Science* 37, 597–607.

Holcombe, D.J., Roland, D.A. and Harms, R.H. (1976) The ability of hens to regulate phosphorus intake when offered diets containing different levels of phosphorus. *Poultry Science* 55, 308–317.

Jessen, C. (1981) Independent clamps of peripheral and central temperatures and their effects on heat production in the goat. *Journal of Physiology* 311, 11–22.

Kutlu, H.R. and Forbes, J.M. (1993a) Effect of changes in environmental temperature on self-selection of ascorbic acid in coloured feeds by broiler chicks. *Proceedings of the Nutrition Society* 52, 29A.

Kutlu, H.R. and Forbes, J.M. (1993b) Self-selection of ascorbic acid in coloured foods by heat-stressed broiler chicks. *Physiology & Behavior* 53, 103–110.

McFarlane, J.M. and Curtis, S.E. (1989) Multiple concurrent stressors in chicks. 3. Effects on plasma-corticosterone and the heterophil-lymphocyte ratio. *Poultry Science* 68, 522–527.

Poppi, D.P., Gill, M. and France, J. (1994) Integration of theories of intake regulation in growing ruminants. *Journal of Theoretical Biology* 167, 129–145.

Rovee-Collier, C.K., Clapp, B.A. and Collier, G.H. (1982) The economics of food choice in chicks. *Physiology & Behavior* 28, 1097–1102.

Rusby, A.A. and Forbes, J.M. (1987) Food intake of intact and vagotomised chickens infused with lysine and glucose into the hepatic portal vein. *Proceedings of the Nutrition Society* 46, 33A.

Shariatmadari, F. and Forbes, J.M. (1992) Diurnal food intake patterns of broiler chickens offered a choice of feed varying in protein content. *Animal Production* 54, 470.

Shariatmadari, F. and Forbes, J.M. (1993) Growth and food intake responses to diets of different protein contents and a choice between diets containing two levels of protein in broiler and layer strains of chicken. *British Poultry Science* 34, 959–970.

Siegel, P.B., Picard, M., Nir, I., Dunnington, E.A., Willemsen, M.H.A. and Williams, P.E.V. (1997) Responses of meat-type chickens to choice feeding of diets differing in protein and energy from hatch to market weight. *Poultry Science* 76, 1183–1192.

Steinruck, U., Kirchgessner, M. and Roth, F.X. (1990) [Selective methionine intake of broilers by changing the position of the diets]. *Archives für Geflügelkunde* 54, 245–250.

Wilcoxon, H.C., Dragoin, W.B. and Kral, P.A. (1971) Illness-induced aversions in rat and quail: relative salience of visual and gustatory cues. *Science* 171, 826–828.

Yasar, S. and Forbes, J.M. (1999) Performance and gastro-intestinal response of broiler chickens fed on cereal grain-based foods soaked in water. *British Poultry Science* 40, 65–76.

Yeates, M.P., Tolkamp, B.J. and Kyriazakis, I. (2002) The relationship between meal composition and long-term diet choice. *Journal of Animal Science* 80, 3165–3178.

Yo, T., Siegel, P.B., Faure, J.M. and Picard, M. (1998) Self-selection of dietary protein and energy by broilers grown under a tropical climate: adaptation when exposed to choice feeding at different ages. *Poultry Science* 77, 502–508.

8 Feed Restriction and Welfare in Domestic Birds

I.C. de Jong[1] and B. Jones[1,2]
[1]Animal Sciences Group of Wageningen UR, Division Animal Resources
Development, Research Group Animal Welfare, PO Box 65, 8200 AB Lelystad,
The Netherlands; [2]Roslin Institute, Roslin BioCentre, Midlothian EH25 9PS,
Scotland, UK

Introduction

This chapter will focus mainly on the effects of feed restriction on the welfare of the broiler breeder chicken. This is a good example because this is where most of the research on feed restriction has been carried out in this species. However, the effects of short- and long-term feed restriction and food deprivation on welfare in other types of domestic birds will also be discussed where relevant. In the first part of the chapter the substantial literature on the effects of feed restriction or food deprivation on parameters indicative of domestic bird welfare will be reviewed. Thereafter, the merits and demerits of the various measures that can be applied to reduce the negative effects of feed restriction on broiler breeder welfare while maintaining target body weight and (re)production at the required level will be discussed.

Broiler Breeders: the Dilemma between Optimal (Re)production and Bird Welfare

The selection of broilers for rapid growth has resulted in increased appetite through the modulation of both central and peripheral mechanisms of hunger regulation. Indeed it has been suggested that food intake in broilers and broiler breeders is driven by a lack of sati-ety signals (Burkhart et al., 1983; Lacy et al., 1985; Denbow, 1999; Bokkers and Koene, 2003), although this hypothesis has since been criticized (Nielsen, 2004). It has also been suggested that broilers eat to their maximum gut capacity when allowed to feed ad libitum (Nir et al., 1978 in Siegel, 1999; Bokkers and Koene, 2003). Overconsumption of feed has resulted in obesity (Siegel, 1999; Mench, 2002), with broilers and broiler breeders that are fed ad libitum suffering from increased mortality (Hocking et al., 2002) and a variety of health problems, including metabolic (e.g. ascites), immune (decreased immune function), muscular (idiopathic myopathy, dystrophies), cardiovascular and skeletal disorders (Dunnington et al., 1987; Scheele, 1997; Mitchell, 1999; Sandercock et al., 2001; Mench, 2002; Whitehead et al., 2003). Skeletal problems include various infectious and non-infectious disorders, porosity and poor mineralization (Williams et al., 2000; Whitehead et al., 2003). The incidence of gait abnormalities increases and the birds' locomotor activity declines markedly with age, probably as a result of the increased incidence and severity of skeletal and/or muscular problems (Estevez et al., 1997; Kjaer and Mench, 2003). In addition to the welfare insult, reproductive performance is severely impaired. Although ad libitum fed broiler breeders enter lay sooner than feed-restricted birds, ad libitum feeding and the resultant obesity are associated with low egg production, erratic oviposition and defective eggs due to the simultaneous development of follicles and excessive ovulations – a condition called 'erratic oviposition and defective egg syndrome' (EODES) (Jaap and Muir, 1968; Hocking et al., 1989; Hocking and White-head, 1990; Siegel, 1999; Heck et al., 2004; Renema and Robinson, 2004).

In order to maintain health and reproduction at acceptable levels severe feed restriction has to be applied to control body weight (Nir et al., 1996; Siegel, 1999; Mench, 2002). In commercially applied restriction programmes feed intake is typically restricted to about 25–33% of the intake of ad libitum fed birds at the same age (or 50% of the intake of ad libitum fed birds at the same weight) during rearing (Savory et al., 1996; de Jong et al., 2002), while it is restricted to 50–90% of ad libitum intake for birds of the same age when the hens are in lay (Bruggeman et al., 1999). The restricted feeding regimes result in a reduction of body weight of adult birds to approximately 40–50% of that of ad libitum fed birds (Katanbaf et al., 1989; Savory et al.,

1996). Figure 8.1 shows the difference in body weight and build between broiler breeders fed ad libitum and those fed at the commercially applied restriction level. The most commonly used commercial restriction programmes for broiler breeders are skip-a-day or everyday feeding, although regimes of 'skip-two-days' or 'feeding for 4 or 5 days followed by 2 or 3 days of deprivation' have also been applied (Mench, 2002). The everyday feeding programme is currently the most common in The Netherlands.

Despite the positive effects of feed restriction on health and reproduction, there is no doubt that it exerts negative effects on other aspects of welfare, as reflected by behavioural and physiological changes. For example, while the domestic fowl and the jungle fowl normally spend most of the day foraging for food (pecking and ground-scratching) (Savory et al., 1978; Dawkins, 1989), feed-restricted broiler breeders usually spend less than 15 min consuming their meal during rearing (Savory et al., 1993b, 1996; de Jong et al., 2002) and 1–2 h during the laying period (de Jong et al., 2005a). Thus, the birds' time budgets

Fig. 8.1. Effects of feeding regimes on body weight and build in broiler breeders at 6 weeks of age. Birds were fed either ad libitum or at the commercially applied restriction level.

are dramatically altered. Perhaps not surprisingly then, during rearing feed-restricted broiler breeders show a number of behaviours that may be indicative of boredom and of frustration of the motivation to feed (e.g. Hocking et al., 1996; Savory and Maros, 1993; de Jong et al., 2002; Picard et al., 2004). Although feed restriction is less severe during the laying period, behaviours indicative of hunger and frustration can also be observed at that time (Picard et al., 2004; de Jong et al., 2005a); indeed the birds show spot-pecking and increased drinking (Hocking et al., 2002). Feed-restricted birds are also thought to be chronically hungry and to show more aggression, particularly during mating, as well as physiological indications of chronic distress (Savory, 1995; King, 2001; Mench, 2002; Kjaer and Mench, 2003). These effects are described more fully below.

Feed Restriction in other Types of Domestic Birds

Although feed restriction is predominantly applied to broiler breeders to safeguard health and reproduction, this procedure is also used for other types of domestic birds for various reasons. For instance, feed restriction programmes are imposed on turkey breeder hens during rearing to minimize the multiple ovulation, low egg production and poor eggshell quality associated with ad libitum feeding (e.g. Crouch et al., 1999, 2002) and on male turkeys during rearing and laying to control body weight and improve their semen production (Hocking et al., 1999). Although growing broilers are primarily fed ad libitum, restriction regimes are sometimes applied to reduce growth rate during the weeks of maximal growth rate to avoid metabolic disorders, improve bird health (especially to reduce skeletal problems), change carcass composition or reduce body fat content (e.g. Nielsen et al., 2003). In the USA laying hens are also subjected to feed deprivation in order to induce moult (Ruszler, 1998; Webster, 2003), though this act is banned in many other countries, such as The Netherlands and the UK. Indeed, induced moulting has been severely criticized

with respect to bird welfare as birds can be food-deprived for up to 16 days and thereby lose up to 25–35% of their body weight (Ruszler, 1998; Webster, 2003). Moreover, mortality rates increase fourfold during the period of moult (Duncan and Mench, 2000).

Effects of Feed Restriction on Bird Behaviour

Feed-restricted broiler breeders show a number of behaviours that are considered to be indicative of frustration and boredom (Savory et al., 1993b). Firstly, for example, they display high rates of stereotypic pecking, which may be directed at the litter, drinker, (empty) feeder or the walls of the pen (Kostal et al., 1992; Savory and Maros, 1993; Hocking et al., 1996; Savory and Kostal, 1996; de Jong et al., 2002; Hocking et al., 2002). Increased pecking at the drinker may result in overdrinking (polydipsia), with the attendant problem of wet litter (Kostal et al., 1992; Hocking et al., 1996). For that reason, water intake is often also restricted in practice in an attempt to maintain litter quality.

Secondly, restricted feeding leads to very high activity levels that largely reflect elevated frequencies of pacing. While stereotypic pecking is predominantly observed after feeding, pacing is mainly apparent before the birds are fed (Savory and Maros, 1993; Savory and Kostal, 1996). The amounts of stereotypic pecking and pacing seem to be positively correlated with the level of feed restriction (Savory and Maros, 1993; Hocking et al., 1996) and may thus conceivably be used to quantify the level of hunger. Feed-restricted turkeys also show similar behavioural responses to broiler breeders, i.e. increased locomotor activity and stereotypic pecking (Hocking et al., 1999). It has since been suggested that the level of hunger may be more accurately reflected by the amount of activity than of stereotypic pecking (Savory et al., 1996). The circadian rhythm of activity differs between ad libitum fed broiler breeders and those fed at commercial levels of restriction (Fig. 8.2). Ad libitum fed birds show low daytime activity with no apparent peaks, whereas a peak in activity is observed around feeding time

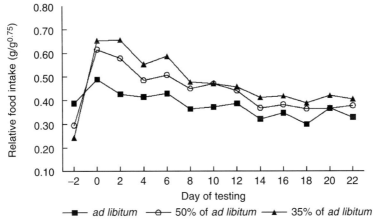

Fig. 8.2. Compensatory feed intake in broiler breeders fed at the commercially applied restriction level and fed at 50% of *ad libitum* intake, as compared to relative feed intake of *ad libitum* fed broiler breeders. (Reprinted from de Jong *et al.*, 2003 with permission from Elsevier.)

in feed-restricted broiler breeders, and they maintain a high level of activity during the day (de Jong *et al.*, 2002). The same circadian activity pattern can be observed in *ad libitum* and feed-restricted broilers (Nielsen *et al.*, 2003). The activity pattern in feed-restricted birds can be explained by the increase in (stereotypic) appetitive foraging activities around feeding, a phenomenon that is also found in feed-restricted mammals (e.g. Day *et al.*, 1995).

The neuropharmacology of stereotypic pecking has received some attention. A report that treatment with nalmefene, an antagonist of central opioid peptide receptors, caused a reduction in stereotypic object-pecking in feed-restricted broiler breeders during rearing suggested that this behaviour might function as a coping strategy because of the de-arousing and positively reinforcing properties of opioids (Savory *et al.*, 1992). Further research revealed that the expression of object-pecking depends on activation of D2 dopamine receptors (Kostal and Savory, 1994) and that feed restriction led to elevated concentrations of dopamine in the basal telencephalon (Kostal *et al.*, 1999). Collectively, these findings indicate a crucial role for dopamine in the control of feed restriction–induced stereotypies in broiler breeders.

Aggression is another potential problem, with different scenarios apparent during rearing and laying. In the weeks following the start of feed restriction at 2 weeks of age, broiler breeder males showed more aggressive pecks and threats but less sparring than did fully fed ones (Mench, 1988). The feed-restricted males were subjected to a skip-a-day feeding regime and aggression peaked at the feed-off mornings, i.e. when food delivery was expected but did not occur (Mench, 1988; Shea *et al.*, 1990). Food deprivation in laying hens was also positively correlated with aggression (Duncan and Wood-Gush, 1971); when deprived of food, the cockerels showed a large increase in overt aggression towards hens that they usually dominate passively (Duncan and Wood-Gush, 1971). Similarly, feather and skin damage in female broiler breeders during rearing has been recently ascribed to the occurrence of aggression around feeding time (I.C. de Jong, 2004, Lelystad, personal observation). In contrast, during laying fully fed broiler breeders displayed more male-to-male and male-to-female aggression than did feed-restricted males (Millman and Duncan, 2000; Millman *et al.*, 2002). Males of laying strains tested at similar ages did not behave aggressively towards females, regardless of whether they were feed-restricted or not (Millman and Duncan, 2000). It has been suggested that the male-to-female aggression commonly observed during laying in broiler breeders – and that constitutes a significant welfare problem for the females – is not a function of feed restriction experienced during rearing; on the other hand, aggression during rearing has been ascribed to restriction-induced frustration (Millman and Duncan, 2000). Clearly, further research is required to illuminate some of these anomalies.

Feed restriction is thought to reduce arousal, as measured by headflick frequency during exposure to events such as silence or sudden noise (Savory et al., 1993a). However, the interpretation of headflicking behaviour is open to debate (Hughes, 1983). The duration of tonic immobility, an anti-predator fear reaction (Jones, 1986), was also found to be shorter in broiler breeders that were feed-restricted than in fully fed ones (Savory et al., 1993a). However, the greater weight and consequently reduced mobility of the fully fed birds may have confounded the results (R.B. Jones, 2004, Roslin, personal observation). Interestingly, feed-restricted birds were more active than fully fed ones when placed in a novel environment (open field) (de Jong et al., 2003). Since fear inhibits other behaviour systems such as feeding and social behaviour (Jones, 1996; Jones et al., 2002) the previous finding may indicate lower fearfulness in feed-restricted birds (de Jong et al., 2003).

Effects of Feed Restriction on Feed Intake Motivation

There is substantial evidence that broiler breeders fed at the commercially applied restriction level are chronically hungry and that they are highly motivated to eat at all times. Using operant conditioning, it was shown that the motivation to eat in broiler breeders kept under commercial restriction programmes was approximately four times as high as that of ad libitum fed birds that had been subjected to food withdrawal for 72 h (Savory et al., 1993b). Furthermore, compared to their food intake during restriction relative compensatory feed intake (feed intake/g metabolic weight)

was more than three times higher when the birds were allowed to eat freely (de Jong et al., 2003). This very high level of feed consumption was maintained for several days and, though a gradual decline was apparent, it was still two times higher than the intake during feed restriction after 3 weeks (de Jong et al., 2003) (see Fig. 8.2).

Effects of Feed Restriction on Physiological Stress Indicators

It has been suggested that feed restriction is a potent physiological stressor in domestic fowl. Feed withdrawal for 24 h resulted in adrenal hypertrophy, increased concentrations of circulating corticosterone and of free fatty acids, as well as decreased plasma glucose concentrations (Freeman et al., 1980; Scanes et al., 1980; Zulkifli et al., 1995). Indeed, compared to ad libitum fed birds, broiler breeders (Hocking et al., 1996; Savory and Mann, 1997; de Jong et al., 2002) and turkeys (Hocking et al., 1999) fed at commercially applied restriction levels showed chronically elevated plasma corticosterone concentrations. It is unclear whether the increase in plasma corticosterone concentrations reflects the metabolic effects of feed restriction, (chronic) psychological stress or both. However, corticosterone helps to regulate blood glucose levels (Schwartz, 1997; Schwartz et al., 2000; Mench, 2002), and glucocorticoids activate the hypothalamic neuropeptide Y (NPY) pathway that stimulates feeding behaviour and increases body energy stores (Schwartz, 1997; Boswell et al., 1999; Schwartz et al., 2000). Furthermore, it has been suggested, at least from other avian studies, that corticosterone might be directly implicated in the initiation of food-searching behaviour (Webster, 2003). Thus, it is likely that the increased plasma corticosterone concentrations found in feed-restricted birds are indicative of altered intermediary metabolism due to a different food intake pattern as well as of chronic stress (de Jong et al., 2002, 2003). In view of the curvilinear relationship between plasma corticosterone concentrations and the level of feed restriction (Fig. 8.3) it may be tentatively suggested that metabolic effects play a predominant role at restriction levels of up to

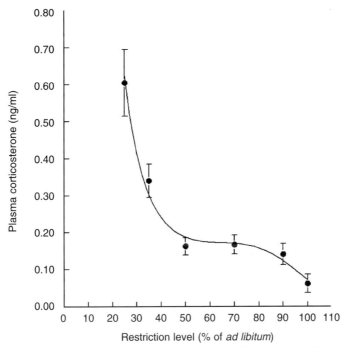

Fig. 8.3. The relationship between feed restriction level and baseline plasma corticosterone concentrations in broiler breeders. (Reprinted from de Jong *et al.*, 2003 with permission from Elsevier.)

50% of *ad libitum* intake, whereas psychological stress may operate when restriction levels exceed 50% (de Jong *et al.*, 2003).

The heterophil/lymphocyte (H/L) ratio, which can be used as an indicator of chronic stress (Gross and Siegel, 1983; Jones, 1989; Maxwell, 1993), was increased in feed-restricted birds in some studies (Hocking *et al.*, 1993, 1996; Savory *et al.*, 1993a) but not in others (Savory *et al.*, 1996; Hocking *et al.*, 1999; de Jong *et al.*, 2002). This apparent inconsistency might reflect differences in sampling strategies. For example, there may be an increase in H/L ratios over the first few days of restriction, followed by normalization after 2–3 weeks (Zulkifli *et al.*, 1993). Increased proportions of basophils were also found when broiler breeders were subjected to feed restriction (Hocking *et al.*, 1993).

Elevated levels of plasma creatine kinase and aspartate accompanied restricted feeding in broiler breeders and turkeys (Hocking *et al.*, 1993, 1999). These findings suggest that the rapid growth of fully fed birds is associated with elevated plasma enzyme activities reflecting possible muscle dysfunction or injury and/or increased muscle protein turnover (Hocking *et al.*, 1993, 1999); in turkeys this was consistent with mortality from cardiovascular disease in *ad libitum* fed birds (Hocking *et al.*, 1999). These harmful effects were reduced by restricted feeding (Hocking *et al.*, 1993, 1999), thus confirming that the regime can have a positive effect on bird health.

A chronically increased heart rate and body temperature may indicate stress but may also have a relationship with metabolic rate (de Jong *et al.*, 2002). Heart rate and body temperature are much lower in feed-restricted broiler breeders, both during the day and night (Fig. 8.4); however, these findings reflect differences in food intake patterns and thus in intermediary metabolism. Moreover, circadian patterns in heart rate, body temperature and activity can be observed in feed-restricted birds, whereas such clear patterns are absent

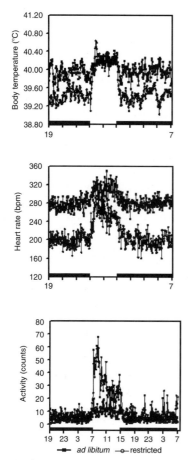

Fig. 8.4. Circadian variation in body temperature, heart rate and activity in 7-week-old broiler breeders fed *ad libitum* or at the commercially applied restriction level. Dark bars indicate the dark period. (Reprinted with permission from de Jong *et al.*, 2002.)

in *ad libitum* fed birds. Although their night-time activity levels are similar, body temperature and heart rate are higher in birds kept under *ad libitum* rather than restricted feeding regimes; the latter observations may reflect increased metabolic rates in the *ad libitum* fed birds due to their very high food intake instead of differences in the level of stress (de Jong *et al.*, 2002).

There is substantial evidence, at least from behavioural observations, that feed restriction is a potent stressor. However, it is more difficult to draw such conclusions from physiological measurements because most physiological stress indicators are sensitive to

the metabolic as well as the psychological effects of feed restriction. It remains to be determined how the effects of psychological stress can be distinguished from those of altered metabolism.

Possible Ways of Reducing the Negative Effects of Long-term Feed Restriction on Broiler Breeder Welfare

Low-density diets

Several studies have evaluated the use of low-density diets as a possible method of reducing distress and the feelings of hunger induced by quantitative feed restriction in broiler breeders. The notion is that the provision of low-density diets prolongs the time spent eating, thereby reducing frustration associated with the conflict between continued motivation to eat and the absence of food (see discussion in Mench, 2002). In addition, a fuller gastrointestinal tract may promote the perception of satiety and thereby reduce feelings of hunger, although inputs from distension receptors of the gastrointestinal tract are not the only factor influencing satiety (Whittaker *et al.*, 1998, 1999). For example, the provision of a diet diluted with 15% or 30% of ground oat hulls from 0 to 56 weeks of age significantly reduced the H/L ratio, at 12 weeks of age but not at 20 and 40 weeks (Zuidhof *et al.*, 1995). It also significantly increased the time required for broiler breeders to eat the food, though this effect was only apparent during the first half of the rearing period. Other behavioural effects were only found in the laying period; birds fed a diluted rather than a standard restricted diet spent less time at the water source, but there were no differences in stereotypic spot-pecking during either rearing or laying. Benefits of dietary dilution included improved flock uniformity, egg production and feed conversion, but the absence of effects on behaviours characteristic of hunger and frustration, like object-pecking and hyperactivity, argues against a stress-reducing property of this feeding strategy (Zuidhof *et al.*, 1995).

Furthermore, when birds were given free access to feed diluted with either unmolassed sugarbeet pulp, two concentrations of oat hulls or softwood sawdust up to 10 weeks of age, they grew faster than those fed the standard restricted diet, thus compromising industry targets (Savory et al., 1996; Savory and Lariviere, 2000). Other disadvantages of feeding these low-density diets included reduced uniformity in body weight gain and an increased production of excreta (Savory et al., 1996). Although the incidence of stereotypic spot-pecking was decreased in one study, activity levels were not affected by dietary dilution, and the use of diluents such as sugarbeet pulp and sawdust caused physiological stress as indicated by elevations in H/L ratios and plasma corticosterone concentrations (Savory et al., 1996). Subsequent studies suggested that neither feeding motivation nor activity levels were affected by feeding both broilers and broiler breeders, diets diluted with 15% oat hulls (Savory and Lariviere, 2000; Nielsen et al., 2003). Thus, the debate over the welfare implications of dietary dilution continues.

In another study, broiler breeders were fed diets diluted with oat hulls, sugarbeet pulp and sunflower meal up to 15 weeks of age (Hocking et al., 2004). Therein, stereotypic object-pecking was significantly decreased in birds fed diets diluted with sugarbeet pulp (50 g/kg) and the highest concentrations of oat hulls (200 g/kg). Standing and walking were also reduced with increasing concentrations of oat hulls, but the opposite was found for sugarbeet pulp (Hocking et al., 2004). Birds fed diets containing the highest fibre concentrations had the lowest H/L ratios, but no other physiological measures were examined (Hocking et al., 2004). Since activity levels and H/L ratios are thought to be positively associated with hunger (Savory et al., 1996) and distress (Jones, 1989; Maxwell and Robertson, 1998), respectively, it may be concluded that low-density diets diluted with either oat hulls or sugarbeet pulp might have beneficial effects on broiler breeder welfare during rearing. The effects of adding sugarbeet pulp may be related to its water-retaining property and the increased sense of satiety that such a 'saturated' feed might induce in the crop and gizzard (Hocking et al., 2004).

A recent study (de Jong et al., 2005a) determined the effects on welfare and produc-tion characteristics of feeding broiler breeders with low-density diets diluted with sunflower seed meal, wheat bran and palm kernel meal or with oats and sugarbeet pulp during rearing and laying. The energy content of the feeds was lowered by 1.28 or 2.51 MJ/kg. Thus, the birds were still fed restricted diets but the amount of feed supplied was increased. In the rearing period, the times spent object-pecking and sitting were reduced and increased, respectively for the diets diluted with wheat bran, palm kernel meal and the highest concentration of sunflower seed meal (de Jong et al., 2005a), suggesting reductions in hunger and in frustration of feed restriction (Savory et al., 1996; de Jong et al., 2003). However, object-pecking was still apparent in a large proportion of observations, indicating that hunger had not been eliminated. Moreover, behavioural effects were only found during the first half of the rearing period (at 6 and 10 weeks of age). Physiological measures of stress and hunger (plasma corticosterone concentrations, H/L ratios and glucose/non-esterified fatty acids (NEFA) ratios) did not differ between birds fed diluted diets or standard commercial ones. Feeding motivation was significantly reduced in birds fed the diet with the lowest energy content (de Jong et al., 2005a), indicating reduced hunger, but surprisingly, in contrast with Hocking's results (Hocking et al., 2004), the observed effects of the diet diluted with sugarbeet pulp were very small. Behavioural and physiological observations were indicative of distress when birds were given the diet with the lowest energy content during laying (de Jong et al., 2005a), but dietary dilution in general had no adverse effects on production characteristics like uniformity, egg production and mortality of the progeny (H. Enting, 2004, Lelystad, personal observation). Collectively, the results indicate that a low-density diet may exert some beneficial effects on broiler breeder welfare during rearing but that feelings of hunger remain (de Jong et al., 2005a).

Table 8.1 summarizes the effects of different types of diluted diets on parameters indicative of broiler breeder welfare during rearing. The results of the various studies suggest that, though small effects may be apparent, feeding diluted diets is unlikely to realize substantial welfare improvements.

Table 8.1. Effects of diet dilution during rearing on behaviour, feed intake motivation and common physiological parameters of distress in broiler breeders. The use of brackets denotes that the effects are concentration-dependent.

Feed diluent	Stereotypic pecking	Activity level	H/L ratio	Plasma corticosterone concentration	Feed intake motivation	References
Oat hulls, 15–30%	↔	↔	↓	ND	ND	Zuidhof et al. (1995)
Oat hulls, 30/60%	↓	↓	↔	↔	↔	Savory et al. (1996); Savory and Lariviere (2000)
Oat hulls, 5/10/20%	↓ (20%)	↓	↓ (10/20%)	ND	ND	Hocking et al. (2004)
Oat hulls, 30–40% and calcium propionate 5–10%	↓	ND	↔	↔	ND	Sandilands et al. (2004)
Sugarbeet pulp, 40%	↓	↓	↑	↑	ND	Savory et al. (1996)
Sugarbeet pulp, 5/10/20%	↓ (5%)	↑	↓	ND	ND	Hocking et al. (2004)
Sugarbeet pulp and oats, 30%	↔	↔	↔	↔	↓	de Jong et al. (2005a)
Sunflower meal, 5/10/20%	↔	↔	↓	ND	ND	Hocking et al. (2004)
Sunflower meal, wheat bran and palm kernel meal, 17/32.5%	↓ (32.5%)	↓ (32.5%)	↔	↔	↓	de Jong et al. (2005a)
Sawdust, 50%	↓	↔	↔	↑	ND	Savory et al. (1996)

Note: ↓ reduced; ↑ increased; ↔ no difference, compared to broiler breeders fed a standard restricted diet during rearing. ND: non-determined.

Anorectic Agents

Self-restriction represents a potential alternative approach. A voluntary reduction in feed intake can be induced by adding various anorexic chemicals, such as calcium propionate or phenylpropanolamine, to a balanced diet. For example, adding propionic acid to the standard diet suppressed behavioural characteristics of hunger and frustration, like stereotypic pecking behaviour, in broiler breeders, but it did not affect feeding motivation (Savory and Lariviere, 2000) or physiological indices of stress, such as plasma corticosterone concentrations and H/L ratios (Savory *et al.*, 1996; Sandilands *et al.*, 2004). Thus, it is not clear if such a self-restricting diet can also reduce feelings of hunger. Moreover, adding propionic acid to the standard diet failed to inhibit voluntary feed intake or to control growth to the extent achieved through conventional feed restriction programmes (Pinchasov *et al.*, 1993; Savory *et al.*, 1996; Savory and Lariviere, 2000). Other disadvantages included the expense of using propionic acid on a large scale and the reduced uniformity in body weight gain (Savory *et al.*, 1996). Therefore, we conclude that self-restriction by the addition of propionic acid to the diet is not currently a useful way of reducing hunger-induced distress in growing broiler breeders.

Fenfluramine may be an alternative to propionic acid. Fenfluramine suppresses feed intake by releasing endogenous serotonin in the brain, thereby mimicking the natural release of serotonin during satiation that terminates feeding. Treatment with this agent depressed body weight and reduced feed intake in a dose-dependent manner while maintaining egg production at a high level, but effects on egg fertility remain to be assessed (Hocking and Bernard, 1993; Rozenboim *et al.*, 1999). However, fenfluramine is also expensive and, like propionic acid, it is not yet known if it suppresses feelings of hunger. Other anorectic agents like furanone and crotonolactone yielded inconsistent decreases in body weight and were therefore not considered likely to be suitable candidates for the control of food intake in broiler breeders (Hocking and Bernard, 1993).

A supplement of jojoba meal in the diet of broiler breeders reduced feed intake and controlled body weight to the extent recommended by the breeding company (Vermaut *et al.*, 1998), probably by its negative taste (Vermaut *et al.*, 1997). However, it has a negative effect on egg production during the laying period (Vermaut *et al.*, 1998, 1999) and is therefore no candidate for the control of food intake in broiler breeders.

feeder' has become a recognized industrial practice in some parts of the world, e.g. Canada and The Netherlands (Van Middelkoop *et al.*, 2000). The assumption is that foraging behaviour is stimulated when a hard pelleted feed is scattered in the litter. Encouragingly, farmers reported an attenuation of leg problems and they attributed this to the increased (foraging) activity of the birds. Moreover, they reported that the birds seemed to show less pacing (I.C. de Jong, 2004, Lelystad, personal observation), which may indicate that scattered feeding can reduce frustration, elicited by thwarting of feeding. It was also previously suggested that the stimulation of foraging behaviour might eradicate the occurrence of stereotypic object-pecking that could otherwise develop due to a restricted nutrient supply or via reduced access to a foraging substrate (Lawrence and Terlouw, 1993).

The effects on welfare of exposing broiler breeders to scattered feeding during rearing have also received substantial attention (de Jong *et al.*, 2005b). For example, behaviour patterns, feeding motivation, and plasma corticosterone, glucose and NEFA concentrations were measured in broiler breeders at 7–8 weeks of age when feed was either delivered in the food trough or scattered in the litter. However, there were no detectable effects of scattered feeding on the physiological parameters or on feeding motivation. The occurrence of stereotypic object-pecking was significantly reduced in birds exposed to scattered feeding, but it remains unclear if this finding can be considered indicative of reduced frustration and improved welfare when the same birds are spending a substantial amount of their time budget on foraging activities. In other words, we cannot rule out the possibility that pecking at the litter where the birds expect to find food may also be compulsive in nature. It was concluded that scattered feeding did not significantly improve broiler breeder welfare during rearing because there was no evidence that feelings of hunger were reduced (de Jong *et al.*, 2005b).

Scattered Feeding

During the rearing of broiler breeders, scattering feed in the litter by using a so-called 'spin-

Meal Feeding

Chickens show increased foraging behaviour just before the dark period, suggesting that they have a need for energy intake at that time (Savory et al., 1978; de Jong et al., 2005b). Therefore, it is conceivable that providing broiler breeders with a second meal at that time might serve to reduce feelings of frustration and, possibly, of hunger. The effects of feeding the birds twice a day, with the first meal early in the morning and the second meal just before the dark period, were therefore examined. However, behavioural observations revealed that stereotypic object-pecking was not reduced and that activity levels were actually elevated. Indeed, substantial pacing was observed before feeding, both in the morning and afternoon (I.C. de Jong, 2004, Lelystad, personal observation; de Jong et al., 2005b). The absence of any detectable differences in feeding motivation or in physiological parameters of distress, like plasma corticosterone, glucose and NEFA concentrations, between birds fed two rather than one meal per day (de Jong et al., 2005b) suggests that this strategy failed to alleviate feelings of hunger and frustration. Savory and Mann (1999) also fed broiler breeders twice a day and found that with increasing meal size the time spent in stereotypic object-pecking after feeding increased while the time spent standing decreased. Thus, paradoxically, though the expression of stereotypic object-pecking is known to correlate negatively with daily total food intake, it appears to correlate positively with meal size within a given level of restriction. Thus, the collective findings (Savory and Mann, 1999; de Jong et al., 2005b) do not support the notion that broiler breeder welfare can be improved by feeding twice a day.

Modified Rearing Programmes

Commercial restriction programmes impose a severe feed restriction between 6 and 16 weeks of age, thereby resulting in the standard flattened growth curve. Hocking et al. (2001) then asked whether modifying the restriction programme so that there was a more linear increase in growth during this period could have beneficial effects on bird welfare. However, they found no detectable effects on welfare indices such as selected behaviours, plasma corticosterone and enzyme concentrations, or H/L ratios (Hocking et al., 2001).

After 15 weeks of age birds that are then fed either ad libitum or a restricted diet can attain good reproductive performance (Bruggeman et al., 1999), but ad libitum access to food at this time was not considered to have significantly improved their overall welfare because the most severe feed restriction had already occurred, i.e. from 6 to 16 weeks of age. Thus, modified restricted feeding programmes, or at least those described above (Bruggeman et al., 1999; Hocking et al., 2001) did not significantly improve broiler breeder welfare.

Genetic Selection

Some production-related problems can be addressed more efficiently through improved management, environmental control or nutrition than via genetic selection (Emmerson, 2003). Furthermore, given the growing societal concerns about farm animal welfare, the use of selective breeding to manipulate animal production now has to be deemed acceptable before it can be applied (Decuypere et al., 2003).

However, within these constraints appropriate selection programmes could conceivably be used to reduce the level of feed restriction while maintaining the required production level (Mench, 2002). Here, we consider just two examples. Firstly, broiler chickens can be produced by mating dwarf hens with normal males. Dwarf hens have good reproductive performance and they do not need to be feed-restricted (de Jong et al., 2003; Kjaer and Mench, 2003). Secondly, selection for lower residual feed consumption, which is the difference between the amount of feed actually consumed and the food consumption expected on the basis of weight gain, egg production and metabolic body weight, may also have positive implications for welfare (Kjaer and Mench, 2003). Of course, we must not lose sight of the fact that decisions for breeding are driven not only in response to the impact of consumer demands for a welfare-friendly product on purchase decisions, i.e. the fork-to-farm component, but also by economic requirements. Thus, there is a need to balance welfare and performance goals in an animal-friendly and economically viable fashion. This may be easier said than done.

Conclusions

Although it is not clear whether the observed effects of feed restriction on physiological parameters reflect (chronic) stress or merely metabolic effects, measures of bird behaviour and feed intake motivation substantiate the conclusion that long-term feed restriction, as is commonly applied in broiler breeder production, has a major negative effect on bird welfare. While long-term feed restriction is not applied on such a large scale in other types of domestic birds, its use in the management of laying hens, meat-type broilers and turkeys is also likely to exert adverse effects on their welfare. Thus far, efforts to improve broiler breeder welfare solely by maintaining good reproductive performance, and thus restricting body weight, have not led to a significant overall improvement. Conversely, *ad libitum* feeding is not a solution because it is related to serious health and mobility problems that, in turn, have a strong negative effect on bird welfare. The ongoing genetic selection for increased growth and meat production in meat-type birds worsens the situation (Rauw *et al.*, 1998; Kjaer and Mench, 2003). Currently, an improvement of the welfare of meat-type birds in terms of reducing their hunger can only be achieved by the use of slower growing strains (Hocking and Jones, 2004; Picard *et al.*, 2004). However, this strategy may be economically unacceptable to many breeding companies and farmers without compensatory inducements such as subsidies and/or premium prices.

References

Bokkers, E.A.M. and Koene, P. (2003) Eating behaviour, and preprandial and postprandial correlations in male broiler and layer chickens. *British Poultry Science* 44, 538–544.

Boswell, T., Dunn, I.C. and Corr, S.A. (1999) Neuropeptide Y gene expression in the brain is stimulated by fasting and food restriction in chickens. *British Poultry Science* 40, S42–S61.

Bruggeman, V., Onagbesan, O., D'Hondt, E., Buys, N., Safi, M., Vanmontfort, D., Berghman, L., Vandesande, F. and Decuypere, E. (1999) Effects of timing and duration of feed restriction during rearing on reproductive characteristics in broiler breeder females. *Poultry Science* 78, 1424–1434.

Burkhart, C.A., Cherry, J.A., Van Krey, H.P. and Siegel, P.B. (1983) Genetic selection for growth rate alters hypothalamic satiety mechanisms in chickens. *Behavior Genetics* 13, 295–300.

Crouch, A.N., Grimes, J.L., Christensen, V.L. and Garlich, J.D. (1999) Restriction of feed consumption and body weight in two strains of large white turkey breeder hens. *Poultry Science* 78, 1102–1109.

Crouch, A.N., Grimes, J.L., Christensen, V.L. and Krueger, K.K. (2002) Effect of physical feed restriction during rearing on large white turkey breeder hens: 2. Reproductive performance. *Poultry Science* 81, 16–22.

Dawkins, M.S. (1989) Time budgets in red junglefowl as a baseline for the assessment of welfare in domestic fowl. *Applied Animal Behaviour Science* 24, 77–80.

Day, J.E.L., Kyriazakis, I. and Lawrence, A.B. (1995) The effect of food-deprivation on the expression of foraging and exploratory-behavior in the growing pig. *Applied Animal Behaviour Science* 42, 193–206.

de Jong, I.C., Van Voorst, A., Ehlhardt, D.A. and Blokhuis, H.J. (2002) Effects of restricted feeding on physiological stress parameters in growing broiler breeders. *British Poultry Science* 43, 157–168. Available at: http://www.tandf.co.uk/journals

de Jong, I.C., Van Voorst, A. and Blokhuis, H.J. (2003) Parameters for quantification of hunger in broiler breeders. *Physiology & Behavior* 78, 773–783.

de Jong, I.C., Enting, H., Van Voorst, A. and Blokhuis, H.J. (2005a) Do low-density diets improve broiler breeder welfare during rearing and laying? *Poultry Science* 84, 194–203.

de Jong, I.C., Fillerup, M. and Blokhuis, H.J. (2005b) Effect of scattered feeding and feeding twice a day during rearing on indicators of hunger and frustration in broiler breeders. *Applied Animal Behaviour Science* 92, 61–76.

Decuypere, E., Bruggeman, V., Barbato, G.F. and Buyse, J. (2003) Growth and reproduction problems associated with selection for increased broiler meat production. In: Muir, W.M. and Aggrey, S.E. (eds) *Poultry Genetics, Breeding and Biotechnology*. CAB International, Wallingford, UK, pp. 13–28.

Denbow, D.M. (1999) Food intake regulation in birds. *Journal of Experimental Zoology* 283, 333–338.

Duncan, I.J.H. and Mench, J. (2000) Does hunger hurt? *Poultry Science* 79, 934.

Duncan, I.J.H. and Wood-Gush, D.G.M. (1971) Frustration and aggression in domestic fowl. *Animal Behaviour* 19, 500–504.

Dunnington, E.A., Martin, A. and Siegel, P.B. (1987) Antibody responses to sheep erythrocytes in early and late feathering chicks in a broiler line. *Poultry Science* 66, 2060–2062.

Emmerson, D. (2003) Breeding objectives and selection strategies for broiler production. In: Muir, W.M. and Aggrey, S.E. (eds) *Poultry Genetics, Breeding and Biotechnology*. CAB International, Wallingford, UK, pp. 113–126.

Estevez, I., Newberry, R.C. and de Reyna, L.A. (1997) Broiler chickens: a tolerant social system? *Etologia* 5, 9–29.

Freeman, B.M., Manning, A.C.C. and Flack, I.H. (1980) Short-term stressor effects of food withdrawal on the immature fowl. *Comparative Biochemistry and Physiology* 67A, 569–571.

Gross, W.B. and Siegel, H.S. (1983) Evaluation of the heterophil/lymphocyte ratio as a measure of stress in chickens. *Avian Diseases* 27, 972–979.

Heck, A., Onagbesan, O., Tona, K., Metayer, S., Putterflam, J., Jego, Y., Trevidy, J.J., Decuypere, E., Williams, J., Picard, M. and Bruggeman, V. (2004) Effects of *ad libitum* feeding on performance of different strains of broiler breeders. *British Poultry Science* 45, 695–703.

Hocking, P.M. and Bernard, R. (1993) Evaluation of putative appetite suppressants in the domestic fowl. *British Poultry Science* 34, 393–404.

Hocking, P.M. and Jones, E.K.M. (2004) Changes in behaviour and the assessment of welfare in broiler breeders. In: *Proceedings of the XXII World's Poultry Congress*, Turkish Branch of the WPSA, p. 1014.

Hocking, P.M. and Whitehead, C.C. (1990) Relationship between body fatness, ovarian structure and reproduction in mature females from lines of genetically lean or fat broilers given different food allowances. *British Poultry Science* 31, 319–330.

Hocking, P.M., Waddington, D., Walker, M.A. and Gilbert, A.B. (1989) Control of the development of the ovarian follicular hierarchy in broiler breeder pullets by food restriction during rearing. *British Poultry Science* 30, 161–174.

Hocking, P.M., Maxwell, M.H. and Mitchell, M.A. (1993) Welfare assessment of broiler breeder and layer females subjected to food restriction and limited access to water during rearing. *British Poultry Science* 34, 443–458.

Hocking, P.M., Maxwell, M.H. and Mitchell, M.A. (1996) Relationships between the degree of food restriction and welfare indices in broiler breeder females. *British Poultry Science* 37, 263–278.

Hocking, P.M., Maxwell, M.H. and Mitchell, M.A. (1999) Welfare of food restricted male and female turkeys. *British Poultry Science* 40, 19–29.

Hocking, P.M., Maxwell, M.H., Robertson, G.W. and Mitchell, M.A. (2001) Welfare assessment of modified rearing programmes for broiler breeders. *British Poultry Science* 42, 424–432.

Hocking, P.M., Maxwell, M.H., Robertson, G.W. and Mitchell, M.A. (2002) Welfare assessment of broiler breeders that are food restricted after peak of lay. *British Poultry Science* 43, 5–15.

Hocking, P.M., Zaczek, V., Jones, E.K.M. and McLeod, M.G. (2004) Different concentrations and sources of dietary fibre may improve the welfare of female broiler breeders. *British Poultry Science* 45, 9–19.

Hughes, B.O. (1983) Headshaking in fowls – the effect of environmental stimuli. *Applied Animal Ethology* 11, 45–53.

Jaap, R.G. and Muir, F.V. (1968) Erratic oviposition and egg defects in broiler-type pullets. *Poultry Science* 47, 417–423.

Jones, R.B. (1986) Fear and adaptability in poultry: insights, implications and imperatives. *World's Poultry Science Journal* 52, 131–174.

Jones, R.B. (1989) Chronic stressors, tonic immobility and leucocytic responses in the domestic fowl. *Physiology & Behavior* 46, 439–442.

Jones, R.B. (1996) Fear and adaptability in poultry: insights, implications and imperatives. *World's Poultry Science Journal* 52, 131–174.

Jones, R.B., Facchin, L. and McCorquodale, C. (2002) Social dispersal by domestic chicks in a novel environment: reassuring properties of a familiar odourant. *Animal Behaviour* 63, 659–666.

Katanbaf, M.N., Dunnington, E.A. and Siegel, P.B. (1989) Restricted feeding in early and late feathering chickens. *Poultry Science* 68, 344–351.

King, L.A. (2001) Environmental enrichment and aggression in commercial broiler breeder production. In: Garner, J.P., Mench, J.A. and Heekin, S.P. (eds) *Proceedings of the 35th Conference of the ISAE*, Centre of Animal Welfare of UC Davis, Davis, California, p. 174.

Kjaer, J.B. and Mench, J. (2003) Behaviour problems associated with selection for increased production. In: Muir, W.M. and Aggrey, S.E. (eds) *Poultry Genetics, Breeding and Biotechnology*. CAB International, Wallingford, UK, pp. 67–82.

Kostal, L. and Savory, C.J. (1994) Influence of pharmacological manipulation of dopamine and opioid receptor subtypes on stereotyped behaviour of restricted fed fowls. *Pharmacology, Biochemistry and Behavior* 48, 241–252.

Kostal, L., Savory, C.J. and Hughes, B.O. (1992) Diurnal and individual variation in behaviour of restricted-fed broiler breeders. *Applied Animal Behaviour Science* 32, 361–374.

Kostal, L., Vyboh, P., Savory, C.J., Jurani, M., Kubikova, L. and Blazicek, P. (1999) Influence of food restriction on dopamine receptor densities, catecholamine concentrations and dopamine turnover in chicken brain. *Neuroscience* 94, 323–328.

Lacy, M.P., Van Krey, H.P., Skewes, P.A. and Denbow, D.M. (1985) Effect of intrahepatic glucose infusions on feeding in heavy and light breed chicks. *Poultry Science* 64, 751–756.

Lawrence, A.B. and Terlouw, E.M.C. (1993) A review of behavioral factors involved in the development and continued performance of stereotypic behaviors in pigs. *Journal of Animal Science* 71, 2815–2825.

Maxwell, M.H. (1993) Avian blood leucocyte responses to stress. *World's Poultry Science Journal* 49, 34–43.

Maxwell, M.H. and Robertson, G.W. (1998) The avian heterophil leucocyte: a review. *World's Poultry Science Journal* 54, 155–178.

Mench, J.A. (1988) The development of aggressive behaviour in male broiler chicks: a comparison with laying-type males and the effects of feed restriction. *Applied Animal Behaviour Science* 21, 233–242.

Mench, J.A. (2002) Broiler breeders: feed restriction and welfare. *World's Poultry Science Journal* 58, 23–30.

Millman, S.T. and Duncan, I.J.H. (2000) Effect of male-to-male aggressiveness and feed-restriction during rearing on sexual behaviour and aggressiveness towards females by male domestic fowl. *Applied Animal Behaviour Science* 70, 63–82.

Millman, S.T., Duncan, I.J.H. and Widowski, T.M. (2002) Male broiler breeder fowl display high levels of aggression toward females. *Poultry Science* 79, 1233–1241.

Mitchell, M.A. (1999) Muscle abnormalities: pathophysiological mechanisms. In: Richardson, M.I. and Mead, G.C. (eds) *Poultry Meat Science*. CAB International, Wallingford, UK, pp. 65–98.

Nielsen, B.L. (2004) Behavioural aspects of feeding constraints: do broilers follow their gut feelings? *Applied Animal Behaviour Science* 86, 251–260.

Nielsen, B.L., Litherland, M. and Noddegaard, F. (2003) Effects of qualitative and quantitative feed restriction on the activity of broiler chickens. *Applied Animal Behaviour Science* 83, 309–323.

Nir, I., Nitsan, Z., Dror, Y. and Shapira, N. (1978) Influence of overfeeding on growth, obesity and intestinal tract in young chicks of light and heavy breeds. *British Journal of Nutrition* 39, 27–35.

Nir, I., Nitsan, Z., Dunnington, E.A. and Siegel, P.B. (1996) Aspects of food intake restriction in young domestic fowl: metabolic and genetic considerations. *World's Poultry Science Journal* 52, 251–266.

Picard, M., Bruggeman, V., Jego, Y., Onagbesan, O. and Hocking, P. (2004) In: *Proceedings of the XXII World's Poultry Congress*, Turkish Branch of the WPSA, p. 1013.

Pinchasov, Y., Galili, D., Yonash, N. and Klandorf, H. (1993) Effect of feed restriction using self-restricting diets on subsequent performance of broiler breeder females. *British Poultry Science* 72, 613–619.

Rauw, W.M., Kanis, E., Noordhuizen-Stassen, E.N. and Grommers, F.J. (1998) Undesirable side effects of selection for high production efficiency in farm animals: a review. *Livestock Production Sciences* 56, 15–33.

Renema, F.R. and Robinson, F. (2004) Defining normal: comparison of feed restriction and full feeding of female broiler breeders. *World's Poultry Science Journal* 60, 508–522.

Rozenboim, I., Kapkowska, E., Robinzon, B. and Uni, Z. (1999) Effects of fenfluramine on body weight, feed intake and reproductive activities of broiler breeder hens. *Poultry Science* 78, 1768–1722.

Ruszler, P.L. (1998) Health and husbandry considerations of induced molting. *Poultry Science* 77, 1789–1793.

Sandercock, D.A., Hunter, R.R., Mitchell, M.A. and Hocking, P.M. (2001) The effect of genetic selection for muscle yield on idiopathic myopathy in poultry: implications for welfare. In: Oester, H. and Wyss, C. (eds) *Proceedings of the 6th European Symposium on Poultry Welfare*, Swiss Branch of the WPSA, Zollikofen, Switzerland, pp. 118–123.

Sandilands, V., Tolkamp, B.J., Savory, C.J. and Kyriazakis, I. (2004) Alternative methods for limiting broiler breeder growth rate during rearing. In: Hanninen, L. and Valros, A. (eds) *Proceedings of the 38th Congress of the ISAE*, ISAE, Helsinki, Finland, p. 49.

Savory, C.J. (1995) Broiler welfare: problems and prospects. *Archives für Geflügelkunde Sonderheft* 1, 48–52.

Savory, C.J. and Kostal, L. (1996) Temporal patterning of oral stereotypies in restricted-fed fowls: 1. Investigations with a single daily meal. *International Journal of Comparative Psychology* 9, 117–139.

Savory, C.J. and Lariviere, J.-M. (2000) Effects of qualitative and quantitative food restriction treatments on feeding motivational state and general activity level of growing broiler breeders. *Applied Animal Behaviour Science* 69, 135–147.

Savory, C.J. and Mann, J.S. (1997) Is there a role for corticosterone in expression of abnormal behaviour in restricted-fed fowls? *Physiology & Behavior* 62, 7–13.

Savory, C.J. and Mann, J.S. (1999) Stereotyped pecking after feeding by restricted-fed fowls is influenced by meal size. *Applied Animal Behaviour Science* 62, 209–217.

Savory, C.J. and Maros, K. (1993) Influence of degree of food restriction, age and time of day on behaviour of broiler breeder chickens. *Behavioural Processes* 29, 179–190.

Savory, C.J., Wood-Gush, D.G.M. and Duncan, I.J.H. (1978) Feeding behaviour in a population of domestic fowls in the wild. *Applied Animal Ethology* 4, 13–27.

Savory, C.J., Seawright, E. and Watson, A. (1992) Stereotyped behaviour in broiler breeders in relation to husbandry and opiod receptor blockade. *Applied Animal Behaviour Science* 32, 349–360.

Savory, C.J., Carlisle, A., Maxwell, M.H., Mitchell, M.A. and Robertson, G.W. (1993a) Stress, arousal and opioid peptide-like immunoreactivity in restricted- and *ad lib*-fed broiler breeder fowls. *Comparative Biochemistry and Physiology* 106A, 587–594.

Savory, C.J., Maros, K. and Rutter, S.M. (1993b) Assessment of hunger in growing broiler breeders in relation to a commercial restricted feeding programme. *Animal Welfare* 2, 131–152.

Savory, C.J., Hocking, P.M., Mann, J.S. and Maxwell, M.H. (1996) Is broiler breeder welfare improved by using qualitative rather than quantitative food restriction to limit growth rate? *Animal Welfare* 5, 105–127.

Scanes, C.G., Merrill, G.F., Ford, R., Mauser, P. and Horowitz, C. (1980) Effects of stress (hypoglycaemia, endotoxin, and ether) on the peripheral circulating concentration of corticosterone in the domestic fowl (*Gallus domesticus*). *Comparative Biochemistry and Physiology* 66C, 183–186.

Scheele, C.W. (1997) Pathological changes in metabolism of poultry related to increasing production levels. *Veterinary Quarterly* 19, 127–130.

Schwartz, M.W. (1997) Regulation of appetite and body weight. *Hospital Practice* 32(7), 109–119.

Schwartz, M.W., Woods, S.C., Porte, D., Seeley, R.J. and Baskin, D.G. (2000) Central nervous system control of food intake. *Nature* 404, 661–671.

Shea, M.M., Mench, J.A. and Thomas, O.P. (1990) The effect of dietary tryptophan on aggressive-behavior in developing and mature broiler breeder males. *Poultry Science* 69, 1664–1669.

Siegel, P.B. (1999) Body weight, obesity and reproduction in meat type chickens. In: *International Congress on Bird Reproduction*. Tours, pp. 209–213.

Van Middelkoop, J.H., Van der Haar, J.W., Robinson, F.E., Luzi, C.A. and Zuidhof, M.J. (2000) The application of scatter feeding in managing feed restriction during the rearing of broiler breeders. In: *Proceedings of the 21st World's Poultry Congress*, Montréal, Canada.

Vermaut, S., DeConinck, K., Flo, G., Cokelaere, M., Onagbesan, M. and Decuypere, E. (1997) Effect of deoiled jojoba meal on feed intake in chickens: satiating or taste effect? *Journal of Agricultural and Food Chemistry* 45, 3158–3163.

Vermaut, S., Onagbesan, O., Bruggeman, V., Verhoeven, G., Berghman, L., Flo, G., Cokelaere, M. and Decuypere, E. (1998) Unidentified factors in jojoba meal prevent oviduct development in broiler breeder females. *Journal of Agricultural and Food Chemistry* 46, 194–201.

Vermaut, S., De Coninck, K., Bruggeman, V., Onagbesan, O., Flo, G., Cokelaere, M. and Decuypere, E. (1999) Evaluation of jojoba meal as a potential supplement in the diet of broiler breeder females during laying. *British Poultry Science* 40, 284–291.

Webster, A.B. (2003) Physiology and behavior of the hen during induced molt. *Poultry Science* 82, 992–1002.

Whitehead, C.M., Fleming, R.H., Julian, R.J. and Sorensen, P. (2003) Skeletal problems associated with selection for increased production. In: Muir, W.M. and Aggrey, S.E. (eds) *Poultry Genetics, Breeding and Biotechnology*. CAB International, Wallingford, UK, pp. 29–52.

Whittaker, X., Spoolder, H.A.M., Edwards, S.A., Lawrence, A.B. and Corning, S. (1998) The influence of dietary fibre and the provision of straw on the development of stereotypic behaviour in food restricted pregnant sows. *Applied Animal Behaviour Science* 61, 89–102.

Whittaker, X., Edwards, S.A., Spoolder, H.A.M., Lawrence, A.B. and Corning, S. (1999) Effects of straw bedding and high fibre diets on the behaviour of floor fed group-housed sows. *Applied Animal Behaviour Science* 63, 25–39.

Williams, B., Solomon, S., Waddington, D., Thorp, B. and Farquharson, C. (2000) Skeletal development in meat-type chickens. *British Poultry Science* 41, 141–149.

Zuidhof, M.J., Robinson, F.E., Feddes, J.J.R., Hardin, R.T., Wilson, J.L., Mckay, R.I. and Newcombe, M. (1995) The effects of nutrient dilution on the well-being and performance of female broiler breeders. *Poultry Science* 74, 441–456.

Zulkifli, I., Dunnington, E.A., Gross, W.B., Larsen, A.S., Martin, A. and Siegel, P.B. (1993) Responses of dwarf and normal chickens to feed restriction, *Eimeria-Tenella* infection, and sheep red-blood-cell antigen. *Poultry Science* 72, 1630–1640.

Zulkifli, I., Siegel, H.S., Mashaly, M.M., Dunnington, E.A. and Siegel, P.B. (1995) Inhibition of adrenal steroidogenesis, neonatal feed restriction, and pituitary–adrenal axis response to subsequent fasting in chickens. *General and Comparative Endocrinology* 97, 49–56.

9   Feeding Ostriches

T.S. Brand[1] and R.M. Gous[2]

[1]*Animal Production Institute: Elsenburg, Department of Agriculture: Western Cape, Private Bag X1, Elsenburg 7607, South Africa;* [2]*Animal and Poultry Science, University of KwaZulu-Natal, Pietermaritzburg 3200, South Africa*

Introduction

The ostrich (*Struthio camelus*) originated in Africa and the Arabian Desert (Smith, 1964). The domestication and farming of ostriches started in South Africa around 1866 (Mosenthal and Harting, 1897), from where it expanded to the rest of the world. Today the number of breeding hens used in commercial farming amounts to about 90,000, these being found in South Africa (67%), Western and Eastern Europe (13%), Middle and Far East (8%), Australasia (7%) and North and South America (5%) (South African Ostrich Business Chamber, 2002). The current efficiency rate of world ostrich production amounts to 5–6 birds slaughtered per breeding hen per year (South African Ostrich Business Chamber, 2002). Ostriches were originally domesticated for the harvesting of their feathers, but nowadays ostriches are produced to provide low-cholesterol meat (~50% of the current commercial income), durable high-quality hides (~45% of the current commercial income) and plumes for the fashion market (~5% of the current commercial income) (South African Ostrich Business Chamber, 2002).

Anatomical Structures

Ostriches are monogastric herbivores with a relatively large digestive tract, which provides an ideal environment for fermentation of fibrous plant material. Like other birds, their digestive tract consists of a beak and mouth, oesophagus, proventriculus (glandular stomach), gizzard (ventriculus/muscular stomach), small intestine (subdivided into the duodenum, jejunum and ileum), large intestine (consisting of two caeca and the proximal, middle and distal colon) and a cloaca (see Fig. 9.1).

The beak of the ostrich is relatively large (without any teeth) and is used to tear off plant material and to pick up food. The ostrich, like other ratites, has no crop; instead, the upper part of the oesophagus is slightly enlarged for the accumulation of food, while the proventriculus acts as the high volume storage organ for ingested food (Holtzhauzen and Kotzé, 1990). As in all birds, hydrochloric acid and enzymes are secreted by the proventriculus to start the digestive process. The gizzard is muscular with thick walls and a horny interior epithelium. Ingested plant material or feed is mechanically ground to a finer form in the gizzard with the help of ingested stones and pebbles and the contractions of the wall of the gizzard (Holtzhauzen and Kotzé, 1990), while digestion is also promoted by the strong acid milieu (pH ~2.2) in the gizzard. The small intestine of the ostrich is relatively long (Table 9.1) and can be divided into the duodenum, jejunum and ileum. Partially digested food is digested here in an alkaline environment by various digestive enzymes like

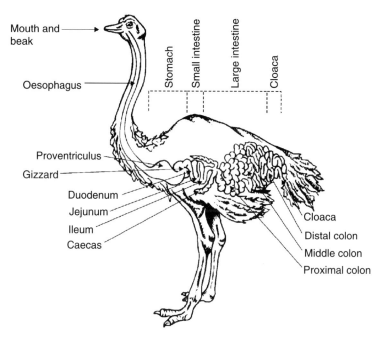

Fig. 9.1. Graphic illustration of the digestive system of the ostrich.

amylase, trypsin, chymotrypsin, lipase, maltase, sucrase, alkaline phosphatase and arginase (Iji *et al.*, 2003) and the absorption of digested nutrients starts in the small intestine.

The large intestine comprises the two caeca and the colon or rectum. Digested food is exposed to microbial fermentation in an alkaline environment in the colon, which enables the bird to digest fibrous material. The large intestine (colon), as a proportion of the

Table 9.1. The relative length of the digestive tract of a mature ostrich (Skadhauge *et al.*, 1984).

Component	Length (cm)	Percentage of total
Oesophagus	110	4.6
Proventriculus and gizzard	35 (30–40)	1.5
Duodenum	150	6.3
Jejunum and ileum	700 (600–800)	29.2
Caecum (total of both branches)	200	8.4
Colon	1200 (1100–1300)	50.0
Total	2395	100

digestive tract, also increases with age, the ratio of colon to small intestine increasing from about 1:1 at hatching, through about 1.5:1 at 3 months of age and reaching the adult ratio of about 2:1 at 6 months of age (Bezuidenhout, 1986). The development of the ostrich gastrointestinal tract switches dramatically from the typical bird neonate (i.e. large postnatal yolk sac through that resembling the typical monogastric animal to that of a hindgut fermenter), all in the space of 70–80 days. The dependence on the yolk sac ends shortly after 7 days, after which the development of the proventriculus, gizzard and small intestine stabilizes after 27 days post hatching. The switch to the hindgut starts to accelerate by day 55 (Van der Walt *et al.*, 2003).

Digestion

The ability of the ostrich to digest fibre distinguishes it from other monogastric herbivores. Ostriches consistently utilize 25% more energy than pigs on the same feeds (Brand *et al.*, 2000a). Ostriches also exhibit considerably higher (30%) metabolizable energy values for

Table 9.2. The metabolizable energy content of feeds determined using ostriches, pigs, poultry and ruminants (*in vitro* values for ruminants) (Brand *et al.*, 2000a).

Diet	Ostrich (TME,[a] MJ/kg)	Pig (ME,[b] MJ/kg)	Poultry (TME,[b] MJ/kg)	Ruminant (IVOMD,[c] %)
High-energy/low-fibre				
Starter diet	14.98	12.84	–	87.90
Grower diet	14.76	12.26	14.70	83.87
Finisher diet	14.88	13.02	14.10	81.78
Medium-energy/medium-fibre				
Starter diet	14.08	10.36	12.30	75.68
Grower diet	13.99	10.39	9.70	67.63
Finisher diet	13.94	11.08	12.20	72.88
Low-energy/high-fibre				
Starter diet	11.96	9.79	8.70	66.61
Grower diet	12.80	8.82	8.20	63.89
Finisher diet	12.44	10.45	9.00	68.36

[a]True metabolizable energy.
[b]Metabolizable energy.
[c]*In vitro* organic matter digestibility.

high-fibre feeds compared with poultry (Table 9.2), confirming the results of Swart (1988), who demonstrated that plant fibres, specifically hemi-celluloses and cellulose, are efficiently digested by ostriches (66% and 38%, respectively).

Table 9.3 shows the metabolizable energy values for both poultry and ostriches of ingredients generally used in ostrich feeds.

Data collected by Cilliers (1994, 1998) revealed a significant linear relationship between the true metabolizable energy (TME) values of individual ingredients determined with ostriches and with poultry. He suggested that energy values for raw materials to be used in ostrich feed formulations could be predicted from values obtained with poultry using the following equation (Cilliers, 1998):

TME ostrich = $6.35 + 0.645$ TME poultry (R^2 = 0.80).

Linear regression analysis of the TME values determined with the different animal species using these mixed diets revealed useful equations to predict ostrich TME values for individual ingredients as well as mixed feeds from the values respectively obtained with poultry, pigs or *in vitro* ruminant values (Brand *et al.*, 2003b):

TME ostrich (MJ/kg DM) = $6.743 + 0.638$ ME pig ($P \leq 0.01$; R^2 = 67.7%).

Table 9.3. Metabolizable energy values (MJ per kg feed on a 90% DM basis) of feedstuffs as determined with ostriches and roosters.

Ingredients	Ostriches	Roosters
Concentrates		
Maize[a]	15.22	14.42
Triticale[a]	13.21	11.82
Oats[a]	12.27	10.63
Barley[a]	13.92	11.33
Maize and cob meal[d]	13.45	–
Molasses meal[d]	7.77	–
Protein sources		
Fish meal[a]	15.13	13.95
Ostrich meat and bone meal[a]	12.81	8.34
Soybean oilcake meal[a]	13.44	9.00
Sunflower oilcake meal[a]	10.79	8.89
Lupins (*Lupinus albus*)[a]	14.61	9.40
Canola oilcake meal[b]	12.38	7.80[c]
Full-fat canola seed[b]	20.25	16.65[c]
High-fibre sources		
Lucerne hay[a]	8.91	4.00
Lucerne hay[b]	9.39	4.48[c]
Common reed (*Phragmites australis*)[a]	8.67	2.79
Saltbush hay (*Artiplex nummular*)[a]	7.09	4.50
Wheat bran[a]	11.91	8.55
Grape residue[d]	7.81	–
Hominy chop[d]	11.49	–
Alivera (*Agave americana*)[d]	12.20	–

[a]Cilliers (1994); [b]Brand *et al.* (2000b); [c]Wiseman (1987); [d]Cilliers (1998).

TME ostrich (MJ/kg DM) = 9.936 + 0.326 ME poultry ($P \leq 0.01$; R^2 = 79.6%).
TME ostrich (MJ/kg DM) = 5.500 + 0.1111 IVOMD ruminant ($P \leq 0.01$; R^2 = 73.7%).

Because the ostrich is capable of extracting energy from fibre, the accuracy of equations used for predicting the TME values of ingredients for ostriches from TME values obtained with poultry is improved by including a term for the dietary crude fibre (CF), neutral detergent fibre (NDF) or acid detergent fibre (ADF) content of the ingredient or feed. The appropriate coefficients are given in Table 9.4.

Ostrich chicks under the age of 10 weeks do not digest fibre well (Cilliers and Angel, 1999; Nizza and Di Meo, 2000), although Angel (1993) found that ostriches were able to digest fibre at 10–17 weeks of age (51.2% and 58.0% NDF digestion) as efficiently as mature birds (61.6% NDF digestion). The lower digestive tract (colon plus caecum) of poultry, pigs (Getty, 1975) and ostriches (Bezuidenhout and van Aswegen, 1990) contributes about 11%, 21% and 61% of the total tract of the different species. The retention time of feed, or the rate of passage of digesta, which determines the time for digestive enzymes and microbes to digest material, also differs between these species, being approximately 10 h for chickens, 30 h for pigs and 40 h for ostriches (Ensminger and Olentine, 1978; Swart, 1988).

The ability of the ostrich to utilize lower-quality raw materials was clearly illustrated by the pioneering work of Swart (1988), who established that ostriches could probably obtain between 12% and 76% of their energy in the form of volatile fatty acids (the end product of the digestion of fibre-rich feed in the large intestine). Pigs are also able to digest fibre fractions (hemicellulose) to a certain extent, but in this case energy supply in the form of volatile fatty acids from the lower digestive tract may contribute between 10% and 30% to their total energy requirements (Eggum et al., 1982). The lower digestive tract of chickens can supply only 8% of the required energy in the form of short-chain fatty acids (Jozefiak et al., 2004).

Baltmanis et al. (1997) reported that the type of feed offered to ostriches affected the relative size of the lower digestive tract as well as the ability to obtain more nutrients from the diet. Fibre-rich feeds increased the capacity of the colon (Salih et al., 1998; Salih, 2000), and it has been shown that the number and type of gut bacteria present in ostriches was largely affected by diet composition (Law-Brown et al., 2004). Viljoen et al. (2004) similarly reported that the length of the large intestine, small intestine and caeca decreased with an increase in dietary energy concentration. The study by Salih et al. (1998) revealed that the number of cellulolytic bacteria found in the large intestine of ostriches was only two- to onefold (10^2–10^1) lower than the amount found in the intestine of ruminants.

A toxic effect of short-chain fatty acids on some enteric bacteria in the large intestine of avian species has been found, while these organic acids do not inhibit beneficial bacteria such as lactobacillus. The high fermentation action in the large intestine resulting from the use of fibrous feed may also result in a lower pH, which may in turn inhibit some pathogenic bacteria (McHan and Shotts, 1993; Van der Wielen et al., 2000 as cited by Jozefiak et al., 2004).

Table 9.4. Regression coefficients relating the ME obtained with poultry, and the fibre content of feed, to the ME determined with ostriches.

	Using CF		Using NDF		Using ADF	
	b	SE	b	SE	b	SE
Poultry ME	0.8844	0.02430	0.8507	0.0363	0.8924	0.0231
CF	0.2165	0.0166				
NDF			0.1251	0.0134		
ADF					0.1663	0.0124
R^2	91.8%		84.6%		90.6%	

Fat digestibility seems to be low in young birds (44.1% at 3 weeks of age), but increases rapidly (74.3%, 85.7%, 91.1% and 92.9% at 6, 10, 17 weeks and 30 months of age) (Angel, 1993).

Cilliers (1994) compared faecal digestibility values between ostriches and poultry, and found that ostriches had marginally higher faecal protein (64.6% vs. 60.9%) and amino acid (83.7% vs. 79.5%) digestibility values than poultry. Retention of dietary protein was also marginally higher in ostriches (64.6% vs. 60.9% for ostriches and poultry, respectively).

However, faecal digestibility values for protein and amino acids may not reveal the true utilization potential of these nutrients, due to the metabolism and synthesis of amino acids in the large intestine of the monogastric animal (Just et al., 1981). This may be especially true in the case of ostriches with their relatively large colon and caecum. Protein digestibility values obtained with normal and ileo-rectum anastomosed ostriches were compared to determine ileal protein digestibilities derived at the end of the ileum (Tables 9.5 and 9.6).

Grazing Behaviour

Ostriches are well adapted to savannah, semi-desert and desert regions (Van Niekerk, 1995). They normally avoid areas with high grass or bushveld that may obstruct their visual field, and normally graze in the wild in large herds of up to 100 individuals (Sauer and Sauer, 1996). Ostriches graze during daytime and sit down and remain inactive throughout the night.

Adult ostriches are totally herbivorous (Kok, 1980; Williams et al., 1993), although young ostriches may consume insects (Milton et al., 1994). Young birds lack the gizzard

Table 9.5. Ileal and faecal DM and protein digestibility values for ostriches (Brand et al., 2005).

Parameter measured	Digestibility (%)	
	Faecal	Ileal
Dry matter	71.6 ± 3.1	64.9 ± 2.9
Crude protein	78.6 ± 5.1	64.7 ± 4.9

Table 9.6. The effect of type of feed on DM and protein digestibility (Brand et al., 2005).

Type of feed	DM digestibility	Protein digestibility (%)
Pre-starter	77.1 ± 4.3	81.0 ± 7.0
Starter	77.4 ± 4.2	75.5 ± 7.0
Grower	73.4 ± 3.9	68.4 ± 6.4
Finisher	59.7 ± 6.0	68.8 ± 9.9
Maintenance	54.1 ± 5.2	64.6 ± 8.6

stones that are able to grind fibrous feed to a paste, as well as the bacteria necessary to digest cellulose and hemicellulose (Milton et al., 1994). The ingestion of the faeces of adult birds (Smith, 1964), as well as stones and pebbles, prepares their digestive system for a high-fibre-containing plant diet (Milton et al., 1994).

The newly hatched chickens are dependent mainly on their yolk sac, which disappears within the first 14 days after hatching (Mushi et al., 2004). It will therefore not start to feed from external sources for the first 24–72 h after hatching. In the wild, chickens will initially feed on the fresh dung of their parents (Osterhoff, 1979), which acts as a source of microorganisms and aids in the digestion of food, especially in the lower digestive tract. It is believed that young birds may be attracted to certain colours like green and white (Bubier et al., 1996). Ostrich chickens are strong imitators of the behaviour of their parents and other chickens, and will follow each other in pecking at the same objects (Kubirskie, 1985), thereby learning from their parents and each other their selection of food. Placing chicks that are already eating with newly hatched chicks may therefore help the latter to start eating. Brightly coloured food ingredients of different sizes, about 3 mm in diameter, may stimulate eating by the young bird (Erasmus and Erasmus, 1993). With juveniles in captivity it has been found that food in bowls is largely ignored, whereas food scattered on the floor is readily taken (Bubier et al., 1996).

Ostriches are very selective, and graze by stripping shrubs or pastures from their leaves by running the branches through their beaks (Van Niekerk, 1995). They will never consume dead or woody material in a natural environment (Milton et al., 1994). Birds of all ages will also continuously consume stones and pebbles, which are normally between 50% and 75% of the size of the large toenail of the bird at the respective age. These pebbles and stones are never excreted, but gradually wear out. Bones and shell are commonly ingested, probably as a source of calcium (Smith, 1964; Kok, 1980). Ostriches generally graze with their head 30 cm above the ground, which enables them to select appropriate plant material. Their round sharp bill, good eyesight and long neck enable them to scan the ground at close range, to see and selectively eat even the smallest green shoots, which they often pluck from the ground, roots and all (Williams et al., 1993). During grazing, the food will accumulate in the gullet, the first part of which is like a pouch, until the bird lifts its head to swallow (Osterhoff, 1979). A bolus of food normally contains the results of 40–100 pecks (Bertram, 1980). Birds normally feed on leaves and short grass blades. Ingestion of grass blades and stems longer than 15 cm may result in compaction in the proventriculus, intestinal blockage and sudden death of the bird (Kok, 1980). This is especially a problem with young growing birds. Older birds have a larger digestive tract and they avoid the excessive intake of plant parts that may cause compaction.

The diet of juvenile ostriches in *Acaci-Themeda* savanna consists of selected forbs and new grass, with some seeds and fruit when available (Cooper and Palmer, 1994). Milton *et al.* (1994) studied adult ostriches in a natural environment in Southern African savannah, desert grassland, arid scrubland and Mediterranean fynbos. They found that the stomach (proventriculus and gizzard) of the adult ostrich contains between 4.5 and 5.5 kg of fresh material. Stone mass in the gizzard amounts to about 0.80% of the total body mass. They observed that ostriches in their natural environment consumed only green non-woody plant material (see Fig. 9.2) containing 70% water and 24% CF, 12% crude protein (CP), 16% ash and 3% lipids on a dry matter (DM) basis.

Williams *et al.* (1993) reported that free-ranging Namibian desert ostriches selected plant material containing 35.2% CF, 11.2% CP and 4.2% lipids. It was clear from the studies both by Milton *et al.* (1994) and Williams *et al.* (1993) that free-ranging ostriches select a diet with less than 12% CP and an excess of 24% CF. It was observed that when ostriches graze on very short pasture, they will repeatedly peck at plants until they are uprooted, in which case sand may comprise 31–46% of the stomach contents of these birds (Robinson and Seely, 1975). It is also known that birds in captivity will ingest sticks and sharp woody material, which may lead to atrophy and eventually death. This divergent behaviour may probably be related to boredom and/or stress. Although not observed by other authors, Farrell *et al.*

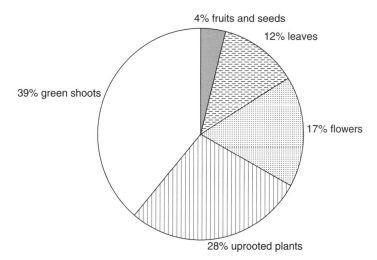

Fig. 9.2. Diet selection of adult ostriches in a natural environment. (From Milton *et al.*, 1994.)

(2000) observed that ostriches practised coprophagy on all feeds used in their studies. They were uncertain as to whether ostriches in the wild recycled their excreta, or whether their observations were a behavioural response to confinement.

Several behavioural studies were performed on ostriches. Bubier et al. (1996) found that juveniles spend over 50% of their time foraging and walking. They spend 27.7% of their time pecking at items on the floor, including eating pellets spread over the floor, and 23.1% of their time walking around. Chicks frequently moved under the brooder lamp (11.2% of their time budget), while 10.2% of their time was consumed in pecking at objects other than food (e.g. food trays, water trays or walls). Only 3.5% of their time was used for feeding from the food bowl (Deeming et al., 1996).

Both Degen et al. (1989) and Brand et al. (2003a) studied the activity budget of slaughter ostriches under feedlot conditions (Fig. 9.3). They found that birds under these captivity conditions spend approximately 45 min/day feeding, 57 min/day foraging (or pecking ground) and 8 min/day to drink water.

In relation to this, Bertram (1980) and Williams et al. (1993) observed that ostriches in their natural environment spent between 65% and 85% of their daytime walking and pecking (~540 min/day). Sambraus (1994) as cited by Mohamed (2001) pointed out that ostriches in their natural habitat spent about 33% of their time in search for food compared to ostriches in a feedlot which spent only about 18% of their time feeding or drinking.

It is clear from these studies that ostriches in their natural habitat spend much more time feeding and foraging compared to their counterparts kept in captivity in feedlots. This may probably be related to the divergent behaviour of feather-pecking and/or ingestion of sticks. In their study, Degen et al. (1989) observed that ostriches pecked at their feed about 2800 times per day and at the ground about 1950 times per day. A lot of the time was spent pecking and studying an object before dropping it again.

Diet Selection

Birds merely rely on their vision in the selection of food (El Boushy et al., 1989; Cooper and Palmer, 1994), although it is believed that even birds are able to distinguish between the five primary taste activities (sweet, bitter, sour, salty and umami) (Roura, 2003). Birds have a small number of taste buds compared to mammals: 8 in a day-old chick and 24 in a 3-month-old cockerel (Lindenmaier and Kare, 1959); and about 300 in chickens compared to about 20,000 in pigs and 7000 in humans (Roura, 2003). Initial indications suggest that there are no taste buds in the oral cavity of the ostrich (J.P. Soley and H.B. Groenewald, Department of Anatomy and Physiology, Onderstepoort Campus, South Africa, personal communication, 2005). Milton et al. (1994) observed that ostriches seldom fed on plant species containing high concentrations of ether extract, phenolics, tannins or Na and Ca oxalate. They postulated that ostriches did

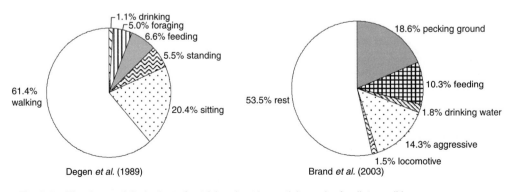

Fig. 9.3.   The time-activity budget of ostriches kept in captivity under feedlot conditions.

not feed on plants that are toxic to mammalian herbivores. It has been suggested that the refusal or acceptance of newly introduced feed by ostriches is determined by the colour and surface texture of the feed, which appear to take precedence over all other qualities (Van Niekerk, 1995; Brand et al., 2004a). Smell and/or odour probably also play a prominent role in the selection of food. In practice, it is well known, especially with young growing and/or finishing birds in feedlots, that birds will refuse to consume feeds that are introduced suddenly, i.e. without exposing the birds to the new feed over a period of time. This phenomenon generally occurs when feeds with a green colour are changed to those of a lesser green colour. A study by Bubier et al. (1996) revealed that juvenile ostriches showed a preference for green-coloured strips (15.5 pecks/min) followed by white-coloured strips (1.5 peck/min), while they practically ignored red, blue, yellow and black strips (all less than 0.5 pecks/min). A recent study by Brand et al. (2005) revealed no preference for green coloured feed by young ostrich chickens. It is also known that, in practice, birds will refuse to consume mouldy feed. Young birds are especially sensitive to micotoxins (Scheideler, 1997). El Boushy and Kennedy (1987) stated that apart from smell and taste, the general factors affecting the acceptability of feed for poultry include memory, physical conditions, alternativeness and physiological effects, which is believed will also generally affect the acceptability of feed for ostriches.

It is believed that commercial farm animals may, under certain circumstances, make a rational choice between feeds according to their nutritional requirements (Ferguson et al., 2002). In an experiment reported by Brand et al. (2003b), male and female ostrich breeders were able to select between a high- and low-quality feed, their feeding patterns being recorded by means of an electronic ostrich-feeding system (Brand and Steyn, 2002). However, no significant difference was found between male (~46% of their time at the high-quality feed and 54% of their time at the low-quality feed) and female breeders (~53% of their time at the high-quality feed and ~47% of their time at the low-quality feed). This result differs from that in which slaughter birds were offered a choice between four feeds differing in both protein and energy (Brand et al., 2005) (Table 9.7). In this case the birds preferred the high-energy feeds throughout the trial period (hatching to 9 months). However, the preference for feeds that are high in dietary protein throughout the three periods of growth is contrary to previous observations with pigs, that the proportion of protein selected is reduced as the animals age, reflecting a change in the requirement for protein as a proportion of the feed (Bradford and Gous, 1991).

It was evident in this experiment that the coarser structure of the high-grain-containing feeds might have had an effect on the preference of the birds (feeds were provided in meal form). In a series of free-choice studies performed by Farrell et al. (2000) ostriches selected a high-energy feed in the ratio of 1.5:1.0 compared to the low-energy feed. When offered a choice of high-protein–high-energy vs. high-protein–low-energy feeds, birds consumed the high-protein feeds in largest quantities, similarly to the results shown in Table 9.7.

Table 9.7. Preference of slaughter birds when provided with feeds differing in energy and protein in a free choice situation (Brand et al., 2005).

Dietary energy (MJ ME)	Dietary protein (%)	Age group		
		Hatching to 2 months of age (%)	2–6 months (%)	6–9 months (%)
13.5	18	50.5	54.0	60.0
13.5	12	30.0	30.0	31.0
8.5	18	13.0	9.0	4.0
8.5	12	6.5	7.0	5.0

Drinking Behaviour

In free-ranging conditions, when ostriches can find sufficient juicy plants, they seldom need to drink water. In spite of this, in the wild they are normally found within a radius of 24 km from water (Berry and Louw, 1982). Degen *et al.* (1989) found the water intake of 5–6-month-old birds (~57 kg) to be about 10 l/day. In a subsequent study by Degen *et al.* (1991) the ratio of water consumption to dry matter intake (DMI) remained relatively constant, at approximately 2.3 l water/kg DMI. Withers (1983) found that the DMI of ostriches was reduced by 84% when water was withheld from birds fed dry feed in captivity. According to Withers (1983), mature birds (95 kg) on dry lucerne hay will drink 7.9 l water/day whilst obtaining 0.4 l/day from the hay. Water loss is mainly through the faeces (2.9 l/day), urine (2.5 l/day) and evaporation (1.4 l/day).

Feeding Strategies

Ostriches are raised extensively (birds are totally dependent on natural veld and/or cultivated pasture), semi-intensively (birds graze on veld or cultivated pastures, and receive a feed concentrate as a supplement) or intensively (a fully balanced feed is provided).

In nature, ostriches may range up to 18.5 km/day, while their grazing field may cover up to 84.3 km^2 (Williams *et al.*, 1993). In natural Karoo veld (semi-desert) of good quality, normal stocking density may vary between 3 and 5 ha per bird (Smith, 1934; Mosenthal and Harting, 1879), although the average carrying capacity of Karoo veld is about 10–12 ha per bird (Osterhoff, 1979). However, little research has been conducted on the utilization of natural veld as a feed source, although it is well known that grazing ostriches may destroy veld when the stocking density is too high. For this reason, ostriches in South Africa may not be raised for commercial slaughtering purposes on natural veld (J. Nel, Oudtshoorn Agricultural Development Centre, Oudtshoorn, South Africa, personal communication, 2005).

The most common cultivated pasture used for grazing ostriches is lucerne. The stocking density on lucerne pasture is approxi-

mately 6.5–12 slaughter birds per hectare, depending on the quality of the pasture as well as the age of the birds (Smith, 1934; Osterhoff, 1979; Nel, 1993). Lucerne is most commonly used as grazing for young birds from about 2 weeks to 3–4 months of age. In succession to this period, they are normally moved to an intensive feeding system (de Kock, 1995). It is, however, advisable to provide the birds with supplementary feed when cultivated pastures are used, to prevent any nutrient deficiencies that may otherwise occur. Because grazing ostriches may be destructive to the pasture, either due to trampling or as a result of their feeding behaviour (stripping the leaves from the stem of the plant), a system of zero grazing is often recommended, in which the lucerne is mechanically harvested and fed to the birds as chopped green feed, in the feedlot as the hay component of a balanced diet or as silage (Nel, 1993).

In South Africa, at present, 80% of the ostriches raised for slaughtering in the Klein Karoo area are kept intensively in feedlots and 20% are on pasture, while in the Southern Cape, 60% of slaughter birds are raised on pasture and 40% intensively in feedlots (J. Nel, Oudtshoorn Agricultural Development Centre, Oudtshoorn, South Africa, personal communication, 2005). It is well known that ostriches do not perform well on pastures dominated by grass. In the wild, ostriches will feed mainly on young grass blades or will consume grass seeds that are stripped from the inflorescence (Cooper and Palmer, 1994). Birds in captivity, however, perform well on pastures dominated by any type of legume, for example, lucerne, clover, *Medicago* spp. and seradella. Ostriches may also be raised successfully on canola pastures or even saltbush plantations (Brand *et al.*, 2004a). Unfortunately very little is known with regard to the intake and selection by ostriches raised on cultivated pasture, or to the substitution effect of feed supplementation on intake by birds on pasture. Generally, in practical feed formulation of supplementary feeds for birds on pasture, the assumed intake of pasture is limited to 0% (pre-starter phase), 30% (starter phase) and 50% (grower phase), after which birds are placed in feedlots on complete feeds, due to the potential destructive effect on the pasture. Studies by Farrell *et al.* (2000)

revealed that supplementation levels below 70% of the total diet had a detrimental effect on growth, and suggest that pasture can only contribute significantly to the nutrition of ostriches when it is of good quality.

It is well known that pelleting of feed may increase animal performance for various reasons. In a study described by Brand et al. (2003b) ostriches (~44–70 kg live weight) in feedlots were fed either a high- and low-energy diet as 6 mm pellets or mash. The pelleting of diets tended to improve feed conversion efficiency by about 13% and growth rate of the birds by about 24%.

Nutrient Requirements

There is consensus amongst nutritionists that information on ostrich nutrition is still lacking compared with that on other domesticated farm animals (O'Malley, 1995; Ullrey and Allen, 1996; Janssens et al., 1997; Farrell et al., 2000). Feed formulators and manufacturers of ostrich diets initially used generous specifications based on the requirements for chickens (Farrell et al., 2000) or turkeys (Ullrey and Allen, 1996). According to O'Malley (1995) most feeds for ostriches have been formulated on limited data, or by the extrapolation of values determined for poultry, which has resulted in the development of feeds that have supplied excessive amounts of the birds' true nutritional requirements. Recent scientifically based information has led to more accurate calculation of the nutrient requirement of ostriches.

Energy Requirements

Swart et al. (1993) found, from colorimetric studies, that the energy requirements for ostriches were 0.44 MJ/kg $W^{0.75}$ per day, and that the efficiency of utilization of ME for tissue synthesis was 0.32.

More recent predicted matter intake (DMI) and TME requirement values for ostriches at different growth stages are presented in Table 9.8.

Several studies have demonstrated that ostriches, like broilers and pigs, overconsume energy when dietary energy contents are increased (Table 9.9). Because feed intake is reduced as dietary energy content increases, the feed conversion ratio (FCR) improves, but because feed price increases with nutrient density, it does not necessarily follow that high-nutrient-density feeds are the most economical to feed to ostriches.

Protein and Amino Acid Requirements

DMI, protein and essential amino acid requirements for ostriches at different growth stages are presented in Table 9.10.

Brand et al. (2005) determined the chemical and amino acid composition of ostriches (body plus feathers) at six growth stages (day-old, 5, 10, 20, 40 and 80 kg). Results revealed a mean CP content of 63.4%, a mean ether extract content of 18.9%, a mean calcium content of 3.35% and a mean phosphorus content of 2.38% on a DM basis. The amino acid

Table 9.8. Predicted mean dry matter intake and energy requirements for ostriches calculated from values published by du Preez (1991), Smith et al. (1995) and Cilliers et al. (2002).

Production stage	Live mass (kg)	Age (months)	Feed intake (g per bird day)	TME (MJ per kg feed)	Predicted growth rate (g per bird day)
Pre-starter	0.85–10.0	0–2	275[a]	14.65	163[a]
Starter	10–40	2–5	875	13.58	296
Grower	40–60	5–7	1603	10.80	387
Finisher	60–90	7–10	1915	9.83	336
Maintenance	90–120	10–20	2440	7.00	115
Female breeder	110	20-plus	2000	11.58	–

[a]Value published by Gandini et al. (1986).

Table 9.9. Dietary energy contents and corresponding dry matter intakes of ostriches.

Reference	Growth interval (kg)	Dietary energy content (MJ ME per kg DM)	Dry matter intake (g per bird day)	FCR (kg feed per kg weight gain)
Brand et al. (2000c)	25–94	12.0	2410	6.8
		10.5	2630	7.5
		9.0	2900	8.8
Brand et al. (2003b)	45–65	12.5	1370	3.9
		10.5	1640	5.0
		8.5	1870	6.2
Brand et al. (2004b)	65–98	11.5	2790	11.7
		9.5	3310	13.7
		7.5	3830	18.7

composition of the body plus feathers of the ostrich is compared to other species in Table 9.11. The amino acid composition of ostriches, relative to lysine, is similar in many respects to that of chickens and turkeys, but less so to pigs.

A formal method of determining the amino acid requirements of ostriches, from which a feeding strategy may be developed, requires the characterization of the growth potential of the body and feather proteins of ostriches. Whereas a few studies have been made from which the parameters of a Gompertz growth equation may be estimated, in no instances has the rate of feather growth been measured separately from the rest of the body. Reasonable estimates of the parameters describing the growth of the body plus feathers have been published by du Preez et al. (1992) and by Cilliers (1994), these values being about 0.83 kg for W_0 (chick weight at hatching), between 100 kg and 120 kg for W_m (the

Table 9.10. Predicted mean dry matter intake, protein and amino acid requirements for ostriches calculated from values published by du Preez (1991), Smith et al. (1995) and Cilliers et al. (2002).

Parameter predicted	Pre-starter	Starter	Grower	Finisher	Maintenance	Layers[a]
Live mass (kg)	0.85–10.0	10–40	40–60	60–90	90–120	110
Age (months)	0–2	2–5	5–7	7–10	10–20	24[a]
Feed intake (g/day)	275	875	1603	1915	2440	2000
Protein (g/100 g feed)	22.89	19.72	14.71	12.15	6.92	10.50
Lysine (g/100 g feed)	1.10	1.02	0.84	0.79	0.58	0.68
Methionine (g/100 g feed)	0.33	0.33	0.29	0.28	0.24	0.26
Cystine (g/100 g feed)	0.23	0.22	0.18	0.17	0.14	–
Total SAA (g/100 g feed)	0.56	0.55	0.47	0.45	0.38	–
Threonine (g/100 g feed)	0.63	0.59	0.49	0.47	0.36	0.59
Arginine (g/100 g feed)	0.97	0.93	0.80	0.78	0.63	0.51
Leucine (g/100 g feed)	1.38	1.24	0.99	0.88	0.59	0.90
Isoleucine (g/100 g feed)	0.70	0.65	0.54	0.51	0.38	0.45
Valine (g/100 g feed)	0.74	0.69	0.57	0.53	0.36	0.55
Histidine (g/100 g feed)	0.40	0.43	0.40	0.40	0.37	0.25
Phenylalanine (g/100 g feed)	0.85	0.79	0.65	0.61	0.45	0.47
Tyrosine (g/100 g feed)	0.45	0.44	0.38	0.38	0.31	0.37
Phenylalanine and tyrosine (g/100 g feed)	1.30	1.23	1.03	0.99	0.76	0.84

[a]Based on a 110 kg breeder laying a 1.4 kg egg every second day.

Table 9.11. Amino acid composition of body and feathers of ostriches, relative to lysine content, compared to pigs, chickens and turkeys.

Amino acid	Ostriches (Brand *et al.*, 2005)	Pigs (Wiseman, 1987)	Chickens (NRC, 1994)	Turkeys (Firman, 2003)
Lysine	100	100	100	100
Methionine	31	–	46	–
Cystine	–	–	36	–
Total SAA	–	60	82	64
Threonine	67	60	73	59
Tryptophan	–	18	18	21
Arginine	101	29	114	105
Isoleucine	74	60	73	72
Leucine	123	72	109	69
Valine	88	70	82	61
Histidine	39	26	32	38
Phenylalanine	69	–	66	–
Tyrosine	42	–	56	–
Phenylalanine and Tyrosine	111	100	122	128

final mature weight), and between 0.009 and 0.013 for B, the rate of maturing. Using these values, and the amino acid composition of the ostrich carcass at various stages of growth (Cilliers, 1994), amino acid requirements may be calculated for each stage of the growth of the ostrich, from which a feeding programme can be designed. An example of the change in the lysine requirement as a proportion of the feed is given in Fig. 9.4, in which a feeding

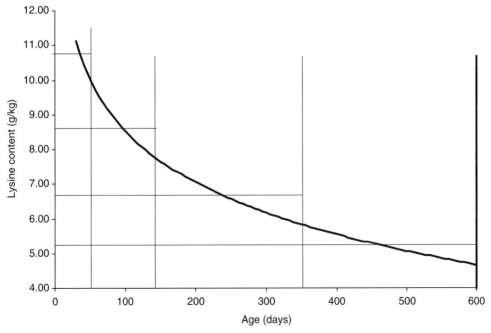

Fig. 9.4. Theoretical change in lysine requirement (g/kg feed) of growing ostriches with time, and a feeding programme to minimize deficiencies and excesses of lysine during the growth period.

programme has been superimposed on the lysine requirement. Until the rates of growth of feathers and the body are determined separately, this technique cannot be regarded as being an accurate method of determining the amino acid requirements but it has the potential to improve considerably the manner in which nutrient requirements of growing animals are estimated (Gous, 1993). In addition, once the potential growth rates of body and feather protein have been accurately characterized for the ostrich, optimization routines may be used to determine the most profitable means of feeding growing ostriches (Gous, 2002).

Growth Performance Studies

Several studies have been performed to determine the effect of dietary energy and protein (amino acids) on the growth performance of slaughter ostriches as well as the production performance of breeding birds.

In a study by Brand (2001), in which ostriches were reared from hatching to slaughter weight, birds consumed about 8 kg of feed during the pre-starter (hatching to 5 kg live weight) and about 62 kg in the starter (5–30 kg) phases, being 14% of their total intake per bird from hatching to slaughter of about 500 kg. Therefore, feeding cost during these two phases has only a minor effect on the total cost of raising the bird. However, chick mortalities during this phase are high (up to ~30% of total chicks hatched according to Cloete et al., 2001). The composition of feed during this phase may have a major effect on productivity, due to the effect of growth rate of chicks on viability and survival rate.

Gandini et al. (1986) found no significant difference in the growth performance of young birds fed iso-energetic feeds (11.5 MJ ME poultry per kg feed) with CP contents of 16%, 18% and 20% (lysine levels of 0.80%, 0.90% and 0.99%). Birds consuming 14% CP (0.7% lysine) feed did not perform as well as those on the other feeds. Angel (1996) fed ostriches two iso-energetic feeds with either 14.5% or 17.0% CP from hatching to 68 days of age. Chickens on the high-protein feed showed superior growth rates up to 56 days, with no difference in growth response from 56 up to

68 days of age due to dietary CP level. Brand et al. (2003b) provided ostrich chickens (~13–34 kg live weight) with feeds with three energy levels (10.5, 12.5 and 14.5 MJ ME ostrich) and increasing protein and amino acid levels (14–22% CP, with all amino acid profiles remaining the same). Feed intake was highest on the feed with about 18% CP (1.07% lysine, 0.68% methionine-cystine, 0.29% tryptophan, 0.75% threonine and 0.95% arginine). Higher growth rates were observed with the higher-energy feeds, while growth rate generally increased with an increase in CP intake. FCR increased with both dietary energy and CP.

During the grower (~30–60 kg) and finisher phases (~60–90 kg), approximately 430 kg feed (150 kg during the grower phase and 280 kg during the finisher phase) is consumed per bird. This represents about 86% of the total feed normally consumed by birds from hatching up to slaughter (Brand, 2001). These two growing phases are therefore of significant economic importance for the ostrich producer. In a study by Swart and Kemm (1985) with slaughter birds (60–110 kg), the highest growth rate (201 g per bird day) and best FCR (11 kg feed per kg gain) were obtained with birds consuming the high-energy feeds (dietary energy levels of ~8.1, 9.5 and 10.7 MJ ME pigs per kg feed were used), while growth rates were, respectively, 210, 146 and 210 g per bird day for birds consuming feeds with either 14%, 16% or 18% CP. Cornetto et al. (2003) fed ostriches (98 up to 146 days of age) diets with three different energy levels (11.71, 12.90, and 14.09 MJ/kg). They recorded an improved growth rate and FCR for birds consuming the high-energy-containing diet.

Brand et al. (2000c) fed ostriches from 4 to 11 months of age (~25–94 kg) diets with increasing CP (13.0–17.0%) and energy (9.0–12.0 MJ ME ostrich per kg) contents. Growth rate was unaffected by energy content, but birds on the low-energy diets consumed more than ostriches on the high-energy diets with a resultant poorer FCR. The CP and amino acid contents (between 0.67% and 0.90% lysine) of the feeds had no effect on performance. In another study by Brand et al. (2003a), where ostriches were given feeds varying in both energy (8.5, 10.5 and 12.5 MJ ME ostrich per kg) and protein (11.5%, 13.5%, 15.5%, 17.5% and 19.5%) during the grower phase (45–65 kg), dietary CP content had no effect on performance in terms of DMI, average daily gain (ADG) or FCR. Lower dietary energy levels again significantly increased feed intake, with a resultant decrease in FCR. Growth rate was reduced by about 15% when dietary energy content was reduced from 12.5 to 8.5 MJ ME ostrich per kg feed. During the finisher phase (~65–95 kg) similar performance results were obtained. DMI increased and both growth rate and FCR decreased with a decrease in dietary energy content, while dietary CP and amino acid contents ranging from 11.5% CP (0.58% lysine) to 19.5% CP (0.98% lysine) had no effect on performance during this production phase (Brand et al., 2004c). It is apparent from these results that ostriches in the finisher period do not benefit from dietary CP contents in excess of 11.5%.

Several studies to evaluate the effect of dietary CP and energy on ostrich breeder performance have been performed (Brand et al., 2002a,b,c, 2004d). Dietary CP contents between 8.5% (0.40% lysine) and 16.5% (0.85% lysine), at a DMI of 2.5 kg per bird day, had no effect on reproductive performance (egg production, egg mass, laying interval and live chicks hatched). Dietary energy contents lower than 8.5 MJ ME ostrich per kg feed at the above daily DMI reduces both egg production and the number of live chicks hatched. There are potential negative carry-over effects from one breeding season to the next when dietary energy contents less than 8.5 MJ ME per kg feed are used. No consistent effect of either dietary CP or energy on the distribution of eggs produced over the breeding season was observed (Brand et al., 2004d).

Mineral and Vitamin Requirements

Very little scientifically proven knowledge exists regarding the vitamin and mineral requirements of ostriches. In practice, data from other species are normally used when formulating premixes for ostriches. Minimum calcium (Ca) requirements for growing birds vary between 0.8% and 1.2%, with maximum dietary levels of 1.5% and 1.8%, while phosphorus (P) requirements vary between 0.5% and 0.6%. Minimum/maximum Ca requirements for breeding birds varied between 2.0% and 2.5% and 3.0% and 3.5% (Department of Agriculture, 2001). Carcass Ca and P contents were 3.35% and 2.38%, respectively (Brand et al., 2005).

Production Norms and Feeding Guidelines

The production rate of growing ostriches is dependent on the nutrient composition of the feed. Table 9.12 illustrates practical production norms based on several production studies for producing ostriches under intensive

Table 9.12. Yearly amount of feed required per slaughter bird and per breeding pair.

Period/Interval	Age (months)	Mass (kg)	DMI (g/ bird day)	ADG (g/ bird day)	FCR (kg feed/ kg gain)	Annual feed intake (kg per bird)[a]	Cumulative feed intake (kg per bird)[a]
Slaughter bird							
Pre-starter period	0–2	1–10	275	150	1.80	16	16
Starter period	2–4.5	10–40	1100	400	2.75	84	100
Grower period	4.5–6.5	40–60	1650	330	5.00	100	200
Finisher period	6.5–10.5	60–90	2500	240	10.00	300	500
Post-finisher period	10.5–12.0	90–100	3000	200	15.00	150	650
Breeding bird			Days	Intake/day (kg)[b]		Annual feed intake (kg)	
Rest period			120	4.0		480	
Breeding period			245	4.0		980	
Total			–	–		1460	

[a]Based on feeding pellets.
[b]Based on a 10:6 female/male ratio production system.

feedlot conditions that may be useful in feed flow programmes (adapted from Brand and Jordaan, 2004).

It is important to keep in mind the effect of the energy concentration of the feed, especially on daily DMI and on the resultant FCR, and the potential effect thereof on the amount of feed required per slaughter bird (see Table 9.9). The major effect of slaughter age on the total amount of feed required to reach slaughter weight is also clear from Table 9.12. To increase live weight from 90 to 100 kg an additional 30% of the total feed normally required to reach 90 kg is needed. The total amount of feed per slaughter bird to reach 12 months of age (when the average skin and nodule size are normally acceptable to the market) is approximately 650 kg per bird, while the total amount of feed required per breeding pair (including the rest period) amounts to about 1500 kg. When the assumption of 25 chickens per breeding pair is applied, the amount of feed required per chicken hatched is about 60 kg.

Production norms for slaughter birds used in Table 9.12 compare well with results published by Degen et al. (1991), who similarly found poor FCRs (17.1 kg feed per kg weight gain) for birds between 10.7 and 11.6 months of age. Commercial guidelines for ostrich feeds in South Africa are presented in Table 9.13. These guidelines regulated by the Animal Feed Act of South Africa (Act 96 of 1947) are regularly adapted to the most recent research findings.

Concluding Remarks and Recommendations for Future Research

Several studies have been done that enable us to determine the nutrient requirements of ostriches in terms of energy, protein and the essential amino acids for the different production stages. Similarly, the nutrient contents of available feed ingredients are known or can be calculated. Growth and performance studies were done either to test these calculated nutrient requirements or to fine-tune them. Studies on the anatomical structures and digestion help us to understand the utilization of ingested feed by the bird. Behavioural and selection studies were important to help us to adjust either complete feeds, supplementary feeds and cultivated pastures to food that is acceptable to the bird.

A range of challenges, however, still exists before ostriches can be produced commercially as successfully as other farm animals. The high mortality rate (30–40%) of chickens that occurs under commercial conditions may be ascribed mainly to disease and stress-related problems, although feeding behavioural and nutritional inadequacies probably also play a role.

The nutritional requirements of birds during all stages of production need to be estimated using a more scientific basis. Current standards are inadequate, and are based on a limited number of studies, none of which has addressed the need to determine separately the growth rates of body and feather protein. With-

Table 9.13. Commercial guidelines (as fed) for the composition of ostrich feeds (From Department of Agriculture, 2001).

Feed type	Minimum crude protein (g/kg)	Minimum lysine (g/kg)	Maximum moisture (g/kg)	Minimum crude fat (g/kg)	Maximum crude fibre (g/kg)	Calcium Min	Max	Minimum phosphorus
Pre-starter	190	10	120	25	100	12	15	6
Starter	170	9	120	25	135	12	15	6
Grower	150	7.5	120	25	175	10	16	5
Finisher	120	5.5	120	25	225	9	18	5
Slaughter	100	4	120	25	250	8	18	5
Maintenance	100	3	120	20	300	8	18	5
Breeder	120	5.8	120	25	240	25	30	5

out such information, simulation modelling cannot produce the results of which this approach is capable. The ostrich, being a monogastric herbivore, has the ability to utilize high-fibre-containing ingredients, which cannot be utilized to the same extent by other monogastric animals. The decision to use predominantly high-fibre or high-concentrate feeds in the feeding system should be based entirely on economic principles. Ileal amino acid digestibility values for ingredients used in ostriches still have to be collected, which will assist feed formulators to determine more accurate diets. The integration of production systems using pastures may play an important role in the long-term sustainability of the ostrich industry. Scientific information on the production of ostriches on different types of pastures, as well as related aspects of grazing systems, is still lacking. The science of providing supplementary feed to birds grazing pasture is limited. Several aspects of feed intake of birds on pasture, level and type of supplementary feed and the accompanying level of substitution still have to be solved. With current ostrich production systems there may be a conflict in production principles, since animals are kept for hides, meat and feathers. The market requires the skin of a bird of at least 12 months of age, while for meat production per se it may be more economical to feed the bird to achieve its maximum growth rate, which will result in the bird being ready for slaughter at too young an age to maximize returns from the sale of the hide. Simulation models will assist the producer in deciding on the most economic feeding strategy (Brand and Gous, 2003).

References

Angel, C.R. (1993) Research update: age changes in the digestibility of nutrients in ostriches and nutrient profiles in hen and chick. In: *Proceedings of the Association of Avian Veterinarians*, Nashville, Tennessee, pp. 275–281.

Angel, C.R. (1996) A review of ratite nutrition. *Animal Feed Science and Technology* 60, 241–246.

Baltmanis, B., Blue-McLendon, A. and Angel, C.R. (1997) Effect of diet on the ostrich gastrointestinal tract size. *American Ostrich: Research Issue* (April), 17–19.

Berry, H.H. and Louw, G.N. (1982) Nutritional balance between grassland productivity and large herbivore demands in the Etosha National Park. *Madoqua* 13, 141–150.

Bertram, B.C.R. (1980) Vigilance and group size in ostriches. *Animal Behaviour* 28, 278–286.

Bezuidenhout, A.J. (1986) The topography of the thoraco-abdominal viscera in the ostrich (*Struthio camelus*). *Onderstepoort Journal of Veterinary Research* 53, 111–117.

Bezuidenhout, A.J. and van Aswegen, G. (1990) Light microscopic and immunocytochemical study of the gastrointestinal tract of the ostrich (*Struthio camelus*). *Onderstepoort Journal of Veterinary Research* 57, 37–48.

Bradford, M.M.V. and Gous, R.M. (1991) The response of growing pigs to a choice of diets differing in protein content. *Animal Production* 52, 185–192.

Brand, T.S. (2001) Ostrich nutrition: cost implications and possible savings. *Elsenburg Journal*, 21–27 (in Afrikaans).

Brand, T.S. and Gous, R.M. (2003) Ostrich nutrition: using simulation models to optimize ostrich feeding. *Feed Technology* 7(9/10), 12–14.

Brand, T.S. and Jordaan, J.W. (2004) Ostrich nutrition: cost implications and possible savings. *Feed Technology* 8(2), 22–25.

Brand, T.S. and Steyn, H.H. (2002) Electronic feeding system for ostriches. *South African Patent* No 2003/3491.

Brand, T.S., Salih, M., van der Merwe, J.P. and Brand, Z. (2000a) Comparison of estimates of feed energy obtained from ostriches with estimates obtained from pig, poultry and ruminants. *South African Journal of Animal Science* 30, 13–14.

Brand, T.S., de Brabander, L., van Schalkwyk, S.J., Pfister, B. and Hayes, J.P. (2000b) The true metabolisable energy content of canola oilcake meal and full-fat canola seed for ostriches (*Struthio camelus*). *British Poultry Science* 41, 201–203.

Brand, T.S., Nel, C.J. and van Schalkwyk, S.J. (2000c) The effect of dietary energy and protein level on the production of growing ostriches. *South African Journal of Animal Science* 80, 15–16.

Brand, T.S., Aucamp, B.B., Kruger, A. and Sebake, Z. (2003a) *Ostrich Nutrition: Progress Report 2003*. Ostrich Research Unit, Private Bag X1, Elsenburg 7607, South Africa, pp. 1–19.

Brand, T.S., Gous, R., Brand, Z., Aucamp, B.B., Kruger, A.C.M. and Nel, J. (2003b) Review: research on ostrich nutrition on South Africa. *Proceedings of the 11th World Ostrich Conference*, 17–19 October, Vienna, Austria.

Brand, T.S., Smith, N., Aucamp, B.B. and Kruger, A. (2004a) Preference of different types of lupins by ostriches. *Book of Abstracts – 2nd Joint Congress of the Grassland Society of Southern Africa Society of Animal Science*, 28 June–1 July, Goudini, South Africa, p. 188.

Brand, T.S., Gous, R.M., Kruger, A., Aucamp, B.B., Nel, C.J. and Horbanczuk, J. (2004b) The effect of dietary energy and protein (amino acid) on the performance of slaughter ostriches. *Book of Abstracts – 2nd Joint Congress of the Grassland Society of Southern Africa and the South African Society of Animal Science*, 28 June–1 July, Goudini, South Africa, p. 189.

Brand, T.S., Nel, C.J., Brand, Z. and Engelbrecht, S. (2004c) The replacement of lucerne hay with saltbush (*Atriplex nummularia*) in diets of grower-finisher ostriches. *Book of Abstracts – 2nd Joint Congress of the Grassland Society of Southern Africa and the South African Society of Animal Science*, 28 June–1 July, Goudini, South Africa, p. 94.

Brand, T.S., Gous, R.M., Kruger, A., Brand, Z., Nel, C.J., Aucamp, B. and Engelbrecht, S. (2004d) Mathematical feeding optimization model for ostriches: recent progress in the prediction of nutrient requirements for slaughter and breeder birds. *Invited Review at the Ostrich Information Day of the South African Ostrich Producers Organisation*, June, pp. 1–13 (in Afrikaans).

Brand, T.S., Aucamp, B.B., Kruger, A. and Brand, Z. (2005) *Ostrich Nutrition: Progress Report 2005*. Ostrich Research Unit, Private Bag X1, Elsenburg 7607, South Africa.

Brand, Z., Brand, T.S. and Brown, C.R. (2002a) The effect of dietary energy and protein levels during a breeding season on production in the following breeding season (*Struthio camelus*). *Journal of Animal Science* 32, 226–230.

Brand, Z., Brand, T.S. and Brown, C.R. (2002b) The effect of dietary energy and protein levels on body condition and production of breeding male ostriches. *South African Journal of Animal Science* 32, 231–239.

Brand, Z., Brand, T.S. and Brown, C.R. (2002c) The effect of dietary energy and protein levels on production in breeding female ostriches. *British Poultry Science* 44, 598–606.

Bubier, N.E., Lambert, M.S., Deeming, D.C., Ayres, L.L. and Sibly, R.M. (1996) Time budget and colour preference (with reference to feeding) of ostrich chicks in captivity. *British Poultry Science* 37, 547–551.

Cilliers, S.C. (1994) Evaluation of feedstuffs and the ME and amino acid requirements for maintenance and growth in ostriches. PhD thesis, University of Stellenbosch, Stellenbosch, South Africa.

Cilliers, S.C. (1998) Feedstuff evaluation and the metabolisable energy and amino acid requirements for maintenance and growth in ostriches. In: *Proceedings of the AFMA Symposium*, Pretoria, South Africa, pp. 1–16.

Cilliers, S.C. and Angel, C.R. (1999) Basic concepts and recent advances in digestion and nutrition. In: Deeming, D.C. (ed.) *The Ostrich: Biology, Production and Health*. CAB International, Wallingford, UK, pp. 105–128.

Cilliers, S.C., Hayes, J.P., Chwalibog, A., Sales, J. and du Preez, J.J. (2002) Determination of energy, protein and amino acid requirements for maintenance and growth in ostriches. *Animal Feed Science and Technology* 72, 283–293.

Cloete, S.W.P., Nel, C.J. and van Schalkwyk, S.J. (2001) *Report on Chicken Mortality for Ostriches in South Africa*. Ostrich Research Unit, Elsenburg Agricultural Research Centre, Private Bag X1, Elsenburg 7607, South Africa, 29 pp. (in Afrikaans).

Cooper, S.M. and Palmer, T. (1994) Observations on the dietary choice of free-ranging juvenile ostriches. *Ostrich* 65(3/4), 251–255.

Cornetto, T., Angel, R. and Estevez, I. (2003) Influence of stocking density and dietary energy on ostrich performance. *International Journal of Poultry Science* 2, 102–106.

Deeming, D.C., Bubier, N.E., Paxton, C.G.M., Lambert, M.S., Magole, I.L. and Silbly, R.M. (1996) A review of recent work on the behavior of young ostrich chicks with respect to feed. In: *Proceedings of a Conference on Improving our Understanding of Ratites in a Farming Environment*, 27–29 March, University of Manchester, UK, pp. 20–21.

Degen, A.A., Kam, M. and Rosenstrauch, A. (1989) Time-activity budget of ostriches (*Struthio camelus*) offered concentrate feed and maintained in outdoor pens. *Applied Animal Behaviour Science* 22, 347–358.

Degen, A.A., Kam, M., Rosenstrauch, A. and Plavnik, P. (1991) Growth rate, total body water volume, dry-matter intake and water consumption of domesticated ostriches (*Struthio camelus*). *Animal Production* 52, 225–232.

Department of Agriculture (2001) *Guidelines for the Composition of Animal Feeds*. Department of Agriculture, Pretoria, South Africa.

de Kock, J.A. (1995) *Natural Rising of Ostrich Chicks*. Klein Karoo Development Centre, Oudtshoorn, South Africa (in Afrikaans).

du Preez, J.J. (1991) Ostrich nutrition and management. In: Farrell, D.J. (ed.) *Recent Advances in Animal Nutrition in Australia*. University of New England, Armidale, Australia, pp. 278–291.

du Preez, J.J., Jarvis, M.J.F., Capatos, D. and de Kock, J.A. (1992) A note on growth curves for the ostrich (*Struthio camelus*). *Animal Production* 54, 150–152.

Eggum, B.O., Thorbek, G., Beames, R.M., Chwalibog, A. and Henckel, S. (1982) Influence of diet and microbial activity in the digestive tract on digestibility, and energy metabolism in rats and pigs. *British Journal of Nutrition* 48, 161–175.

El Boushy, A.R. and Kennedy, D.A. (1987) Flavoring agents improve feed acceptability. *Poultry* (June/July), pp. 32–33.

El Boushy, A.R., van der Poel, A.F.B., Verhaart, J.C.J. and Kennedy, D.A. (1989) Sensory involvement controls feed intake in poultry. *Feedstuffs* (June), 16, 18, 19, 41.

Ensminger, M.E. and Olentine, C.G. (1978) *Feeds and Nutrition Complete*. The Ensminger Publishing Company, Clovis, California.

Erasmus, J. and Erasmus, E de V. (1993) Ostrich odyssey: a guide to ostrich farming in South Africa. Review Printers, Pietersburg, South Africa.

Farrell, D.J., Kent, P.B. and Schermer, M. (2000) *Ostriches: Their Nutritional Needs under Farming Conditions*. Rural Industries Research and Development Corporation, Barton ACT, 2600, p. 55.

Ferguson, N.S., Bradford, M.M.V. and Gous, R.M. (2002) Diet selection priorities in growing pigs offered a choice of feeds. *South African Journal of Animal Science* 32, 136–143.

Firman, J.D. (2003) Computer formulation of low protein diets for turkeys. *Multi-State Poultry Meeting*, 20–22 May, pp. 1–13.

Gandini, G.C.M., Burroughs, R.E.J. and Ebedes, H. (1986) Preliminary investigation into the nutrition of ostrich chicks (*Struthio camelus*) under intensive conditions. *Journal of the South African Veterinary Association* 57, 39–42.

Getty, R. (1975) *The Anatomy of the Domestic Animals*, 5th edn. W.B. Saunders, London.

Gous, R.M. (1993) The use of simulation models in estimating the nutritional requirements of broilers. *Australian Poultry Science Symposium* 5, 1–9, University of Sydney, Sydney, Australia.

Gous, R.M. (2002) Modelling energy and amino acid requirements as a means of optimising the feeding of commercial broilers. *New Zealand Poultry Industry Conference*, Palmerton North, New Zealand.

Holtzhausen, A. and Kotzé, M. (1990) *The Ostrich 56p*. C.P. Nel Museum, Oudtshoorn, South Africa.

Iji, P.A., Van der Walt, J.G., Brand, T.S., Boomker, E.A. and Booyse, D. (2003) Development of the digestive function in the ostrich. *Archives of Animal Nutrition* 57, 217–228.

Janssens, G.D.J., Seynaeve, M., de Wilde, R.O. and Rycke, H. (1997) Voedingsaspecten van de struisvogel (*Struthio camelus*). *Vlaams Dieregeneeskundig Tijdschrift* 66, 153–160.

Jozefiak, D., Rutkaski, A. and Martin, S.A. (2004) Carbohydrate fermentation in the avian ceca: a review. *Animal Feed Science and Technology* 113, 1–15.

Just, A., Jorgensen, H. and Fernandez, J.A. (1981) The digestive capacity of the caecum-colon and value of nitrogen absorbed from the hindgut of protein synthesis in pigs. *British Journal of Nutrition* 32, 479–483.

Kok, O.B. (1980) Feed intake of ostriches in the Namib-Naukluft park, Suidwes-Afrika. *Madoqua* 12(3), 155–161 (in Afrikaans).

Kubirskie, K.R. (1985) The intensive raising of wild ostriches (*Struthio camelus australis*). *Agricola* 2, 69–71.

Law-Brown, J., Reid, S.J., Thomson, J.A. and Brand, T.S. (2004) The effect of diet on gut bacteria in ostriches, *Struthio camelus*. *Book of Abstracts – 2nd Joint Congress of the Grassland Society of Southern African and the South African Society of Animal Science*, 28 June–1 July, Goudini, South Africa, p. 194.

Lindenmaier, P. and Kare, M.R. (1959) The taste end-organs of the chicken. *Poultry Science* 38, 545–550.

McHan, F. and Shotts, E.B. (1993) Effect of short-chain fatty acids on the growth of *Salmonella typhimurium* in an *in vitro* system. *Avian Disease* 37, 396–398.

Milton, S.J., Dean, W.R.J. and Siegfried, W.R. (1994) Food selection by ostrich in South Africa. *Journal of Wildlife Management* 58, 234–248.

Mohamed, A.E. (2001) Nutritional, behavioral and pathological studies on captive red-neck ostriches. PhD thesis, University of Khartoum, Sudan.

Mosenthal, J. and Harting, J.E. (1879) *Ostriches and Ostrich Farming.* Trubner & Co, London.

Mushi, E.Z., Binta, M.G. and Chabo, R.G. (2004) Yolk sac utilization in ostrich (struthio camelus) chicks. *Onderstepoort Journal of Veterinary Research* 71, 247–249.

Nel, J.C. (1993) Pasture utilization by ostriches. *Report on the Ostrich Industry in South Africa.* Oudtshoorn Development Centre, Private Bag, Oudtshoorn, South Africa, pp. 20–21 (in Afrikaans).

Nizza, A. and Di Meo, C. (2000) Determination of the apparent digestibility coefficients in 6-, 12- and 18-week-old ostriches. *British Poultry Science* 41, 518–520.

NRC (1994) *Nutrient Requirements of Poultry*, 9th edn. National Academy Press, Washington, DC.

O'Malley, P. (1995) Nutrient requirements of ratites: comparison of emu and ostrich requirements. *Recent Advances in Animal Nutrition in Australia* (July 1995), 53–61.

Osterhoff, D.R. (1979) Ostrich farming in South Africa. *World Review of Animal Production* 15, 19–30.

Robinson, E.R. and Seely, M.K. (1975) Some food plants of ostriches in the Namib Desert Park, South West Africa. *Madoqua* 4, 99–100.

Roura, E. (2003) Recent studies on the biology of taste and olfaction in mammals and aves. *Congress Latinoamericano de Avicultura y Suinocultura.* Florianopolis, Brazil, 12–16 May, pp. 1–12.

Salih, M.E. (2000) The effects of high-fibre diets on the digestion, growth and performance of Black African ostriches. PhD thesis, Ankara University, Turkey (in Turkish).

Salih, M.E., Brand, T.S., van Schalkwyk, S.J., Blood, J.R., Pfister, B., Brand, Z. and Akbay, R. (1998) The effect of dietary fibre level on the production of growing ostriches. *Proceedings of the Second International Ratite Congress*, Oudtshoorn, South Africa, 21–25 September, pp. 31–37.

Sambraus, H.H. (1994) Circadian rhythm in the behavior of ostriches kept in pens. *Berliner-und-Muchener-Tierarztliche-Wochenschft* 107, 339–341.

Sauer, E.G.F. and Sauer, E.M. (1996) The behaviour and ecology of the South African ostrich. *Living Bird* 5, 45–75.

Scheideler, S.E. (1997) Effects of vomitoxin on ostrich growth. *American Ostrich* (April), 14–16.

Skadhauge, E., Waruei, C.N., Kamau, J.M.Z. and Maloiy, G.M.O. (1984) Function of the lower intestine and osmoregulation in the ostrich: preliminary anatomical and physiological observation. *Quarterly Journal of Experimental Physiology* 69, 809–818.

Smith, D.J.v.Z. (1964) *Ostrich farming in the Little Karoo.* South African Department of Technical Services, Bulletin No 358.

Smith, W.A., Cilliers, S.C., Mellet, F.D. and van Schalkwyk, S.J. (1995) Nutrient requirements and feedstuff values in ostrich production. *Feed Compounder* (September), 2–29.

South African Ostrich Business Chamber (2002) *Report of the South African Ostrich Business Chamber.* Oudtshoorn, South Africa.

Swart, D. (1988) Studies on the hatching, growth and energy metabolism of the ostrich chick (*Struthio camelus*). PhD thesis, University of Stellenbosch, South Africa.

Swart, D. and Kemm, E.H. (1985) The effect of diet protein and energy levels on the growth, performance and feather production of slaughter ostriches under feedlot conditions. *South African Journal of Animal Science* 15, 146–150.

Swart, D., Siebrits, F.K. and Hayes, J.P. (1993) Utilization of metabolizable energy by ostrich (struthio camelus) chicks at two different concentrations of dietary energy and crude fibre originating from lucerne. *South African Journal of Animal Science* 23, 136–141.

Ullrey, D.E. and Allen, M.E. (1996) Nutrition and feeding of ostriches. *Animal Feed Science and Technology* 59, 27–36.

Van der Walt, J.G., Iji, P.A., Brand, T.S., Boomker, E.A. and Booyse, D. (2003). Early development of the digestive function in the ostrich. *International Herbivore Symposium*, Mexico.

Van der Wielen, P.W., Biesterveld, S., Notermans, S., Hofstra, H., Urlings, B.A.P. and van Knapen, F. (2000) Role of volatile fatty acids in development of cecal micro flora in broiler chickens during growth. *Applied Environmental Biology* 66, 2536–2540.

Van Niekerk, B.D.H. (1995) The science and practice of ostrich nutrition. *Proceedings of the AFMA Conference*, June, Pretoria, South Africa, pp. 1–8.

Viljoen, M., Brand, T.S. and Van der Walt, J.G. (2004) The effect of different dietary energy and protein levels on the digestive anatomy of ostriches. *Book of Abstracts – 2nd Joint Congress of the Grassland Society of Southern Africa and the South African Society of Animal Science*, 28 June–1 July, Goudini, South Africa, p. 95.

Williams, J.B., Siegfried, W.R., Milton, S.J., Adams, N.J., Dean, W.R.J., du Plessis, M.A., Jackson, S. and Nagy, K.A. (1993) Field metabolism, water requirements and foraging behaviour of wild ostriches in the Namib. *Ecology* 74, 390–403.

Wiseman, J. (1987) *Feeding of Non-ruminant Livestock*. Butterworths, London, 208 pp.

Withers, P.C. (1983) Energy, water and solute balance of ostrich *Struthio*. *Physiological Zoology* 56, 568–579.

10 Feeding Behaviour in Pigs

B.L. Nielsen, K. Thodberg, L. Dybkjær and E.-M. Vestergaard
Danish Institute of Agricultural Sciences, Department of Animal Health, Welfare and Nutrition Research Centre Foulum, PO Box 50, DK-8830 Tjele, Denmark

Feeding Behaviour of Piglets

The pig is a rather unique member of the ungulates. Pigs are omnivorous as opposed to their herbivorous relatives and are the only ungulates that give birth to litters, which they hide in a nest. The physiology and behaviour associated with nursing and suckling are relatively complex. Milk is only released in discrete ejections after stimulation from the whole litter and is not present in a cisterne as in ruminants (Wakerley et al., 1994). Ten days after giving birth in a carefully constructed nest, the sow and her litter rejoin the female group, in which the sow usually lives (Jensen and Redbo, 1987; Jensen et al., 1993). Within this group, litters are born within a short time period and the sows attempt to synchronize their nursing. Together with the necessary stimulation from the whole litter this is an evolutionary good adaptation to group living as all piglets are engaged in suckling their own mother at the same time and are not able to poach milk from other sows.

The importance of colostrum

Immediately after birth the newborn piglet moves to the udder to ingest its first meal. The time from birth to first suckle is highly variable ranging from 10 to 80 min (e.g. Bünger, 1985; Damm et al., 2002; Pedersen et al., 2003).

Piglets are born without immune protection and therefore need access to the udder, not only to get nourishment but also to ensure the transfer of immunoglobulins from the sow. Six hours after the onset of farrowing, the concentration of immunoglobulins in the colostrum drops to 70%; after 12 h it is 40% and only 10% is left 48 h after the onset of farrowing (Klobasa et al., 1987). The relatively short availability of colostrums, together with the piglets' inability to absorb immunoglobulins after 24–48 h (Klobasa et al., 1990; Blecha, 1998), makes it very important for the piglets to reach the udder soon after birth. The longer the interval from birth until first suckle, the lower the concentration of antibodies in the blood. The first piglets born visit more teats, consume more meals and have longer duration of suckling leading to higher levels of immunoglobulins (De Passillé et al., 1988b; Damm et al., 2002).

Effects of the behaviour of the sow on the first sucklings

Failure to obtain milk in general and colostrum in particular is very serious for piglets. Apart from the obvious consequences of hunger it also leads to a lower body temperature and reduced vitality. This puts the piglets at higher risk of being crushed by the sow, since cold and hungry piglets tend to stay close to the sow for heat and milk (Weary *et al.*, 1996) and may be less able to get away when the sow changes position (Bauman *et al.*, 1966).

The behaviour of the sow during and immediately after farrowing plays a role in the progress of the piglets' first suckle. It is easier and less dangerous for the piglets to get access to the udder if the sow is lying quietly with the udder exposed (Fig. 10.1), but sows show a large individual variation in the time spent lying laterally both during farrowing and over the following 24 h. The variations between sows can be attributed to their reaction pattern or temperament (Thodberg *et al.*, 2002a,b) as well as to the farrowing environment. Provision of biologically relevant nest-building materials in late pregnancy affects the subsequent maternal behaviour during farrowing and early lactation, probably through feedback from the nest built by the sow (Jensen, 1993). Some of the positive effects are reduced nest-building activity during farrowing (Thodberg *et al.*, 1999; Damm *et al.*, 2000), higher responsiveness towards piglets (Cronin and van Amerongen, 1991; Herskin *et al.*, 1998) and longer nursing bouts in the first week of lactation (Herskin *et al.*, 1999).

Nursing pattern and teat order

Nursing and suckling behaviour is a cooperation between sow and piglets involving sounds, hormones and tactile stimulation. During the first hours after the onset of farrowing the pattern of nursing has not yet become cyclic. The very first milk is more freely available to the piglets than at later stages, and can be removed from the teats without the stimulation that is required later, where the milk is only available in short and less frequent ejections (Fraser, 1984; Fraser and Rushen, 1992). In the first hours after farrowing, ejections of milk last 1–4 min and occur in intervals of 5–30 min (Fraser, 1984; Kent *et al.*, 2003). Milk ejections are usually associated with a peak in the concentration of the oxytocin hormone, which is also responsible for the contractions of the uterus during delivery. Milk ejections can occur, however, without an oxytocin peak within the first hours of lactation due to a high baseline concentration of oxytocin during and immediately after farrowing (Castrén *et al.*, 1993).

The transition from the early nursing to a cyclic nursing pattern occurs during the first

Fig. 10.1. A sow in lateral recumbency suckling a litter of piglets.

24 h after farrowing. According to Lewis and Hurnik (1985) this change in nursing pattern takes place around 11 h after onset of farrowing, whereas Castrén *et al.* (1989b) found a change already after 5 h, based partly on the grunt peak pattern of the sow. It is debatable whether this change in nursing pattern is an abrupt or a more gradual process (De Passillé and Rushen, 1989).

During the first days the piglets develop a preference for a specific teat or teat pair, which they fight to protect from their littermates (e.g. Hemsworth *et al.*, 1976; Rosillon-Warnier and Paquay, 1984; De Passillé *et al.*, 1988a). These fights are severe and occur most frequently on the first day *post partum*. Piglets with a high birth weight win most fights and the number of fights won is correlated to the number of teats suckled (De Passillé *et al.*, 1988a; De Passillé and Rushen, 1989). There is great variation between litters as to when the teat order becomes stable, ranging from 2 to 10 days after birth. This variation is related, for example, to litter size and variation in teat quality.

Once established, the cyclic nursings follow the same pattern every time. A nursing is usually divided into five descriptive phases (Fig. 10.2; Whittemore and Fraser, 1974; Fraser, 1980; Algers and Jensen, 1985; review by Algers, 1993). In phase 1 the nursing is announced by the sow. The piglets gather and find their individual teat. Early in lactation the sow usually takes the initiative to nurse, but later the piglets take over this role and make the sow expose the udder by contacting her and nuzzling the udder. The onset of the udder massage by the piglets marks the beginning of phase 2, with the sow grunting slowly. The massaging of individual teats stimulates the secretion of oxytocin. The oxytocin peak is accompanied by a peak in the grunt rate of the sow, which is highly dependent on the rise in oxytocin. Castrén *et al.* (1993) found that milk ejections during and immediately after farrowing can occur without an oxytocin peak, but then no grunt peak is observed. As the grunting of the sow gets faster the piglets stop massaging and take the teat into the mouth and start sucking slowly, which characterizes phase 3. It takes 20–30 s for oxytocin to stimulate an increase in the intramammary pressure and start the contraction of the myoepithelial cells (Ellendorf *et al.*, 1982) after which milk is ejected. At this time piglets make rapid sucking movements as they ingest the milk during phase 4, which lasts only 10–20 s. In phase 5 the piglet resumes the udder massage, which can last from a few seconds up to several minutes (Jensen, 1988). This whole sequence of events is a fine interplay between the sow and her piglets involving both tactile and auditory communication, which can easily be disturbed by the surroundings. The noise from a ventilation fan can disturb the transition between the phases and lead to a decrease in milk

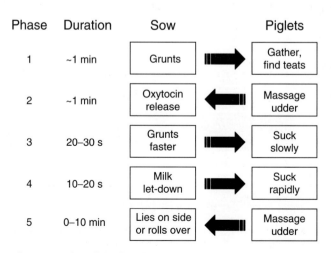

Phase	Duration	Sow		Piglets
1	~1 min	Grunts	➡	Gather, find teats
2	~1 min	Oxytocin release	⬅	Massage udder
3	20–30 s	Grunts faster	➡	Suck slowly
4	10–20 s	Milk let-down	➡	Suck rapidly
5	0–10 min	Lies on side or rolls over	⬅	Massage udder

Fig. 10.2.   Schematic presentation of the five phases that constitute the pattern of nursing.

production of the sow, underlining the significance of the sow grunts as a transition signal (Algers and Jensen, 1985, 1991).

Internursing interval

There is a huge variation in the reported internursing interval with durations from 30 to 80 min during the first week of lactation (e.g. Jensen et al., 1991; Wechsler and Brodman, 1996; Spinka et al., 1997). This variation is due to differences in the experimental conditions, breeds used, observation methods, definitions and stage of lactation. Several studies found a decreasing frequency of nursing with the highest frequency 6–8 days after farrowing when 23–31 daily nursing rounds were reported (e.g. Puppe and Tuchscherer, 2000; Valros et al., 2002; Damm et al., 2003). Already after 2 weeks of lactation the sow starts the natural weaning process. The sow restricts the udder exposure by assuming sternal recumbency more frequently (De Passillé and Robert, 1989), and gradually initiates less nursing and ends nursing more frequently (Valros et al., 2002; Damm et al., 2003). Concurrently the piglets try more actively to make the sow nurse, creating a so-called parent–offspring or weaning conflict. The sow becomes

more and more restrictive, saving energy for future generations, whereas the piglets try to manipulate the sow into providing more milk here and now (Fraser et al., 1995). The time of the natural weaning process is different for individual sows and litters, and can be affected by housing (Bøe, 1993; Pajor et al., 2000; Thodberg et al., 2002b), individual reaction pattern (Thodberg et al., 2002b) and litter features (Puppe and Tuchscherer, 2000).

An important aspect of the transfer of milk from the sow to the piglets is the sow's motivation to nurse. The longer the interval since last nursing, the shorter the latency to assume the nursing posture. A large individual variation has been found between sows when faced with the choice between eating and nursing (Thodberg and Jensen, 2005). The sows were kept on two hunger levels and the hungrier sows did not respond to the increased nursing interval (Fig. 10.3).

Factors affecting milk production

The amount of milk produced changes during lactation. From its lowest level during the first week, it increases until the third week, after which it is relatively constant for a further 3 weeks (Toner et al., 1996; review by Etienne

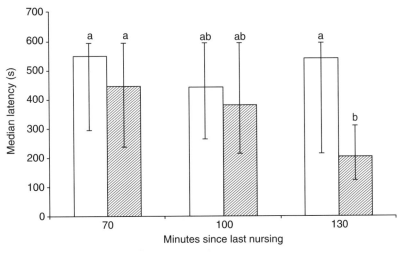

Fig. 10.3. Latency to assume nursing posture in relation to the interval since last nursing for sows kept on a high (white bars) or a low hunger level (hatched bars). From Thodberg and Jensen (2005).

et al., 1998; Valros et al., 2002). The milk
production is influenced by the management
and feeding of the sow during rearing, gestation and lactation, but the piglets can also play
a role (review by Hurley, 2001). It is essential
that the milk is removed from the teats to prevent them from drying out (Hurley, 2001).
King et al. (1997) showed that piglet weight
can influence the milk production of the sow.
In a cross-fostering experiment they found that
sows, which received piglets at 3–4 weeks of
age within the first week after farrowing, produced 26% more milk than those receiving
newborn piglets. Several studies reported a linear relationship between milk production and
litter size, mainly because more teats are suckled in big litters (e.g. Toner et al., 1996; Kim
et al., 1999). Other investigations have
focused on the length of the nursing interval as
a regulator of milk production (Illmann and
Madlafousek, 1995; Spinka et al., 1997; Valros et al., 2002). Milk yield is highest in sows
with most frequent nursing (Valros et al.,
2002). Milk yield can be increased by shortening the interval between nursing either by
manipulating access of the sow's own litter to
the udder (Spinka et al., 1997) or by using two
groups of pigs nursing alternately (Auldist
et al., 2000). The results indicate that the milk
for the next nursing is 'ready' after 35 min
(Spinka et al., 1997; Auldist et al., 2000), and
that more frequent suckling will increase the
daily milk intake of the piglets and thus the
yield of the sow (Etienne et al., 1998). Play-
back of nursing grunts can stimulate the sows
to start nursing (Wechsler and Brodman,
1996). This method has been used to increase
the nursing frequency by exploiting the tendency of sows to synchronize their nursing,
resulting in a higher piglet weight gain (e.g.
Contreras et al., 2001; Cronin et al., 2001).
The function of the final udder massage is
under debate. Algers and Jensen (1985) proposed the 'Restaurant Hypothesis', suggesting
that the piglets – at least in the first days of lactation – 'order' the size of their next
meal according to the duration of the final
massage. A recent study found a weak but
significant positive relationship between intensity of udder massage and amount of udder tissue at slaughter (Thodberg and Sørensen,
2006).

Non-nutritive nursings

Regularly, one nursing takes place without
milk ejection. This occurs once in about three
nursing cycles during the first 3 days after farrowing and subsequently in every fifth nursing
(Ellendorf et al., 1982; Castrén et al., 1989a;
Illmann and Madlafousek, 1995). The non-
nutritive nursing starts with no visible sign that
it will not succeed (Illmann et al., 1999). The
procedure occurs until the point when the
grunt rate of the sow usually increases. This
does not occur in non-nutritive nursing and
neither do the rapid sucking movements of the
piglets (phases 3 and 4 in Fig. 10.2). Instead
the piglets continue to massage the udder.
Non-nutritive nursing has been observed both
in production and under natural conditions and
is therefore not a phenomenon found only in
production systems (Castrén et al., 1989a).
The function of the non-nutritive nursing has
been suggested to be a tool for the sow to regulate the frequency of milk ejection, thereby
controlling the milk intake of the piglets
(Illmann and Madlafousek, 1995).

The pre-weaning intake of solid feed under semi-natural conditions

Studies of free-ranging domestic pigs kept in
two semi-natural enclosures of 7 and 13 ha with
several different biotopes have revealed that the
natural weaning process, from the piglets rely-
ing exclusively on milk to when they cease to
suckle, is a long and gradual one (Jensen and
Recén, 1989; Jensen and Stangel, 1992).
From 7 to 10 days of age the piglets follow the
sow to the feeding place (Jensen, 1986).
Although rooting, chewing and sniffing at substrate (which are exploratory elements all
included in foraging behaviour) can be observed
during the first week *post partum*, the piglets
do not appear to ingest much food at that time
(Petersen, 1994). The exploratory behaviour
gradually increases during the first 4–5 weeks of
life as the piglets familiarize themselves with
their surroundings (Newberry and Wood-Gush,
1988; Petersen, 1994) and gradually begin to
eat small amounts of solid food. They start grazing and eating pellets from 4 weeks of age, and
increase time spent grazing from 7% to 42% of

the daytime in the subsequent 4 weeks (Petersen et al., 1989; Petersen, 1994).

The piglets are weaned when they cease to suckle and feed exclusively on solids. This usually occurs when the piglets are around 17 weeks of age (Jensen, 1986; Petersen, 1994), but the time of the last suckling can vary from 9 weeks (Newberry and Wood-Gush, 1985) to 22 weeks after birth (Jensen and Stangel, 1992). Seasonal changes in availability of food sources probably explain some of the variations between litters with regard to weaning age (Newberry and Wood-Gush, 1985; Jensen and Recén, 1989) in natural environments.

Individual piglets will sometimes miss milk let-down during the first week *post partum*, after which the proportion of piglets missing suckling increases up to 12–15 weeks of age (Jensen, 1988; Jensen and Recén, 1989). The speed of the transition from exclusive intake of milk to exclusive intake of solid food and water differs both between litters and between littermates. According to the so-called 'Fast-food Hypothesis' proposed by Jensen and Recén (1989), the differences between individual piglets are related to the milk productivity of the teat occupied by the piglet. The piglet will optimize its foraging behaviour by ceasing to suckle and compensate by eating solid food once the milk from its teat cannot meet its nutritional demand. The hypothesis is supported by Jensen (1995), who found a negative correlation between attachment to the sow and time spent eating solid food in piglets from 4 to 10 weeks of age.

The pre-weaning intake of feed and water in commercial pig production

In commercial pig production, the sow and her litter are most often housed in a farrowing pen of limited size until weaning. The sow is confined and provided with feed and water from a trough and a drinker designed to minimize waste. The piglets can be offered water from a water nipple (Fig. 10.4) or bowl, and creep fed in a small trough placed somewhere in the pen.

Piglets have been observed to drink water from a bowl within the first 2 days after birth, and to discover the water sooner, and drink more, if the water is presented as an exposed water surface in a facility wide enough for several piglets to drink simultaneously (Phillips and Fraser, 1990, 1991). This early water intake may prevent dehydration, especially in warm environments, and improve survival rates in litters with a temporary inadequate access to milk (Fraser et al., 1988, 1990). During the following weeks the nursing piglets spend increasingly more time at the water facility (Petersen, 2004). The intake of water and creep feed is related, especially after the first weeks of the nursing period, as some litters eat significantly more creep feed if provided with water (Friend and Cunningham, 1966).

Although the overall effects of offering creep feed to piglets are not consistent (see Pluske et al., 1995), many studies indicate some positive effects of giving the piglets the possibility to familiarize themselves physiologically and behaviourally with solid feed before weaning and to compensate for insufficient access to milk (e.g. English et al., 1980; Aherne et al., 1982). The amount and time of first feed intake depend on the age at first presentation. In many studies creep feed is offered from 14 days of age, but some piglets have been observed to eat at 10–11 days of age (Barnett et al., 1989; Pajor et al., 1991). Giving the piglets access to creep feed from day 7 after farrowing may even encourage them to start eating creep feed earlier (Labayen et al., 1994).

The amount of feed consumed is low during the first 3–5 weeks after birth but then increases considerably during the following weeks (see Fig. 10.5). As seen in semi-natural environments the behaviour of the piglets directed at the food source shifts gradually from being exclusively exploratory at first into actual consumption of the creep feed (Delumeau and Meunier-Salaün, 1995). In addition to the age of the piglets, several other factors affect the intake of creep feed. These include trough space (higher intake if sufficient space; Appleby et al., 1991, 1992) and quality of the creep feed (higher intake if complex diet; Fraser et al., 1994; Pajor et al., 2002). Allowing the sow to leave the piglets for short periods can also stimulate the intake of creep feed. In a sow-controlled housing system, where the sow was able to leave the litter, the piglets ate 65% more solid feed before weaning at

Fig. 10.4. Example of a water nipple drinker used for weaned piglets.

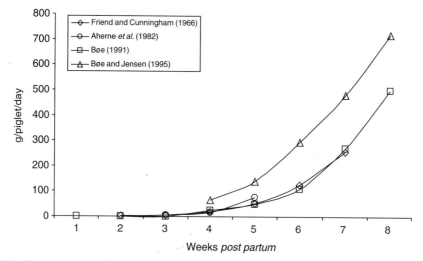

Fig. 10.5. Examples of development of creep feed intake during the first weeks *post partum*.

28 days compared to piglets in pens where the sow was confined (Pajor *et al.*, 2002).

Large variations between littermates with regard to the intake of creep feed are well described. As an example, Pajor *et al.* (1991) found that the total creep feed intake from 10 to 28 days of age varied from 118 to 1385 g/piglet within a litter consisting of ten piglets. In some studies the small piglets start consuming solid feed at a younger age than their larger littermates, whereas in other studies the opposite occurred. Piglets with low weight gain during the first weeks of lactation (indicating access to less productive teats) may start eating solid feed sooner and/or eat more solid food as compensation for the lack of milk than their larger littermates (e.g. Algers *et al.*, 1990; Appleby *et al.*, 1992; Fraser *et al.*, 1994). Heavier individuals, however, may have a more developed gastrointestinal tract than their smaller littermates, and therefore may show more interest in creep feed at an earlier age (De Passillé *et al.*, 1989; Pajor *et al.*, 1991).

Intake of feed and water in relation to commercial weaning

In some parts of the world, piglets are weaned as early as 2–3 weeks of age to improve productivity. In many European Union (EU) countries the average weaning age is closer to 4 weeks of age. Other countries and production systems such as organic pig production have weaning ages of 7 weeks or more. In all cases the whole litter is weaned simultaneously and abruptly, independent of the developmental stage of the individual piglet.

Despite attempts to stimulate early feed intake of piglets during lactation, considerable variations exist in feed intake within and between litters. Thus many piglets weaned at 3–4 weeks of age have not consumed any solid feed before the weaning (Bruininx *et al.*, 2002). The lower the weaning age, the less developed is their eating and drinking behaviour and the lower the number of individuals familiar with consuming solid feed and water. These piglets face a sudden transition from getting all their nourishment and liquid from a known source – the sow – to having the nourishment and the liquid usually offered separately in two different facilities: the feed trough and the water nipple or bowl.

When piglets are weaned early and moved to weaning pens, they fast for various lengths of time. The majority of the piglets start to eat solid feed within 5 h or earlier, but some piglets take more than 50 h post weaning before they eat for the first time (Bruininx *et al.*, 2002). Eating time and feed intake are low on the first day after weaning and then increase as illustrated in Fig. 10.6 (e.g. Bark *et al.*, 1986; Gardner *et al.*, 2001; Dybkjær

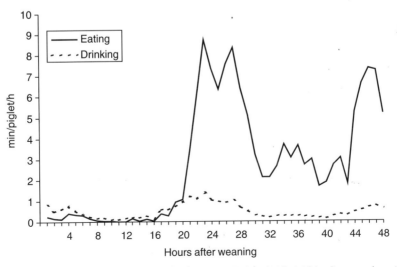

Fig. 10.6. Mean time spent eating and drinking by piglets during the first 48 h after weaning at 27 days of age. (Modified from Dybkjær *et al.*, 2006.)

et al., 2006). The amount of feed ingested during the first day(s) after weaning is important to the growth and health of the piglets. A long fasting period or a low feed consumption has detrimental effects on the digestive tract and disposes the piglet to diarrhoea and weight loss (McCracken and Kelly, 1993; Pluske *et al.*, 1997; Madec *et al.*, 1998). Individuals that have already consumed creep feed in the nursing period start to eat solid feed earlier after weaning (Bruininx *et al.*, 2002). They also eat more solid feed (Aherne *et al.*, 1982; Delumeau and Meunier-Salaün, 1995; Bruininx *et al.*, 2002, 2004) and/or gain more weight during the early post-weaning period (Aherne *et al.*, 1982; Appleby *et al.*, 1991; Fraser *et al.*, 1994; Bruininx *et al.*, 2002, 2004) than individuals that have not eaten creep feed. Some authors, however, found the relationship between creep feed intake and post-weaning growth of the piglets to be weak and confounded with other variables, with no documented causal link (e.g. Appleby *et al.*, 1991; Pajor *et al.*, 1991). However, given the potential benefit of creep feed, and the lack of documented negative effects to piglets, creep feeding appears to be a worthwhile strategy provided weaning age is not too low.

The intake of dry feed after weaning must obviously be accompanied by water intake. Piglets spend more time at the water facility on the first day after weaning compared to before weaning. Drinking time then decreases (Algers *et al.*, 1990). This may be due to a high motivation for exploring the new post-weaning environment, including the water facility. Alternatively, some piglets may be trying to achieve a feeling of satiety by filling their stomachs with water (Yang *et al.*, 1981). The amount of water released from a water nipple within the first week after weaning reaches a maximum on the second day. In some instances no water is consumed until day 2 after weaning, and the water/food ratio is not stabilized until 2–3 weeks after weaning (Brooks *et al.*, 1984). Insufficient intake of water, sometimes caused by a low water delivery rate from the drinker, reduces feed intake after weaning (Barber *et al.*, 1989), resulting in an increased risk of health problems and poor growth. Limitations in access to water should be avoided and all efforts made to help

the piglets establish a good balance between intake of feed and water as soon as possible after weaning. Mixing feed and water may provide several advantages (see Brooks *et al.*, 2001), but due to hygiene problems this practice is not often used in large intensive herds during the first days or weeks after weaning.

Feeding Behaviour of Growing Pigs

Under natural conditions growing pigs will explore their environment using their snouts to sniff, root and find items to be tasted, chewed and eaten (Fig. 10.7). This appetitive foraging behaviour is somewhat different to the behaviour observed in the restricted and often relatively barren environment offered in conventional housing. The description of feeding behaviour in the following section is based on commercially housed and reared growing pigs.

Methods and measurements used to describe feeding behaviour

The daily food intake of growing pigs – like that of most free-feeding animals – is a summation of discrete bursts of feeding behaviour interspersed with periods where no food is ingested (Collier *et al.*, 1972; Auffray and Marcilloux, 1980; Bigelow and Houpt, 1988). The way in which these bursts of eating are distributed across time depends in part on the way feeding behaviour is measured (e.g. bite of food, visit to trough). In order to describe the feeding behaviour in a biologically meaningful way, it may be appropriate to collapse feeding activity (e.g. visits to the feeding trough) into behavioural bouts (e.g. meals) through some systematic and objective method (e.g. Lehner, 1979; Sibly *et al.*, 1990; Berdoy, 1993). However, most methods assume the occurrence of feeding to be a random process, which it is not. Tolkamp *et al.* (1998) developed a method which takes into account that the probability of feeding occurring will increase with the time since it last occurred. The method describes the frequency distribution of log-transformed interval lengths, and has been successfully applied to data from

Fig. 10.7.   Free-ranging pigs rooting in a field.

growing pigs (Morgan *et al.*, 2000). Thus, growing pigs consume their food in discrete meals, and show a bimodal distribution of feeding activity (Kersten *et al.*, 1980; De Haer and Merks, 1992) with high levels of feeding behaviour displayed at the beginning and towards the end of the active period. Most studies report that approximately one-third of the feed is consumed during the night (e.g. Labroue *et al.*, 1999; Collin *et al.*, 2001).

Meal frequency, meal size and meal duration are three base variables from which can be derived all other parameters used to describe the feeding behaviour of an animal offered a given food (Nielsen, 1999; Fig. 10.8). Breaking down daily food intake into its component behaviours reveals a greater underlying variation between individuals. It also offers a way to describe how different aspects of the environment affect the feeding behaviour of growing pigs.

The size of a meal – however one defines it – is dependent on the physical characteristics of the food offered, and is limited by the size of the stomach and controlled largely by inhibitory signals from gastrointestinal sites to the central nervous system (i.e. satiety feedback; Houpt, 1985). Important factors in this

context are the physiological constraints on feeding behaviour, which may limit the range within which an animal can express its feeding pattern.

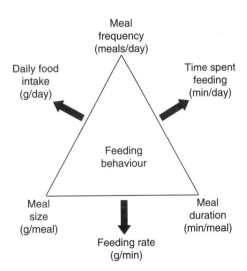

Fig. 10.8.   Schematic presentation of the relationship between short-term feeding behaviour parameters.

Influence of animal characteristics on feeding behaviour

The feeding behaviour of a pig in a given environment is dependent on various individual characteristics of the animal. Differences in feeding behaviour have been found between breeds of pigs, with Large White having more and Meishan having fewer feeder visits compared to Landrace (Labroue *et al.*, 1994; Quiniou *et al.*, 1999). Differences have also been found in the feeding behaviour of males, females and castrates that cannot be attributed directly to differences in growth rate or daily food intake. Across breeds, boars eat less frequently and more per visit than gilts (De Haer and de Vries, 1993), and castrates spend longer eating, and eat more, both per visit and per day, than boars (Labroue *et al.*, 1994; Hyun and Ellis, 2000).

The feeding pattern of pigs changes over time, from frequent visits to the trough by newly weaned piglets (Bigelow and Houpt, 1988) to few, large meals ingested by sows (Auffray and Marcilloux, 1980). It is not possible, however, to disentangle the simultaneous effects of increases in live weight, degree of maturity and level of experience.

Food accessibility, number of troughs and feeder design

The accessibility to the food can be reduced through increases in the pig/trough ratio, through modifications to the design of the feeder by making it more difficult to access the trough or by introducing some form of arbitrary response (e.g. bar-pressing) in order to obtain food. Whatever the reason, reduced accessibility to the food results in fewer, but larger, meals, most often with no affect on the level of daily food intake (Morrow and Walker, 1994; Nielsen *et al.*, 1995a,b, 1996a).

Although growing pigs are most often fed *ad libitum*, the feeders used in modern pig production do not all allow the same level of food accessibility. The way in which the different feeders may reduce the accessibility to the food can be divided into two main groups: feeders that make it difficult to gain access to the food (procurement cost, e.g. by placing a door in front of the trough), and feeders that make it difficult to retain access to the food (consumption cost, e.g. intermittent bar presses required). Some feeders incorporate both kinds of reduced accessibility. A procurement cost will reduce the number of meals, whereas a consumption cost is mostly found to increase the number of daily meals (e.g. Johnson *et al.*, 1984). A feeding trough, which contains separate food and water dispensers operated by the pig (e.g. Morrow and Walker, 1994), may effectively introduce a consumption cost and elevate the number of visits overall, especially if this is also the only source of water available in the pen.

Social constraints

For growing pigs kept in groups, the social environment impinges on almost every action performed by the individual animals. Individually housed pigs have been found to obtain their daily food intake by a large number of small meals, whereas group-housed pigs eat less frequent, but larger, meals (De Haer, 1992; Bornett *et al.*, 2000b). This is only found, however, when groups of growing pigs are sharing one trough space. When groups of ten pigs were given access to a four-space feeding trough (Nielsen *et al.*, 1996a), they displayed a feeding pattern similar to that of individually housed animals (i.e. many visits with a low feed intake per visit; De Haer, 1992), indicating that a low feeding frequency is not a result of the social environment *per se*, but is mainly attributable to increases in the pig/trough ratio. This effectively corresponds to the change in feeding behaviour when access to the trough is made more difficult.

Over and above changes with time and age, the feeding rate of individual animals in a given environment appears to be relatively stable (Auffray and Marcilloux, 1983; Nielsen, 1999). Within-meal variation in feeding rate is relatively small; it may be an artefact of the meal criterion used, and will generally have little or no influence on the overall feeding rate of an individual (Le Magnen, 1985). Nielsen (1999) suggested that augmented rates of feeding result from higher feeding motivation (i.e. hunger) and/or increased social constraint, when environmental constraints force the animal to eat faster in order to obtain a sufficient level of daily food intake. Changes in the environment, which result in an increased feeding rate, may be stressful to the animals. Thus, changes in the feeding rate may be used as an index of the social pressure imposed by a change in the given environment. Pigs may also change their feeding pattern in order to synchronize their behaviour with the rest of the group, even to the extent that daily feed intake is reduced compared with individually housed pigs (e.g. Gomez et al., 2000).

Group-housed pigs given access to a single-space feeder have no opportunity to feed simultaneously, but do show the bimodal feeding pattern also observed in individually housed animals (De Haer and Merks, 1992; Nielsen et al., 1995a). Pigs feeding from a multispace trough display a high level of simultaneous feeding (61% of all visits; Nielsen et al., 1996a; Fig. 10.9), and also show a preference for troughs directly adjacent to troughs occupied by another feeding pig. Morgan et al. (2003) found that pigs would more readily learn to select an appropriate diet when housed with a pig trained to select between two foods. Pigs also place a higher value on food in the presence of a companion pig than when tested in isolation (Pedersen et al., 2002).

A number of experiments have reported positive correlations between social rank and performance (McBride et al., 1964; Beilharz and Cox, 1967; Hansen et al., 1982), especially when access to feed has been restricted. Georgsson and Svendsen (2002) found that the smaller pigs in groups with access to only one feeding trough would eat less, and more often during the night, compared to the smaller pigs in groups with two trough spaces. This indicates that they are more constrained by the pig/trough ratio than their larger group-mates. However, the causal relationship between dominance value and production performance (e.g. live weight) is not clear, and some of the reported correlations are weak (McBride et al., 1964) or non-existent (Nielsen et al., 1995a). Some data suggest that dominant individuals may spend more time defending their food rather than eating it (Brouns and Edwards, 1994). Also, dominance assessed in a non-feeding situation may show little correlation with that displayed in a feeding context. Factors such as high space allowance and straw provision may disguise or eliminate any

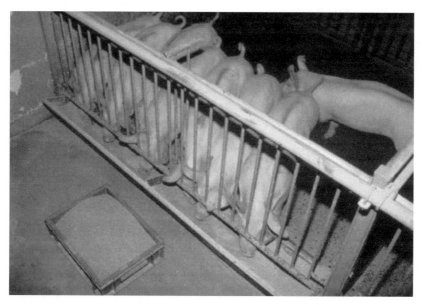

Fig. 10.9. Pigs feeding together from a long trough.

significant correlation between dominance rank and performance (Hansen and Hagelsø, 1980).

Physical form of the feed and nutrient density

The physical form of the feed most profoundly influences feeding rate. When calculated in grams (but not necessarily joules) per minute, the feeding rate will be substantially higher on liquid diets than on pelleted food, which in turn can be eaten faster than dry meal (Laitat *et al.*, 1999). Also, provision of water in the feed trough will alter the rate of ingestion as well as eliminate the need to interrupt a meal in order to drink. Pigs make rapid changes in feeding behaviour to accommodate changes in feed bulk content (Whittemore *et al.*, 2002). Pigs with previous experience of bulky feeds are able to consume more of a high-fibre diet through enlargements of the gastrointestinal tract (Kyriazakis and Emmans, 1995). When pigs are fed an amino acid-deficient diet, a reduction in feeding rate is seen, and daily food intake decreases through smaller meal sizes (Montgomery *et al.*, 1978).

Ambient temperature

Ferguson *et al.* (1994) suggest that it is unlikely that fast-growing pigs will be able to reach their growth potential when kept at temperatures above 20°C, especially when high-nutrient-density feeds are offered. Lowering the ambient temperature increases the heat demand of the animals. Ingram and Legge (1974) found a sustained increase in intake when changing the ambient temperature for growing pigs from 25°C to 5°C. When temperature fluctuates around a comfort value over the day, growing pigs will have the lowest feed intake during the warmest periods of the day compensated by a higher feed intake during the cold periods (Quiniou *et al.*, 2000). Nienaber *et al.* (1991) found that the usual age/time-dependent decrease in meal frequency and increase in meal size disappeared when group-housed pigs were kept under cool (13°C) and severe cold (5°C) stress conditions,

compared to thermoneutral (21°C) conditions, without significant differences in feed intake or final live weight.

The effects of environmental factors on feeding behaviour are non-linear. A set of thresholds appears to operate where a change from 'one trough' to 'more than one trough' and from 'one pig' to 'more than one pig' have major influences on the feeding behaviour observed. We do not yet know enough about the effects on the feeding behaviour of individuals when they approach maximum feeder capacity of 11–17 pigs per trough (Korthals, 2000). Nielsen *et al.* (1996b) found that pigs which display a relatively low number of feeder visits when group-housed had a lower daily food intake than pigs with a high number of daily feeder visits. This difference subsequently disappeared as the pigs were moved to individual housing. However, such a positive correlation between number of daily feeder visits and daily food intake is not always found (e.g. Nielsen *et al.*, 1995a,b). A high feeding frequency is likely to be a more flexible feeding pattern (Fig. 10.10), and pigs appear to change their feeding pattern only during periods when restrictions are imposed (Bornett *et al.*, 2000a).

Feeding Behaviour of Sows and Boars

Under semi-natural conditions male and female adult pigs use 50–60% of their daily active time grazing and rooting. This foraging activity takes place diurnally during the morning and afternoon (Stolba and Wood-Gush, 1989). Commercially kept fully grown pigs with *ad libitum* access to food obtain their daily food intake in three large meals, one of which occurs during the night (Auffray and Marcilloux, 1980).

Restrictive feeding and hunger in sows

Modern slaughter pigs are genetically selected for their ability to optimize growth and food utilization in order to reach slaughter weight as fast as possible. The breeding stock, however, is fed restrictively during most of their life in

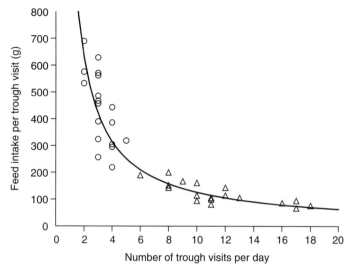

Fig. 10.10. Number of daily trough visits (NDV) recorded over 21 days and plotted against mean feed intake per visit (FIV) for two pigs of similar size housed within a group of 20. Both pigs had similar levels of daily food intake (1255 g/day (circles) and 1264 g/day (triangles), respectively), and the pictured isoline represents all combinations of NDV and FIV resulting in a daily intake of 1260 g. (From Nielsen, 1995.)

order to prolong their longevity. During rearing, gilts and boars are fed at a level of 75–100% of appetite depending on the strategy to meet the correct balance of genetics, age, body weight, body condition and sexual maturity at first mating (Close and Cole, 2000). During the pregnancy period gilts are restricted to a feeding level that accommodates demands for maintenance and body growth, which often is around 2 kg of feed (26.8 MJ ME/day) for a gilt with a body weight of 125 kg. When sows and sexually active boars are fully grown, their daily food allowance is restricted further to 1.5 times the level of maintenance, equal to 60% of the *ad libitum* intake. For a sow with a body weight of 200 kg this is equivalent to 1.8 kg of feed per day (24.6 MJ ME/day), when the anticipated number of pigs in a litter is 12 and the gestation weight gain is 30 kg (Anonymous, 1998). Such feeding practices are generally accepted in most of the major pig-producing countries (Danielsen and Vestergaard, 2001). Independent of the housing system used, the feed ration during gestation (2 kg/day) can be eaten by the sow in about 5 min, if the food is composed of energy dense low dietary fibre grain products, such as wheat. This leaves the ani-

mals in a situation of unsatisfied feeding motivation (hunger), which may enhance frustration, aggression, stereotypies and excessive rooting behaviour after feeding and sham-chewing (Vestergaard and Danielsen, 1998; Whittaker *et al.*, 1998; Danielsen and Vestergaard, 2001). During the lactation period sows are fed *ad libitum* with high-energy diets in order to maximize milk production. The estimated feed intake will often be between 4.3 and 6.4 kg/day depending on the number of piglets in the litter and the anticipated weight changes of the sow during lactation (Anonymous, 1998).

Effects of housing system on sow feeding behaviour

The feeding behaviour of gestating sows is highly influenced by the housing system in which they are kept. The normal diurnal feeding activity of sows kept under semi-natural conditions is not observed in systems with electronic sow feeding (ESF) stations, where the feeding activity is affected by the accessibility of the feeding station, leading to feeder occupation over most of the 24-h period. If sows are protected from interference from other sows whilst eating, such as in a feeding box system, this leads to a diurnal feeding activity resembling the situation under semi-natural conditions (Krause *et al.*, 1997). The frequency of agonistic interactions between group-housed sows is highest when they are competing for food, and this aggression is especially caused by the high-ranking, dominant animals (Csermely and Wood-Gush, 1986). Indeed, dietary deviations from nutritional requirements such as high-density diets, extended feed or water deprivation, and insufficient roughage (Hurnik, 1993) are among the most commonly mentioned concerns regarding the welfare of farm animals in modern production systems.

Methods to prevent the feeling of hunger in pregnant sows

The short-term regulation of meal size and frequency is determined by satiety signals, which consist of several elements including palatability of the food, activation of stretch- and chemoreceptors in the stomach and duodenum, plasma glucose, insulin, glucagon, glycerol, body temperature, stress and conditioned responses (Vander *et al.*, 1986; Westerterp-Platenga *et al.*, 1997). The long-term regulation of the total body energy balance is controlled by multiple physiological systems, of which the hormonal and neuronal signals from the gut to the brain play a key role (Schwartz *et al.*, 1992; Riedy *et al.*, 1995). One way to relieve the feeling of hunger in pregnant sows (without overfeeding them) is to allow the animals access to low-energy, high-fibre feedstuffs in the form of roughage. It has been shown that inclusion of soluble viscous polysaccharides in carbohydrate meals curbs the postprandial glucose and insulin responses in humans (Vahouny and Cassidy, 1985) as well as in pigs (Vestergaard, 1998). The soluble dietary fibres appear to delay the transport of nutrients across the intestinal epithelium (Asp, 1995).

   The dietary fibre content varies greatly among the grain products normally used in pig diets (Knudsen, 1997). While maize and wheat are both very low in dietary fibre

(maize: 108 ± 4 g/kg; wheat: 138 ± 10 g/kg), rye, barley and oats all contain considerably higher dietary fibre levels (rye: 174 ± 10 g/kg; barley: 221 ± 13 g/kg; oats: 298 ± 19 g/kg). Oats in particular contain welfare-beneficial soluble dietary fibres (beta-glucans), which can delay stomach emptying after feeding (Johansen *et al.*, 1996). The capacity of sows to utilize dietary fibre is much higher than for growing pigs, due to their more developed digestive tract (especially the caecum and colon). In addition, the low feeding level practised for the adult animals leads to a slow transit of the feed in the intestine (Fernández *et al.*, 1986; Noblet and Knudsen, 1997). Roughage and fibre-rich feed can therefore be added to the diet of sows without negative effects on their energy balance. Pigs appear to prefer types of roughage that they can utilize, such as leguminous plants, which are shown to have a higher digestibility and a more beneficial composition of dietary fibre than different grasses (Ohlsson, 1987; Andersson, 1997). Olsen *et al.* (2000) showed that whole-crop silage of oats, vetch and lupin was preferred by pigs compared to whole-crop silage of barley and peas, whole-crop silage of clover and grass, green grass meal, hay of clover and grass or chopped fodder beets. Sows on pasture will consume white clover rather than grass when both are available (Sehested *et al.*, 2000).

Feed-related diseases and injuries

Stomach ulcers

The stomach of the pig has developed to receive small amounts of food mainly from vegetable origin with high levels of dietary fibre. The last-eaten feed will be introduced into the central parts of the stomach, while feed already present will be positioned along the stomach walls. From these walls digestive secretions will gradually penetrate into the feed, making it more viscous so that it is transported against the fundus and duodenum by peristaltic movements. The digesta gradually becomes more and more acidic as it reaches the fundus, which leaves the anterior part of the stomach next to the oesophagus in a relatively alkaline condition, reinforced also by the alkaline saliva introduced in connection with each meal. This normal function, processing and passage of feed in the stomach do not take place when very concentrated and intensive feed items are used in the diets for sows or boars. Diets composed of low dietary fibre products and finely ground grain (especially wheat) will lead to a quick drop in acidity throughout the stomach (Nielsen and Ingvartsen, 2000), which leaves the oesophageal part at a higher risk of ulcer development. If the condition becomes chronic, there is a high risk of vessel ruptures, which can lead to acute bleeding in the stomach and sudden death.

Torsion of abdominal organs

Torsion of the abdominal organs occurs suddenly and can involve the spleen, stomach or a part of the small intestines or the liver. It will always be fatal to the animal either due to disturbances in the circulation or internal bleeding. Restricted feeding or missed meals increases the risk of torsion (Morin *et al.*, 1984; Sanford *et al.*, 1984). This is mainly due to fighting in connection with feeding, or rapid consumption of large amounts of food, which is more likely to occur when sows compete for a restricted quantity of food. Chagnon *et al.* (1991) found increased risk of intestinal torsion in sows the day after weaning, which coincides with the time when sows are drastically food-restricted after *ad libitum* feeding during lactation.

Vulva biting

Vulva lesions caused by biting are only found in loose sow herds, and is more common in systems with ESF. In ESF systems only one sow is able to eat at a time. This causes a highly competitive feeding situation, thereby increasing the risk of vulva biting, and the mechanism of gate control is suggested to be one of the major determinants (Jensen *et al.*, 1995). The relative risk of vulva lesions was found to increase 2.6 times in systems without roughage feeding (Gjein and Larssen, 1995).

Concluding Remarks

Food is the most important short-term commodity for pigs, showing an inelastic demand no matter how much work the animals have to do in order to get their food reward (Ladewig and Matthews, 1996). It is therefore not surprising that the different characteristics of feeding behaviour play a pivotal role in the life of pigs at all stages of their development. From the first attempts to suckle to the restricted feeding of pregnant sows, this chapter has highlighted a number of issues related to the feeding behaviour of pigs. Other aspects of feeding are dealt with elsewhere, such as detailed descriptions of suckling (German *et al.*, Chapter 4, this volume), mastication (Popowics and Herring, Chapter 5, this volume), feeding of extensively kept pigs (Glatz *et al.*, Chapter 18, this volume), as well as additional welfare concerns associated with feeding (Koene, Chapter 6, this volume).

References

Aherne, F.X., Danielsen, V. and Nielsen, H.E. (1982) The effects of creep feeding on pre- and post-weaning pig performance. *Acta Agriculturae Scandinavica* 32, 155–160.

Algers, B. (1993) Nursing in pigs: communicating needs and distributing resources. *Journal of Animal Science* 71, 2826–2831.

Algers, B. and Jensen, P. (1985) Communication during suckling in the domestic pig. Effects of continuous noise. *Applied Animal Behaviour Science* 14, 49–61.

Algers, B. and Jensen, P. (1991) Teat stimulation and milk production during early lactation in sows: effects of continuous noise. *Canadian Journal of Animal Science* 71, 51–60.

Algers, B., Jensen, P. and Steinwall, L. (1990) Behaviour and weight changes at weaning and regrouping of pigs in relation to teat quality. *Applied Animal Behaviour Science* 26, 143–155.

Andersson, C. (1997) Forages for growing pigs – partition of digestion and nutritive values. PhD thesis, Swedish University of Agricultural Sciences, Uppsala, Sweden, 77 pp.

Anonymous (1998) *Nutrient Requirements of Swine.* National Research Council, National Academy Press, Washington, DC, 189 pp.

Appleby, M.C., Pajor, E.A. and Fraser, D. (1991) Effect of management option on creep feeding by piglets. *Animal Production* 53, 361–366.

Appleby, M.C., Pajor, E.A. and Fraser, D. (1992) Individual variation in feeding and growth of piglets: effects of increased access to creep food. *Animal Production* 55, 147–152.

Asp, N.-G.L. (1995) Classification and methodology of food carbohydrates as related to nutritional effects. *American Journal of Clinical Nutrition* 61(suppl.), 930S–937S.

Auffray, P. and Marcilloux, J.C. (1980) Analyse de la séquence alimentaire du porc, du sevrage à l'état adulte. *Reproduction, Nutrition, Developpement* 20(5B), 1625–1632.

Auffray, P. and Marcilloux, J.C. (1983) Etude de la sequence alimentaire du porc adulte. *Reproduction, Nutrition, Developpement* 23, 517–524.

Auldist, D.E., Carlson, D., Morrish, L., Wakeford, C.M. and King, R.H. (2000) The influence of suckling interval on milk production of sows. *Journal of Animal Science* 78, 2026–2031.

Barber, J., Brooks, P.H. and Carpenter, J.L. (1989) The effects of water delivery rate on the voluntary food intake, water use and performance of early-weaned pigs from 3 to 6 weeks of age. In: Forbes, J.M., Varley, M.A. and Lawrence, T.L.J. (eds) *The Voluntary Feed Intake of Pigs.* Occasional Publication No. 13, British Society of Animal Production, Edinburgh, UK, pp. 103–104.

Bark, L.J., Crenshaw, T.D. and Leibbrandt, V.D. (1986) The effect of meal intervals and weaning on feed intake of early weaned pigs. *Journal of Animal Science* 62, 1233–1239.

Barnett, K.L., Kornegay, E.T., Risley, C.R., Lindemann, M.D. and Schurig, G.G. (1989) Characterization of creep feed consumption and its subsequent effects on immune response, scouring index and performance of weanling pigs. *Journal of Animal Science* 67, 2698–2708.

Bauman, R.H., Kadlec, J.E. and Powlen, P.A. (1966) *Some Factors Affecting Death Loss in Baby Pigs.* Research Bulletin No. 810, Purdue University Agriculture Experiment Station, Lafayette, Indiana, 9 pp.

Beilharz, R.G. and Cox, D.F. (1967) Social dominance in swine. *Animal Behaviour* 15, 117–122.

Berdoy, M. (1993) Defining bouts of behaviour: a three-process model. *Animal Behaviour* 46, 387–396.

Bigelow, J.A. and Houpt, R.T. (1988) Feeding and drinking patterns in young pigs. *Physiology & Behavior* 43, 99–109.

Blecha, F. (1998) Immunological aspects: comparison with other species. In: Verstegen, N.W.A., Moughan, P.J. and Schrama, J.W. (eds) *The Lactating Sow.* Wageningen Pers, Wageningen, The Netherlands, pp. 23–44.

Bøe, K. (1991) The process of weaning in pigs: when the sow decides. *Applied Animal Behaviour Science* 30, 47–59.

Bøe, K. (1993) Maternal behaviour of lactating sows in a loose-housing system. *Applied Animal Behaviour Science* 35, 327–338.

Bøe, K. and Jensen, P. (1995) Individual differences in suckling and solid food intake by piglets. *Applied Animal Behaviour Science* 42, 183–192.

Bornett, H.L.I., Morgan, C.A., Lawrence, A.B. and Mann, J. (2000a) The flexibility of feeding patterns in individually housed pigs. *Animal Science* 70, 457–469.

Bornett, H.L.I., Morgan, C.A., Lawrence, A.B. and Mann, J. (2000b) The effect of group housing on feeding patterns and social behaviour of previously individually housed growing pigs. *Applied Animal Behaviour Science* 70, 127–141.

Brooks, P.H., Russell, S.J. and Carpenter, J.L. (1984) Water intake of weaned piglets from three to seven weeks old. *Veterinary Record* 115(20), 513–515.

Brooks, P.H., Morgan, C.A., Beal, J.D., Demeckova, V. and Cambell, A. (2001) Liquid feeding for the young piglet. In: Varley, M.A. and Wiseman, J. (eds) *The Weaner Pig: Nutrition and Management.* CAB International, Wallingford, UK, pp. 153–178.

Brouns, F. and Edwards, S.A. (1994) Social rank and feeding behaviour of group-housed sows fed competitively or *ad libitum*. *Applied Animal Behaviour Science* 39, 225–235.

Bruininx, E.M.A.M., Binnendijk, G.P., van der Peet-Schwering, C.M.C., Schrama, J.W., den Hartog, L.A., Everts, H. and Beynen, A.C. (2002) Effect of creep feed consumption on individual feed intake characteristics and performance of group-housed weanling pigs. *Journal of Animal Science* 80, 1413–1418.

Bruininx, E.M.A.M., Schellingerhout, A.B., Binnendijk, G.P., van der Peet-Schwering, C.M.C., Schrama, J.W., den Hartog, L.A., Everts, H. and Beynen, A.C. (2004) Individually assessed creep food consumption by suckled piglets: influence on post-weaning food intake characteristics and indicators of gut structure and hind-gut fermentation. *Animal Science* 78(1), 67–75.

Bünger, B. (1985) Eine ethologische methode zur vitalitätseinschätzung neugeborener Ferkel. *Monatsheft Veterinär-Medizin* 40, 519–524.

Castrén, H., Algers, B. and Jensen, P. (1989a) Occurrence of unsuccessful sucklings in newborn piglets in a semi-natural environment. *Applied Animal Behaviour Science* 23, 61–73.

Castrén, H., Algers, B., Jensen, P. and Saloniemi, H. (1989b) Suckling behaviour and milk consumption in newborn piglets as a response to sow grunting. *Applied Animal Behaviour Science* 24, 227–238.

Castrén, H., Algers, B., De Passillé, A.M., Rushen, J. and Uvnäs-Moberg, K. (1993) Early milk ejection, prolonged parturition and periparturient oxytocin release in the pig. *Animal Production* 57, 465–471.

Chagnon, M., D'Allaire, S. and Drolet, R. (1991) A prospective study of sow mortality in breeding herds. *Canadian Journal of Veterinary Research* 55, 180–184.

Close, W.H. and Cole, D.J.A. (2000) *Nutrition of Sows and Boars*. Nottingham University Press, UK, 377 pp.

Collier, G., Hirsch, E. and Hamlin, P.H. (1972) The ecological determinants of reinforcement in the rat. *Physiology & Behavior* 9, 705–716.

Collin, A., van Milgen, J., Dubois, S. and Noblet, J. (2001) Effect of high temperature on feeding behaviour and heat production in group-housed young pigs. *British Journal of Nutrition* 86, 63–70.

Contreras, C., Mujica, F. and Bate, L. (2001) Effects of increasing the frequency of suckling, induced by maternal vocalizations, on the live weight of piglets. *Avances en Producción Animal* 26, 165–173.

Cronin, G.M. and van Amerongen, G. (1991) The effect of modifying the farrowing environment on sow behaviour and survival and growth of the piglets. *Applied Animal Behaviour Science* 30, 287–298.

Cronin, G.M., Leeson, E., Cronin, J.G. and Barnett, J.L. (2001) The effect of broadcasting sow suckling grunts in the lactation shed on piglets growth. *Asian–Australasian Journal of Animal Science* 14, 1019–1023.

Csermely, D. and Wood-Gush, D.G.M. (1986) Agonistic behaviour in grouped sows. I. The influence of feeding. *Biology of Behaviour* 11, 244–252.

Damm, B.I., Vestergaard, K.S., Schrøder-Petersen, D.L. and Ladewig, J. (2000) The effects of branches on prepartum nest building in gilts with access to straw. *Applied Animal Behaviour Science* 68, 113–124.

Damm, B.I., Friggens, N.C., Nielsen, J., Ingvartsen, K.L. and Pedersen, L.J. (2002) Factors affecting the transfer of porcine parvovirus antibodies from sow to piglets. *Journal of Veterinary Medicine* 49A, 487–495.

Damm, B.I., Pedersen, L.J., Jessen, L.B., Thamsborg, S.M., Mejer, H. and Ersbøll, A.K. (2003) The gradual weaning process in outdoor sows and piglets in relation to nematode infections. *Applied Animal Behaviour Science* 82, 101–120.

Danielsen, V. and Vestergaard, E.-M. (2001) Dietary fibre for pregnant sows: effect on performance and behaviour. *Animal Feed Science and Technology* 90, 71–80.

De Haer, L.C.M. (1992) Relevance of eating pattern for selection of growing pigs. PhD thesis, Wageningen University, Wageningen, The Netherlands, 159 pp.

De Haer, L.C.M. and de Vries, A.G. (1993) Effects of genotype and sex on the feed intake pattern of group-housed growing pigs. *Livestock Production Sciences* 36, 223–232.

De Haer, L.C.M. and Merks, J.W.M. (1992) Patterns of daily food intake in growing pigs. *Animal Production* 54, 95–104.

Delumeau, O. and Meunier-Salaün, M.C. (1995) Effect of early familiarity on the creep feeding behaviour in suckling piglets and after weaning. *Behavioural Processes* 34, 185–196.

De Passillé, A.M.B. and Robert, S. (1989) Behaviour of lactating sows: influence of stage of lactation and husbandry practices at weaning. *Applied Animal Behaviour Science* 23, 315–329.

De Passillé, A.M.B. and Rushen, J. (1989) Suckling and teat disputes by neonatal piglets. *Applied Animal Behaviour Science* 22, 23–38.

De Passillé, A.M.B., Rushen, J. and Hartsock, T.G. (1988a) Ontogeny of teat fidelity in pigs and its relation to competition at suckling. *Canadian Journal of Animal Science* 68, 325–338.

De Passillé, A.M.B., Rushen, J. and Pelletier, G. (1988b) Sucking behaviour and serum immunoglobulin levels in neonatal piglets. *Animal Production* 47, 447–456.

De Passillé, A.M.B., Pelletier, G., Ménard, J. and Morisset, J. (1989) Relationships of weight gain and behaviour to digestive organ weight and enzyme activities in piglets. *Journal of Animal Science* 67, 2921–2929.

Dybkjær, L., Jacobsen, A.P., Poulsen, H.D. and Tøgersen, F.A. (2006) Eating and drinking activity of newly weaned piglets: Effects of individual characteristics, social mixing, and addition of extra zinc to the feed. *Journal of Animal Science* (in press).

Ellendorf, F., Forsling, M.L. and Poulain, D.A. (1982) The milk ejection reflex in the pig. *Journal of Physiology* 333, 577–594.

English, P.R., Robb, C.M. and Dias, M.F.M. (1980) Evaluation of creep feeding using a highly-digestible diet for litters weaned at 4 weeks of age. *Animal Production* 30, 496.

Etienne, M., Dourmad, J.Y. and Noblet, J. (1998) The influence of some sow and piglet characteristics and of environmental conditions on milk production. In: Verstegen, N.W.A., Moughan, P.J. and Schrama, J.W. (eds) *The Lactating Sow.* Wageningen Pers, Wageningen, The Netherlands, pp. 285–299.

Ferguson, N.S., Gous, R.M. and Emmans, G.C. (1994) Preferred components for the construction of a new simulation model of growth, feed intake and nutrient requirements of growing pigs. *South African Journal of Animal Science* 24, 10–17.

Fernández, J.A., Jørgensen, H. and Just, A. (1986) Comparative digestibility experiments with growing pigs and adult sows. *Animal Production* 43, 127–132.

Fraser, D. (1980) A review of the behavioural mechanism of milk ejection of the domestic pig. *Applied Animal Ethology* 6, 247–255.

Fraser, D. (1984) Some factors influencing the availability of colostrum to piglets. *Animal Production* 39, 115–123.

Fraser, D. and Rushen, J. (1992) Colostrum intake by newborn piglets. *Canadian Journal of Animal Science* 72, 1–13.

Fraser, D., Phillips, P.A., Thompson, B.K. and Peeters Weem, W.B. (1988) Use of water by piglets in the first days after birth. *Canadian Journal of Animal Science* 68, 603–610.

Fraser, D., Patience, J.F., Phillips, P.A. and McLeese, J.M. (1990) Water for piglets and lactating sows: quantity, quality and quandaries. In: Haresign, W. and Cole, D.J.A. (eds) *Recent Advances in Animal Nutrition.* Butterworths, London, pp. 137–160.

Fraser, D., Feddes, J.J.R. and Pajor, E.A. (1994) The relationship between creep feeding behaviour of piglets and adaptation to weaning: effect of diet quality. *Canadian Journal of Animal Science* 74, 1–6.

Fraser, D., Kramer, D.L., Pajor, E.A. and Weary, D.M. (1995) Conflict and cooperation: sociobiological principles and the behaviour of pigs. *Applied Animal Behaviour Science* 44, 139–157.

Friend, D.W. and Cunningham, H.M. (1966) The effect of water consumption on the growth, feed intake, and carcass composition of suckling piglets. *Canadian Journal of Animal Science* 46, 203–209.

Gardner, J.M., de Lange, C.F.M. and Widowski, T.M. (2001) Belly-nosing in early-weaned piglets is not influenced by diet quality or the presence of milk in the diet. *Journal of Animal Science* 79, 73–80.

Georgsson, L. and Svendsen, J. (2002) Degree of competition at feeding differentially affects behavior and performance of group-housed growing-finishing pigs of different relative weights. *Journal of Animal Science* 80, 376–383.

Gjein, H. and Larssen, R.B. (1995) Housing of pregnant sows in loose and confined systems – a field study 1. Vulva and body lesions, culling reasons and production results. *Acta Veterinaria Scandinavica* 36, 185–200.

Gomez, R.S., Lewis, A.J., Miller, P.S. and Chen, H.Y. (2000) Growth performance and digestive and metabolic responses of gilts penned individually or in groups of four. *Journal of Animal Science* 78, 597–603.

Hansen, L.L. and Hagelsø, A.M. (1980) A general survey of environmental influence on the social hierarchy function in pigs. *Acta Agriculturae Scandinavica* 30, 388–392.

Hansen, L.L., Hagelsø, A.M. and Madsen, A. (1982) Behavioural results and performance of bacon pigs fed *ad libitum* from one or several self-feeders. *Applied Animal Ethology* 8, 307–333.

Hemsworth, P.H., Winfield, C.G. and Mullaney, P.D. (1976) A study of the development of the teat order in piglets. *Applied Animal Ethology* 2, 225–233.

Herskin, M.S., Jensen, K.H. and Thodberg, K. (1998) Influence of environmental stimuli on maternal behaviour related to bonding, reactivity and crushing of piglets in domestic sows. *Applied Animal Behaviour Science* 58, 241–254.

Herskin, M.S., Jensen, K.H. and Thodberg, K. (1999) Influence of environmental stimuli on nursing and suckling behaviour in domestic sows and piglets. *Animal Science* 68, 27–34.

Houpt, T.R. (1985) The physiological determination of meal size in pigs. *Proceedings of the Nutrition Society* 44, 323–330.

Hurley, W.L. (2001) Mammary gland growth in the lactating sow. *Livestock Production Sciences* 70, 149–157.

Hurnik, J.F. (1993) Ethics and animal agriculture. *Journal of Agricultural and Environmental Ethics* 6, 21–35.

Hyun, Y. and Ellis, M. (2000) Relationships between feed intake traits, monitored using a computerized feed intake recording system, and growth performance and body composition of group-housed pigs. *Asian–Australasian Journal of Animal Sciences* 13, 1717–1725.

Illmann, G. and Madlafousek, J. (1995) Occurrence and characteristics of unsuccessful nursings in mini-pigs during the first week of life. *Applied Animal Behaviour Science* 44, 9–18.

Illmann, G., Spinka, M. and Stetkova, Z. (1999) Predictability of nursings without milk ejection in domestic pigs. *Applied Animal Behaviour Science* 61, 303–311.

Ingram, D.L. and Legge, K.F. (1974) Effects of environmental temperature on food intake in growing pigs. *Comparative Biochemistry and Physiology* 48A, 573–581.

Jensen, K.H., Pedersen, B.K., Pedersen, L.J. and Jørgensen, E. (1995) Well-being in pregnant sows: confinement versus group housing with electronic sow feeding. *Acta Agriculturae Scandinavica Section A, Animal Science* 45, 266–275.

Jensen, P. (1986) Observations on the maternal behaviour of free-ranging domestic pig. *Applied Animal Behaviour Science* 16, 131–142.

Jensen, P. (1988) Maternal behaviour and mother–young interactions during lactation in free-ranging domestic pigs. *Applied Animal Behaviour Science* 20, 297–308.

Jensen, P. (1993) Nest building in domestic sows: the role of external stimuli. *Animal Behaviour* 45, 351–358.

Jensen, P. (1995) The weaning process of free-ranging domestic pigs: within- and between-litter variations. *Ethology* 100, 14–25.

Jensen, P. and Recén, B. (1989) When to wean – observations from free-ranging domestic pigs. *Applied Animal Behaviour Science* 23, 49–60.

Jensen, P. and Redbo, I. (1987) Behaviour during nest leaving in free-ranging domestic pigs. *Applied Animal Behaviour Science* 18, 355–362.

Jensen, P. and Stangel, G. (1992) Behaviour of piglets during weaning in a semi-natural enclosure. *Applied Animal Behaviour Science* 33, 227–238.

Jensen, P., Stangel, G. and Algers, B. (1991) Nursing and suckling behaviour of semi-naturally kept pigs during the first 10 days postpartum. *Applied Animal Behaviour Science* 31, 195–209.

Jensen, P., Vestergaard, K. and Algers, B. (1993) Nestbuilding in free-ranging domestic sows. *Applied Animal Behaviour Science* 38, 245–255.

Johansen, H.N., Knudsen, K.E.B., Sandström, B. and Skjøth, F. (1996) Effects of varying content of soluble dietary fibre from wheat flour and oat milling fractions on gastric emptying in pigs. *British Journal of Nutrition* 75, 339–351.

Johnson, D.F., Ackroff, K.M., Collier, G.H. and Plescia, L. (1984) Effects of dietary nutrients and foraging costs on meal patterns of rats. *Physiology & Behavior* 33, 465–471.

Kent, J.C., Kennaugh, L.M. and Hartman, P.E. (2003) Intramammary pressure in the lactating sow in response to oxytocin and during natural milk ejections throughout lactation. *Journal of Dairy Science* 70, 131–138.

Kersten, A., Strubbe, J.H. and Spiteri, N.J. (1980) Meal patterning of rats with changes in day length and food availability. *Physiology & Behavior* 25, 953–958.

Kim, S.W., Osaka, I., Hurley, W.L. and Easter, R.A. (1999) Mammary gland growth as influenced by litter size in lactating sows: impact on lysine requirement. *Journal of Animal Science* 77, 3316–3321.

King, R.H., Mullan, B.P., Dunshea, F.R. and Dove, H. (1997) The influence of piglets body weight on milk production of sows. *Livestock Production Sciences* 47, 169–174.

Klobasa, F., Werhahn, E. and Butler, J.E. (1987) Composition of sow milk during lactation. *Journal of Animal Science* 64, 1458–1466.

Klobasa, F., Habe, F. and Werhahn, E. (1990) Untersuchungen über die absorption der kolostralen immunoglobuline bei neugeborener Ferkeln. I. Mitteilung: einfluss der zeit von der geburt bis zur ersten nahrungsaufnahme. *Berliner Münchner Tierärztliche Wochenschrift* 103, 335–340.

Knudsen, K.E.B. (1997) Carbohydrate and lignin contents of plant materials used in animal feeding. *Animal Feed Science and Technology* 67, 319–338.

Korthals, R.L. (2000) Evaluation of space requirements for swine finishing feeders. *Transactions of the American Society of Agricultural Engineers* 43, 395–398.

Krause, M., van't Klooster, C.E., Buré, R.G., Metz, J.H.M. and Sambraus, H.H. (1997) The influence of sequential and simultaneous feeding and the availability of straw on the behaviour of gilts in group housing. *Netherlands Journal of Agricultural Science* 45, 33–48.

Kyriazakis, I. and Emmans, G.C. (1995) The voluntary feed intake of pigs given feeds based on wheat bran, dried citrus pulp and grass meal, in relation to measurements of feed bulk. *British Journal of Nutrition* 73, 191–207.

Labayen, J.P., Batungbacal, M.R. and Lopez, P.L. (1994) The performance of suckling piglets given creep feed at seven and fourteen days of age. *Philippine Journal of Veterinary Medicine* 31, 38–41.

Labroue, F., Gueblez, R., Sellier, P. and Meunier-Salaun, M.C. (1994) Feeding behaviour of group-housed Large White and Landrace pigs in French central testing stations. *Livestock Production Sciences* 40, 303–312.

Labroue, F., Gueblez, R., Meunier-Salaun, M.C. and Sellier, P. (1999) Feed intake behaviour of group-housed Pietrain and Large White growing pigs. *Annales de Zootechnie* 48, 247–261.

Ladewig, J. and Matthews, L.R. (1996) The role of operant conditioning in animal welfare research. *Acta Agriculturae Scandinavica Section A, Animal Science* 27(suppl.), 64–68.

Laitat, M., Vandenheede, M., Desiron, A., Canart, B. and Nicks, B. (1999) Comparison of performance, water intake and feeding behaviour of weaned pigs given either pellets or meal. *Animal Science* 69, 491–499.

Lehner, P.N. (1979) *Handbook of Ethological Methods.* Garland STPM Press, New York and London, 403 pp.

Le Magnen, J. (1985) *Hunger: Problems in the Behavioural Sciences.* Cambridge University Press, Cambridge, UK, 157 pp.

Lewis, N.J. and Hurnik, J.F. (1985) The development of nursing behaviour in swine. *Applied Animal Behaviour Science* 14, 225–232.

Madec, F., Bridoux, N., Bounaix, S. and Jestin, A. (1998) Measurement of digestive disorders in the piglet at weaning and related risk factors. *Preventive Veterinary Medicine* 35, 53–72.

McBride, G., James, J.W. and Hodgens, N.W. (1964) Social behaviour of domestic animals. IV. Growing pigs. *Animal Production* 6, 129–140.

McCracken, K.J. and Kelly, D. (1993) Development of digestive function and nutrition/disease interactions in the weaned pig. In: Farrell, D.J. (ed.) *Recent Advances in Animal Nutrition in Australia.* University of New England, Armidale, Australia, pp. 182–192.

Montgomery, G.W., Flux, D.S. and Carr, J.R. (1978) Feeding patterns in pigs: the effects of amino acid deficiency. *Physiology & Behavior* 20, 693–698.

Morgan, C.A., Emmans, G.C., Tolkamp, B.J. and Kyriazakis, I. (2000) Analysis of the feeding behavior of pigs using different models. *Physiology & Behavior* 68, 395–403.

Morgan, C.A., Kyriazakis, I., Lawrence, A.B., Chirnside, J. and Fullam, H. (2003) Diet selection by groups of pigs: effect of a trained individual on the rate of learning about novel foods differing in protein content. *Animal Science* 76, 101–109.

Morin, M., Sauvageau, R., Phaneuf, J.-B., Teuscher, E., Beauregard, M. and Lagacé, A. (1984) Torsion of abdominal organs in sows: a report of 36 cases. *Canadian Veterinary Journal* 25, 440–442.

Morrow, A.T.S. and Walker, N. (1994) A note on changes to feeding behaviour of growing pigs by fitting stalls to single-space feeders. *Animal Production* 59, 151–153.

Newberry, R.C. and Wood-Gush, D.G.M. (1985) The suckling behaviour of domestic pigs in a semi-natural environment. *Behaviour* 95, 11–25.

Newberry, R.C. and Wood-Gush, D.G.M. (1988) Development of some behavioural patterns in piglet under semi-natural conditions. *British Society of Animal Production* 46, 103–109.

Nielsen, B.L. (1995) Feeding behaviour of growing pigs: Effects of the social and physical environment. Ph.D. Thesis, The University of Edinburgh, United Kingdom, 123 pp.

Nielsen, B.L. (1999) On the interpretation of feeding behaviour measures and the use of feeding rate as an indicator of social constraint. *Applied Animal Behaviour Science* 63, 79–91.

Nielsen, B.L., Lawrence, A.B. and Whittemore, C.T. (1995a) Effect of group size on feeding behaviour, social behaviour, and performance of growing pigs using single-space feeders. *Livestock Production Sciences* 44, 73–85.

Nielsen, B.L., Lawrence, A.B. and Whittemore, C.T. (1995b) Effects of single-space feeder design on feeding behaviour and performance of growing pigs. *Animal Science* 61, 575–579.

Nielsen, B.L., Lawrence, A.B. and Whittemore, C.T. (1996a) Feeding behaviour of growing pigs using single or multi-space feeders. *Applied Animal Behaviour Science* 47, 235–246.

Nielsen, B.L., Lawrence, A.B. and Whittemore, C.T. (1996b) Effect of individual housing on the feeding behaviour of previously group housed growing pigs. *Applied Animal Behaviour Science* 47, 149–161.

Nielsen, E.K. and Ingvartsen, K.L. (2000) Effects of cereal disintegration method, feeding method and straw as bedding on stomach characteristics including ulcers and performance in growing pigs. *Acta Agricultura Scandinavica Section A, Animal Science* 50, 30–38.

Nienaber, J.A., McDonald, T.P., Hahn, G.L. and Chen, Y.R. (1991) Group feeding behavior of swine. *Transactions of the American Society of Agricultural Engineers* 34, 289–294.

Noblet, J. and Knudsen, K.E.B. (1997) Comparative digestibility of wheat, maize and sugar beet pulp non-starch polysaccharides in adult sows and growing pigs. In: Laplace, J.P., Feverier, C. and Barbeau, A. (eds) *Digestive Physiology in Pigs.* Saint Malo, INRA, France, pp. 571–574.

Ohlsson, C. (1987) Phenological stage comparisons between American and Swedish cultivars of red clover and timothy. MSc thesis, Iowa State University, Ames, Iowa, 138 pp.

Olsen, A.W., Vestergaard, E.-M. and Dybkjær, L. (2000) Roughage as additional rooting substrates for pigs. *Animal Science* 70, 451–456.

Pajor, E.A., Fraser, D. and Kramer, D.L. (1991) Consumption of solid food by suckling pigs: individual variation and relation to weight gain. *Applied Animal Behaviour Science* 32, 139–155.

Pajor, E.A., Kramer, D.L. and Fraser, D. (2000) Regulation of contact with offspring by domestic sows: temporal patterns and individual variation. *Ethology* 106, 37–51.

Pajor, E.A., Weary, D.M., Caceres, C., Fraser, D. and Kramer, D.L. (2002) Alternative housing for sows and litters. Part 3. Effects of piglet diet quality and sow-controlled housing on performance and behaviour. *Applied Animal Behaviour Science* 76, 267–277.

Pedersen, L.J., Jensen, M.B., Hansen, S.W., Munksgaard, L., Ladewig, J. and Matthews, L. (2002) Social isolation affects the motivation to work for food and straw in pigs as measured by operant conditioning techniques. *Applied Animal Behaviour Science* 77, 295–309.

Pedersen, L.J., Damm, B.I., Marchant-Forde, J.N. and Jensen, K.H. (2003) Effects of feed-back from the nest on maternal responsiveness and postural changes in primiparous sows during the first 24 h after farrowing onset. *Applied Animal Behaviour Science* 83, 109–124.

Petersen, H.T. (2004) Stimulation of feed intake of piglets during lactation. MSc thesis, The Royal Veterinary and Agricultural University, Copenhagen, Denmark, 31 pp.

Petersen, H.V. (1994) The development of feeding and investigatory behaviour in free-ranging domestic pigs during their first 18 weeks of life. *Applied Animal Behaviour Science* 42, 87–98.

Petersen, H.V., Vestergaard, K. and Jensen, P. (1989) Integration of piglets into social groups of free-ranging domestic pigs. *Applied Animal Behaviour Science* 23, 223–236.

Phillips, P.A. and Fraser, D. (1990) Water bowl size for newborn pigs. *Applied Engineering in Agriculture* 6(1), 79–81.

Phillips, P.A. and Fraser, D. (1991) Discovery of selected water dispensers by newborn piglets. *Canadian Journal of Animal Science* 71, 233–236.

Pluske, J.R., Williams, I.H. and Aherne, F.X. (1995) Nutrition of the neonatal pig. In: Varley, M.A. (ed.) *The Neonatal Pig: Development and Survival*. CAB International, Wallingford, UK, pp. 187–238.

Pluske, J.R., Hampson, D. and Williams, I.H. (1997) Factors influencing the structure and function of the small intestine in the weaned pig: a review. *Livestock Production Sciences* 51, 215–236.

Puppe, B. and Tuchscherer, A. (2000) The development of suckling frequency in pigs from birth to weaning of their piglets: a sociobiological approach. *Animal Science* 71, 273–279.

Quiniou, N., Dubois, S., Le Cozler, Y., Bernier, J.F. and Noblet, J. (1999) Effect of growth potential (body weight and breed/castration combination) on the feeding behaviour of individually kept growing pigs. *Livestock Production Sciences* 61, 13–22.

Quiniou, N., Renaudeau, D., Collin, A. and Noblet, J. (2000) Influence of high ambient temperatures and physiological stage on feeding behaviour of pigs. *Productions Animales* 13, 233–245.

Riedy, C.A., Chavez, M., Figlewicz, D.P. and Woods, S.C. (1995) Central insulin enhances sensitivity to cholecystokinin. *Physiology & Behavior* 58(4), 755–760.

Rosillon-Warnier, A. and Paquay, R. (1984) Development and consequences of teat-order in piglets. *Applied Animal Behaviour Science* 13, 47–58.

Sanford, S.E., Waters, E.H. and Josephson, G.K.A. (1984) Gastrosplenic torsions in sows. *Canadian Veterinary Journal* 25, 364.

Schwartz, M.W., Figlewicz, D.P., Baskin, D.G., Woods, S.C. and Porte, D. Jr (1992) Insulin in the brain: a hormonal regulator of energy balance. *Endocrine Review* 13(3), 387–414.

Sehested, J., Søegaard, K., Danielsen, V. and Kristensen, V.F. (2000) Mixed grazing with sows and heifers: effects on animal performance and pasture. In: Hermansen, J.E., Lund, V. and Thuen, E. (eds) *Ecological Animal Husbandry in the Nordic Countries*. Proceedings from NJF-Seminar No. 303, Horsens, Denmark, pp. 35–39.

Sibly, R.M., Nott, H.M.R. and Fletcher, D.J. (1990) Splitting behaviour into bouts. *Animal Behaviour* 39, 63–69.

Spinka, M., Illman, G., Algers, B. and Stetkova, Z. (1997) The role of nursing frequency in milk production in domestic pigs. *Journal of Animal Science* 75, 1223–1228.

Stolba, A. and Wood-Gush, D.G.M. (1989) The behaviour of pigs in a semi-natural environment. *Animal Production* 48, 419–425.

Thodberg, K., Jensen, K.H., Herskin, M.S. and Jørgensen, E. (1999) Influence of environmental stimuli on nest building and farrowing behaviour in domestic sows. *Applied Animal Behaviour Science* 63, 131–144.

Thodberg, K., Jensen, K.H. and Herskin, M.S. (2002a) Nest building and farrowing in sows: relation to the reaction pattern during stress, farrowing environment and experience. *Applied Animal Behaviour Science* 77, 21–42.

Thodberg, K., Jensen, K.H. and Herskin, M.S. (2002b) Nursing behaviour, postpartum activity and reactivity in sows – effects of farrowing environment, previous experience and temperament. *Applied Animal Behaviour Science* 77, 53–76.

Thodberg, K. and Jensen, K.H. (2005) A test of sows' willingness to nurse. *Applied Animal Behaviour Science* 94, 49–58.

Thodberg, K. and Sørensen, M.T. (2006) Mammary development and milk production in the sow: Effects of udder massage, genotype and feeding in late gestation. *Livestock Science* (in press).

Tolkamp, B.J., Allcroft, D.J., Austin, E.J., Nielsen, B.L. and Kyriazakis, I. (1998) Satiety splits feeding behaviour into bouts. *Journal of Theoretical Biology* 194, 235–250.

Toner, M.S., King, R.H., Dunshea, F.R., Dove, H. and Atwood, C.S. (1996) The effect of exogenous somatotropin on lactation performance in first-litter sows. *Journal of Animal Science* 74, 167–172.

Vahouny, G.V. and Cassidy, M.M. (1985) Dietary fibres and absorption of nutrients. *Proceedings of the Society for Experimental Biology and Medicine* 180, 432–446.

Valros, A.E., Rundgren, M., Spinka, M., Saloniemi, H., Rydhmer, L. and Algers, B. (2002) Nursing behaviour of sows during 5 weeks lactation and effects on piglets growth. *Applied Animal Behaviour Science* 76, 93–104.

Vander, A.J., Sherman, J.H. and Luciano, D.S. (1986) *Human Physiology. The Mechanisms of Body Function.* McGraw-Hill, New York, pp. 533–541.

Vestergaard, E.-M. (1998) The effect of dietary fibre on welfare and productivity of sows. PhD thesis, The Royal Veterinary and Agricultural University, Copenhagen, Denmark, 104 pp.

Vestergaard, E.-M. and Danielsen, V. (1998) Dietary fibre for sows: effects of large amounts of soluble and insoluble fibres in the pregnancy period on the performance of sows during three reproductive cycles. *Animal Science* 68, 355–362.

Wakerley, J.B., Clarke, G. and Summerlee, A.J.S. (1994) Milk ejection and its control. In: Knobil, E. and Neill, J.D. (eds) *The Physiology of Reproduction.* Raven Press, New York, pp. 1131–1177.

Weary, D.M., Pajor, E.A., Thompson, B.J. and Fraser, D. (1996) Risky behaviour by piglets: a trade off between feeding and risk of mortality by maternal crushing. *Animal Behaviour* 51, 619–624.

Wechsler, B. and Brodman, N. (1996) The synchronization of nursing bouts in group-housed sows. *Applied Animal Behaviour Science* 47, 191–199.

Westerterp-Platenga, M.S., Wijckmans-Duijsens, N.E.G., Verboeket-van de Venne, W.P.H.G., de Graaf, K., Weststrate, J.A. and van het Hof, K.H. (1997) Diet-induced thermogenesis and satiety in humans after full-fat and reduced-fat meals. *Physiology & Behavior* 61(2), 343–349.

Whittaker, X., Spooler, H.A.M., Edwards, S.A., Lawrence, A.B. and Corning, S. (1998) The influence of dietary fibre and the provision of straw on the development of stereotypic behaviour in food restricted pregnant sows. *Applied Animal Behaviour Science* 61, 89–102.

Whittemore, C.T. and Fraser, D. (1974) The nursing and suckling behaviour of pigs. II. Vocalization of the sow in relation to suckling behaviour and milk ejection. *British Veterinary Journal* 130, 234–356.

Whittemore, E.C., Kyriazakis, I., Tolkamp, B.J. and Emmans, G.C. (2002) The short-term feeding behavior of growing pigs fed foods differing in bulk content. *Physiology & Behavior* 76, 131–141.

Yang, T.S., Howard, B. and MacFarlane, W.V. (1981) Effects of food on drinking behaviour of growing pigs. *Applied Animal Ethology* 7, 259–270.

11 Feeding Behaviour in Rabbits

T. Gidenne[1] and F. Lebas[2]

[1]INRA, Centre de Toulouse, Station de Recherches Cunicoles, BP 27, 31326
Castanet-Tolosan, France; [2]Cuniculture, 87a Chemin de Lassère, 31450
Corronsac, France

Introduction

The rabbit is a monogastric herbivore, belonging to the Lagomorpha order (Leporidae family: rabbits and hares; Grassé and Dekeuser, 1955). Thus, it is not a rodent although one of its main feeding behaviour features is to gnaw. Information about the feeding behaviour has been mainly obtained on the domestic rabbit, either bred for meat or fur production, or as a laboratory animal. It basically involves rabbits receiving *ad libitum* a balanced complete pelleted feed, supplemented or not with dry forages or straw, but most generally without a real free choice of food. One of the most original features of the rabbit feeding behaviour is the caecotrophy, which involves an excretion and an immediate consumption of specific faeces named soft faeces or 'caecotrophes'. Consequently, daily intake behaviour of the rabbit is constituted of two meals: feeds and caecotrophes.

We choose first to recall some basics of digestive anatomy and physiology, since it governs the feeding behaviour and allows us a better understanding. Regulation of the intake behaviour will be reviewed according to several factors: age, type of feed, etc. The last part of the chapter will be devoted to feeding behaviour of the wild and domestic rabbit in a situation of free choice.

Anatomy of the Alimentary Tract, Basics of Digestive Physiology, Caecotrophy

The digestive system of the rabbit is adapted to a herbivorous diet, including specific adaptations, from teeth to an enlarged hindgut for fermentation, and the separation of caecal digesta particles allowing for caecotrophy.

Anatomy

In an adult (4–4.5 kg) or semi-adult (2.5–3 kg) rabbit, the total length of the alimentary canal is 4.5–5 m. The general organization of the digestive tract is presented in Fig. 11.1 together with the main characteristics of each segment.

Mouth and oesophagus

The rabbit's dental formula is 2/1 0/0 3/2 3/3. The 28 teeth grow continuously throughout its life (1–2.4 mm/week). All teeth of the upper jaw are mainly worn out by those of the lower jaw and vice versa, without any real relationship with feed's hardness. In practice, incisors cut raw feeds and molars shred them coarsely; altogether, mastication efficiency is poor. Salivary glands produce saliva with low amylase concentration (10–20 times lower than

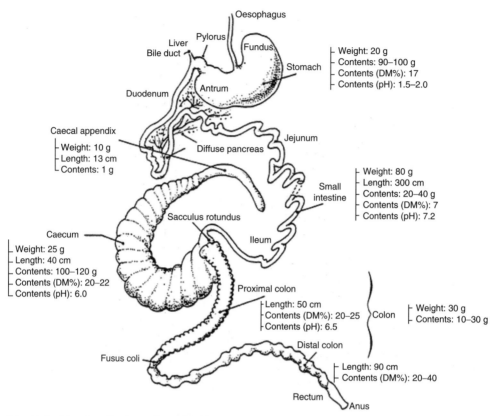

Fig. 11.1. The digestive tract of a rabbit. Numerical values are those observed in a 2.5 kg New Zealand White fed a pelleted balanced diet *ad libitum* (Lebas *et al.*, 1997).

that of pancreatic juice). The time between feed intake and swallowing is only a few seconds.

The oesophagus is short and acts exclusively to transport feed material from the mouth to the stomach. Regurgitation is impossible.

Stomach

After a quick transit through the oesophagus, feeds arrive in the simple stomach, which stores about 90–100 g of a rather pasty mixture of feedstuffs. In the stomach, the blind part (great curve) is named fundus and the opposite part is the antrum, i.e. the opening to the small intestine through the pylorus. The latter has a powerful sphincter, which regulates the entrance into the duodenum (the first part of the small intestine). The glands in the stomach wall secrete hydrochloric acid, pepsin and some minerals (Ca, K, Mg and Na). The stomach pH, which is always acidic, varies along the day

mainly in the fundus (in relation with soft faeces storage) (Fig. 11.2); the average pH is 1.5–2.0.

In a 9-week-old rabbit the stomach contents vary from 90 to 120 g of fresh material, depending on daytime. Its dry matter (DM) varies from 16% to 21%.

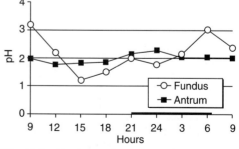

Fig. 11.2. Nycthemeral variations of the pH in the two main parts of the stomach in 10-week-old rabbits (Gidenne and Lebas, 1984).

Small intestine

The small intestine is about 3 m long and 0.8–1 cm in diameter. It is classically divided into three parts: duodenum, jejunum and ileum. The biliary duct opens immediately after the pylorus but the pancreatic duct opens 40 cm farther away in the duodenum. The contents are liquid, especially in the upper part. Their pH is slightly basic in the upper part (pH 7.2–7.5) and more acidic at the end of the ileum (pH 6.2–6.5). Normally there are small segments, about 10 cm long, which are empty.

Caecum

The small intestine ends at the base of the caecum in the ileocaecal valve or sacculus rotundus. This second storage segment contains about 40% of the whole digestive content, and is about 40–45 cm long with an average diameter of 3–4 cm. It contains 100–120 g of a uniform pasty mix, with a DM content of about 22–24%. The pH varies at around 6.0 depending on daytime (Fig. 11.3). All along the caecum the organ wall enters partially in the direction of the caecum lumen following a spiral (22–25 spires), increasing the possibility of contact between the inner surface and the caecal contents. The caecal appendix situated at the end of the caecum is 10–12 cm long and has a much smaller diameter. Its walls are composed of lymphoid tissues.

Colon

At the base of the caecum begins the colon, which is 1.5 m long. The first 50 cm are called the proximal or haustrated colon. This ends

with the short fusus coli (1–2 cm) – the only part of the intestine with red muscle. After this zone begins the distal colon, which is 1.0 m long, ending with the rectum and the anus. The proximal colon has a diameter of about 2–3 cm and the distal one a diameter of 1 cm.

General development of the digestive tract

The digestive tract is relatively more developed in a young rabbit than in an adult. In a breed like the New Zealand White, the most commonly studied rabbit, the definitive dimension of the digestive tract is observed when live rabbits are 2.6–2.7 kg, i.e. only 60–70% of the adult weight. In addition, development of the end part of the digestive tract (caecum and colon) is clearly later than that of the upper part.

Digestive physiology

A classical digestion process in the upper part of the digestive tract

Feed eaten by the rabbit quickly reaches the stomach. There it finds an acid environment and remains in the stomach for a few hours (2–4 h), undergoing little chemical change. Thus the rabbit stomach could be considered a short-term storage compartment. The contents of the stomach are gradually 'injected' into the small intestine in short bursts, by strong stomach contractions. As the contents enter the small intestine they are diluted by the flow of bile, the first intestinal secretions and finally the pancreatic juice.

After enzymatic action from these last two secretions the feed's elements that can easily be broken down are freed and pass through the intestinal wall to be carried by the blood to the cells after collection by the portal vein system and a passage through the liver.

The particles that are not broken down after a total stay of about 1½ h in the small intestine enter the caecum. There they stay from 2 to 18 h (mean: 6–12 h), while they are attacked by bacterial enzymes. Thus the caecum is the second storage compartment of the digestive tract, but a long-term one. Elements that can be broken down by this bacterial attack (producing mainly volatile fatty acids and ammonia) are freed and most of them

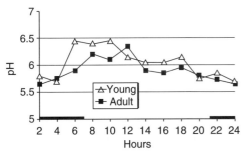

Fig. 11.3. Nycthemeral variations of pH in the caecal contents, in 6-week-old and adult rabbits (18 weeks old) (Bellier *et al.*, 1995).

pass through the wall of the digestive tract and into the bloodstream or are metabolized by some other bacteria.

The contents of the caecum are then evacuated into the colon. Approximately half consists of both large and small food particles not already broken down mixed with partially degraded intestinal secretions, while the other half consists of bacteria that have developed in the caecum, fed on matter from the small intestine.

The dual functioning of the colon and caecotrophy

So far, the functioning of the rabbit's upper digestive tract is virtually the same as that of other monogastric animals. The uniqueness of the rabbit species (and of Lagomorpha in general) lies in the dual function of its proximal colon. If the caecum contents enter the colon in the early part of the morning, they undergo a few biochemical changes. The colon wall secretes mucus, which gradually envelops the pellets formed by the wall contractions. These pellets gather in elongated clusters and are called soft pellets (more scientifically, caecotrophes).

If the caecal contents enter the colon at any other time of the day, the activity of the proximal colon is entirely different. Successive waves of contractions in alternating directions begin: the first to evacuate the contents normally and the second to push them back into the caecum. Under the varying pressure and rhythm of these contractions the contents are squeezed like a sponge. Most of the liquid part, containing soluble products and small particles of less than 0.1 mm, is forced back into the caecum (Björnhag, 1972). The solid part, containing mainly large particles over 0.3 mm, forms hard pellets, which are then expelled. In fact, as a result of this dual action, the colon produces two types of excrement: hard and soft. Table 11.1 shows the chemical composition of these pellets.

The hard pellets are expelled, but the soft pellets are recovered by the rabbit directly upon being expelled from the anus. To do this the rabbit twists itself round, sucks in the soft faeces as they emerge from the anus and then swallows them without chewing. The rabbit can retrieve the soft pellets easily, even from a mesh floor. By the end of the morning there are large numbers of these pellets inside the stomach, where they may comprise three-quarters of the total contents. The intriguing presence of these soft pellets in the stomach was at the origin of the first correct description of caecotrophy by Morot (1882), i.e. production of two types of faeces and systematic ingestion of one of the two types (the soft ones). This makes caecotrophy different from the coprophagy classically described for rats or pigs where only one type of faeces is produced.

After a stay in the stomach for 4–6 h (longer than that of normal feed) and breaking up of the aggregative structure, the soft pellets follow the same digestive process as normal feed. Considering that some parts of the intake may be recycled once, twice and even three or four times, and depending on the type of feed, the rabbit's digestive process lasts from 18 to 30 h in all, averaging 20 h.

The soft pellets consist of few modified caecal contents, i.e. half of imperfectly broken down food residues and what is left of the

Table 11.1. Average composition of hard and soft pellets (Proto, 1980). Means and dispersion correspond to a study with ten feedstuffs including complete pelleted feeds, green and dry forages.

	Hard pellets		Caecotrophes	
	Average	Extremes	Average	Extremes
Dry matter (%)	53.3	48–66	27.1	18–37
Proteins (%DM)	13.1	9–25	29.5	21–37
Crude fibre (%DM)	37.8	22–54	22.0	14–33
Lipids (%DM)	2.6	1.3–5.3	2.4	1.0–4.6
Minerals (%DM)	8.9	3–14	10.8	6–18

intestinal secretions and half of bacteria. The latter contains an appreciable amount of high-value proteins and water-soluble vitamins. The practice of caecotrophy therefore has a real nutritional value. In normally fed rabbits caecotrophy provides 15–20% of the total daily nitrogen entering the stomach. It also provides all C and B group vitamins necessary for the rabbit.

The composition of the soft pellets and the quantity expelled daily are relatively independent of the type of feed ingested, since the bacteria remain constant. In particular, the amount of DM recycled daily through caecotrophy is independent of the fibre content of the feed. As a consequence, suppression of soft faeces ingestion has very little influence on fibre digestibility. The higher the crude fibre content of the feed and/or the coarser the particles, the sooner it passes through the digestive tract. From the feeding behaviour point of view, this acceleration of the speed of digestive transit with the increase of fibrous materials in the diet makes possible a higher daily ingestion of forages than that of concentrates.

On the other hand, this particular functioning requires roughage. If the feed contains few large particles and/or it is highly digestible, most of the colon contents are pushed back to the caecum and lose elements that nourish the 'normal' bacteria living in the caecum. This would appear to increase the risk of undesirable bacteria developing in this impoverished environment, some of which might be harmful. It is thus advisable to include a minimum of roughage in the feed providing enough undigested raw particles in the colon during hard faeces production and then enabling the rabbit's digestive process to be completed fairly rapidly.

Feeding Behaviour in the Domestic Rabbit

From birth to weaning

The feeding pattern of the newborn rabbits is imposed by the dam. A doe in fact feeds her young only once every 24 h. However, recent studies suggest that some does (either in wild or domestic rabbit) nurse their young twice a day (more frequent in the second week of lactation: mean suckling 1.3 per 24 h; Hoy and Selzer, 2002). Suckling lasts only 2–3 min for a litter of 8–11 kits. The first suckling (colostrum) normally occurs during the parturition and within the first hour after the birth, and is essential to ensure the kits' survival and growth (Coureaud et al., 2000). The newborn rabbits do not appropriate one nipple (contrary to piglets), but are able to pass from one to another nipple frequently within one suckling (Hudson et al., 2000). Kits can drink up to 25% of their live weight of milk in one nursing session, and their nipple-searching behaviour is very stereotyped and controlled by a pheromonal signal (Schaal et al., 2003).

If there is not enough milk, the young try to feed every time the doe enters the nest box, but she holds back her milk. The young are able to suckle twice a day or more, and from two different does within a day, leading to a higher growth rate (Gyarmati et al., 2000).

During the first post-natal week (between 4 and 6 days of age) the young also consume hard faeces deposited by the doe in the nest, thus stimulating the caecal flora maturation (Kovacs et al., 2004).

From 1 to 3 weeks of age, the young increase their milk intake from 10 to 30 g of milk per day (or suckling) (Fig. 11.4), and the doe's milk production decreases (more sharply if pregnant). A young rabbit, reared in a litter of 7–9 kits, therefore consumes about 360–450 g of milk between birth and day 25 (vs. 100–150 g from 26 to 32 days). Individual milk intake patterns are relatively variable and depend partly on the live weight of the kit (Fortun-Lamothe and Gidenne, 2000). The dry feed intake begins significantly when the young rabbit is able to move easily to access a feeder (with pelleted feed) and a drinker, i.e. around 17–20 days (Fig. 11.4). In classical breeding conditions, the total dry feed intake is about 25–30 g/head for the period 16–25 days. Then, the food intake increases 25-fold from 20 to 35 days (Gidenne and Fortun-Lamothe, 2002). However, large variations among litters have been observed in the starting time of the dry feed intake. For example, an increase in the competition for milk, depending on litter size, stimulates the dry feed intake of the young rabbit (Fortun-Lamothe and Gidenne, 2000). Reversely, offering a second milking to the young kit (using a second doe) delays the dry feed intake (Gyarmati *et al.*, 2000). Besides, the young rabbits prefer to suckle at the mother feeder rather than a young specific feeder (Fortun-Lamothe and Gidenne, 2003), probably because of an imitation of the feeding behaviour of their mother. Age at weaning is obviously an efficient factor to modulate the dry feed intake. For instance, in wild conditions the young rabbit could be weaned at about 3 weeks of age, when the doe is pregnant and preparing a new nest for her next litter, and thus is forced to consume dry food rapidly.

The 25–30-day period is particularly important, since the intake of solid feed and water will exceed milk intake. During this period the changes in feeding behaviour are remarkable: the young rabbit goes from a single milk meal per day to a large number of alternating solid and liquid meals distributed irregularly throughout the day. Additionally, the caecotrophy behaviour probably starts at about 22–25 days of age, when a significant dry feed intake occurs that leads to a caecal and colon filling and to the dual motility pattern of the proximal colon described earlier. However, the individual feeding behaviour of the young remains largely unknown (regulation factors, number of meals, etc.), since no method is presently available to assess the intake level of young reared collectively (till weaning).

Feeding behaviour of the growing and adult rabbit

From weaning (classically between 4 and 5 weeks) the daily feed intake of the domestic rabbit (fed a complete pelleted feed) increases correlatively to the metabolic live weight (Fig. 11.5) and levels up at about 5 months of age. Taking as reference an adult fed *ad libitum* (140–150 g DM/day, e.g. for a 4 kg New Zealand White), the following observations are made: at 4 weeks a young rabbit eats a quarter of the amount an adult eats, but its live weight is only 14% of the adult's; at 8 weeks the relative proportions are 62% and 42%; at 16 weeks they are 100–110% and 87%. Between the weaning (4–5 weeks) and 8 weeks of age, the weight gain reaches its highest level (Table 11.2) while the feed conversion is optimal. Then the feed

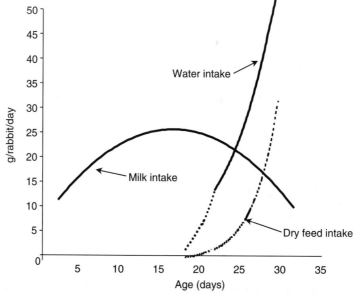

Fig. 11.4. Milk, water and dry feed intake of the young rabbit. Note: mean values, for litters of 7–9 kits, with pelleted dry feed, nipple drinker, and weaned at 30 days (doe re-mated 11 days after kindling). (Adapted from Szendrö *et al.*, 1999 and Fortum-Lamothe and Gidenne, 2000.)

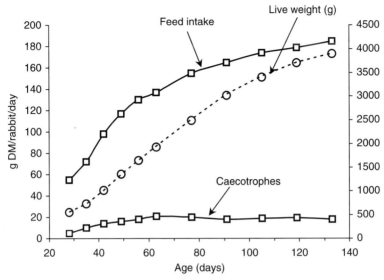

Fig. 11.5. Dry matter intake from pelleted feed, caecotrophes, and live weight from weaning (28 days) till adulthood. Values are for a domestic rabbit fed *ad libitum* (Gidenne and Lebas, 1987). Note: data of caecatrophes excretion are obtained from a rabbit wearing a collar.

intake increases less quickly as well as the growth rate, and the intake levels up at around 12 weeks of age for current hybrid lines of domestic rabbit. A rabbit regulates its feed intake according to energy need, just as other mammals do. Chemostatic mechanisms are involved, by means of the nervous system and blood levels of compounds used in energy metabolism. However, in monogastric animals the glycemia plays a key role in food intake regulation, while in ruminants the levels of volatile fatty acids in blood play a major role. Since the rabbit is a monogastric herbivore, it is not clear as to which is the main blood component regulating feed intake, but it is likely to be the blood glucose level. Voluntary intake, proportional to metabolic live weight ($LW^{0.75}$), is about 900–1000 kJ DE/day/kg $LW^{0.75}$ (DE: digestible energy), and the chemostatic regulation appears only with a dietary DE concentration higher than 9–9.5 MJ/kg (Parigi-Bini and Xiccato, 1998). Below this level, a physical-type regulation is prevalent and linked to gut fill.

The intake of soft faeces increases only till 2 months of age and then remains steady (Fig. 11.5). Expressed as fresh matter, the soft faeces intake evolves from 10 g/day (1 month old) to 55 g/day (2 months), thus representing

Table 11.2. Feeding behaviour of the domestic rabbit after weaning.

	Periods of age	
	5–7 weeks	7–10 weeks
Solid feed (pellets, 89% DM)		
Solid feed intake (g/day)	100–120	140–170
Weight gain (g/day)	45–50	35–45
Food conversion	2.2–2.4	3.4–3.8

Note: Mean values from rabbits (current commercial lines) fed *ad libitum* a pelleted diet (89% DM), and having free access to drinkable water.

15–35% of the feed intake (Gidenne and Lebas, 1987).

The rabbit fractionates its voluntary solid intake into numerous meals: about 40 at 6 weeks of age, and a slightly lower number in adulthood (Table 11.3). This meal fractionation is probably linked to the relatively weak storage capacity of the stomach discussed earlier, particularly when compared to herbivorous animals or even carnivorous or omnivorous ones (such as the dog or pig).

For 6-week-old rabbits, fed a pelleted diet, the time spent on feeding every 24 h exceeds 3 h. Then it drops rapidly to less than 2 h. If a ground non-pelleted diet is proposed to rabbits, the time spent to eat is doubled (Lebas, 1973). The number of liquid meals evolve in parallel to that of feed, and the time spent to drink is lower than that spent to eat. Furthermore, at any age, feed containing over 70% water, such as green forage, will provide rabbits with sufficient water at temperatures under 20°C, and in this case rabbits may not drink at all. In growing rabbits fed with pellets, the normal ratio of water to DM is about 1.6:1.8. In the adult or the breeding doe it is increased up to 2.0:2.1.

The solid intake fluctuates over a 24-h period, as shown in Fig. 11.6. Over 60% of the solid feed (excluding soft faeces meals) is consumed in the dark period for a domestic rabbit submitted to a 12L/12D light schedule. The nycthemeral changes of liquid meals are strictly parallel to those of solid meals for the domestic rabbit fed with pellets (Prud'hon et al., 1975), but no correlation can be established between time or intervals of solid and liquid meals. With ageing, the nocturnal feeding behaviour becomes more pronounced. The feeding habits of wild rabbits are even more nocturnal than those of domesticated rabbits. In fact, the domestic rabbit is no longer without eating, since it has more than 20 meals of dry feed a day, and it also has meals of caecotrophes (early morning). Moreover, Hirakawa (2001) pointed out that leporids (including rabbits) also consume a part of their own hard faeces, which are masticated, contrary to soft faeces that are swallowed. Meals of soft faeces (and sometime hard) increase in proportion when food availability is insufficient for rabbits.

Obviously, the feed intake level is modulated by the physiological status of the animal. For instance, a doe's voluntary intake varies greatly during the reproduction cycle (Fig. 11.7). The intake during the final days of pregnancy drops off markedly. Some does refuse solid food just before kindling. Water intake, however, never stops completely. After kindling, the feed intake increases very rapidly and can exceed 100 g DM/kg live weight per day. Water intake is also increased at that time from 200 to 250 g/day/kg live weight. When a doe is both pregnant and lactating, she eats amounts similar to those observed in a doe that is only lactating.

External Factors Modulating the Feeding Behaviour of the Domestic Rabbit

Feed composition and presentation form

One of the main dietary components implicated in feed intake regulation, after weaning, is the DE concentration. The domestic rabbit (fed a pelleted balanced diet) is able to regulate its DE intake (and thus its growth) when the dietary DE concentration is between 9 and 11.5 MJ/kg, or when the dietary fibre level is between 10% and 25% acid detergent fibre

Table 11.3. Feeding and drinking behaviour of the domestic rabbit from 6 to 18 weeks. (From Prud'hon et al., 1975.)

	Age in weeks		
	6	12	18
Solid feed (pellets, 89% DM)			
Solid feed intake (g/day)	98	194	160
Number of meals per day	39	40	34
Average quantity per meal (g)	2.6	4.9	4.9
Drinking water			
Water intake (g/day)	153	320	297
Number of drinks per day	31	28.5	36
Average weight of one drink (g)	5.1	11.5	9.1

Note: Mean values from nine New Zealand White rabbits fed *ad libitum* a pelleted diet (89% DM), and having free access to drinkable water.

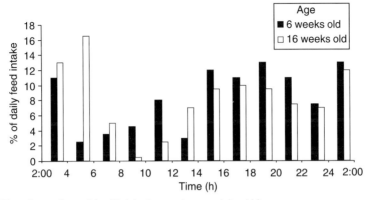

Fig. 11.6. Circadian pattern of feed intake in growing or adult rabbits.
Note: mean values for domestic rabbits (*n* = 6) fed *ad libitum* a pelleted feed (daily feed intake = 80 and 189 g, respectively for 6- and 16-week-old rabbits) and bred under a 7:00–19:00 light schedule. (From Bellier *et al.*, 1995.)

(ADF). The intake level is thus better correlated with the dietary fibre level, compared to the dietary DE content (Fig. 11.8). However, the incorporation of fat in the diets, while maintaining the dietary fibre level, increases the dietary DE level, but leads to a slight reduction of the intake. Reversely, before weaning, the young seem not to regulate their feed intake according to the DE level (Gidenne and Fortun-Lamothe, 2002), since litters consume preferentially a diet with a higher DE content.

Other nutrients in the diets are able to modify the food intake, such as protein and amino acids (Tome, 2004). For example, an

excess of methionine reduced by at least 10% the feed intake of the growing rabbit (Colin *et al.*, 1973; Gidenne *et al.*, 2002).

The diet presentation is an important factor modulating the feeding behaviour in the rabbit. Compared to meals, pelleted feeds are preferred in 97%, when offered in free choice (Harris *et al.*, 1983). Furthermore, meals seem to disturb the circadian cycle of feed intake (Lebas and Laplace, 1977). Pellet size and quality (hardness, durability) are also able to affect the feeding behaviour (Maertens and Villamide, 1998). A reduction in pellet diameter, which also increases the hardness, lessens

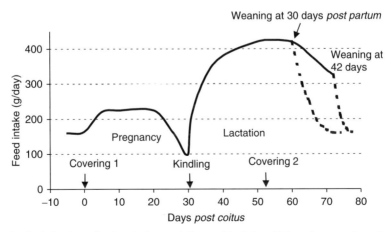

Fig. 11.7. Intake behaviour of a doe during gestation and lactation. Note: values are for a domestic rabbit fed a balanced pelleted feed (89% DM). (From Lebas, 1975.)

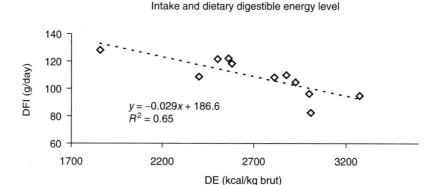

Intake and dietary digestible energy level

$$y = -0.029x + 186.6$$
$$R^2 = 0.65$$

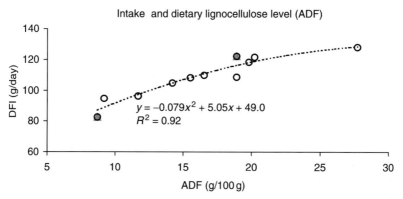

Intake and dietary lignocellulose level (ADF)

$$y = -0.079x^2 + 5.05x + 49.0$$
$$R^2 = 0.92$$

Fig. 11.8. Feed intake prediction in the domestic rabbit after weaning. DFI: daily feed intake measured between weaning (4 weeks) and 11 weeks of age.

the feed intake of the young (Gidenne *et al.*, 2003) or growing rabbit (Maertens, 1994) although the time budget for feeding is increased.

Environmental factors affecting the feeding behaviour of the rabbit

The rabbit's energy expenditure depends on ambient temperature. Feed intake to cope with energy needs is therefore linked to temperature. Studies on growing rabbits showed that at temperatures between 5°C and 30°C intake of pelleted feed dropped from 180 to 120 g/day and water intake rose from 330 to 390 g/day (Table 11.4). A closer analysis of feeding behaviour shows that as temperature rises the number of solid meals eaten in 24 h drops. From 37 solid feeds at 10°C the num-

ber drops to only 27 at 30°C (young New Zealand White rabbits). The amount eaten at each meal drops with high temperatures (5.7 g/meal from 10°C to 20°C down to 4.4 g at 30°C) but the water intake goes up from 11.4 to 16.2 g/meal between 10°C and 30°C.

The feeding and drinking behaviours were also studied for the doe and her litters according to the climatic conditions, as reviewed by Cervera and Fernandez-Carmona (1998).

If drinking water is not provided and if the only feed available is dry with a moisture content of less than 14%, dry matter intake (DMI) drops to nil within 24 h. With no water at all, and depending on temperature and humidity, an adult rabbit can survive for 4–8 days without any irreversible damage, though its weight may drop by 20–30% in less than a week (Cizek, 1961). Rabbits with access to drinking water but no solid feed can survive for 3–4

Table 11.4. Feeding behaviour of the growing rabbit according to ambient temperatures. (From Eberhart, 1980.)

Ambient temperature	5°C	18°C	30°C
Relative humidity (%)	80	70	60
Pelleted feed eaten (g/day)	182	158	123
Water drunk (g/day)	328	271	386
Water/feed ratio	1.80	1.71	3.14
Average weight gain (g/day)	35.1	37.4	25.4

weeks. Within a few days they will drink 4–6 times as much water as normal. Sodium chloride in the water (0.45%) reduces this high intake, but potassium chloride has no effect (sodium loss through urination). The rabbit is therefore very resistant to hunger and relatively resistant to thirst; but any reduction in the water supply, in terms of water requirements, causes a proportional reduction in DMI, with a consequent drop in performance.

Other environmental factors have also been studied in the domestic rabbit, such as the light schedule or the housing systems. In the absence of light (0L/24D), the feed intake of fattening rabbits is increased when compared with rabbits submitted to a sunlight programme (Lebas, 1977). In the absence of light, rabbits organized their feeding pattern in a regular 23.5–23.8-h programme, with about 5–6 h devoted to soft faeces ingestion and the remaining part of the cycle to feed intake. In continuous lighting, the feeding pattern is organized in an about 25-h programme (Jilge, 1982; Reyne and Goussopoulos, 1984). For breeding does, reduction of the lighting duration during a 24-h cycle by introducing two 4-h periods of dark during the normal 12 h of lighting in a 12L/12D programme (intermittent lighting) does not modify the average daily feed intake despite an increase of the milk production leading to a better feed efficiency for milk production (Virag et al., 2000).

As previously mentioned, type of caging also influences the daily feed intake and the feeding pattern of rabbits. For instance, the feed intake is affected by the density of rabbits in the cage. Increasing the density, which seems to lead to a higher competition for feeders among the animals, results in a reduction of feed intake (Aubret and Duperray, 1993). But this is not necessarily a result of a compe-tition for feeders since it is also observed with rabbits in individual cages (Xiccato et al., 1999). In comparisons of cage and pen housing of rabbits, enlarging the cage size for a group (with or without variation of the density) allows more movements to the rabbits and reduces the daily feed intake (Maertens and Van Herck, 2000). At the same density, rabbits caged by two or by six have the same daily feed intake, but in cages of two, rabbits spend a lower proportion of their time budget for feed consumption: 5.8% vs. 9.9% of the 10 h of the lighting period during which they were observed (Mirabito et al., 1999). Finally according to the feeding pattern, the number of places at the feeder (1–6) for a group of 10 rabbits does influence the daily feed intake (Lebas, 1971).

Feeding Behaviour in a Situation of Choice

All the above results are based on studies that were conducted with domestic rabbits, generally fed pelleted and more or less balanced diets. In the wild or in a situation of free choice for caged rabbits, another dimension must be added to the feeding behaviour: how do rabbits select the feeds?

Feeding behaviour of wild rabbits in an open situation (grazing rabbits)

First of all, the feed resources available for wild rabbits are mostly constituted by a great range of plant material. Rabbits clearly prefer graminaceous plants (*Festuca* sp., *Brachypodium* sp. or *Digitaria* sp.) and graze only few dicotyledons if sufficient grasses are available (Williams et al., 1974). Within the dicotyledons rabbits graze especially some leguminous plants and some compositae. But it could be underlined that grazing pressure on carrots (*Daucus carotta*) is very light, this plant being those preferred by rabbits (CTGREF, 1978).

Proportion of dicotyledonous species and even mousses may increase during some seasons depending on the availability of plants (Bhadresa, 1977). In winter and early spring, grazing of cultivated cereals by rabbits may completely compromise the crop, especially up to a distance of 30–100 m from the warren (Biadi and Guenezan, 1992). When rabbits can choose between winter cereals cultivated with or without mineral fertilization (phosphorus and/or nitrogen), they clearly prefer the cereals without artificial fertilization (Spence and Smith, 1965).

Grazing rabbits may be very selective and, for example, choose one part of the plant or the type of plant with the highest nitrogen concentration (Seidenstücker, 2000). Similarly, wild rabbits have grazed more intensively on one variety of spring barley than four others in a test performed in Ireland, probably in relation with the plant's composition. But differences in sugar content of varieties did not fully explain this varietal selection by grazing rabbits (Bell and Watson, 1993).

The great winter appetence of rabbits for buds and young stems of some woody plants must be underlined. Grazing of very young trees or shoots may completely compromise regeneration of some forests (CTGREF, 1980), or more specifically the regeneration of different shrubs like juniper (RSPB, 2004) or common broom. In winter rabbits like to eat the bark of some cultivated trees (not only young stems), especially that of apple trees and to some extent that of cherry and peach trees. Barks of pear, plum or apricot trees are generally less attacked (CTGREF, 1980). In forests, rabbits clearly prefer broad-leaved trees but may also attack the bark of conifers (mainly spruce and some types of pines); on the contrary, when very young trees are available, rabbits prefer to eat apical or lateral sprouts of spruces or firs instead of that of oaks (CTGREF, 1978).

So basic reasons for the choices remain unclear, even if they are constant. It could only be said that it is under the regulation of the hypothalamus since hypothalamic lesions modify clearly the choice pattern of rabbits (Balinska, 1966).

Many experiments were conducted, especially in Australia and New Zealand, to study the wild rabbit's comportment when different more or less manufactured baits are proposed (the final objective being the eradication of imported wild rabbits). Many variations were observed depending on the type of bait, but also of season. For example pollard plus bran pellets (5:1 in weight) are well consumed throughout the year. In contrast, the acceptability of carrots or oats varies seasonally. Addition of salt (1% or 5% NaCl) or of lucerne meal (15%) to the pollard plus bran pellets significantly reduces the baits' consumption (Ross and Bell, 1979).

Free choice for domestic caged rabbit

When a choice is proposed between a control diet and the same diet plus an appetizer, rabbits generally prefer the diet with the appetizer. But when the same two diets are offered separately to rabbits, the daily feed intake is exactly the same as well as the growth performance (Fekete and Lebas, 1983). This means that the pleasant smell of the proposed food is not essential for the feed intake regulation. This was also proved with a repellent diet (addition of formalin), which was clearly rejected in the free choice test but consumed in the same quantity in the long-term single-food test (Lebas, 1992, unpublished data).

In the same way Cheeke et al. (1977) have demonstrated that rabbits prefer lucerne with saponin, a bitter component, up to 3 mg/g of the diet, whereas rats always prefer the control diet without saponin in the range of 0.4–5 mg/g (Fig. 11.9). But when single feeds with different levels of saponin are offered to rabbits (saponin from 1.8 to 6.4 mg/g of complete diet), the feed intake and growth rate are independent of the saponin level (Auxilia et al., 1983).

On the contrary, when a toxin is present such as aflatoxins, rabbits refuse completely to consume the diet or consume it in very small quantities (Fehr et al., 1968; Morisse et al., 1981; Saubois and Nepote, 1994). This regu-

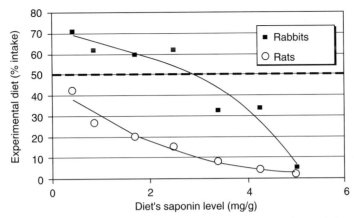

Fig. 11.9. Relative feed intake of a lucerne-based diet with various levels of saponin in rats and rabbits in a situation of free choice between this diet and a control diet without saponin. (From Cheeke *et al.*, 1977.)

lation may be considered pertinent to protect the animal against food injuries. When a concentrate (low-fibre diet or compound diet) and a fibrous material are proposed as free choice to rabbits, they prefer the concentrate. The fibrous material is consumed in only small quantities and the growth rate may be reduced (Lebas *et al.*, 1997). The consequence is also an immediate increase in the sanitary risk of rabbits with digestive disorders due to lack of fibre (Gidenne, 2003). This is the consequence of the specific search of the rabbit for energetic sources (scarce in the wild), which are the dominant regulation system of feed intake in rabbits.

Effectively, when two energetic concentrates are proposed as free choice, as was done by Gidenne (1985) with a complete diet and fresh green bananas, the growth rate is equivalent to that of the control and the digestible energy daily intake is identical. Nevertheless, it must be underlined that in this study the proportion of bananas in the DMI decreased from 40% at weaning (5 weeks) to 28% at the end of the experiment 7 weeks later.

In another study, rabbits receiving a diet deficient in one essential amino acid (lysine or sulphur amino acid) and drinking water with or without the missing amino acid in solution, preferred clearly the solution with the missing amino acid (Lebas and Greppi, 1980).

To add a last constituent to this chapter on free choice, it could be remembered that in a free choice situation, a simple variation of humidity in one component may change the equilibrium in the rabbit's choice. For example, when dehydrated lucerne and normally dried maize grains (11% DM) are offered *ad libitum* to rabbits, the result of the choice is 65% lucerne:35% maize. But if the DM content of the maize grains is increased to 14–15%, the proportion of maize becomes 45–50% (Lebas, 2002). In this case, the choice of rabbits seems motivated more by the immediate palatability of the feeds than by their nutritive value.

As described above, regulation of intake in a free choice situation is delicate to predict. Thus, in most practical situations of rabbit production the utilization of a complete balanced diet is advisable.

Conclusion

Rabbit feeding behaviour is very peculiar compared to that of other mammals, with special features, such as caecotrophy, associated to a particular digestive physiology, intermediate between the monogastric and the herbivorous. As herbivorous, the feeding strategy of the rabbit is almost opposite to that of ruminants. The feeding strategy of the latter consists of retaining food particles in the rumen till they reach a sufficiently small size. The rabbit has adopted a reverse strategy characterized by a preferential retention of fine digesta particles in the fermentative segment (caecum and proximal colon), with a quick removal of the coarse particles (such as low-digested fibres) in hard faeces. This is associated to numerous meals, thus favouring a quick digesta rate of passage and the digestion of the most digestible fibre fractions. Therefore, the rabbit is adapted to various feeding environments, from desert to temperate or even cold climates, and is able to consume a very wide variety of feeds, from seeds to herbaceous plants.

References

Aubret, J.M. and Duperray, J. (1993) Effets d'une trop forte densité dans les cages d'engraissement. *Cuniculture* 109, 3–6.

Auxilia, M.T., Bergoglio, G., Masoero, G., Mazzocco, P., Ponsetto, P.D. and Terramoccia, S. (1983) Feeding meat rabbits. Use of lucerne with different saponin content. *Coniglicoltura* 20(3), 51–58.

Balinska, H. (1966) Food preference in rabbits with hypothalamic lesions. *Revue Roumaine de Biologie* 11, 243–247.

Bell, A.C. and Watson, S. (1993) Preferential grazing of five varieties of spring barley by wild rabbits (*Oryctolagus cuniculus*). *Annals of Applied Biology* 122, 637–641.

Bellier, R., Gidenne, T., Vernay, M. and Colin, M. (1995) *In vivo* study of circadian variations of the cecal fermentation pattern in postweaned and adult rabbits. *Journal of Animal Science* 73, 128–135.

Bhadresa, R. (1977) Food preferences of rabbits *Oryctolagus cuniculus* (L.) at Holkham Sand Dunes, Norfolk. *Journal of Applied Ecology* 14, 287–291.

Biadi, F. and Guenezan, M. (1992) Le lapin de garenne. *Bulletin Technique d'Information* 3, 89–95.

Björnhag, G. (1972) Separation and delay of contents in the rabbit colon. *Swedish Journal of Agricultural Research* 2, 125–136.

Cervera, C. and Fernandez-Carmona, J. (1998) Climatic environment. In: De Blas, C. and Wiseman, J. (eds) *The Nutrition of the Rabbit.* CAB International, Wallingford, UK, pp. 273–295.

Cheeke, P.R., Kinzell, J.H. and Pedersen, M.W. (1977) Influence of saponins on alfalfa utilisation by rats, rabbits and swine. *Journal of Animal Science* 46, 476–481.

Cizek, L.J. (1961) Relationship between food and water ingestion in the rabbit. *American Journal of Physiology* 201, 557–566.

Colin, M., Arkhurst, G. and Lebas, F. (1973) Effet de l'adition de methionine au régime alimentaire sur les performances de croissance chez le lapin. *Annales de Zootechnie* 22, 485–491.

Coureaud, G., Schaal, B., Coudert, P., Rideaud, P., Fortun-Lamothe, L., Hudson, R. and Orgeur, P. (2000) Immediate postnatal suckling in the rabbit: its influence on pup survival and growth. *Reproduction, Nutrition, Developpement* 40, 19–32.

CTGREF (1978) Observations sur les préférences alimentaires du lapin de garenne et les dégâts causés aux plantations forestières. *Le Saint-Hubert* (February), 74–76.

CTGREF (1980) Protection des cultures agricoles et des régénérations forestières contre le lapin de garenne. *CTGREF Informations Techniques*, cahier 39, note 4, p. 5.

Eberhart, S. (1980) The influence of environmental temperatures on meat rabbits of different breeds. In: *Proceedings of the 2nd World Rabbit Congress*, Barcelona, April, 1, 399–409.

Fehr, P.M., Delage, J. and Richir, C. (1968) Répercutions de l'ingestion d'aflatoxine sur le lapin en croissance. *Cahiers de Nutrition et Diététique* 5, 62–64.

Fekete, S. and Lebas, F. (1983) Effect of a natural flavour (thyme extract) on the spontaneous feed ingestion, digestion coefficients and fattening parameters. *Magyar Allatorvosok Lapja* 38(2), 121–125.

Fortun-Lamothe, L. and Gidenne, T. (2000) The effect of size of suckled litter on intake behaviour, performance and health status of young and reproducing rabbits. *Annales de Zootechnie* 49, 517–529.

Fortun-Lamothe, L. and Gidenne, T. (2003) Les lapereaux préfèrent manger dans la même mangeoire que leur mère. In: Bolet, G. (ed.) *Proceedings of the 10èmes Journées de la Recherches Cunicoles France*, 19 and 20 November. ITAVI Publications, Paris, France, pp. 111–114.

Gidenne, T. (1985) Effet d'un apport de banane en complément d'un aliment concentré sur la digestion des lapereaux à l'engraissement. *Cuni-Sciences* 3, 1–6.

Gidenne, T. (2003) Fibres in rabbit feeding for digestive troubles prevention: respective role of low-digested and digestible fibre. *Livestock Production Sciences* 81, 105–117.

Gidenne, T. and Fortun-Lamothe, L. (2002) Feeding strategy for young rabbit around weaning: a review of digestive capacity and nutritional needs. *Animal Science* 75, 169–184.

Gidenne, T. and Lebas, F. (1984) Evolution circadienne du contenu digestif chez le lapin en croissance relation avec la cæcotrophie. In: *Proceedings of the 3rd World Rabbit Congress.* WRSA Publications, Rome, Italy, vol. 2, 494–501.

Gidenne, T. and Lebas, F. (1987) Estimation quantitative de la caecotrophie chez le lapin en croissance: variations en fonctions de l'âge. *Annales de Zootechnie* 36, 225–236.

Gidenne, T., Jehl, N., Segura, M. and Michalet-Doreau, B. (2002) Microbial activity in the caecum of the rabbit around weaning: impact of a dietary fibre deficiency and of intake level. *Animal Feed Science and Technology* 99, 107–118.

Gidenne, T., Lapanouse, A. and Fortun-Lamothe, L. (2003) Comportement alimentaire du lapereau sevré précocement: effet du diamètre du granulé. In: Bolet, G. (ed.) *Proceedings of the 10èmes Journées de la Recherche Cunicole*, 19 and 20 November. ITAVI Publications, Paris, France, pp. 17–19.

Grassé, P.P. and Dekeuser, P.L. (1955) Ordre des lagomorphes. In: Grassé, P.P. (ed.) *Traité de Zoologie. Anatomie, Systématique, Biologie. Tome XVII Mammifères Les Ordres: Anatomie, Ethologie, Systématique.* Masson et Cie Editeur, Paris, France, pp. 1285–1320.

Gyarmati, T., Szendrõ, Z., Maertens, L., Biró-Németh, E., Radnai, I., Milisits, G. and Matics, Z. (2000) Effect of suckling twice a day on the performance of suckling and growing rabbits. In: Blasco, A. (ed.) *Proceedings of the 7th World Rabbit Congress*, 5–7 July. Polytechnic University of Valence Publications, Valence, Spain, *World Rabbit Science* 8(suppl.1C), 283–290.

Harris, D.J., Cheeke, P.R. and Patton, N.M. (1983) Feed preference and growth performance of rabbits fed pelleted versus unpelleted diets. *Journal of Applied Rabbit Research* 6, 15–17.

Hirakawa, H. (2001) Coprophagy in leporids and other mammalian herbivores. *Mammal Review* 31, 61–80.

Hoy, S. and Selzer, D. (2002) Frequency and time of nursing in wild and domestic rabbits housed outdoors in free range. *World Rabbit Science* 10, 77–84.

Hudson, R., Schaal, B., Martinez-Gomez, M. and Distel, H. (2000) Mother–young relations in the European rabbit: physiological and behavioral locks and keys. *World Rabbit Science* 8, 85–90.

Jilge, B. (1982) Monophasic and diphasic patterns of the circadian caecotrophy rhythm of rabbits. *Laboratory Animals* 16, 1–6.

Kovacs, M., Szendrõ, Z., Csutoras, I., Bota, B., Bencsne, K.Z., Orova, Z., Radnai, I., Birone, N.E. and Horn, P. (2004) Development of the caecal microflora of newborn rabbits during the first ten days after birth. In: Becerril, C. and Pro, A. (eds) *Proceedings of the 8th World Rabbit Congress*, 7–10 September. Colegio de Postgraduados for WRSA Publications, Puebla, Mexico, 5, 1091–1096.

Lebas, F. (1971) Nombre de postes de consommation pour des groupes de lapins en croissance. *Bulletin Technique d'Information* 260, 561–564.

Lebas, F. (1973) Possibilités d'alimentation du lapin en croissance avec des régimes présentés sous forme de farine. *Annales de Zootechnie* 22, 249–251.

Lebas, F. (1975) Etude chez les lapines de l'influence du niveau d'alimentation durant la gestation. I. Sur les performances de reproduction. *Annales de Zootechnie* 24, 267–279.

Lebas, F. (1977) Faut-il éclairer les lapins durant l'engraissement? *Cuniculture* 5, 233–234.

Lebas, F. (2002) Biologie du lapin. 4.4 Comportement alimentaire. Available at: http://www.cuniculture.info/Docs/Biologie/biologie-04-4.htm

Lebas, F. and Greppi, G. (1980) Ingestion d'eau et d'aliment chez le jeune lapin disposant d'un aliment carencé en méthionine ou en lysine et pour boisson, en libre choix, d'une solution de cet acide aminé ou d'eau pure. *Reproduction, Nutrition, Developpement* 20, 1661–1665.

Lebas, F. and Laplace, J.P. (1977) Le transit digestif chez le lapin. VI. Influence de la granulation des aliments. *Annales de Zootechnie* 26, 83–91.

Lebas, F., Coudert, P., Rochambeau, de, H. and Thébault, R.G. (1997) *The Rabbit – Husbandry, Health and Production*, 2nd edn. FAO Publications, Rome, Italy, 223 pp.

Maertens, L. (1994) Infuence du diamètre du granulé sur les performances des lapereaux avant sevrage. In: Coudert, P. (ed.) *Proceedings of the VIèmes Journées de la Recherches Cunicole France*, 6 and 7 December. ITAVI Publications, Paris, France, vol. 2, 325–332.

Maertens, L. and Van Herck, A. (2000) Performances of weaned rabbits raised in pens or in classical cages: first results. In: Blasco, A. (ed.) *Proceedings of the 7th World Rabbit Congress*, 4–7 July, Valence, Spain. Polytechnic University Publications, *World Rabbit Science* 8(suppl. 1B), 435–440.

Maertens, L. and Villamide, M.J. (1998) Feeding systems for intensive production. In: De Blas, C. and Wiseman, J. (eds) *The Nutrition of the Rabbit*. CAB International, Wallingford, UK, pp. 255–271.

Mirabito, L., Galliot, P., Souchet, C. and Pierre, V. (1999) Logement des lapins en engraissement en cages de 2 ou 6 individus: etude du budget-temps. In: Perez, J.M. (ed.) *Proceedings of the 8èmes Journées de la Recherche Cunicole*, 9–10 June. ITAVI Publications, Paris, France, pp. 55–58.

Morisse, J.P., Wyers, M. and Drouin, P. (1981) Aflatoxicose chronique chez le lapin. Essai de reproduction expérimentale. *Recueil de Médecine Vétérinaire* 157(4), 363–368.

Morot, C. (1882) Mémoire relatif aux pelotes stomacales des léporidés. *Recueil de Médecine Vétérinaire* 59, 635–646.

Parigi-Bini, R. and Xiccato, G. (1998) Energy metabolism and requirements. In: De Blas, C. and Wiseman, J. (eds) *The Nutrition of the Rabbit*. CAB International, Wallingford, UK, pp. 103–131.

Proto, V. (1980) Alimentazione del coniglio da carme. *Coniglicoltura* 17, 17–32.

Prud'hon, M., Cherubin, M., Goussopoulos, J. and Carles, Y. (1975) Evolution au cours de la croissance des caractéristiques de la consommation d'aliments solides et liquides du lapin domestique nourri ad libitum. *Annales de Zootechnie* 24, 289–298.

Reyne, Y. and Goussopoulos, J. (1984) Caractéristiques du système endogène responsable des rythmes circadiens de la prise de nourriture et d'eau de boisson chez le lapin de garenne: étude en lumière permanente et en obscurité permanente. In: *Proceedings of the 3rd World Rabbit Congress*, Rome, Italy, WRSA publications, vol. 2, 473–480.

Ross, W.D. and Bell, J. (1979) A field study on preference for pollard and bran pellets by wild rabbits. *New Zealand Journal of Experimental Agriculture* 7, 95–97.

RSPB (The Royal Society for the Protection of Birds) (2004) Salisbury plain project. Available at: http://www.rspb.org.uk/england/southwest/conservation/salisbury_life_project.asp

Saubois, A. and Nepote, M.C. (1994) Aflatoxins in mixed feeds for rabbits. *Boletin Micologico* 9, 115–120.

Schaal, B., Coureaud, G., Langlois, D., Ginies, C., Semon, E. and Perrier, G. (2003) Chemical and behavioural characterization of the rabbit mammary pheromone. *Nature* 424, 68–72.

Seidenstücker, S. (2000) Rabbit grazing and N-fertilization on high grass-encroachment in dry coastal grassland in 'Meijendel', The Netherlands. Available at: www.coastalguide.org/dune/meijen2.html

Spence, T.B. and Smith, A.N. (1965) Selective grazing by rabbits. *Agriculture Gazette of N.S. Wales* 76, 614–615.

Szendrö, Z., Papp, Z. and Kustos, K. (1999) Effect of environmental temperature and restricted feeding on production of rabbit does. *CIHEAM, Cahiers Options Méditeranéennes* 41, 11–17.

Tome, D. (2004) Protein, amino acids and the control of food intake. *British Journal of Nutrition* 92, S27–S30.

Virag, G., Papp, Z., Rafai, P., Jakab, L. and Kennessy, A. (2000) Effects of an intermittent lighting schedule on doe and suckling rabbit's performance. In: Blasco, A. (ed.) *Proceedings of the 7th World Rabbit Congress*. Polytechnic University of Valence publications, Valence, Spain. *World Rabbit Science* 8(suppl. 1B), 477–481.

Williams, O.B., Wells, T.C.E. and Wells, D.A. (1974) Grazing management of Woodwalton Fen: seasonal changes in the diet of cattle and rabbits. *Journal of Applied Ecology* 11, 499–516.

Xiccato, G., Verga, M., Trocino, A., Ferrante, V., Queaque, P.I. and Sartori, A. (1999) Influence de l'effectif et de la densité par cage sur les performances productives, la qualité bouchère et le comportement chez le lapin. In: Perez, J.M. (ed.) *Proceedings of the 8èmes Journées de la Recherche Cunicole*, 9–10 June. ITAVI Publications, Paris, France, pp. 59–62.

12 Mastication and Feeding in Horses

K.A. Houpt

*Cornell University, College of Veterinary Medicine, Ithaca,
New York 14853-6401, USA*

Introduction

Horses differ from most of the domestic species considered in this volume because they are usually kept for recreational rather than production purposes. Although horse meat is consumed in some European and Japanese cultures, the vast majority of horses are used for racing or pleasure riding, including competitive jumping or dressage as well as long-distance endurance rides and weekend canters. Partly because many horses exist in a suburban environment lacking pastures and where hay and straw are difficult and inconvenient to obtain, and because of the high caloric demands of racing performance, horses are fed high-concentrate, low-roughage diets. This chapter will review the feeding behaviour of free-ranging and stabled horses, the controls of feeding behaviour and the behavioural consequences of feeding horses high-concentrate, low-roughage diets.

The Gastrointestinal Tract

In order to understand feeding in the horse, it is important to consider its gastrointestinal anatomy and physiology. The stomach of the horse is simple and relatively small; its capacity is 8–15 l. The anterior third is lined with stratified squamous epithelium. There is a narrow band of cardiac mucosa. Proper gastric and pyloric mucosa occupies the remaining two-

thirds of the stomach. The small intestine is 22 m long and holds 40–50 l in the average 500 kg horse. The major difference between the horse and carnivores is that the former has a greatly enlarged large intestine and caecum. The large intestine is 8 m in length. The caecum is 1.25 m in length and has the capacity to hold 25–30 l. The large colon is 3–3.7 m long and its capacity is twice that of the caecum (Sisson and Grossman, 1953). These structures serve as fermentation chambers in which bacteria break down cellulose to volatile fatty acids, primarily acetic, butyric, propionic and lactic. Within the large intestine, the concentration of bacteria is 10^{10}/g and of protozoa, 10^6/g.

The strategy of the horse is to increase total intake and output, extracting digestible material in the small intestine. This allows for efficient digestion of soluble materials by mammalian enzymes yielding glucose prior to fermentation in the hindgut. Fibre and undigested starch are digested in the large intestine (Janis, 1976; Demment and Van Soest, 1985). Hindgut fermentors can extract more from plants with low (<30% cell wall content) rather than high cell wall content. Timothy hay contains 65–75% cell wall content and lucerne, 40–50% (Duncan *et al.*, 1990). In order to have high intake rates, they must devote the major portion of their time to eating. As fibre concentration increases, digestibility decreases. In comparison, ruminants take in a smaller amount of high cell wall content forage, but derive more energy and protein from it through foregut fermentations and subsequent digestion of the microbes and their products in the small intestine.

Chewing

The purpose of the herbivore's teeth is twofold: prehension of fodder (biting) and chewing. The function of chewing is to break down cell walls because in most grasses the potential nutrients are enclosed in a wall of cellulose. Cellulose cannot be digested by mammals, so physical rather than enzymatic destruction is necessary. The evolution of the horse's teeth from the simple structures, such as the triangular molar teeth of most eutherian mammals found in *Hyracotherium*, to the complex teeth with high crowns found in *Equus* indicate the importance of these structures. The premolars became molarized (square or quadrate). The function of the molars of more primitive horses was to crush, but the molars of the modern horse grind by lateral shear.

In most mammals, the cheek teeth are composed of enamel and dentin and, in the high crowned teeth of the horse, cementum. Enamel consists of 95% hydroxyapatite microcrystals, which are resistant to wear. One way to increase durability is by augmenting the amount of enamel in the exposed surface of the tooth. Horses do this by infolding the enamel. The change in the horses' teeth presumably reflects their change from browsers to grazers (MacFadden, 1992). Grazers not only ingest more fibrous food than browsers but they also ingest, albeit inadvertently, abrasive material from the soil with the plants they eat. The teeth must be able to resist the abrasion. They wear at the rate of 3–4 mm/year.

The skull and musculature of the horse have also adapted for chewing. The pre-orbital skull has lengthened so that the horse has become dolichocephalic, and there have been several additional biomechanical changes. The mandible has become progressively larger and the cheek teeth have shifted anteriorly and have also become more ventrally located. The masseter and pterygoid muscles have become very well developed.

Chewing Rate

In a study of stabled horses fed hay *ad libitum*, we observed 40,000 chews/day, but only 10,000 chews/day if they were eating a complete pelleted feed (Elia, 2002). The grazing horse chews at a rate of 30–50 bites/min for 8–12 h/day. Mayes and Duncan (1988) observed free-ranging Camargue horses that chew at a rate of 30,000 bites/day. Horses grazing in the Himalayas displayed a bite rate of 51 bites/min – greater than the sheep and goats in the same environment. Bite size was 120 g when eating grasses, but half that when grazing forbs (Negi *et al.*, 1993). Horses exert 5000–6000 N/min in bite force or 141 N/bite (Hongo and Akimoto, 2003). The bite area is 290 cm^2 with 10 g of dry matter consumed per bite.

Horses prehend at a rate of 15–30 bites/min (Duren *et al.*, 1989; Marinier and Alexander, 1992). Prehending bites should be differentiated from chewing. Shingu *et al.* (2001) found that Hokkaido native ponies eating hay had a bite rate of 5/min and a chew rate of 68/min.

Horses, sheep and cattle each have different modes of prehending grass: horses use their lips; cattle, their tongues; and sheep, their teeth. Presumably, the ability to be selective varies with the method of prehension, as well as with the size of the muzzle. Sheep should be more selective than horses. The reason why selectivity is important is that the grazer can choose a more nutritious plant or plant part or can avoid a toxic one. Horses choose sward heights greater than 7 cm (Naujeck *et al.*, 2005). Bite depth, weight, volume and area all increase with sward height (Naujeck and Hill, 2003). When a pasture contains both long grass and lawns, horses choose the lawns that contain a higher crude protein level (Fleurance *et al.*, 2001).

Mechanics of Ingestion

The initial step in feeding is selection of a feeding patch. The horse uses both avoidance and preference behaviour in making a choice. They tend to avoid faecally contaminated areas (Francis-Smith and Wood-Gush, 1977). The next step in feeding is prehension of the food. Horses use their prehensile upper lip to gather food, and then bite off a clump with their incisors. Once the food has been prehended, the tongue moves grass to the molar area. The horse chews, insalivates and swallows it. Horses seem to be able to chew with their molars while biting a new mouthful with their incisors. The upper lip is used to brush a thin layer of snow from the grass, but when the snow is thicker, the horse paws to uncover it (Salter and Hudson, 1979).

Phases of Grazing

Horses feed, lie, stand and travel, but the main activity is feeding, which, in the case of free-ranging horses, is grazing. Grazing is not just ingesting grass, but consists of appetitive and consumptive phases. The appetitive phase is the seeking of food in general and a specific type of food in particular. The consummatory phase is the process of ingesting the food. Grazing, therefore, is not a homogeneous activity. It consists of a number of components that enable the horse to adjust to the forage availability and to its own energy needs. Grazing involves walking, but the horse does not move forward at a uniform rate. Rather, it eats several bites, and then takes a step or two forward or sideways. The area where a horse grazes without moving is called a feeding station. The horse remains at a feeding station for about 12 s, and then takes several steps. The horse takes about 8 bites at each feeding station or 51 bites/min. The time spent at each feeding station will vary with the density of the forage.

Horses rest for 5 h/day and spend 3 h in other activities, but spend the majority of their time – 50–80% of the 24 h – grazing. This is true of free-ranging horses, whether they are living in the American West (Salter and Hudson, 1979; Berger, 1986), the barrier islands of eastern USA (Keiper and Keenan, 1980; Rubenstein, 1981), the Camargue region of France (Mayes and Duncan, 1986), the New Forest of England (Tyler, 1972) or the steppes of Mongolia (Boyd and Bandi, 2002). The amount of time spent grazing varies with the season, age and sex of the horse, and with the herbage availability. The lower the availability of pasture as measured in yield (kg/ha), the more time the horse spends grazing.

Seasonal and Temperature Effects

Grazing takes place both during the day and night. The time of day when grazing takes place varies with the presence of biting insects. More time is spent grazing during the day in winter when forage is scarcer (more steps between bites of food) and when biting insects do not drive the horses to refuge in snow, water or barren areas (Duncan and Cowtan, 1980; Keiper and Berger, 1982). Horses graze 15 min less for every extra hour of sunlight per day in the spring. During the winter in the New Forest, ponies spend nearly all of the daylight hours grazing, interrupted for only 40 min in mid-day, whereas in the summer more time is spent resting (Tyler, 1972). Table 12.1 lists the percentage of time each day spent grazing by different populations of free-ranging or pastured horses and ponies.

In the summer, grazing activity has a nadir during the warmest hours (08:00–16:00), but in winter, the nadir occurs in the early morning (03:00–06:00) in both western Australia (low 3°C) (Arnold, 1984) and Japan (low 10°C) (Arnold, 1984).

Duncan (1985) analysed grazing time in free-ranging horses in the Camargue region of southern France. He, like Tyler (1972), found that grazing decreased as the environmental temperature increased, but that there was no effect of insect abundance, wind speed, relative humidity, faecal crude protein or green plant matter.

Effect of lactation

Lactating mares graze more than barren or pregnant mares, reflecting the greater energy

Table 12.1. Grazing and eating.

Population	Environment	Time	Diet	Feed %	Author
Pony	Pasture	Night	Grass	55	Houpt *et al.* (1986)
Pony	Pasture	Day	Grass	70	Crowell-Davis *et al.* (1985)
Horse	Pasture	24 h	Grass	60	Fleurance *et al.* (2001)
Horse	Pasture	24 h	Grass		Menard *et al.* (2002)
Przewalski	Pasture	24 h	Grass	46	Boyd *et al.* (1988)
Camargue horse	Free-range	24 h	Grass	59	Duncan (1980)
Misaki horse	Free-range	24 h	Grass	76	Kaseda (1983)
Horse	Metabolism cage	Day	Limited hay concentrate	50	Willard *et al.* (1977)
Horse	Corral	Day	Limited hay and grain	43	Houpt *et al.* (1986)
Pony	Box stall	Night	Limited hay and grain	15	Houpt *et al.* (1986)
Horse	Box stall	Night	Limited hay and grain	27	Shaw *et al.* (1988)
Pony	Pen	24 h	*Ad libitum* grain	17	Laut *et al.* (1985)
Pony	Pen	24 h	*Ad libitum* pellets	31	Ralston *et al.* (1979)
Pony	Box stall	Day	Hay	76	Sweeting *et al.* (1985)
Horse	Tie stall	24 h	*Ad libitum* hay	32	McDonnell *et al.* (1999)
Horse	Tie stall	24 h	*Ad libitum* hay	43	Houpt *et al.* (2001)
Horse	Box stall	Day	*Ad libitum* hay/ restricted oats	45	Doreau (1978)
Przewalski	Small corral	Day	Limited hay	68	Boyd (1988)
Przewalski	Large corral	Day	Limited hay	44	Boyd (1988)

demands of milk production. Intake of lactating mares is 150–200 gW$^{0.75}$/day (Boulot in Duncan *et al.*, 1990). Lactating pony mares on pasture during the summer spend 70% of the daylight hours grazing (Duncan *et al.*, 1984). They supplement with geophagia and browsing on leaves, twigs and bark (Crowell-Davis *et al.*, 1985).

Plant preferences while grazing

Horses are generally monocotyledon specialists (Hansen and Clark, 1977; Olsen and Hansen, 1977; Krysl *et al.*, 1984) but can broaden their diet and include many species of dicot, especially when food is sparse. Grazing behaviour is selective; unless the pasture is composed of only one species of plant, the animal has the opportunity to be selective as to the species ingested. Different parts of the plants can be eaten – stem, leaf, fruit or inflorescence (Negi *et al.*, 1993). Of the 60 plants in a central Himalayan meadow, each of three horses con-

sumed an average of 20 (Negi *et al.*, 1993). There have been several studies of the plants that ponies and horses choose to eat on pasture. For example, in the Mediterranean climate of the Camargue, horses consume graminoids in the marshy areas, moving to the less preferred long grasses in winter. Archer (1973, 1978) found that of 29 species of grass, horses and ponies preferred timothy, white clover (but not red clover) and perennial rye grass. Dandelions were the most preferred herbs. There are seasonal differences in the usage of habitat. For example, usage of marsh areas was much lower in winter than in summer on the barrier island of Shackleford (Rubenstein, 1981). Similarly, meadow and shrubland were the vegetation types grazed by feral horses in the Great Basin of Nevada, but food preferences and nutritional needs had to be weighed against the dangers of exposure in winter and the irritation of insects in summer (Berger, 1986).

A good example of feeding choices in free-ranging horses is that of Feist and McCullough (1976). They observed horses in the Pryor Mountains Wild Horse Range where grasses, the preferred food, were scarce. The horses consumed marsh grasses, reeds and numerous forbs. Where these were not available, they ate new growth of woody plants (*Atriplex* spp., *Sarcobatus vermiculatus*, *Chrysothamnusya-mus*) and sagebrush (*Artemisia tridentata* and *nova*). They also pawed up the roots of *Erotia lanata* and *Astralgus kentrophyta* and *gilvi-florus*. In south-eastern Oregon, 88% of the horses' diet was grass, primarily *Satanion hys-trix* (bottlebrush squirreltail grass), *Agropyron spicatum* (bearded bluebunch wheatgrass) and *Stipa thurberiana* (Thurber needlegrass). Grass consumption peaked in spring and species grazed varying seasonally with *Bromus tecto-rum* (cheat grass brome) and bearded blue-bunch wheatgrass comprising 40% of the diet (McInnis and Vavra, 1987). On Cumberland Island off the coast of Georgia the feral horse diet was composed primarily of only five grass genera: *Eremochloa*, *Spartina*, *Sporobolus*, *Unolia* and *Eragrostix* (Lenarz, 1985).

Putnam *et al.* (1987) found that cattle in the New Forest were less flexible in their diet than ponies. Cattle grazed on improved and streamside grasslands. Ponies used improved grasslands too, but made use of browse during the winter. Menard *et al.* (2002) obtained different results in the Camargue, where the cattle ate more broad-leafed plants, herb-rich swards and clover in winter, whereas horses continued to eat grass.

Based on a study of Icelandic horses, dry areas with high grass cover and biomass are preferred to wet areas with relatively low biomass. Hummocks are preferred to depressions. These preferences are exhibited only at low to moderate stocking density; at higher densities, grazing activity is evenly distributed over biomass type and topography (Magnús-son and Magnússon, 1990). Ponies on the island of Rhum off the British Isles select *Agrostis–Festuca* grasslands throughout the year, and the marsh community to an increasing extent in winter (Gordon, 1989).

Tyler (1972) found that the New Forest ponies of England eat eight different species of plants but avoid the poisonous *Senecio*. These ponies will eat acorns, an overconsumption of which leads to acorn poisoning. Horses have been reported to become ill after repeatedly eating buttercups (*Ranunculus*) (Zahorik and Houpt, 1981). Horses can fall victim to poi-

sonous plants and are particularly at risk from plants containing secondary compounds, which are cumulative toxins; examples are *Senecio* and locoweed (*Oxytropis sericea*). One way to prevent deaths is to teach horses to avoid the plants most likely to poison them. Horses can learn to avoid a feed they associate with gastrointestinal malaise (Houpt *et al.*, 1990). Using this principle of learned taste aversion, Pfister *et al.* (2002) taught horses to avoid *Oxytropis sericea* by giving them lithium chloride (190 mg/kg) by gavage soon after they ingested locoweed. In some cases the lithium treatment had to be repeated before the horses avoided the locoweed. The horses ate very little of the plant when it was presented to them as cut pieces in their pens or was growing in their pasture.

Horses differ individually in their selectivity. Marinier and Alexander (1991) used two choice preference tests of all possible combinations of five different plants to determine if horses were consistent, strong, constant and stringent in their choices. They also determined whether the horse could sort preferred plants from other plants in a trough or in a clump consisting of a mixture of two plants. They found that horses could be grouped into efficient, semi-efficient and inefficient grazers. Inefficient grazers may be at greater risk of plant poisoning.

Eating Hay and Grain

Horses are kept in three general management systems: (i) on pasture; (ii) on pasture during the day, but in a stall at night; and (iii) stall confinement most of the time with some hours outside in a grassless paddock. What the horse eats and how long it spends eating depends on its environment and the types of feed provided in the stall. Most horses are provided with some hay and variable amounts of grain.

When the feeding time of stalled horses is compared with the grazing time of free-ranging horses, there can be large differences or almost none, depending on the amount of hay available to the horse. Stabled horses fed hay *ad libitum* spend almost as much time eating as free-ranging ones (Houpt *et al.*, 2001), but reduction of hay and substitution of grain or pelleted feed reduces feeding time considerably and may lead to physical and behavioural abnormalities such as colic wood-chewing, and cribbing (Ralston *et al.*, 1979). When fed hay *ad libitum*, horses in box stalls or tie (straight) stalls spend approximately 60% of their time eating. Eating hay differs from grazing in several ways: the horse does not move while eating hay; it does not have to bite to prehend the forage; depending on the manner in which the hay is provided, the horse may have its head up (hay rack or hay bag), at chest height (manger) or down in the natural position (floor). When given the opportunity, horses will remove the hay from a manger and place it on the floor (Sweeting *et al.*, 1985), indicating their preference. Time spent eating hay increases when hay is available in several locations in the stack and when several types of hay are fed (Ninomiya *et al.*, 2004).

The greatest difference between hay and natural forage is that hay is dry. Large ponies (350 kg) will consume about 11 kg/day of freshly harvested grass, which is 88% water, and 8 kg of grass hay (15% water). The horse must obtain more fluid as water when eating hay rather than grazing. Some horses will dunk their hay in their water, which may be an attempt to rehydrate it. The closer the hay is to the water source, the more likely the horse is to place the hay in the water. Horses do not apparently carry hay even a few steps to reach water, so their motivation is not very high.

Vigilance and Social Facilitation

As prey animals, horses should be vigilant for predators. They demonstrate this behaviour by raising their heads and scanning the environment. Horses eat when other horses eat, and eat more if they can see another horse eating (Sweeting *et al.*, 1985). This is important when creep feeding a foal and when encouraging an anorexic horse to eat. Horses appear to prefer to eat from the floor and from shallow buckets or mangers. This enables the horse to see in all directions between its legs and may have evolved as an anti-predator strategy. Grazing is, of course, from the ground rather than from the usual chest height of a manger. The problem that arises when feeding confined horses is to prevent ingestion of parasite ova while still encouraging intake.

Holmes *et al.* (1987) demonstrated a similar effect on social faciliation or vigilance when they allowed pairs of horses to eat when separated by a solid partition, by a wire partition or not separated. The dominant horse of each pair ate for a longer period when there was no partition, but neither horse spent much time eating when a solid partition separated them. The total time spent eating by both horses was greatest when they could see one another through the wire partition. In that case, vigilance for predators and for a competing horse were probably major factors.

Meal Patterns

The normal feeding pattern of horses is to graze continuously for several hours and then to rest for longer or shorter periods, depending on the weather conditions and distances that must be travelled to obtain water and sufficient forage. The circadian rhythms of most domestic animals are influenced by the provision of meals. This is especially true of herbivores, such as horses and ruminants, who normally would have access to grass at all times and whose hour-to-hour behaviour would not depend on access to food. The large amount of time occupied by oral behaviour indicates the reasons for the appearance of oral stereotypies, such as wood-chewing and cribbing, in stalled horses on low-roughage diets.

When offered a pelleted complete diet, a similar feeding pattern emerges: many long meals. It is interesting to observe that whether the horse is grazing or eating pellets, the statistical definition of a meal (feeding with breaks no more than 10 min duration) is the same (Ralston, 1984).

Horses spend about 1 h/day eating grain, varying with the form, amount and probably type of grain. Chewing rate is higher per minute when grain is consumed, but because eating time is so short, the number of total chews per day is much lower (Elia, 2002). This may have implications for gastrointestinal function because, in contrast to many species, horses have no psychic phase of salivation; instead they salivate when they chew. The psychic phase of digestion refers to stimulation of the vagus branch of the parasympathetic system by visual, olfactory and auditory cues before feed is tasted. Saliva is released by the parotid salivary gland on one side with each jaw movement (Alexander, 1966). This indicates that the horse consuming most of its calories as grain adds less saliva to its upper gastrointestinal tract.

Horses prefer lucerne to Matua bromegrass hay, and both are preferred to timothy hay (Guay et al., 2002) based on the amount consumed when fed separately. When three hays were fed simultaneously to yearling horses, lucerne was chosen over Matua bromegrass, which was chosen over coastal Bermuda grass (LaCasha et al., 1999). When offered four stages of grass hay, lucerne and barley straw, less time was spent eating straw (Dulphy et al., 1997). Horses eat 20 g DM/kg body weight, spend 84 min/kg consuming hay and devote 800 min/day to chewing (Dulphy et al., 1997). Ninomiya et al. (2004) found that horses would eat lucerne hay first, then timothy and finally orchard grass.

Effects of the form of feed

Complete pelleted diets have been devised for horses. Pellets are convenient, easily stored and may not trigger inhalant allergies in horse or owners. The pellets can be fed in addition to hay or as a complete diet. Complete pelleted diets contain fibre, but it is ground. Therefore, the horse can consume its caloric requirement in 1–2 h/day. Horses fed a ration of barley, ground lucerne and wheat spent 110 min eating their National Research Council requirement, but spent 30 min less if it were pelleted (Hintz and Loy, 1966). Grain can be consumed at a rate of 160 g/min (Hintz et al., 1989).

The question is: What does the horse do when it is not grazing or eating hay? To a certain extent, the horse continues to graze, i.e. it searches the bedding, ingesting some wood shavings, and ingests more if the bedding is straw, finding stray pellets or consuming faeces. Coprophagia is a common response to any dietary insufficiency – calories, protein, fibre, etc. Although the horse forages in the stall, he does not spend as much time as he would spend eating hay. Standing time, the default behaviour, also increases in the horse fed with pellets.

The natural horse diet is high in long stem roughage, but many contemporary equine diets are not. One way to assess whether the horse 'cares' about roughage is to see how hard he will work for it. Horses can be trained to push a switch with their muzzles to obtain access to a bucket containing 100 g (a handful) of hay. To determine the strength of their motivation, the horses had to push the switch more times for each 100 g reward. A progressive ratio was used so that the horse pressed once, twice, four times, seven times, 11 times and so on for each successive reward. If they refused to complete the required number of presses or would not eat the reward, they were considered extinguished. If the horses had hay ad libitum, they would not press the switch to obtain hay. If they were being fed only a complete pelleted diet, they would press 13 times (Elia, 2002). Apparently horses do have a motivation for hay. In the absence of hay or of adequate hay, the horse may 'browse' on wood. Wood-chewing is a common horse behaviour problem that can be caused by lack of roughage. Johnson et al. (1998) found that the addition of virginiamycin to the diet of concentrate-fed horses reduced stall walking, probably because of a decline in the number of lactic acid-producing bacteria in the large intestine.

Flavour Preferences

Randall et al. (1978) have reported the basic taste preferences of immature horses. They show a strong preference for sucrose at concentrations of 1.25–10 g/ml, but no preferences for sour (hydrochloric acid), bitter (quinine) or even salt (sodium chloride) tastes. They rejected salt solutions at concentrations greater than 0.63 g/100 ml, acetic acid at concentrations greater than 0.16 mg/100 ml and quinine at concentrations of 20 mg/100 ml. In other words, at high concentrations, all the solutions except sucrose were rejected. In a study of adult ponies, 9 of 10 showed a strong preference for sucrose (Hawkes et al., 1985). Salt intake varies from 19 to 143 g/day depending upon the individual horse (Schryver et al., 1987). Horses no doubt use smell as well as taste to identify food, especially to reject it. Ott et al. (1979) found that horses appeared to find orange odour aversive and Ödberg and Francis-Smith (1977) have noted that horses avoid areas contaminated by faeces, presumably on the basis of odour.

Fats have been advocated for equine diets to reduce the risk of myopathies, and lower reactivity. Holland *et al.* (1996, 1998) demonstrated that not all fats are equally acceptable to horses. Using a cafeteria of concentrates, each containing a different oil, they determined that horses preferred maize oil to tallow, groundnut oil or safflower oil. There was no difference in acceptance of maize and soy lectins. Horses prefer oats to maize and both of these to barley, rye and wheat (Hintz, 1980). Sometimes adding flavourings, such as apple, caramel and especially anis, to feed can slow down consumption (Hintz *et al.*, 1989).

Physiological Controls of Feeding

There has been only limited research on the controls of ingestive behaviour in horses, much of it performed by Sarah Ralston.

Measurements of feeding

There are various ways to measure feeding. Total amount consumed daily is the simplest, but size of each meal and intermeal interval can reveal information on satiety that amount consumed cannot. Another measure is the satiety ratio, the grams of food consumed divided by the postmeal interval.

Defence of Energy Input and Body Weight

Horses can regulate their energy balance by controlling the amount they ingest. In other words, horses do not eat a fixed amount or eat until they are ill (although they may overeat grain and later become ill). They increase or decrease the amount they eat to compensate for caloric dilution or enrichment, indicating that they are able to determine the caloric density of feed. This has been shown by two different approaches. In the earlier study, Laut *et al.* (1985) fed ponies *ad libitum* grain. The density of the diet was raised by adding indigestible sawdust. Molasses had to be added to all diets; otherwise the ponies were able to sep-

arate the sawdust from the grain. The ponies consumed 5 kg of 3.4 Mcal feed. Two dilutions were used: 25% dilution (2.6 Mcal/kg) and 50% dilution (1.7 Mcal/kg). The ponies were able to ingest 5 kg and compensate for the 25% deletion by eating 6 kg of feed, but, although they increased their intake to 7.5 kg, they could not increase it enough to compensate for 50% dilution. In order to compensate, the ponies ate longer meals; they spent 246 min eating the 3.4 Mcal diet and 408 min eating the 1.7 Mcal diet. A later study by Cairns *et al.* (2002) tested the ability of horses to learn the post-ingestive consequences associated with a particular flavour. Most of the horses initially preferred mint to garlic, but when garlic was paired with a calorically dense diet, the preference for mint decreased and that for garlic increased. After 40 meals of a diet of intermediate caloric density, the horses were fed a high-caloric-density mint-flavoured diet and a low-caloric-density garlic-flavoured diet, which resulted in a reversal of consequences. The horses' preferences for mint increased.

Pregastric factors

Pregastric stimuli appear to be more important in horses than in the other species studied – rats, dogs and monkeys. Ponies with oesophageal fistulas that were sham fed (food dropping from the fistula) consumed the meals of the same size as they would have eaten normally (Ralston and Baile, 1982a). The other species sham feed larger and longer meals. The taste of the food, mastication and swallowing are enough to satiate the horse. They will initiate another meal much sooner after sham feeding – presumably in response to internal assessment of body energy stores via leptin or central glucostatic mechanisms.

Glucostatic

The role of blood glucose in producing satiety in horses was investigated by Ralston and Baile (1982b). Intragastric loads of glucose (300 g) suppressed feeding in fasted ponies. Plasma glucose levels are inversely correlated with the size of the meal and the speed of food intake; however, intravenous administration of 0.2–1 g glucose/BW[75] did not suppress intake. The fact that intragastric, but not intravenous, glucose suppressed feeding indicates that the receptors must be either in the gastrointestinal tract or it was due to some post-absorptive mechanism.

Gastrointestinal Factors

Chemoreceptors in the gastrointestinal tract are not very important controls. When administered either intragastrically or intracaecally, neither glucose, cellulose – the main digestible constituent of grass – nor its breakdown products – the volatile fatty acids solutions – depress intake until they have been absorbed. Intake is suppressed, but only after a long latency. For example, Ralston et al. (1983) infused either propionate or acetate into the caecum of hungry (4 h food-deprived) ponies 15 min before feed was made available. A low dose (0.4 nmol/kg body weight of propionate) increased food intake, but a higher dose (0.75 nmol/kg and 1–1.25 nmol/kg acetate) prolonged the intermeal interval. The highest dose of propionate (1 nmol/kg) reduced the size of the first meal. When these salts were infused intragastrically, only 75 nmol/kg acetate affected feeding – it shortened the intermeal interval. There are mechanisms for controlling caloric intake, but they appear to be post-absorptive (Ralston et al., 1979). Cellulose (300 g in 2 l of water), which is digested to volatile fatty acids by the caecal microorganism, suppressed intake of 200 kg ponies only 3–18 h later. Intragastric administration of kaolin (300 g in 2 l of water) did not affect feeding at all (Ralston and Baile, 1982a). Infusions of maize oil prefeeding did not delay the onset of feeding or the size of the meal, but the onset of the next meal (intermeal interval) and the size of that meal were reduced, indicating a post-absorptive effect (Ralston and Baile, 1983a).

Central Nervous System Depressants

Horses, in common with other more thoroughly studied species, increase their food intake when the brain, presumably the area involved in satiety, is depressed. In a preliminary study, Brown et al. (1976) found that diazepam 0.02–0.03 mg/kg i.v. increased intake of either pelleted feed or hay 50–75% above control levels. Promazine (0.5 mg/kg) also increased intake by about 25%.

The first sign of colic in most horses is anorexia. This clinical observation, as well as controlled studies, indicates that pathological distention of the gastrointestinal tract inhibits feeding in the horse as in other species. This anorexia is probably mediated through pain receptors that travel in the sympathetic nerves. Analgesics intended for use in horses are often tested by determining whether a horse with a dilated caecum will eat when treated with the drug. In this case, the drug is probably working peripherally, so pain signals are not relayed to the brain, whereas diazepam works by inhibiting a centrally acting inhibitor of appetite.

Feeding-related Behaviour Problems

Equine stable problems, the so-called vices, can be divided into oral and locomotory behaviours. These problems may be classified as stereotypies because they are repetitive, serve no known function and occupy a large part (>10%) of the animal's time. Although confinement is common to all these problems, cribbing and wood-chewing appear to be influenced by diet. McGreevy et al. (1995c) found that the type of bedding and the type of feed, as well as the social contacts of thoroughbreds, influenced the incidence of stereotypic behaviours. Horses fed more roughage and roughage of more than one kind, horses bedded on straw, horses fed twice or more than three times a day and horses with more visual contact with other horses had fewer stereotypic behaviours. Although toys may not help with oral problems, provision of a simple foraging device that delivers food as the horse rolls it might help (Malpass and Weigler, 1994; Winskill et al., 1996).

Cribbing

Cribbing is an oral behaviour in which the horse grasps a horizontal surface, such as the rim of a bucket or the rail of a fence, with its incisors, flexes its neck and aspirates air into its oesophagus. Some horses aspirate air without grasping an object. This is called aerophagia or windsucking; the latter term can also be used to refer to pneumovaginitis. It was thought that the horse swallows the air, but this does not usually occur (McGreevy et al., 1995a). The one undeniable consequence of cribbing is excessive wear of the incisor teeth.

Cribbing occurs in association with eating, in particular eating grain or other highly palatable food (Kusunose, 1992). The relation of cribbing to eating is similar to that of non-nutritive suckling in calves that occurs after drinking milk. Cribbing occurs in 2.5–5% of thoroughbreds (McBane, 1987; Vecchiotti and Galanti, 1987; McGreevy et al., 1995b; Luescher et al., 1998; Redbo et al., 1998). Cribbing occurs less frequently in endurance horses (3%) than in those used for dressage or eventing (8%), who are confined in their stall for much longer periods (McGreevy et al., 1995c). It is a clinical impression that cribbing occurs more frequently in confined horses, but once established, it may persist even when the horse is on pasture. Cribbing may be the result, rather than the cause, of gastrointestinal problems. Aspirating air or inflating the oesophagus may be a pleasurable sensation to an animal experiencing gastrointestinal discomfort. Although the opiate blockers such as naloxone will inhibit cribbing (Dodman et al., 1988), opiates do not rise in the blood when horses crib. In fact, horses that crib have lower blood levels of opiates than non-cribbing horses, but the blood levels may not reflect brain levels (Gillham et al., 1994).

There are a variety of surgical treatments for cribbing. These treatments include buccostomy; cutting the ventral branch of the spinal accessory nerve (ninth cranial); myotomy of the ventral neck muscles; or a combination of partial myectomy of the omohyoideus, sternohyoideus and sternothyrohyoideus and neurectomy of the ventral branch of the spinal accessory nerve. The success rates of these treatments vary from 0% to 70% (Owen et al., 1980; Greet 1982; Turner et al., 1984).

The simplest method used to prevent cribbing is to place a strap around the throat just behind the poll so that pressure is exerted when the horse arches its neck and even more pressure is exerted when the animal attempts to swallow. The horse is, in effect, punished for cribbing. If a plain strap does not suffice, a spiked strap or metal collar can be used. A common observation of a horse wearing a cribbing strap is that it continues to grasp horizontal objects with its teeth, but does not swallow as much air. Many stables are designed or modified so that there are few horizontal surfaces available, but water and feed containers usually provide the horse some opportunity to crib. Muzzles may also be used. These are wire baskets that permit the horse to eat and drink, but not grasp a horizontal surface.

Is it necessary to prevent cribbing? Unless the otherwise healthy horse is losing weight or suffering from colic as a result of swallowing air or flatulent colic, the behaviour is not interfering with the horse's well-being. Epiploic foramen entrapment is a type of colic that occurs more frequently in horses that crib (Archer et al., 2004). The noise of cribbing often annoys the owner, but that is not a good reason to subject an animal to the risk of surgery and the possible side effects of infection and disfigurement that may interfere much more with the animal's function than cribbing would. Cribbing is considered an unsoundness, but this may not be justified. Cribbing is also considered to be contagious, but this has never been proven. It is possible that the environment that causes one horse to crib causes the other horse in the same environment to do so. Young horses may be more likely to learn the habit from an adult than are other adults.

Wood-Chewing

Both cribbing and wood-chewing (lignophagia) horses grasp horizontal surfaces with their teeth, but the wood-chewing horse actually ingests the wood, whereas the only damage the cribbing horse does is to mark the wood with its incisors. In contrast to cribbing, wood-chewing appears to have a definite cause – a lack of roughage in the diet. Wood-chewing is more common in dressage and eventing horses than in endurance horses who spend more time outside their stalls. Several investigators have noted that high-concentrate diets or pelleted diets increase the incidence of wood-chewing. Feral horses, as well as well-fed pastured ponies, have been observed to ingest trees and shrubs; so it would seem that there is some need or appetite for wood even when grasses are freely available. Farm managers are well aware that trees, especially young trees, must be protected from horses on pasture. Horses cannot digest wood; nevertheless, there may be some role for indigestible roughage in equine digestion. Jackson et al. (1984) have found that wood-chewing increases in cold, wet weather.

Eliminating edges, covering edges with metal or wire, and painting the surface with taste repellents are the traditional methods for preventing horses from wood-chewing, but providing more roughage is a better practice, both behaviourally and nutritionally. If roughage is provided, the horse's motivation to chew wood is reduced rather than thwarted. An increase in exercise reduces the rate of wood-chewing (Krzak et al., 1991).

Another form of abnormal equine feeding behaviour is coprophagia, ingestion of faeces. Horses may eat their own or other horses' faeces. Coprophagia is normal in foals (Crowell-Davis et al., 1985; Marinier and Alexander, 1995), but usually indicates nutrient or roughage deficiency in adult horses. Foals eat their dams' faeces preferentially.

Conclusion

Horses spend much of their time eating when grass feed is available. They graze selectively and have definite hay and grain preference. Horses seem to orally meter their intake in that they do not continue to eat when sham fed as other species (dogs and rats) do. Horses can meter calories in that they will eat more kilograms of a calorically diluted diet and keep their caloric intake constant. Gastrointestinal factors are important. Clinically, the suppression of appetite due to gastrointestinal pain is well recognized. Physiologically, glucose levels in the blood appear to be more important in predicting meals, i.e. horses eat when their plasma glucose levels fall below 90 mg/dl (Ralston and Baile, 1982a and b). Fat appears to be important, not in controlling meal size, but in controlling intrameal interval (Ralston and Baile, 1983a). Volatile fatty acid concentrations in the caecum may be important as satiety factors (Ralston et al., 1983).

References

Alexander, F. (1966) A study of parotid salivation in the horse. *Journal of Physiology* 184, 646–656.

Archer, D.D., Freeman, D.E., Doyle, A.J., Proudman, C.J. and Edwards, G.B. (2004) Association between cribbing and entrapment of the small intestine in the epiploic foramen in horses: 68 cases (1991–2002). *Journal of American Veterinary Medical Association* 224(4), 562–564.

Archer, M. (1973) The species preferences of grazing horses. *Journal of British Grassland Society* 28, 123–128.

Archer, M. (1978) Further studies on palatability of grasses to horses. *Journal of British Grassland Society* 33, 239–243.

Arnold, G.W. (1984) Comparison of the time budgets and circadian patterns of maintenance activities in sheep, cattle and horses grouped together. *Applied Animal Behaviour Science* 13, 19–30.

Berger, J. (1986) *Wild Horses of the Great Basin*. University of Chicago Press, Chicago, Illinois.

Boyd, L.E. (1988) Time budgets of adult Przewalski horses: effects of sex, reproductive status and enclosure. *Applied Animal Behaviour Science* 21, 19–39.

Boyd, L.E. and Bandi, N. (2002) Reintroduction of takh, *Equus ferus przewalskii*, to Hustai National Park, Mongolia: time budget and synchrony of activity pre- and post-release. *Applied Animal Behaviour Science* 78, 87–102.

Boyd, L.E., Carbonaro, D.A. and Houpt, K.A. (1988) The 24-hour time budget of Przewalski horses. *Applied Animal Behaviour Science* 21, 5–17.

Brown, R.F., Houpt, K.A. and Schryver, H.F. (1976) Stimulation of food intake in horses by diazepam and promazine. *Pharmacology, Biochemistry and Behavior* 5(4), 495–497.

Cairns, M.C., Cooper, J.J., Davidson, H.P.B. and Mills, D.S. (2002) Association in horses of orosensory characteristics of foods with their post-ingestive consequences. *Animal Science* 75, 257–265.

Crowell-Davis, S.L., Houpt, K.A. and Carnevale, J.M. (1985) Feeding and drinking behavior of mares and foals with free access to pasture and water. *Journal of Animal Science* 60(4), 883–889.

Demment, M.W. and Van Soest, P.J. (1985) A nutritional explanation for body-size patterns of ruminant and nonruminant herbivores. *American Naturalist* 125, 641–672.

Dodman, N.H., Shuster, L., Court, M.H. and Patel, J. (1988) Use of a narcotic antagonist (nalmefene) to suppress self-mutilative behavior in a stallion. *Journal of the American Veterinary Medical Association* 192, 1585–1587.

Doreau, M. (1978) Comportement alimentaire du cheval à l = écurie. *Annales de Zootechnie* 3, 291–302.

Dulphy, J.P., Martin-Rosset, W., Dubroeucq, H., Ballet, J.M., Detour, A. and Jailler, M. (1997) Compared feedings in *ad libitum* intake of dry forages by horses and sheep. *Livestock Production Sciences* 52, 49–56.

Duncan, P. (1980) Time-budgets of Camargue horses. II. Time-budgets of adult horses and weaned sub-adults. *Behaviour* 72, 26–49.

Duncan, P. (1985) Time-budgets of Camargue horses. III. Environmental influences. *Behaviour* 92(3–4), 188–208.

Duncan, P. and Cowtan, P. (1980) An unusual choice of habitat helps Camargue horses to avoid blood-sucking horse-flies. *Biology of Behavior* 5, 55–60.

Duncan, P., Harvey, P.H. and Wells, S.M. (1984) On lactation and associated behaviour in a natural herd of horses. *Animal Behaviour* 32, 255–263.

Duncan, P., Foose, T.J., Gordon, I.J., Gakahu, C.G. and Lloyd, M. (1990) Comparative nutrient extraction from forages by grazing bovids and equids: a test of the nutritional model of equid/bovid competition and coexistence. *Oecologia* 84, 411–418.

Duren, S.E., Dougherty, C.T., Jackson, S.G. and Baker, J.P. (1989) Modification of ingestive behavior due to exercise in yearling horses grazing orchardgrass. *Applied Animal Behaviour Science* 22, 335–345.

Elia, J.B. (2002) The effects of diet differing in fiber content on equine behaviour and motivation for fiber. Thesis, Cornell University, Ithaca, New York.

Feist, J.D. and McCullough, D.R. (1976) Behavior patterns and communication in feral horses. *Zeitschrift für Tierpsychologie* 41, 337–371.

Fleurance, G., Duncan, P. and Mallevaud, B. (2001) Daily intake and the selection of feeding sites by horses in heterogeneous wet grasslands. *Animal Research (Annales de Zootechnie)* 50, 149–156.

Francis-Smith, K. and Wood-Gush, D.G.M. (1977) Coprophagia as seen in thoroughbred foals. *Equine Veterinary Journal* 9, 155–157.

Gillham, S.B., Dodman, N.H., Shuster, L., Kream, R. and Rand, W. (1994) The effect of diet on cribbing behavior and plasma β-endorphin in horses. *Applied Animal Behaviour Science* 41, 147–153.

Gordon, I.J. (1989) Vegetation community selection by ungulates on the isle of Rhum. II. Vegetation community selection. *Journal of Applied Ecology* 26, 53–64.

Greet, T.R.C. (1982) Windsucking treated by myectomy and neurectomy. *Equine Veterinary Journal* 14, 299–301.

Guay, K.A., Brady, H.A., Allen, V.G., Pond, K.R., Wester, D.B., Janecka, L.A. and Heninger, N.L. (2002) Matua bromegrass hay for mares in gestation and lactation. *Journal of Animal Science* 80(11), 2960–2966.

Hansen, R.M. and Clark, R.C. (1977) Foods of elk and other ungulates at low elevation in northwestern Colorado. *Journal of Wildlife Management* 41, 76–80.

Hawkes, J., Hedges, M., Daniluk, P., Hintz, H.F. and Schryver, H.F. (1985) Feed preferences of ponies. *Equine Veterinary Journal* 17(1), 20–22.

Hintz, H.F. (1980) Feed preferences of horses. In: *Proceedings of Cornell University Nutrition Conference for Feed Manufacturers*, Syracuse, New York, pp. 113–116.

Hintz, H.F. and Loy, R.G. (1966) Effects of pelleting on the nutritive value of horse rations. *Journal of Animal Science* 25(4), 1059–1062.

Hintz, H.F., Schryver, H.F., Mallette, J. and Houpt, K.A. (1989) Factors affecting rate of grain intake by horses. *Equine Practice* 11(4), 38–42.

Holland, J.L., Kronfeld, D.S. and Meacham, T.N. (1996) Behavior of horses is affected by soy lecithin and corn oil in the diet. *Journal of Animal Science* 74, 1252–1255.

Holland, J.L., Kronfeld, D.S., Rich, G.A., Kline, K.A., Fontenot, J.P., Meacham, T.N. and Harris, P.A. (1998) Acceptance of fat and lecithin containing diets by horses. *Applied Animal Behaviour Science* 56(2–4), 91–96.

Holmes, L.N., Song, G.K. and Price, E.O. (1987) Head partitions facilitate feeding by subordinate horses in the presence of dominant pen-mates. *Applied Animal Behaviour Science* 19, 179–182.

Hongo, A. and Akimoto, M. (2003) The role of incisors in selective grazing by cattle and horses. *Journal of Agricultural Science* 140, 469–477.

Houpt, K.A., O'Connell, M.F., Houpt, T.A. and Carbonaro, D.A. (1986) Night-time behavior of stabled and pastured peri-parturient ponies. *Applied Animal Behaviour Science* 15, 103–111.

Houpt, K.A., Zahorik, D.M. and Swaratzman-Andert, J.A. (1990) Taste aversion learning in horses. *Journal of Animal Science* 68, 2340–2344.

Houpt, K.A., Houpt, T.R., Johnson, J.L., Erb, H.N. and Yeon, S.C. (2001) The effect of exercise deprivation on the behaviour and physiology of straight stall confined pregnant mares. *Animal Welfare* 10, 257–267.

Jackson, S.A., Rich, V.A. and Ralston, S.L. (1984) Feeding behavior and feed efficiency in groups of horses as a function of feeding frequency and use of alfalfa hay cubes. *Journal of Animal Science* 1(suppl.), 152–153.

Janis, C.M. (1976) The evolutionary strategy of the Equidae and the origins of rumen and cecal digestion. *Evolution* 30, 757–774.

Johnson, K.G., Tyrell, J. and Rowe, J.B. (1998) Behavioural changes in stabled horses given nontherapeutic levels of virginiamycin as Founderguard. *Equine Veterinary Journal* 30, 139–143.

Kaseda, Y. (1983) Seasonal changes in time spent grazing and resting of Misaki horses. *Japanese Journal of Zootechnology Science* 54(7), 464–469.

Kawai, M., Juni, K., Yasue, T., Ogawa, K., Hata, H., Kondo, S., Okubo, M. and Asahida, Y. (1995) Intake, digestibility and nutritive value of *Sasa nipponica* in Hokkaido native horses. *Journal of Equine Science* 6(4), 121–125.

Keiper, R.R. and Berger, J. (1982) Refuge-seeking and pet avoidance by feral horses in desert and island environments. *Applied Animal Ethology* 9, 111–120.

Keiper, R.R. and Keenan, M.A. (1980) Nocturnal activity patterns of feral horses. *Journal of Mammalogy* 61, 116–118.

Krysl, L.J., Hubbert, M.E., Sowell, B.F., Plumb, G.E., Jewett, T.K., Smith, M.A. and Waggoner, J.W. (1984) Horses and cattle grazing in the Wyoming Red Desert. I. Food habits and dietary overlap. *Journal of Range Management* 37(1), 72–76.

Krzak, W.E., Gonyou, H.W. and Lawrence, L.M. (1991) Wood chewing by stabled horses: diurnal pattern and effects of exercise. *Journal of Animal Science* 69, 1053–1058.

Kusunose, R. (1992) Diurnal pattern of cribbing in stabled horses. *Japanese Journal of Equine Science* 3, 173–176.

LaCasha, P.A., Brady, H.A., Allen, V.G., Richardson, C.R. and Pond, K.R. (1999) Voluntary intake, digestibility, and subsequent selection of Matua bromegrass, coastal bermudagrass, and alfalfa hays by yearling horses. *Journal of Animal Science* 77(10), 2766–2773.

Laut, J.E., Houpt, K.A., Hintz, H.F. and Houpt, T.R. (1985) The effects of caloric dilution on meal patterns and food intake of ponies. *Physiology & Behavior* 35, 549–554.

Lenarz, M.S. (1985) *Lack of Diet Segregation Between Sexes and Age Groups in Feral Horses*. NPS Cooperative Research Unit, Institute of Ecology, University of Georgia, Athens, Georgia, pp. 2583–2585.

Luescher, U.A., McKeown, D.B. and Dean, H. (1998) A cross-sectional study on compulsive behaviour (stable vices) in horses. *Equine Veterinary Journal* 27(suppl.), 14–18.

MacFadden, B.J. (1992) *Fossil Horses: Systematics, Paleobiology, and Evolution of the Family Equidae*. Cambridge University Press, New York.

Magnússon, S.H. and Magnússon, B. (1990) Studies in the grazing of a drained lowland fen in Iceland. II. Plant preferences of horses during summer. *Búvísindi Iceland Agricultural Science* 4, 109–124.

Malpass, J.P. and Weigler, B.J. (1994) A simple and effective environmental enrichment device for ponies in long-term indoor confinement. *Contemporary Topics* 33, 74–76.

Marinier, S.L. and Alexander, A.J. (1991) Selective grazing behaviour in horses: development of methodology and preliminary use of tests to measure individual grazing ability. *Applied Animal Behaviour Science* 30, 203–221.

Marinier, S.L. and Alexander, A.J. (1992) Use of field observations to measure individual grazing ability in horses. *Applied Animal Behaviour Science* 33, 1–10.

Marinier, S.L. and Alexander, A.J. (1995) Coprophagy as an avenue for foals of the domestic horse to learn food preferences from their dams. *Journal of Theoretical Biology* 173, 121–124.

Martin-Rosset, W. and Dulphy, J.P. (1987) Digestibility interaction between forages and concentrates in horses: influence of feeding level – comparison with sheep. *Livestock Production Sciences* 17, 263–276.

Mayes, E. and Duncan, P. (1988) Temporal patterns of feeding in free-ranging horses. *Behavior* 96, 105–129.

McBane, S. (1987) *Behaviour Problems of Horses*. David and Charles, North Pomfret, Vermont, pp. 115–122.

McDonnell, S.M., Freeman, D.A., Cymbaluk, N.F., Schott, H.C. II, Hinchcliff, K. and Kyle, B. (1999) Behaviour of stabled horses provided continuous or intermittent access to drinking water. *American Journal of Veterinary Research* 60(11), 1451–1456.

McGreevy, P.D., Richardson, J.D., Nicol, C.J. and Lane, J.G. (1995a) Radiographic and endoscopic study of horses performing an oral based stereotypy. *Equine Veterinary Journal* 27, 92–95.

McGreevy, P.D., French, N.P. and Nicol, C.J. (1995b) The prevalence of abnormal behaviours in dressage, eventing and endurance horses in relation to stabling. *The Veterinary Record* 137, 36–37.

McGreevy, P.D., Cripps, P.J., French, N.P., Green, L.E. and Nicol, C.J. (1995c) Management factors associated with stereotypic and redirected behaviour in the thoroughbred horse. *Equine Veterinary Journal* 27, 86–91.

McInnis, M. and Vavra, M. (1987) Dietary relationships among feral horses, cattle and pronghorn in southeastern Oregon. *Journal of Range Management* 40(1), 60–66.

Menard, C., Duncan, P., Fleurance, G., Georges, J.Y. and Lila, M. (2002) Comparative foraging and nutrition of horses and cattle in European wetlands. *Journal of Applied Ecology* 39, 120–133.

Naujeck, A. and Hill, J. (2003) Influence of sward height on bite dimensions of horses. *Journal of Science* 77, 95–100.

Naujeck, A., Hill, J. and Gibb, M.J. (2005) Influence of sward height on diet selection by horses. *Applied Animal Behaviour Science* 90, 49–63.

Negi, G.C.S., Rikhari, H.C., Ram, S.P., Jeet, S.P. and Singh, S.P. (1993) Foraging nice characteristics of horses, sheep and goats in an alpine meadow of the Indian central Himalaya. *The Journal of Applied Ecology* 30(3), 383–394.

Ninomiya, S., Kusunose, R., Sato, S., Terada, M. and Sugawara, K. (2004) Effects of feeding methods on eating frustration in stabled horses. *Animal Science Journal* 75, 465–469.

Ödberg, F.O. and Francis-Smith, K. (1977) Studies on the formation of ungrazed eliminative areas in fields used by horses. *Applied Animal Ethology* 3, 27–34.

Olsen, F.W. and Hansen, R.M. (1977) Food relations of wild free-roaming horses to livestock and big game, Red Desert Wyoming. *Journal of Range Management* 30(1), 17–20.

Ott, E.A., Feaster, J.P. and Lieb, S. (1979) Acceptability and digestibility of dried citrus pulp by horses. *Journal of Animal Science* 49(4), 983–987.

Owen, R.R., McKeathing, F.J. and Jagger, D.W. (1980) Neurectomy in windsucking horses. *The Veterinary Record* 106, 134–135.

Pfister, J.A., Stegelmeier, B.L., Cheney, C.D., Ralphs, M.H. and Gardner, D.R. (2002) Conditioning taste aversions to locoweed (*Oxytropis sericea*) in horses. *Journal of Animal Science* 80, 79–83.

Putnam, R.J., Pratt, R.M., Ekins, J.R. and Edwards, P.J. (1987) Food and feeding behaviour of cattle and ponies in the New Forest, Hampshire. *Journal of Applied Ecology* 24, 369–380.

Ralston, S.L. (1984) Controls in feeding in horses. *Journal of Animal Science* 59(5), 1354–1361.

Ralston, S.L. and Baile, C.A. (1982a) Gastrointestinal stimuli in the control of feed intake in ponies. *Journal of Animal Science* 55(2), 243–253.

Ralston, S.L. and Baile, C.A. (1982b) Plasma glucose and insulin concentrations and feeding behavior in ponies. *Journal of Animal Science* 54(6), 1132–1137.

Ralston, S.L. and Baile, C.A. (1983a) Effects of intragastric loads of xylose, sodium chloride and corn oil on feeding behavior of ponies. *Journal of Animal Science* 56(2), 302–308.

Ralston, S.L. and Baile, C.A. (1983b) Factors in control of feed intake of horses and ponies. *Neuroscience and Biobehavioral Reviews* 7, 465–470.

Ralston, S.L., Van den Broek, G. and Baile, C.A. (1979) Feed intake patterns and associated blood glucose, free fatty acid and insulin changes in ponies. *Journal of Animal Science* 49, 838–845.

Ralston, S.L., Freeman, D.E. and Baile, C.A. (1983) Volatile fatty acids and the role of the large intestine in the control of feed intake in ponies. *Journal of Animal Science* 57(4), 815–825.

Randall, R.P., Schurg, W.A. and Church, D.C. (1978) Response of horses to sweet, salty, sour and bitter solutions. *Journal of Animal Science* 47(1), 51–54.

Redbo, I., Redbo-Torstensson, P., Ödberg, F.O., Hedendahl, A. and Holm, J. (1998) Factors affecting behavioural disturbances in race-horses. *Animal Science* 66, 475–481.

Rubenstein, D.I. (1981) Behavioural ecology of island feral horses. *Equine Veterinary Journal* 13(1), 27–34.

Salter, R.E. and Hudson, R.J. (1979) Feeding ecology of feral horses in western Alberta. *Journal of Range Management* 32, 221–225.

Schryver, H.F., Parker, M.T., Daniluk, P.D., Pagan, K.I., Williams, J., Soderholm, L.V. and Hintz, H.F. (1987) Salt consumption and the effect of salt on mineral metabolism in horses. *Cornell Veterinarian* 77, 122–131.

Shingu, Y., Kondo, S., Hata, H. and Okubo, M. (2001) Digestibility and number of bites and chews on hay at fixed level in Hokkaido native horses and light half-bred horses. *Journal of Equine Science* 12(4), 145–147.

Sisson, S. and Grossman, J.D. (1953) *The Anatomy of Domestic Animal Behavior*, 4th edn. W.B. Saunders, Philadelphia, Pennsylvania, pp. 387–447.

Sweeting, M.P., Houpt, C.E. and Houpt, K.A. (1985) Social facilitation of feeding and time budgets in stabled ponies. *Journal of Animal Science* 160, 369–374.

Turner, A.S., White, N. II and Ismay, J. (1984) Modified Forssell's operation for crib biting in the horse. *Journal of the American Veterinary Medical Association* 184, 309–312.

Tyler, S.J. (1972) The behaviour and social organization of the New Forest ponies. *Animal Behaviour Monograph* 5, 85–96.

Vecchiotti, G.G. and Galanti, R. (1987) Evidence of heredity of cribbing, weaving and stall walking. *Livestock Production Sciences* 14, 91–95.

Willard, J.G., Willard, J.C., Wolfram, S.A. and Baker, J.P. (1977) Effect of diet on cecal pH and feeding behaviour of horses. *Journal of Animal Science* 45, 87–93.

Winskill, L.C., Waran, N.K. and Young, R.J. (1996) The effect of a foraging device (a modified 'Edinburgh Foodball') on the behaviour of the stabled horse. *Applied Animal Behaviour Science* 48, 25–35.

Zahorik, D.M. and Houpt, K.A. (1981) Species differences in feeding strategies, food hazards, and the ability to learn food aversions. In: Kamil, A.C. and Sargent, T.D. (eds) *Foraging Behavior*. Garland STPM Press, New York, pp. 289–310.

13 Foraging in Domestic Herbivores: Linking the Internal and External Milieux

F.D. Provenza and J.J. Villalba

Department of Forest, Range and Wildlife Sciences, Utah State University, Logan, UT 84318-5230, USA

Introduction

Food and sex are basic for life, but of the two, food comes first. Without adequate nutrition, animals do not reproduce. Selecting the right diet, then, is the foundation for the survival of individuals who ensure the persistence of species. During the several million years a species exists, biophysical and social environments change tremendously. How do the individuals who create the populations that make up a species cope with such dynamism and complexity?

Consider the following challenge. Imagine an animal foraging in an environment with 25–50 plant species. These plants differ in their concentrations of energy, protein, minerals and vitamins. They all contain toxins of one sort or another, but at the appropriate dose, many of these toxins have medicinal benefits. Envision further that how much of any one food an animal can eat depends on the other foods it selects because at the biochemical level, nutrients and toxins interact one with another – nutrients with nutrients, nutrients with toxins and toxins with toxins (Provenza *et al.*, 2003a). These challenges are further amplified because cells, organs, individuals, and social and biophysical environments change constantly. Now imagine that 3–7 foods will make up the bulk of the diet at any point in time. Which plants should an animal choose? Clearly, given 25–50 species and their interactions, there are a great many pos-

sibilities to mix and match different plant species.

Certainly the foods an animal can eat and the possibilities to live in an environment are influenced by how the animal is built (its morphology) and how it functions (its physiology) – the expression of its genome. Morphology and physiology affect the need for nutrients, the ability to cope with toxins and the value of medicines, thereby creating the bounds within which animals can use different foods and habitats. But genes do not operate in isolation. They are expressed, beginning at conception, through the interplay with the social and biophysical environments where an individual is reared (Lewontin, 2000; Moore, 2002). Learning itself is a genetically expressed trait, a kind of fixed plasticity, that much like evolution is continually shaped by the environment (Skinner, 1984). Thus, learning in concert with the genome influences the combinations of plants an animal will eat.

Animals function and maintain well-being within the internal milieu through behavioural interactions with the external milieu. These interactions are complex. They involve an ongoing dynamic influenced by history, necessity and chance such that at any time an animal's foraging behaviour is a function: (i) of its evolutionary history, genetically expressed, in concert with its uniquely individualistic history of the social and biophysical environments where it was conceived and reared; (ii) of necessity due to its current nutritional, toxicological and medicinal state relative to the biochemical characteristics of foods it can potentially consume at any moment; and (iii) of chance occurrences that involve gene expression and environmental variability (Lewontin, 2000; Moore, 2002). Unfortunately, we rarely stop to consider that the events we observe are processes, rather than snapshots in time, that reflect this ongoing dynamic influenced by history, necessity and chance.

Behaviour by Consequences

While difficult to realize fully and embrace graciously, the only constant in life is change. Things never were the way they were and they never will be again. To survive in a world where the internal and external milieux change constantly requires that individuals experience the consequences of their behaviours. That necessitates ceaseless monitoring by cells and organs and ongoing changes in behaviour in ways that integrate basic metabolic processes with the experiences of pleasure and pain, drives and motivation, emotions and feelings (Damasio, 2003). Monitoring is essential because behaviours at all levels – cells and organs to social and biophysical environments – alter 'landscapes' at every level as they unfold. At all these levels, behaviour is a function of the consequences that ensue from behaviours and which in turn change the likelihood of future behaviours.

Consequences that increase the likelihood of behaviours recurring are called reinforcement, and they can be either positive (positive reinforcement) or negative (negative reinforcement). Creatures seek positive reinforcers. When a hungry animal searches for a particular nutritious food, or a thirsty animal walks to water, or a hot animal seeks shade, they do so because food, water and shade are positive reinforcers – they are things the body wants. Conversely, animals avoid negative reinforcers. When a hungry animal searches for a nutritious food, or a thirsty animal walks to water, or a hot animal seeks shade, they also do so to get relief from aversive stimuli – lack of food, water and shade.

Consequences that decrease the likelihood of behaviours recurring are called punishment, and they are based on either presentation of an aversive stimulus (positive punishment) or the removal of a positive stimulus (negative punishment). Positive punishment results from presenting an aversive stimulus. For example, when livestock experience toxicosis after eating a plant, they stop eating the plant. Negative punishment results from the removal of a positive reinforcer. For instance, when an animal eats a plant that was once nutritious but is no longer so, the animal decreases its rate of responding (eating the plant) because a positive reinforcer (nutrients) has been removed.

Behaviour by consequences and foraging

In the case of foraging, behaviour by consequences (preference) is manifest as the interrelationship between a food's flavour (behaviour) and its postingestive feedback (consequences) (Provenza, 1995a; Fig. 13.1). Flavour is the integration of odour, taste and texture with postingestive feedback from cells and organs in response to concentrations of biochemicals in foods. The senses of smell, taste and sight enable animals to discriminate among foods and provide pleasant sensations – liking for a food's flavour – associated with eating. Postingestive feedback calibrates sensory experiences – like or dislike – according to a food's utility to the body. Thus, flavour–feedback interactions emanate from an animal's physiological state and a food's chemical characteristics, and feedback affects liking for flavour (palatability) as a function of need and past experience with a food.

Feedback from the 'body' to the 'senses' is critical for health and well-being. Bodies are integrated societies of cells. They interact one with another and with the external environment through feedback mediated by nerves, neurotransmitters and hormones. In the case of flavour–feedback interactions, nerves for taste converge with nerves from the body in the brainstem and from there they synapse and relay to the limbic system and the cortex (Provenza, 1995a). Feedback from the body to

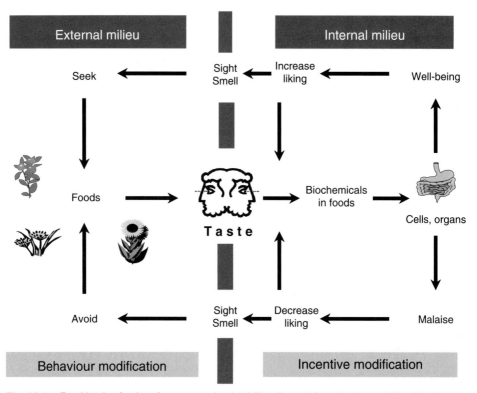

Fig. 13.1. Food intake, food preferences and palatability all result from the interrelationship between a food's flavour and its postingestive consequences. The integration of the internal (cells and organs) and external (foods) milieux occurs through two interrelated systems: affective (implicit or associative) and cognitive (explicit or declarative). Taste plays a critical role in both systems, and receptors for taste are situated at the junction between the internal and external milieux, like a Janus head – the Roman god of beginnings – placed at the gateway to the body, with one face looking outward and the other looking inward. The affective system integrates the taste of food with postingestive feedback from cells and organs in response to levels of ingested substances: nutrients, toxins, medicines. This system causes changes in the intake of food items that depends on whether the effect on the internal milieu is aversive or positive. The net result is incentive modification due to changes in well-being. The cognitive system integrates the odour and sight of food with its taste. Animals use the senses of smell and sight to differentiate among foods, and to select or avoid foods whose effect on the internal milieu is either positive or aversive. The net result is behaviour modification. Together, affective and cognitive processes enable animals to maintain fluidity given ongoing changes in the internal and external milieux and given that nutrients at too high levels are toxic, toxins at appropriate levels may be therapeutic and medicines in suitable doses can ameliorate excessive intakes of nutrients or toxins.

the palate is how societies of cells and organs influence certain foods, and how much of those foods are eaten by an animal. Feedback also influences the senses – smell, taste, touch – that are the interfaces between the internal milieu of the body and the external milieu where animals learn to forage.

Looking for Spinoza: experiencing the consequences of behaviours

While much research has related physiology and behaviour, there has been virtually no work to understand how livestock actually experience the consequences of their behaviours, e.g.

increased well-being following a nutrient boost or illness following toxicosis. Our mechanical notions of animals have inhibited us from exploring the issues Damasio (2003) raises in *Looking for Spinosa: Joy, Sorrow, and the Feeling Brain*. Damasio argues that body mapping of life-sustaining processes by the central nervous system is linked to well-being and feelings in animals. At the most fundamental levels, this involves metabolic regulation, basic reflexes and immune responses. At somewhat higher levels, organisms experience the consequences of their behaviours on the internal milieu as a continuum from pain (aversive) to pleasure (positive). At yet higher levels, drives and motivations (incentives) emanating from the consequences of past behaviours influence the likelihood of future behaviours. And at the highest levels, the experience of emotions and ultimately feelings arises from different degrees of well-being emanating from ongoing interactions among cells, organs, individuals, and social and biophysical environments (Fig. 13.2).

The notion that animals experience and respond to different physiological states by altering their ingestive behaviour has not been widely accepted historically or in the present time. Hence, we know little about the processes that enable animals to ingest appropriate amounts of nutrients, toxins and medicines (Provenza, 1995a). Consider, however, that feedback provides detailed mapping simultaneously about the current states of the living cells throughout the body because every region of the body contains nerves that provide information to the central nervous system (Damasio, 2003). The signalling is complex. It is not a matter of 'zeros' or 'ones' indicating, for example, that a living cell is on or off. The signals are highly variegated and related to well-being at the cellular level. Among other functions – including feedback reflecting pH, oxygen and carbon dioxide in the vicinity of a cell as well as internally generated compounds signalling distress, disease and pleasure – they can indicate the concentrations of nutrients or toxins in the internal (gut) and external (skin) milieux, and thus indicate to the brain what the gut or skin is experiencing at any given moment. In addition, the central nervous system is also directly informed about variations in the concentrations of a myriad of chemical

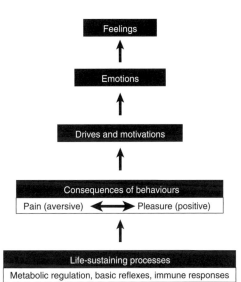

Fig. 13.2. Foraging behaviour is a function of its consequences, and ingesting foods results in postingestive consequences that affect the kinds and likelihoods of future foraging behaviours. Animals probably experience the consequences of their behaviours, for instance well-being following a nutrient boost or illness following nutrient excesses or toxicosis, through body mapping by the central nervous system that links life-sustaining processes to feelings (Damasio, 2003). Nerve endings in every region of the internal milieu provide detailed mapping simultaneously of the current state of cells that feedback to the central nervous system. At the most fundamental levels, this involves metabolic regulation, basic reflexes and immune responses. At somewhat higher levels, organisms experience the consequences of their behaviours as a continuum from pain (aversive) to pleasure (positive). At yet higher levels, drives and motivations (incentives) emanate from the consequences of past behaviours that influence the likelihood of future behaviours. At the highest levels, the experience of emotions and ultimately feelings arise from different degrees of well-being emanating from ongoing interactions among cells, organs, individuals, and social and biophysical environments.

molecules in the bloodstream via non-neural routes.

Thus, feelings arise from collective representations of life in the process of being aligned for survival in a state of optimal performance. When bodily processes are engaged due to internal and external stimuli and situations, the flow of the life process is made either more efficient, unimpeded and easier, or less so. The intensity of feelings depends upon the degree of corrections necessary when individuals experience negative states, and the level to which an animal is able to achieve positive states, both of which are directly related to sustaining life processes. Thus, to paraphrase Damasio (2003), the particular way a feeling is experienced depends on: basic life-sustaining processes in multicellular organisms with central nervous systems; the current functioning – positive or negative – of life-sustaining processes; the corrective responses particular life states beget and the behaviours organisms exhibit given the presence of certain biophysical and social stimuli and situations; and the nature of the neural medium in which all of these structures and processes are mapped.

From Homeostasis to Homeodynamics

Homeostasis

Claude Bernard first introduced the notion that internal organs are protected by the constancy of the internal milieu. The need to maintain a hospitable environment where cells and organs can perform optimally was attributed to continuous engagement at all levels – cells, organs, individuals – in an active self-preservation (Bernard, 1957). Homeostasis is the tendency of an organism to maintain a uniform and beneficial physiological stability within and among its cells and organs, leading to organic equilibrium. The term 'homeostatic regulation' was created to describe physiological processes and mechanisms that keep the body 'in balance' (Cannon, 1966). Set points and thermostats, terms often used to portray homeostasis, connote regulation to maintain dynamic equilibrium, which implies a fixed point of balance around which a body oscillates.

These metaphors taken literally have led to static views of homeostasis and mechanical views of animals. With foraging behaviour, we observe animals selecting particular foods and habitats, and we assume their behaviours are

optimized by predetermined morphological structures and physiological processes. We consider their needs for different nutrients to be rigidly set, and we change physical environments to fit our perception of what we think animals require – we attempt to change the landscape to suit the animal – rather than changing the animal to suit the landscape. In the case of conservation biology, we strive to maintain 'pristine' environments we believe best meet the needs of animals. All of these notions, rooted in views of homeostasis tempered by rigid ideas of genes that affect feedbacks, set points and thermostats, have come to connote rigidity rather than fluidity. They do not consider how history, necessity and chance interact to influence behaviour.

Homeodynamics

Ongoing regulation of the internal milieu requires ceaselessly engaging the external milieu, and both the internal and external milieux are best explained as strands in a web characterized by change. If we consider feedback among cells, organs, individuals, and social and biophysical environments, at all levels smaller-scale parts exist for and by means of larger-scale wholes, which in turn exist for and by means of smaller-scale parts. This is the essence of the holon, which is an autonomous entity when viewed from the perspective of its constituent subsystems, such as an animal from the viewpoint of a cell or an organ. The same holon, viewed from a larger scale, is merely a part of a larger system, such as an animal within a social group or an ecosystem. Interrelationships among holons make identity at any scale real and meaningful, as it is identity that makes the ongoing interrelationships possible. Interactions within and among holons – behaviour by consequences – involve the ongoing exchange of energy and matter at all levels from cells and organs to individuals and biophysical environments. While we often view these interactions in linear, hierarchical ways, in reality there is no one central controlling force, but only a large number of holons, all interacting and adapting to one another and to their local environments. Ultimately, kaleidoscopic patterns emerge from ongoing interactions among all of the parts (Provenza *et al.*, 1998).

As in quantum physics, the critical attribute of any holon – an elementary particle, an individual, a social group – is the fact that the dynamics of its behaviour cannot be defined with certainty because the holon is inseparable from its history (Provenza *et al.*, 1999). An individual animal's behaviour reflects its evolutionary history (gene-expressed morphology and physiology), its cultural history (experiences of the social and biophysical environments where an individual is conceived, born and reared) and its ongoing interactions with those environments. This does not mean behaviours occur in an arbitrary fashion. It means only that the behaviour of any organism is determined by connections to a larger historical whole. Because we do not know all of these connections precisely, and because their interrelationships lead to emergent properties, the classical notion of cause and effect must be supplemented with notions of self-organization (Provenza and Cincotta, 1993; Provenza *et al.*, 1998).

Ultimately, each individual's behaviour is unique, and behavioural processes can be understood only in a dynamic context, in terms of movement, interaction and transformation. Thus, animals are not machines – set points never are – and feedback changes animal behaviour from conception to the death of the individual and among individuals across generations. For these reasons, Rose (1998) coined the term homeodynamics, as opposed to homeostasis, to connote ongoing adjustments to ever-changing internal and external milieux, rather than to oscillations about some fixed point of balance.

Thus, behaviour emerges from the functioning of holons and the integrity of the holon is the variable that is maintained (Maturana and Varela, 1980; Provenza *et al.*, 1998). Deviations in the internal milieu promote behavioural responses intended to maintain the holon's functioning and well-being. Behavioural regulation is maintained via the functioning of feedback loops involving circular arrangements of interconnected holons. Feedback influences behaviour through actual, not expected, performance, and it involves more than the linear cause-and-effect links between

inputs and outputs, such as temperature control by a thermostat (Capra, 1996). The inherent non-linear nature of interconnected loops, which involves thresholds and varying degrees of temporal delays between behaviour and consequences (Glass and MacKey, 1988), creates multiple arrays and levels of complexity that increase the flexibility and adaptability of responses to environmental challenges.

Flavour–Feedback Associations: Linking State and Substance

The amount of food eaten and preferences for foods are typically thought to be influenced by palatability. But what is palatability? It is a narrowly defined term with many meanings. *Webster's* defines palatable as pleasant or acceptable to the taste and hence fit to be eaten or drunk. Animal scientists usually explain palatability as the hedonic liking or affective responses from eating that depend on a food's flavour and texture, or the relish an animal shows when consuming a food or ration. Conversely, plant scientists describe palatability as attributes of plants that alter preference such as chemical composition, growth stage and associated plants. Thus, all popular definitions focus on either a food's flavour or its physical and chemical characteristics.

Historically, researchers thought palatability depended on the species of animals, and they attempted to rank food preferences based on their presumed palatabilities. Nutritionists understood that intake regulation involved feedback from receptors – chemoreceptors, osmoreceptors, mechanoreceptors – in the body (Van Soest, 1982). They also came to appreciate the fact that different concentrations of compounds used to assess the preferences of cattle, sheep and goats for sweet, sour, salty and bitter affect more than just taste – they have postingestive effects (Grovum, 1988). Even so, palatability remained a mystery, and its relationship to intake and preference unknown (Arnold and Dudzinski, 1978; Grovum, 1988), largely because we did not appreciate the complex interrelationships among taste, smell, sight and the postingestive effects of foods, as well as their relationship to past experiences with foods. In essence, we did not consider history, individuality or feedback.

Behavioural regulation in the internal milieu

To maintain fluidity and well-being while foraging, cells and organs must integrate neural and hormonal signals corresponding with specific internal states, and animals must then associate changes in well-being with ingesting specific substances (Figs 13.1 and 13.2). The exchange of information between the internal and external milieux is processed through two interrelated systems: affective (implicit or associative) and cognitive (explicit or declarative). Taste plays a critical role in both systems. The affective system integrates the taste of food with postingestive feedback from cells and organs in response to levels of ingested substances – nutrients, toxins, medicines. This system causes changes in the intake of food items that depend on whether the effect on the internal milieu is aversive or positive. The net result is incentive modification due to changes in well-being. On the other hand, the cognitive system integrates the odour and sight of food with its taste. Animals use the senses of smell and sight to differentiate among foods, and to select or avoid foods whose effect on the internal milieu is either positive or aversive. The net result is behaviour modification. Together, affective and cognitive processes enable animals to maintain fluidity given ongoing changes in the internal and external milieux and given that nutrients at too high levels are toxic, toxins at appropriate levels may be therapeutic and medicines in suitable doses can ameliorate excessive intakes of nutrients or toxins.

Affective (non-cognitive) changes in palatability through flavour–feedback interactions occur automatically. Animals do not need to think about, or even remember, the feedback event, just as none of us need to consider which enzymes to release to digest the foods we eat. Even when animals are anaesthetized, postingestive feedback still changes palatability. When sheep eat a nutritious food and then receive a toxin dose during deep anaesthesia, they become averse to the food because the negative feedback from the toxin occurs even when animals are deeply asleep (Provenza et al., 1994a). Thus, feedback changes palatability without a bit of thought, and often in spite of rationality. For instance, people often acquire strong aversions to foods eaten just prior to becoming nauseated even when they know that the flu or seasickness – rather than the food – was responsible for the nausea.

Historically, few believed that cells and organs could influence the palate to select foods that meet their needs – that bodies have nutritional wisdom. In part, we assumed nutritional wisdom was not possible because it involved more cognitive, rational, analytical thought. We did not appreciate that the wisdom of the body depends strongly on non-cognitive, intuitive, synthetic processes mediated by cells and organs in response to nutrients, toxins and medicines (Provenza, 1995a, 1996). In considering the origins of human behaviour as well, we typically emphasize the cognitive, rational and analytical aspects, but the non-cognitive, intuitive and synthetic facets of 'thinking' most strongly influence our behaviours, and this occurs without a bit of 'thought' (Gladwell, 2005).

Feedback increases preference for substances that enhance well-being

Combinations of nutrients in appropriate doses enhance well-being. For example, livestock benefit from maintaining a balance of energy and protein in their diets and by discriminating specific flavours and nutrient-specific feedbacks. Lambs fed diets low in energy and protein prefer flavoured low-quality foods previously paired with intraruminal infusions of energy (starch, propionate, acetate) or nitrogen (urea, casein, gluten) (Villalba and Provenza, 1996, 1997a,b,c). Given preloads of energy or nitrogen, lambs prefer flavours previously paired with nitrogen or energy, respectively, during the ensuing meals (Villalba and Provenza, 1999). Thus, animals maintain a balance of energy to protein that meets their nutritional needs, and in the process, they recognize different internal states and discriminate among different nutrients (Egan, 1980; Wang and Provenza, 1996).

People learn to take aspirin for headaches, antacids for stomachaches and ibuprofen to relieve pain, and we obtain prescriptions from doctors for medications. Many of the drugs we use come from plants in nature. But what about other animals; can they too learn to write prescriptions from nature's pharmacy? While little is known about the abilities of animals to self-medicate, and many of the observations are anecdotal and equivocal (Clayton and Wolfe, 1993; Lozano, 1998; Houston et al., 2001), there is evidence of self-medication in animals (Engel, 2002). Sheep ingest 'medicines' such as polyethylene glycol (PEG), a substance that attenuates the aversive effects of tannins, when they eat foods high in tannins, and they titrate the dose of PEG according to the amount of tannin in their diet (Provenza et al., 2000). They discriminate the medicinal benefits of PEG from non-medicinal substances by selectively ingesting PEG after eating a meal high in tannins (Villalba and Provenza, 2001). They also forage in locations where PEG is present, rather than where it is absent, when offered nutritious foods high in tannins in different locations (Villalba and Provenza, 2002). Likewise, cattle foraging on endophyte-infected tall fescue high in alkaloids readily use lick tanks that contain FEB-200 (Altec™), but they ignore lick tanks without FEB-200 (C. Bandyck, Dodgeville, Wisconsin, personal communication, 2004). FEB-200 contains the cell walls of yeast, which adsorb the alkaloids in tall fescue, thus acting as a medicine that enhances consumption of tall fescue by cattle. Sheep fed acid-producing substrates such as grains subsequently ingest foods and solutions that contain sodium bicarbonate, which attenuates acidosis (Phy and Provenza, 1998). In the most elaborate studies to date, sheep learned to selectively ingest three medicines – sodium bentonite, polyethylene glycol, dicalcium phosphate – that lead to recovery from illness due to eating too high amounts of grain, tannins and oxalic acid, respectively (Villalba et al., 2006a). This first demonstration of multiple malaise–medicine associations supports the notion that herbivores can learn to self-medicate.

Animals may also learn to overcome internal parasite burdens by eating foods high in tannins and nutrients (Hutchings et al., 2003). Livestock feeding on plants with tannins show lower nematode burdens, lower faecal egg counts and higher body gains than those eating similar plants without tannins (Athanasiadou et al., 2000; Coop and Kyriazakis, 2001; Min and Hart, 2003). Tannins also increase the supply of bypass protein (Reed, 1995; Foley et al., 1999), which enhances immune responses to intestinal parasites (Min and Hart, 2003). Sheep with parasite infections also ingest needed nutrients (Hutchings et al., 2003), thereby better coping with the nutrient drain and correcting the infection through increased immunity (Min and Hart, 2003). Finally, sheep with high parasite loads avoid parasite-rich pastures, even though those pastures offer higher nutrient rewards, to a greater extent than sheep with lower levels of parasite infection (Hutchings et al., 2002).

Feedback decreases preference for substances that diminish well-being

Deficits or excesses of nutrients or toxins cause cells and organs to deviate from well-being and self-preservation. In this case, behaviours are geared towards limiting intakes of particular foods. Feedback from foods inadequate in nutrients decreases intake and preference. For example, sheep are reluctant to eat poorly nutritious foods such as straw; their intake and preference for straw increase only with feedback from more nutritious food (Greenhalgh and Reid, 1971) or starch (Villalba and Provenza, 1997a, 2000a) infused into the rumen immediately after a meal of straw. Animals detect and respond to an amino acid deficit within minutes of eating a diet low in an amino acid (Hao et al., 2005). Preruminant and ruminant lambs become averse to diets deficient in specific amino acids and they readily sample other foods that may help them correct the deficiency (Rogers and Egan, 1975; Egan and Rogers, 1978). Finally, exposure even to a nutritionally balanced food for as little as a day can decrease preference for that food, and the decrease in preference is much more pronounced when the food is low in fermentable protein relative to energy (Early and Provenza, 1998).

Intake and preference also decline when nutrients exceed needs. For instance, when needs for NaCl are met, and lambs are fed flavoured straw previously associated with intraruminal infusions of NaCl, they avoid the flavoured straw because their needs for salt are met (Villalba and Provenza, 1996). Likewise, sheep avoid sulphur when their requirements for sulphur are met (Hills *et al.*, 1999), and cattle stop eating bones when their blood Pi (inorganic phosphate) levels are within normal or excessive ranges (Denton *et al.*, 1986; Blair-West *et al.*, 1992). The same is true with excesses of energy or protein. Lambs prefer a flavour of straw paired with low to moderate doses of energy (propionate, acetate) or nitrogen (urea, casein), but at higher levels of energy or nitrogen, they become averse to that flavour of straw (Villalba and Provenza, 1996, 1997a,b). Dairy cows fed high levels of protein in the barn subsequently avoid eating plants with higher nitrogen concentrations when given choices while they are foraging on pasture (D. Emmik, Cortland, New York, unpublished data, 2004).

Finally, excesses of toxins in foods cause food avoidance because they move the body away from normal functioning. Goats limit intakes of otherwise nutritious foods too high in tannins or lithium chloride (Provenza *et al.*, 1990). Oral gavage of toxins causes dose-dependent decreases in intake of toxin-containing foods (Wang and Provenza, 1997; Dziba and Provenza, 2006). Limits on intake are set by the rates at which toxins can be eliminated from the body (Foley and McArthur, 1994). At critical thresholds, toxins satiate the detoxification capabilities of herbivores (Provenza *et al.*, 2003a). At these levels, animals quit feeding, and resume eating only after toxin concentrations in the body decline due to detoxification and elimination (Pfister *et al.*, 1997; Dziba and Provenza, 2006; Dziba *et al.*, 2006). These processes cause cyclic patterns of intakes of particular foods with peak intakes at the lowest concentration of toxins in the body (Pfister *et al.*, 1997; Foley *et al.*, 1999).

Behavioural regulation when substances interact

Despite the diversity of chemicals herbivores ingest, most studies focus only on single compounds. Nevertheless, biochemical diversity is

essential for homeodynamics and all biochemical interactions depend on dosages (Provenza *et al.*, 2003a). For instance, tannins at high levels adversely affect animals, but sufficient protein can mitigate the effects of excess tannins; conversely, tannins in moderate amounts are beneficial as they reduce nitrogen loss in the rumen by decreasing the breakdown of protein into ammonia, thus increasing the 'bypass value' of proteins and amino acids, especially sulphur-containing amino acids such as methionine and cystine (Reed, 1995). Tannins eaten in modest amounts also decrease internal parasites (Hutchings *et al.*, 2003; Min and Hart, 2003).

Nutrient–nutrient interactions

When different nutrients interact, behavioural responses depend on the specific characteristics of the interaction. For instance, preference increases for diets with appropriate ratios of energy and protein, whereas preference decreases with an excess of either (Kyriazakis and Oldham, 1997; Villalba and Provenza, 1997c). Asynchronous releases of by-products of energy and nitrogen metabolisms cause a build-up of organic acids and ammonia that diminishes food preference (Cooper *et al.*, 1995; Francis, 2003). Balancing the supply of fermentable carbohydrates and nitrogen optimizes microbial protein synthesis and maximizes retention of rumen-degradable nitrogen (Sinclair *et al.*, 1993). Conversely, when the rate of ammonia formation exceeds the rate of carbohydrate fermentation, nitrogen is used inefficiently by microbes, and much nitrogen is lost in urine (Russell *et al.*, 1992). Excessive nitrogen/energy ratios cause ammonia toxicity (Lobley and Milano, 1997), whereas excessive energy relative to nitrogen produces acidosis (Francis, 2003). Supplements high in starch depress the intake of fibrous foods (Mertens and Loften, 1980).

Toxin–toxin interactions

All plants contain toxins, including the vegetables we grow in our gardens, the grasses and forbs we plant in pastures and the plants that grow naturally on rangelands. None the less, herbivores seldom consume enough toxins to be poisoned because they regulate their intake of toxins. While we know little about this topic, we do know that interactions among toxins can cause aversions or preferences, depending on the specific characteristic of the interaction (Provenza et al., 2003a). Ingesting foods with a variety of different toxins, which act upon different organs and detoxification pathways, is likely to be less harmful than a large dose of any one toxin (Freeland and Janzen, 1974). Indeed, sheep eat more when offered choices of foods with various toxins that affect different detoxification mechanisms, and thus are complementary (Burritt and Provenza, 2000; Villalba et al., 2004). In contrast, when toxins impact the same detoxification pathway or are antagonistic, ingestion of toxins decreases (Burritt and Provenza, 2000). Interestingly, sheep and goats maintain high levels of intakes when they can select a variety of shrubs that contain different toxins, and the effect is far greater than that due to medicines such as PEG and activated charcoal; these medicines have a pronounced effect only when the number of shrubs in the diet is reduced to less than two or three (Rogosic et al., 2006a,b).

Nutrient–toxin interactions

Rates of detoxification are influenced by the nutritional status of an animal. The general mechanism of detoxification involves converting more toxic lipophilic compounds to less toxic water-soluble compounds that can be excreted in the urine (Cheeke and Shull, 1985; Cheeke, 1998). Biotransformation of toxins is carried out largely in the liver and usually occurs in two steps. The first step (phase I) introduces a reactive group – such as OH, NH_2, COOH or SH – into the structure of the toxin; those interactions typically produce a less toxic compound. During the second step (phase II), the newly formed compound is conjugated with a small molecule such as glucuronic acid, amino acids (e.g. glycine), sulphates, acetates or methyl groups (Osweiler et al., 1985). Importantly, these transformations require nutrients such as protein and

energy (Illius and Jessop, 1995, 1996). Thus, detoxification processes reduce the protein and energy that otherwise would be available for maintenance and production (Freeland and Janzen, 1974; Illius and Jessop, 1996).

When animals ingest adequate amounts of energy and protein, they can eat more foods that contain toxins. Lambs ingest more of the toxin LiCl as the energy content of their diet increases (Wang and Provenza, 1997). Likewise, sheep offered terpene-containing diets with increasing concentrations of energy or protein consume terpenes in a graded fashion with a positive relationship between energy and protein intake of foods with toxins (Villalba and Provenza, 2005). Supplemental energy and protein increase the ability of sheep and goats to eat foods that contain toxins such as terpenes (Banner et al., 2000; Villalba et al., 2002a), tannins (Villalba et al., 2002b) and saponins (Williams et al., 1992; Martinez et al., 1993). In contrast, herbivores eat less food with toxins when levels of nutrients such as sodium are low (Freeland et al., 1985; Freeland and Choquenot, 1990).

Ingesting specific toxins also influences selection of nutrients by animals, presumably behaviour aimed at correcting the disturbed internal state. Lambs infused with terpenes, nitrates, tannins or lithium chloride select diets with higher protein/energy ratios than animals that do not receive those toxins. In contrast, following infusions of cyanide, lambs prefer foods with lower protein/energy ratios than controls (Villalba et al., 2002c). In every case, the needs for nutrients increase, but the preferred protein/energy ratio depends on the specific toxin involved. Thus, there is not likely to be a set proportion of protein/energy needed to counterbalance a toxin challenge. Rather, these proportions vary on a toxin-by-toxin basis depending on physiological state. Many toxins promote formation of organic acids that disrupt acid/base status, which has led to the proposal that metabolic acidosis is a common effect of absorbed toxins (Foley et al., 1995). The selective effects of different toxins on preferred protein/energy ratios discussed above suggest that toxins cause other physiological effects in addition to acidosis, and that herbivores can discriminate among the postingestive effects of different toxins (Provenza, 1996; Villalba *et al.*, 2005a). Thus, toxins impose different metabolic costs and consequences that modify homeodynamic behaviour.

Finally, feeding decisions depend on an animal's capacity to detoxify plant toxins. Thus, a herbivore that can detoxify a toxin more quickly should be able to eat more. The aforementioned findings are consistent with this thesis, but they do not provide direct tests, which have been difficult because we generally do not know the specific mechanisms the body uses to detoxify toxins. Recently, more direct tests have been conducted with brushtail possums (Marsh *et al.*, 2005). Possums supplemented with glycine metabolize benzoic acid faster and in response eat more; animals detoxify benzoic acid primarily by conjugating it with glycine to form benzoyl glycine (hippuric acid). Moreover, when given a choice, possums select a diet containing both benzoate and glycine over diets with a high concentration of just one of these supplements. The ability of possums to regulate intake of benzoate and glycine when these compounds are offered separately or mixed together suggests they experience excesses of amino acid or benzoate and modify their feeding behaviour accordingly.

Integrating the Internal and External Milieux

Maintaining fluidity while foraging can be viewed as an ongoing series of bifurcations, or choices, in the face of varying degrees of uncertainty. Behaviour at bifurcations is influenced by history, necessity and chance, all of which influence the relationship – preference to aversion – between the internal and the external milieux. These interactions begin *in utero* and continue through life. Thus, more static views of organisms as machines, of innate appetites and of 'wisdom of the body' originally developed by Bernard, Cannon and Richter must be expanded to include homeodynamic notions of behaviour as multifaceted, flexible and organic (Schulkin, 2001; Provenza *et al.*, 2003a).

Learned responses and multiple flavour–feedback associations

Given the dynamic nature of foods and landscapes, common sense suggests that nature did not confer specific recognition, through the senses of smell and taste, of every nutrient (Provenza and Balph, 1990; Schulkin, 2001). It is not enough for animals to have specific and static preferences for 'nutritious' foods, or even to possess an odour–taste system organized to discriminate nutrients from toxins. Such organization would not be sensitive enough to ongoing changes in the internal and external milieux. To maintain well-being throughout the life of the individual and the species, these systems must be plastic (Scott, 1990) and sensitive to ongoing interactions among cells and organs in play with ever-changing social and biophysical environments (Provenza, 1995a). Thus, beyond the ability to sense by odour or taste specific nutrients such as sodium (Richter, 1976), animals evolved mechanisms that enabled them to learn about interactions among nutrients, toxins and medicines.

The plasticity of flavour–feedback associations provides a mechanism for generating, through experience, limitless and very specific flavour–feedback associations. For example, sheep can discriminate among three flavours associated with rumen infusions of three different nutrients – starch (flavour 1), casein (flavour 2) and water (flavour 3) (Villalba and Provenza, 1999), and among three flavours associated with NaCl (flavour 1), NaH_2PO_4 (flavour 2) and water (flavour 3) (Villalba *et al.*, 2006c). Furthermore, the ability to learn multiple associations suggests that animals learn to seek or avoid substances that rectify deviations in their internal milieux. Indeed, experienced animals increase their preference for a buffer, PEG or dicalcium phosphate when challenged, respectively, with excessive amounts of grain, tannins or oxalates (Villalba *et al.*, 2006a). Buffers, PEG and dicalcium phosphate attenuate, respectively, malaise due to over-ingesting grain, tannins and oxalates.

The ability to learn preferences does not negate the possibility of more 'hard-wired' appetites that act in concert with learning. For instance, animals recognize and consume more salt when deficient, and they also learn to associate various flavours with intragastrically (untasted) administered sodium (Hill and Mistretta, 1990; Villalba and Provenza, 1996). The preference for bones, typically observed when animals experience deficits of phosphorous, occurs automatically (Blair-West et al., 1992). However, animals experiencing phosphorus deficiencies can also learn to eat new forms of the element (Reynolds et al., 1953). These learned responses are influenced by the postingestive effects of phosphorus (Blair-West et al., 1992; Villalba et al., 2006c). Likewise, recognizing and rejecting amino acid-deficient diets occurs through a general control system, triggered by accumulation of amino acid-depleted transfer RNA in cells of the anterior piriform cortex in the brain. This internal nutrient sensor involved in maintaining amino acid homeodynamics appears to be conserved across evolution from single-cell organisms to mammals (Hao et al., 2005). Animals use flavour cues to develop aversions or preferences for diets that induce or restore amino acid imbalances, but only after they have experienced the postingestive effects of those diets (Gietzen, 1993).

The ability to learn multiple flavour–nutrient feedback associations, as well as to select different proportions of protein and energy to ameliorate the effects of different toxins (Villalba et al., 2002c) or a parasite load (Kyriazakis et al., 1994), suggests that animals experience multiple internal states as opposed to general states of need (nutrients) or malaise (toxins). This is consistent with findings that feedback from every region of the body provides detailed mapping simultaneously as to the current state of the living cells in particular regions because every region of the body contains nerve endings that provide feedback to the central nervous system (Damasio, 2003). Thus, well is not well is not well, and sick is not sick is not sick. Animals discriminate among multiple internal states.

Learning multiple flavour–feedback associations within and among meals

If palatability is more than a matter of taste, and it is, then how does the body discriminate among different foods, based on specific flavour–feedback interactions, within and among meals? There is some debate concerning how ruminants learn multiple flavour–feedback associations within a meal (Provenza et al., 1998; Duncan and Young, 2002). This debate has arisen in part due to a lack of knowledge of the interactions – from conception until death – among history, necessity and chance in the ongoing evolution of flavour–feedback associations. Here, we discuss separately, for the sake of simplicity, several factors that enable multiple flavour–feedback associations. However, just as a river is in all places simultaneously – at the source, along the course, in the ocean and in the atmosphere – so too these behaviours reflect past and ongoing experiences that are at the same time uniquely individualistic, cultural and environmental. The uniqueness of these interactions makes each individual different. The plasticity of these processes lets animals adapt to ever-changing environments, and enables people to use behaviour to transform systems (Provenza, 2003a).

Historically, we believed that ruminants could not associate a food's flavour with its postingestive consequences for two reasons: (i) the presumed long delays between food ingestion and postingestive consequences; and (ii) the complexity of making multiple flavour–feedback associations within and among meals. More recent research shows ruminants can learn with delays between food ingestion and consequences of up to 8 h for toxins (Burritt and Provenza, 1991) and up to 1 h for calories (Villalba et al., 1999). In many cases, feedback from nutrients and toxins occurs within minutes as opposed to hours. Moreover, the lack of feedback from less nutritious food influences behaviour as much as rapid feedback from nutritious foods because the greater the delay between food ingestion and consequences, the less likely a food is to be nutritious and preferred (Provenza et al., 1992). We are also learning how multiple flavour–feedback associations mediated by cells and organs help the body assess which foods provide which benefits as enteric (gut) and central (brain) nervous systems interact with one another to integrate a food's flavour with its postingestive effects (Provenza, 1995a,b; Provenza et al., 1998; Fig. 13.3).

These interactions begin early in life, and because nervous systems have long memories, flavour–feedback interactions do not have to be relearned each time an animal eats a food, any more than a human has to relearn when different garden vegetables are ripe. Flavour–feedback relationships merely need to be updated as flavours and feedbacks change.

Familiar–novel dichotomy

Pasture and rangeland researchers, as well as nutritionists and ecologists, typically consider foraging only in terms of how plant structural and nutritional characteristics influence nutrient intake. The social environment, if it is considered at all, is typically viewed as a 'nuisance variable' that may modify processes that are physically and chemically driven. This has been an unfortunate oversight. The social environment influences selection of foods and habitats (Provenza and Balph, 1988; Howery et al., 1996, 1998; Mosley, 1999), and creates patterns of behaviour that make sense only in light of social history and culture (Provenza, 2003a,b; Provenza et al., 2003a).

Social organization creates culture, the knowledge and habits acquired by ancestors and passed from one generation to the next about how to survive in an environment (De Waal, 2001). Cultures develop when learned practices contribute to the group's success in solving problems. Cultures evolve as individuals in groups discover new ways of behaving, as with finding new foods or habitats and better ways to use them (Skinner, 1981). Accordingly, interactions with the mother and peers markedly influence what a young animal prefers to eat and where it prefers to live (Provenza, 1994, 1995a); an individual reared in shrub-dominated deserts of Utah will behave differently – and is morphologically and physiologically different – from one reared on grass in the bayous of Louisiana (Distel and Provenza, 1991; Distel et al., 1994, 1996; Wiedmeier et al., 2002).

Socializing enhances learning efficiency because each animal no longer has to discover everything through trial and error. When sheep and goats must learn to drink from a water device that requires pressing a lever, it

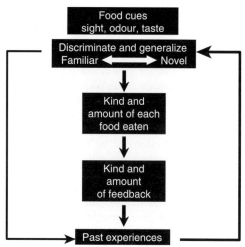

Fig. 13.3. Animals use food cues – sight, odour, taste – to discriminate and generalize among different foods based on the dichotomy between the familiar and the unfamiliar, which ranges along a continuum influenced by the degree of similarity between past and ongoing experiences. That dichotomy affects the kinds and amounts of foods eaten, which in turn impacts the kinds and amounts of feedback an animal experiences. These interactions begin *in utero* and continue throughout life, and because nervous systems have long memories, flavour–feedback interactions do not have to be relearned each time an animal eats a food. They merely need to be updated as flavours and feedbacks change in response to an animal's physiological conditions, a food's chemical characteristics and the biochemical characteristics of other foods in the diet. Collectively, these updates can be viewed as an ongoing series of bifurcations, or choices, influenced by history, necessity and chance. When nutritional state is adequate, familiarity breeds content, novelty breeds contempt and animals are neophobic – they are more cautious of novel foods. Conversely, as nutritional state becomes inadequate, familiarity breeds contempt, novelty breeds content and animals become neophyllic – they sample novel foods more readily.

takes only one individual to learn how to do it, and in no time all the others are drinking. Thus, when an individual discovers how to use a new resource, the group benefits, but discovering new resources is not inevitable (Provenza, 2003a). For example, goats browsing blackbrush-dominated rangelands experience energy and protein deficiencies. Of 18 different groups of goats foraging on six separate blackbrush pastures during three different winters, goats in only one group discovered that the interior chambers of woodrat houses provide a good source of supplemental nitrogen, and they performed much better than their counterparts. One goat probably learned the value of eating woodrat houses, and the other goats learned from it. Animals have similar difficulties learning through trial and error about the medicinal effects of substances, especially if behaviour and consequences (flavour–feedback) are not contingent (paired consistently) and contiguous (paired closely in time), as illustrated in studies that either did or did not closely pair medicine (flavour) and benefit (feedback) (Provenza et al., 2000; Villalba and Provenza, 2001; Huffman and Hirata, 2004; Villalba et al., 2006a). In all of these cases, social models increase the efficiency of learning over trial-and-error learning by individuals. When a mother's behaviour (eat or avoid) is subsequently reinforced by postingestive feedback (positive or negative), her offspring respond strongly (eat or avoid) to a food (Provenza et al., 1993b). Such knowledge then becomes a part of the culture, wherein young animals learn from their ancestors through their mothers.

Critically, lessons learned early in life from a mother create a dichotomy between the familiar and the unfamiliar (novel) essential for survival. Of the many factors that interact during updates about foods in which an animal's past experiences with a food are integrated with new information about the food, none is more important than novelty (Provenza et al., 1998). Novelty includes anything from a complete lack of acquaintance with the flavour of a food never before eaten, to a change in the flavour of a familiar food, to a novel food whose flavour is somewhat similar to a familiar food. The body evaluates new foods and flavours very carefully for their potential nutritional or toxicological effects.

While temporal contiguity – the proximity in time between a behaviour and its consequences – is of utmost importance in shaping behaviour, novelty even trumps temporal contiguity. For example, if an animal eats two foods in sequence, and then gets sick, the animal will strongly avoid the food eaten just prior to illness, unless the food eaten first was novel, in which case the animal will avoid the novel food (Provenza et al., 1993a). Even more critically, novel foods are evaluated cautiously by the animal and the body (internal milieu) within a meal. Sheep eat small amounts of a novel food, and they acquire an aversion only to the novel food when toxicosis follows a meal of several familiar foods and a novel food (Burritt and Provenza, 1989, 1991). Conversely, when sheep eat the same low-energy food in two different flavours, one familiar and the other novel, and then receive intraruminal infusions of starch directly proportional to the amount of the novel food consumed, they form preferences for the novel flavour (Villalba and Provenza, 2000b).

Past experiences

Past experiences influence behaviour, as illustrated in studies of the effects of prior illness on preference. In one study, sheep first made averse to a particular food, and then allowed to eat the food until the aversion extinguished, avoided the food that made them ill in the past when toxicosis followed a meal of several foods (Burritt and Provenza, 1996). In another study, when toxicosis followed a meal of five foods, including one novel and one that made them ill in the past, sheep subsequently would not eat the novel food and markedly decreased their intake of the food that made them ill in the past; they did not avoid the other three familiar foods (Burritt and Provenza, 1991).

Experiences early in life have life-long influences on food intake and animal performance, as illustrated in a 3-year study where 32 beef cattle 5–8 years of age were fed ammoniated straw from December to May (Wiedmeier et al., 2002). Although the cows were similar genetically and were fed the same diet, some cows performed poorly, while others maintained themselves. Researchers were baffled until they examined the dietary histories of the animals. Half of the cows were exposed to ammoniated straw with their mothers for 2 months early in life, while the other half had never seen straw. Throughout the study, the experienced cows had higher body weight and condition, and for the first 2 years of the study, they also produced more milk and bred back sooner than cows with no exposure to straw, even though they had not seen straw for 5 years prior to the study.

Thus, past experiences have life-long influences on behaviour, but we seldom notice because we know or remember so little about the history of any animal. Nutrient-conditioned food preferences, which may have occurred as a result of brief exposure to a food 3–5 years previously, cause some animals (experienced) to readily eat a food other animals (naive) avoid (goats – Distel and Provenza, 1991; Distel et al., 1994, 1996; sheep – Green et al., 1984; cattle – Wiedmeier et al., 2002). Likewise, food aversions often extinguish in the absence of toxicosis. However, if after an aversion has extinguished a sheep eats a meal of familiar foods, one of which previously made it ill, and then experiences toxicosis, the sheep will avoid the food that previously made it ill, not the other familiar foods (Burritt and Provenza, 1991, 1996). This point was highlighted when an adult ewe previously trained – conditioned with toxicosis – to avoid a tree (Russian Olive) began to eat the tree during meals 3 years later. When a meal of foods that included Russian Olive was followed by toxicosis, the ewe subsequently avoided the Russian Olive, the food that made her ill 3 years previously, not the other familiar foods eaten in that meal (F.D. Provenza, Logan, Utah, unpublished data, 2000). Without knowledge of the ewe's dietary history with Russian Olive, her behaviour would not have made sense.

Discriminating and generalizing based on past experiences

The familiar–novel dichotomy – the first line of defence for evaluating foods – also causes animals to discriminate and generalize based on their past experiences. For example, if the flavour of a familiar, nutritious substance such as molasses occurs in a novel food, the likelihood increases that animals will eat the food, provided they had positive experiences with molasses previously. Conversely, if the flavour of a familiar, toxic substance occurs in a novel food, the likelihood that the food will be eaten decreases if the animals previously experienced toxicosis after eating that food.

The ability to discriminate and generalize based on past experience helps animals identify potentially toxic foods quickly. Animals generalize aversions from past experiences, thus reducing the risks associated with toxic novel foods or familiar foods whose flavours have changed. When lambs eat cinnamon-flavoured rice and then experience toxicosis, their preference for cinnamon-flavoured rice declines. When they are subsequently offered wheat, which they prefer, but with cinnamon flavour added, they refuse to eat the cinnamon-flavoured wheat. Thus, lambs generalize an aversion from rice to wheat based on a common flavour – cinnamon (Launchbaugh and Provenza, 1993).

Sheep and goats also generalize preferences. Lambs experienced in eating grains such as milo, barley and wheat, which are about 80% starch, subsequently prefer novel foods such as grape pomace with added starch (Villalba and Provenza, 2000c). Sheep and goats prefer hay sprayed with extracts of preferred high-grain concentrates (Dohi and Yamada, 1997), and sheep more readily eat novel foods such as rice bran when they have been sprayed with extracts from familiar grass (Tien et al., 1999). Livestock also generalize preferences to weeds sprayed with molasses (Provenza, 2003a).

Sheep and goats discriminate among foods based on the concentration of the flavour, and they typically avoid flavour concentrations most different from what they have eaten in the past. For instance, when lambs familiar with unflavoured barley eat a meal of barley with a low and a high concentration of an added flavour and then receive a mild dose of toxin, they subsequently avoid the barley with the highest flavour concentration because it is most different from unflavoured barley (Launchbaugh et al., 1993). The same is true for goats foraging on shrubs. Current season growth (CSG) and older growth (OG) twigs from the blackbrush shrub likely share a common flavour, but the flavour is much stronger in CSG. When goats first eat a meal of CSG and OG and then experience toxicosis, they subsequently avoid CSG because it is most different from the familiar food (Provenza et al., 1994b).

Animals also discriminate based on changes in flavour. Lambs decrease intakes of a familiar food (rice) when rice contains a novel flavour (onion) (Provenza et al., 1995). Similarly, sheep routinely fed elm from one location would not eat elm of the same species from another site, evidently because the smell and taste of elm differed between two locations (Provenza et al., 1993b). By reducing their intakes of familiar foods with novel flavours animals reduce the likelihood of overingesting toxins.

Sheep generalize preferences based on the quantitative interrelationship between flavour and feedback. Lambs exposed to low or high concentrations of a flavour, and then given low or high amounts of energy (starch) by intraruminal infusion, learn to prefer the flavour concentration – low or high – associated with the highest amount of starch infused. When the lambs are then given a gradient of flavour concentrations, they prefer the flavour intensity most similar to the intensity of the flavour associated with the highest amount of starch previously infused (Villalba and Provenza, 2000d).

Animals also generalize across broad classes of experience. Lambs that experience toxicosis, even if it occurs only once after eating a novel food, become more reluctant to eat novel foods (F.D. Provenza, Logan, Utah, unpublished data, 1998). Conversely, when lambs experience positive nutritional consequences every time they eat a novel food, they more readily eat novel foods (Launchbaugh et al., 1997).

Discriminating nutritious from toxic foods within and among meals

DISCRIMINATING AMONG NOVEL FOODS BASED ON THE AMOUNT OF EACH NOVEL FOOD EATEN IN A MEAL.   The amount of food eaten and the novelty of a food's flavour interact with postingestive consequences to enable animals to discriminate nutritious from toxic foods within a meal. When goats naive to the blackbrush shrub first eat a meal of blackbrush – twigs of both CSG and OG – and then experience toxicosis due to the high tannin concentrations in CSG, they subsequently avoid CSG in part because it is most different from the more

familiar OG (Provenza et al., 1994b). The amount of each twig eaten within a meal also influences their behaviour. As soon as a goat eats a meal that contains more CSG than OG, and it eats enough CSG to experience malaise from the tannins in CSG, it subsequently avoids CSG, the twig type it ate in the greatest amount (Provenza et al., 1994b). This behaviour is learned very quickly but not simultaneously by all goats. Some naive goats eat more CSG than OG initially in a meal, and others do the opposite, but all goats learn about the consequences of eating CSG and OG within 1–2 days of beginning to forage on blackbrush.

Animals also must ingest a threshold amount of a novel food for the body to assess the specific biochemical characteristics of the food, and nutritional state influences the response. Sheep exposed for only 20 min/day to two novel foods, one more nutritious than the other, surprisingly preferred the less nutritious of the two foods because it was more familiar and they ate little of the more nutritious food. This response occurred only when sheep were fed a basal diet adequate in energy and protein. When offered only the two novel foods for 8 h/day, they quickly learned to prefer the more nutritious of the two foods (Villalba and Provenza, 2000b). Importantly, animals will include a familiar but less nutritious food in their diet when the food is only a minor part of their diet.

Thus, sheep discriminate based on familiarity, nutritional state and amount of food eaten, factors that undoubtedly influence preferences as nutritional qualities, toxicities and abundances of foods change daily and seasonally. More generally, when nutritional state is adequate, familiarity breeds content, novelty breeds contempt and animals are neophobic – they are more cautious of novel foods. Conversely, as nutritional state becomes inadequate, familiarity breeds contempt, novelty breeds content, and animals become neophyllic – they sample novel foods more readily.

DISCRIMINATING DIFFERENT DOSAGES AND COMBINATIONS OF NUTRIENTS AND TOXINS.   How animals discriminate among foods varying in nutrients and toxins also has been investigated with goats conditioned with different combinations of nutrient–toxin (postingestive feedback) as they ate the branches of four species of conifers (flavours) (Ginane *et al.*, 2005). On conditioning days, animals were fed a conifer species, and as they ate, they were dosed with a mixture of two stimuli: one nutritious (sodium propionate) and the other toxic (lithium chloride). For each goat, four dosage combinations of nutrient–toxin were each paired with a different species of conifer, and a different species was offered each day on 4 successive days per week for 5 weeks.

The goats reacted strongly to the toxin, and the effect was evident early in the study. One encounter with a food plant paired with the toxin was sufficient for the goats to perceive the variation in the four different intensities (doses) of feedback, to associate each conifer species with each intensity and to decrease preference progressively for each species paired with increasingly higher doses of toxins. When all four conifer species were offered simultaneously, animals selected based on the intensity of their previously experienced postingestive effects, indicating that the links between postingestive effects and degree of avoidance of toxic foods are well developed in goats, allowing them to make appropriate choices when faced with a series of foods of varying potential toxicity.

Although the goats responded quickly to the toxin, they reacted less strongly to the energy and they did not respond until the last conditioning period, suggesting that they required more time to perceive the energy signal and respond to it within the context of this study. However, their basal diet provided adequate nutrition to meet their maintenance needs for energy, and the conifer species provided additional energy. This combination probably made them much less responsive to the energy infusions than if they had been more deprived of energy (Villalba and Provenza, 1996). Conversely, other studies show that sheep, even fed *ad libitum*, behave based on input from both positive and negative feedback when both signals are associated with the same food cue (Wang and Provenza, 1996, 1997). Thus, in complex situations with many stimuli, animals may need preingestive as well as postingestive clues to discriminate among foods. As argued previously, learning from the mother plays a critical role with regard to learning preferences and aver-

sions to pre-ingestive cues (Provenza *et al.*, 1992).

Initial conditions, individuality, context and homeodynamic utility

Prevailing theories of food (Optimal Foraging Theory) and habitat (Ideal Free Distribution) selection assume that single 'optimal' solutions exist for the challenge of foraging (Fretwell and Lucas, 1970; Stephens and Krebs, 1986). In other words, given an array of food alternatives and locations a herbivore will find an optimal diet or habitat, which will be similar for all individuals of a species. However, the value of a food or any combinations of foods varies among similar individuals of the same species due to context – the kinds and numbers of foods available – and due to each animal's unique experiences with the foods (Villalba *et al.*, 2004). Thus, while animals undoubtedly optimize, an animal's optimization will depend on its history, necessity and chance, each of which will differ by individual and situation. The kinds and amounts of foods eaten will depend on the utility of those foods within the context of the local environment and the combinations of foods consumed. Thus, animals in the same environment may follow different 'foraging paths' depending on their foraging histories and current conditions, and the greater the alternatives, the greater the potential combinations of foods and habitats that can meet individual needs (Provenza *et al.*, 2003a).

Individuality in form and function influences foraging behaviour

Food intake and preference depend on how individuals are built morphologically and how they function physiologically, and marked variation is common even among closely related animals in need for nutrients and abilities to cope with toxins. Foraging decisions are affected by differences in organ mass and how animals metabolize nutrients and toxins (Konarzewski and Diamond, 1994). From the standpoint of nutrients, for instance, the same dose of sodium propionate that conditions preferences in some lambs conditions aversions in others (Villalba and Provenza, 1996). Likewise, lambs given a choice of barley (high energy) and lucerne (high protein) vary greatly in their preferences (Scott and Provenza, 1999). The same is true with toxins. Doses of tannins that condition an aversion in some goats do not deter others (Provenza et al., 1990). Likewise, some sheep fed high levels of Galega officinalis failed to show any symptoms of toxicosis, whereas others were killed by a low dose (Keeler et al., 1988). Such individual variation in ability to tolerate the same dose of different toxins and in metabolism of nutrients and toxins guarantees that no one diet will be ideal for every individual of a species. On the contrary, the degree of benefit or harm a specific food can induce in an animal will vary as a function of the individual's morphological and physiological tolerances and susceptibilities, which in turn will influence diet selection.

Initial conditions and context influence foraging paths

Life flourishes in a liquid region at the boundary of order and chaos, where systems forever push their way into novelty – molecular, biochemical, morphological, physiological, behavioural and organizational (Kauffman, 1995, 2000). The process of maintaining fluidity in the face of uncertainty is much like a hillclimber on a foggy day: animals may not always be able to perceive the optimum so much as the fact that they are travelling uphill (Provenza and Cincotta, 1993). Learning to mix sub-optimal but complementary foods may enable animals to create diets that are unique due to the particular biochemical contexts offered by the foods and the variety of equally successful ways different individuals may meet their nutritional requirements and tolerate toxins. This leads to multiple 'foraging paths' instead of a single solution to the challenge of foraging.

Initial conditions affect the evolution of systems (Glass and MacKey, 1988), and in the case of foraging, first impressions matter. Initial experience and the availability of alterna-

tives influence the preferences of lambs with 3 months of experience mixing foods with tannins, terpenes and oxalates compared with those of lambs naive to these foods (Villalba et al., 2004). During the studies, all lambs were offered five foods, two of them familiar to all of the lambs (ground lucerne and a 50:50 mix of ground lucerne:ground barley) and three of them familiar only to experienced lambs (a ground ration with either tannins, terpenes or oxalates). Within each group, half of the experienced and half of the naive lambs were offered the familiar foods ad libitum, while the remaining lambs were offered only 200 g of each familiar food daily. Throughout the study, naive lambs ate less of the foods with toxins if they were fed ad libitum as opposed to restricted access to the nutritious alternatives (66 vs. 549 g/day), and experienced lambs did likewise (809 vs. 1497 g/day). In both cases, however, lambs with experience ate significantly more of the foods containing the toxins, whether access to the lucerne–barley alternatives was ad libitum (811 vs. 71 g/day) or restricted (1509 vs. 607 g/day). These differences in food preferences and intake persisted during trials a year later. In a companion study, when access to familiar foods was restricted to 10%, 30%, 50% or 70% of ad libitum, animals ate more of the foods with toxins along a continuum (10% = 30% > 50% = 70%), which illustrates that animals must be encouraged to learn to eat unfamiliar foods that contain toxins (Shaw et al., 2006). Thus, initial conditions – past experiences and contexts – that encourage animals to learn to mix diets that contain toxins and nutrients help explain the partial preferences of herbivores, and they provide implications for managing plant–herbivore interactions (Provenza, 2003a).

In a related study, sheep learned to eat a low-quality food with toxins and a high-quality food in two different temporal arrangements (Villalba et al., 2006b). In one case, sheep were fed the high-quality food for 12 days followed by food with toxins for 12 days such that their synergistic effects were dissociated temporally. In the other case, sheep were fed both foods simultaneously for 12 days so their effects were associated within the same meal. Subsequently, all sheep could forage at locations containing both foods, only the high-quality food or only the food with toxins. Sheep that initially ate both foods in a meal always ate more food with toxins than those that initially experienced the foods in two distinct feeding periods, even when the high-quality food was available *ad libitum*. As the high-quality food decreased in abundance, lambs that learned to mix both foods foraged more opportunistically and remained longer at locations with both foods or with just the food with toxins. Even when both groups spent about the same amount of time at locations with both foods, lambs that initially ate both foods in a meal ate more food with toxins and thus consumed more food.

Finally, nutritional context influences what lambs learn about foods with tannins and terpenes. In one study, for instance, a group of lambs was fed a low-quality food containing tannins while on a basal diet low in nutrients; several weeks later they ate the same low-quality food containing terpenes while on a basal diet adequate in nutrients. Conversely, lambs in another group first ate terpenes and then ate tannins under the same regime described above (Baraza *et al.*, 2005). When offered a choice between the two foods, lambs consumed more of the food – tannin or terpene – they ate while on the basal diet high in nutritional quality. Thus, preference for plants high in toxins is affected by the nutritional state of an animal during exposure to the plant, which is influenced by the choices in the landscape (Provenza *et al.*, 2003a), which in turn influences future choices (Provenza *et al.*, 2003b).

Nutritional Wisdom Revisited: From Instinct to Experience

Looking back

During the last century, nutritional wisdom came to imply that animals 'instinctively' selected specific substances to maintain homeostasis (Fig. 13.4). This notion was referred to as 'genetic programming' of ingestive behaviour (Schmidt-Nielsen, 1994), and as the 'subconscious but irresistible desire' to restore biochemical equilibrium (Katz, 1937). With livestock, the archetypal example of nutritional wisdom is the well-characterized appetite for sodium (Richter, 1976), linked to both a specific gustatory transduction mechanism (Schulkin, 1991) and humoral signals acting in the central nervous system (Stricker and Verbalis, 1990). The sodium model of nutritional wisdom was extrapolated to livestock by scientists attempting to explore 'instinctive appetites' for other minerals and vitamins.

These efforts made researchers doubt that livestock possessed nutritional wisdom. Lactating dairy cows did not instinctively ingest recommended levels of calcium and phosphorus when offered dicalcium phosphate; indeed, many of the calcium-deficient animals never approached the novel source of calcium, and for those that did, intakes of calcium varied greatly (Coppock *et al.*, 1976). Moreover, some animals consumed large amounts of dicalcium phosphate even when neither calcium nor phosphorus was needed (Coppock, 1970). These results were consistent with earlier findings that sheep did not rectify a phosphorus deficit by consuming supplemental dicalcium phosphate (Gordon *et al.*, 1954). Nor did dairy cows offered choices consistently select appropriate minerals and vitamins, though the cows fed different diets did not perform differently during the 16-week trials; the researchers concluded that longer studies were needed due to the ability of cows to store many minerals and vitamins in the body (Muller *et al.*, 1977). Finally, lambs did not eat sufficient amounts of needed minerals, and because they tended to overconsume some minerals, researchers recommended feeding a complete ration, or if that is impossible, to offer *ad libitum* a complete mineral mix (Pamp *et al.*, 1977). Collectively, these studies fostered the notion that domestication had produced animals more responsive to food flavour than to nutritive value, and that acceptability rather than appetite or craving for minerals and vitamins influences free-choice consumption (Pamp *et al.*, 1976). In other words, domestication erased 'nutritional wisdom' and the 'innate ability' to select needed nutrients, a trait that through evolution still confers survival value to wild herbivores.

To add to the confusion, other research suggested that livestock have specific appetites for minerals, and observations by livestock

Nutritional wisdom

Historical views Innate appetites – instinct	Contemporary views Learned appetites – experience
Animals are 'genetically programmed' to recognize instinctively needed nutrients through sodium-like mechanisms.	Animals are 'genetically programmed' to learn flavour–feedback associations. These experiences shape food preferences.
Animals have 'innate desires' to restore biochemical equilibrium–homeostatic.	Animals learn as a function of contexts that change constantly – homeodynamic.
Animals ingest nutrients in exact amounts needed to meet their daily requirements – no under- or overconsumption.	Animals respond to excesses, deficits and imbalances in their diet – they may under- and overconsume needed nutrients.
No appreciation of individual variation. All individuals select nutrients in amounts that match tabular values for requirements.	Individuals all vary. Responses depend on unique morphologies and physiologies, physiological states and experiences.
Social learning and culture are not considered in nutritional wisdom.	Social learning and culture are critical for learning nutritional wisdom.

Fig. 13.4. Historical views of nutritional wisdom were based on innate appetites and instinctive drives to restore homeostasis through selection of the 'right' substance – homeostatic behaviour imbedded in the genes through evolution. When researchers attempted to extrapolate these views and the 'sodium model of nutritional wisdom' to study specific appetites for other minerals and vitamins in livestock, their efforts largely failed to show nutritional wisdom because they did not consider the many factors that enable the wisdom of the body. More recently, the concept of nutritional wisdom has been expanded to the idea that food intake, food preferences and palatability involve interrelationships among a food's flavour and postingestive feedback emanating from cells and organs in response to nutrients and toxins in a food, the biochemical characteristics of other foods in the diet, an animal's current physiological condition and its past experiences with the food.

producers and veterinarians were not always consistent with the research findings, which suggested that livestock lacked nutritional wisdom. Deficits of phosphorus and calcium were linked directly with bone ingestion to restore phosphorus and calcium levels in the body (Denton *et al.*, 1986; Blair-West *et al.*, 1992; Schulkin, 2001). Likewise, practitioners and veterinarians such as Holliday (2003) recounted experiences such as the following from an occasion when Holliday was working with a client, a dairy producer named Carl:

One year inclement weather made planting and harvesting hay and grain crops a great gamble with the result that feedstuffs that fall and winter looked good but had low nutritional

value. By late winter Carl consulted me with two seemingly unrelated problems: (i) his cattle were eating almost 2 lb of a mixed mineral per head per day! (ii) about 10 days before they were due to calve, his heifers would abort a live calf. The calf, with some care, would live, but in spite of all we could do the heifer would die within 2 or 3 days. After the third one in a row had died, I did what *every* smart vet would do. . . . I passed the buck and sent a dying heifer to the University Vet School for autopsy. Their diagnosis came back *starvation*! Carl took good care of his animals and was feeding them all they could eat. This diagnosis was like an insult to Carl and difficult for either of us to accept. We could have accepted a diagnosis of malnutrition because of the poor crops that year but starvation seemed a little too harsh.

We then turned our attention to the mineral consumption problem. Available in that area at that time was a 'cafeteria' mineral programme in which each mineral was fed separately on the theory that each animal could then eat only what it needed to balance it's own needs. Carl decided to try this programme. His mineral feeder was in the middle of his cow lot and he had to carry each bag of minerals through the lot to empty into the feeder. Things went well for the first few trips and then several of the normally docile cows suddenly surrounded him, tore a bag of mineral from his arms, chewed open the bag and greedily consumed every bit of the mineral, the bag and even some mud and muck where the mineral had spilled out. . . astounding behaviour for a bunch of tame dairy cows!

What was in the bag, you ask? . . . a source of the trace mineral, zinc. During the next several days they ate several bags of this zinc source while completely ignoring all other minerals. Gradually they began eating normal amounts of the regular mineral. From that day on his heifers calved normally and things gradually returned to normal. . . . Apparently, the difficult growing season had resulted in crops that were deficient in zinc or perhaps high in zinc antagonists. The basic mineral mix had a small amount of zinc in it but to get the zinc they needed, they had to consume large amounts. This gave them too much calcium. Calcium interferes with zinc absorption, which in turn increased their need for zinc. Even though their quest for zinc impelled them to eat the mixed mineral, every mouthful they took increased the imbalance. Inevitably, symptoms began to show up in the most vulnerable group . . . young heifers, still growing and in the last stages of pregnancy. . . . Finally they just gave up and checked out all for want of a few grams of zinc. The decrease in feed conversion associated with zinc deficiencies coupled with the poor quality feed would result in malnutrition even when feed intake appeared to be adequate. I realize that other secondary factors may have been involved here, but the main factor was a zinc deficiency as evidenced by the remission of symptoms when zinc was supplied.

Holliday concludes with the following:

For me this incident epitomizes the concept that, given the chance, animals can balance rations better than computers or nutritionists can. Many nutritionists tend to discount the ability of animals to balance their ration, asserting that by the time they feel the need to eat a certain item they are already in a deficient state. From their point of view, I suppose they have a point. The fallacy in their reasoning may be that they expect the animal to choose for the level of production that man desires while the animal chooses only what it needs to be healthy.

Looking forward

Undoubtedly, in each case above the researchers and the practitioners made accurate observations of the behaviours they observed. Thus, the issue is not who is right and who is wrong, but how contexts and contingencies influenced the behaviours different people observed. All of these interpretations should be reconsidered in light of new understanding of how nutritional wisdom is likely to be manifest, given the many factors that interact to influence food selection. To do so, we must consider how animals learn flavour–feedback associations, including the roles of past experience and the familiar–novel dichotomy, discrimination and generalization, initial conditions and all the dynamic contingencies that apply when animals learn flavour– feedback associations.

It is highly unlikely that several million years of evolution have been erased by a few thousand years of domestication, especially regarding nutritional wisdom. Acquiring nutrients and avoiding toxins is every bit as important as breathing, which has not changed due to domestication, and is similarly influenced by non-cognitive feedback mechanisms. Indeed, feedback mechanisms for detecting and correcting amino acid imbalances appear to be conserved in animals ranging from single-cell organisms such as yeast, to invertebrates, to humans (Hao *et al.*, 2005). Domesticated herbivores forage on rangelands worldwide, and they must choose from the same plant species available to wild herbivores. Given the way livestock are routinely moved from familiar to unfamiliar environments, it is a testament to their ability to quickly adapt that they survive at all (Provenza, 2003a).

As emphasized throughout this review, the concept of nutritional wisdom has been expanded during the last two decades due to research which shows that food preferences are learned through processes involving complex interrelationships between a food's flavour and its postingestive effects, which emanate from cells and organs in response to nutrients and toxins (Provenza, 1995a; Fig. 13.4). The sense of taste 'manages' diet selection by qualitatively and quantitatively analysing foods in conjunction with the visceral and central nervous systems, all of which evokes current and past experiences with food. The receptors for taste are situated at the junction between the internal and external milieux, like a Janus head placed at the gateway to the body, one face looking at what is outside and the other looking at what is inside (Scott, 1990). Such dynamic integration of taste with the internal and external milieux offers a new dimension to nutritional wisdom: the ability to modify diet selection as a function of the consequences of food experienced throughout the lifetime of the individual. This is critical from an evolutionary standpoint, given that the average lifetime of a species is several million years and the kinds of foods the species is likely to encounter will vary tremendously over that time frame. What animals need – nutrients, medicines – is relatively constant, but how the various foods are packaged is not. The solution to this challenge was to create animals that learn based on flavour–feedback interactions, and that requires researchers to rethink notions of how nutritional wisdom is likely to work.

Past research with livestock established that the 'sodium model of nutritional wisdom' is not a particularly good example for energy, protein, minerals and vitamins because animals do not instinctively recognize through odour or taste all of these nutrients; nor do they necessarily recognize all of the various configurations of any particular nutrient (Provenza and Balph, 1990). Rather, they must learn based on flavour–feedback associations to ingest foods that contain these substances. Hence, it is not surprising that calcium- or sulphur-deficient animals offered choices of 10 novel minerals – $CaCO_3$, K_2CO_3, Na_2CO_3, $ZnCO_3$, $4MgCO_3 \cdot Mg(OH)_2 \cdot 4H_2O$, $CuCO_3$, $MnCO_3$, NaH_2PO_4, Na_2SO_4, $NaCl$ –

did not consume appropriate amounts of $CaCO_3$ or Na_2SO_4, respectively (Pamp et al., 1977). Sulphur-deficient lambs had to discriminate among four anions – $CO_3^=$, $PO_4^=$, $SO_4^=$, Cl^- – that all contained sodium. As discussed previously, animals generalize among substances that share common flavours and feedbacks (sodium), thus making the task difficult both from the standpoint of a similar flavour and excessive feedback from it (sodium). Equally problematic, sodium is often used as a carrier to mix with other minerals in cafeteria offerings to encourage and/or limit intakes of other minerals as well. The same is true for the calcium-deficient lambs challenged to discriminate $CaCO_3$ from the six other carbonates. Appropriate learning is more likely when animals are offered substances that differ in both flavour and feedback, and when recovery is paired with eating the substance that rectifies the deficiency. The latter can be learned in part from social models that have learned appropriate behaviours, though that was never considered, despite the fact that that is how humans have learned to prevent deficiencies: long before scientists knew of amino acids, individuals learned to mix rice with beans to get a full complement of indispensable amino acids, and cultures maintained the practice.

In addition, past studies of nutritional wisdom were based on the erroneous assumption that animals eat to meet what people considered to be their needs for minerals. Thus, to demonstrate an appetite for calcium or phosphorus, researchers expected ruminants to consume minerals from concentrate salts according to National Research Council (NRC) requirements. Recommendations for minerals typically are set higher than animals' needs, and individuals, even within uniform groups, vary greatly in their needs for nutrients such that some animals will consume much more or less than others. Thus, it is not surprising that animals varied greatly in their consumption of minerals in previous studies, that many did not consume minerals at all and that some 'underconsumed' while others 'overconsumed'. Indeed, requirements may be exceeded even with a bite/day of a concentrate salt. The daily phosphorus requirement of a 40 kg lamb of moderate growth potential is 3.9 g (NRC, 1985), which amounts to only 15 g of NaH_2PO_4. Preferences for a mineral supplement also depend on the type of mineral offered. For instance, calcium-deficient animals actually avoid phosphorus (Tordoff, 2001). Thus, calcium phosphate salts are not a good choice when studying specific appetites for calcium. For these reasons alone, individuals are unlikely to ingest a recommended daily allowance of minerals.

Finally, researchers did not appreciate that animals respond more strongly to excesses, deficits and imbalances than to daily nutrient requirements. The degree of the mineral deficiency induced in livestock during most early studies to explore nutritional wisdom was likely not enough to induce a response. Learned preferences for nutrients are manifest when animals are in a physiological state of need, and that depends on the specific nutrient in question (Mehiel, 1991; Provenza, 1995a). The ongoing need for energy is much greater than the need for protein due to the high amounts of energy required daily, and preferences for flavours paired with energy are higher than preferences for flavours paired with protein (Villalba and Provenza, 1999). In contrast, calcium and phosphorus deficiencies develop slowly because daily requirements are low relative to body reserves (NRC, 1985). Animals respond more strongly to daily requirements for energy followed by protein, and then, if at all, to daily requirements for calcium and phosphorus. Responses to calcium and phosphorus are likely to be seen only after long periods of ingesting deficient diets, when a strong need for these minerals develops due to depletion of the ample buffer supplied by body reserves (Ternouth, 1991). When need for a nutrient is not high, animals probably respond more to the novelty or familiarity of the odour or taste than to postingestive effects of the mineral and vitamin supplements, as cattle and sheep evidently did in early studies (Coppock et al., 1976; Pamp et al., 1977). Sheep preferred a less nutritious food in a two-food choice test when their energy requirements were satisfied by a basal diet, and the choice lasted only 20 min/day. However, when their need for energy increased and the choice lasted 8 h/day, lambs quickly learned to prefer the more nutritious of the two foods (Villalba and Provenza, 2000b). Sheep respond likewise to the specific postingestive effects of phosphorus. They avoid flavours previously paired with NaH_2PO_4 when their requirements for phosphorus are met or exceeded, and they prefer flavours paired with phosphorus during periods of need (Villalba et al., 2006c).

Conclusion

Historically, researchers have been sceptical of the notion of nutritional wisdom. Few believed that cells and organs of the body could influence the palate to select foods that meet needs for nutrients and prevent toxicosis. For nutrients, that scepticism has been perpetuated for several reasons. Cafeteria feeding studies provided no evidence for nutritional wisdom because they were based on the erroneous assumption that animals eat to meet needs for minerals; we did not appreciate that animals respond most strongly to excesses, deficits and imbalances. Recommendations for minerals typically are set higher than animal needs, and individuals, even within uniform groups, vary greatly in their needs for minerals such that some animals will consume much less or more than others. In addition, there has been little appreciation for how animals learn flavour–feedback associations. Studies did not take into account past experiences, familiar–novel dichotomies, amounts ingested within a meal, generalization or initial conditions, and they did not consider the contexts necessary for animals to learn specific flavour–feedback associations. Finally, for toxins people focused on a subset of plants that cause problems – the poisonous plants. We did not understand why animals over-ingest poisonous plants (Provenza et al., 1992); nor did we realize that all plants contain toxins and that animals learn to limit intakes of most of the plants they encounter (Provenza et al., 2003a).

Any substance can be harmful or beneficial depending on the dose, which is influenced by the chemical characteristics of all foods in the diet relative to the physiological condition of an animal. In some cases, the combination of substance and amount will be beneficial, whereas in others, the same combination may be harmful. At too high doses, nutrients are toxic, whereas at the appropriate dose, toxins can have medicinal benefits. This is so because at the biochemical level, nutrients and toxins interact with one another – nutrients with nutrients, nutrients with toxins and toxins with toxins. These interactions affect basic metabolic processes, and they cause organisms to experience the consequences of their foraging behaviours as a continuum from pain (aversive) to pleasure (positive). Incentives emanating from the consequences of past behaviours influence the likelihood of future behaviours, and they generate the experiences of emotions and ultimately feelings that arise from different degrees of well-being that emanate from ongoing interactions among cells, organs, individuals, and social and biophysical environments (Damasio, 2003).

The distinction between nutrients, toxins and medicines is artificial because ingesting or avoiding substances are means to the same end – stay well – and that depends on compounds, dosages and their interactions. Nutrients, toxins and medicines are merely labels we use to catagorize various phenomena – nutrition, toxicosis and medication. In reality, bodies do not respond to labels, but to various 'substances' that benefit or harm the internal milieu, and these homeodynamic endeavours promote states of well-being (Damasio, 2003) and sustain functional integrity (Maturana and Varela, 1980). The active selection or avoidance of substances, regardless of their labels as a function of internal state, supports the notion that bodies have nutritional wisdom.

Nutritional wisdom notwithstanding, predicting the behaviour of individuals is elusive, not because it cannot be well understood, but because it is so multifaceted and dynamic. An animal's foraging behaviour is a function of its evolutionary history, genetically expressed, in concert with its uniquely individualistic history of the social and biophysical environments where it was conceived and reared. Genes are expressed as a function of interactions with biophysical and social environments, and

because both change constantly and often unpredictably, so too do expressed morphology, physiology and behaviour. Thus, social and biophysical influences that shape the development of individuals from conception to death depend critically on time and timing – the unforeseen and indiscernible role of chance in life (Taleb, 2001). The net result is that food intake and preference depend on differences in how animals are built morphologically and how they function physiologically, and marked variation is common even among closely related animals in need for nutrients and abilities to cope with toxins. Past experience also influences an animal's propensity to eat different foods. Changes in biochemical contexts spatially and temporally affect what herbivores learn, thereby creating additional variability in behaviour. Experienced animals that have learned to eat a variety of foods differing in nutrients and toxins do so even when nutritious alternatives are available, whereas naive animals familiar only with the nutritious alternatives eat just that subset of foods (Villalba et al., 2004). Nutrients and toxins both cause animals to satiate, and excesses of nutrients, nutrient imbalances and toxins all limit intakes of foods. The amount of a toxin an animal can ingest depends on the kinds and amounts of nutrients and toxins in the forages on offer. Individuals can better meet their needs for nutrients and regulate their intake of toxins when offered a variety of foods that differ in nutrients and toxins than when constrained to a single food, even if the food is 'nutritionally balanced'. Transient food aversions compound the inefficiency of single-food diets – whether in confinement, on pastures or on rangelands – by depressing intake among individual animals, even if they are suited 'on average' to that nutrient or toxin profile (Provenza et al., 2003a). The biochemical diversity of landscapes influences foraging in ways that cannot necessarily be predicted solely by the isolated effect of any single biochemical in the body. Thus, the conventional univariate focus mainly on energy must be replaced with multivariate approaches that recognize multiple biochemical interactions (Simpson and Raubenheimer, 2002; Simpson et al., 2004).

References

Arnold, G.W. and Dudzinski, M.L. (1978) *Ethology of Free-Ranging Domestic Animals*. Elsevier/North-Holland, New York.

Athanasiadou, S., Kyriazakis, I., Jackson, F. and Coop, R.L. (2000) Effects of short-term exposure to condensed tannins on adult *Trichostrongylus colubriformis*. *The Veterinary Record* 146, 728–732.

Banner, R.E., Rogosic, J., Burritt, E.A. and Provenza, F.D. (2000) Supplemental barley and activated charcoal increase intake of sagebrush by lambs. *Journal of Range Management* 53, 415–420.

Baraza, E., Villalba, J.J. and Provenza, F.D. (2005) Nutritional context influences preferences of lambs for foods with plant secondary metabolites. *Applied Animal Behaviour Science* 92(4), 293.

Bernard, C. (1957) *An Introduction to the Study of Experimental Medicine, Trans H.C. Greene*. Drover, New York.

Blair-West, J.R., Denton, D.A., McKinley, M.J., Radden, B.G., Ramshaw, E.H. and Wark, J.D. (1992) Behavioral and tissue response to severe phosphorous depletion in cattle. *American Journal of Physiology* 263, R656–R663.

Burritt, E.A. and Provenza, F.D. (1989) Food aversion learning: ability of lambs to distinguish safe from harmful foods. *Journal of Animal Science* 67, 1732–1739.

Burritt, E.A. and Provenza, F.D. (1991) Ability of lambs to learn with a delay between food ingestion and consequences given meals containing novel and familiar foods. *Applied Animal Behaviour Science* 32, 179–189.

Burritt, E.A. and Provenza, F.D. (1996) Amount of experience and prior illness affect the acquisition and persistence of conditioned food aversions in lambs. *Applied Animal Behaviour Science* 48, 73–80.

Burritt, E.A. and Provenza, F.D. (2000) Role of toxins in intake of varied diets by sheep. *Journal of Chemical Ecology* 26, 1991–2005.

Cannon, W.B. (1966) *The Wisdom of the Body*. Norton, New York.

Capra, F. (1996) *The Web of Life*. Anchor Books, Doubleday, New York.

Cheeke, P.R. (1998) *Natural Toxicants in Feeds, Forages, and Poisonous Plants*. Interstate, Danville, Illinois.

Cheeke, P. and Shull, L.R. (1985) *Natural Toxicants in Feeds and Poisonous Plants*. Avi Publishing, Westport, Connecticut.

Clayton, D.H. and Wolfe, D. (1993) The adaptive significance of self-medication. *Trends in Ecology and Evolution* 8, 60–63.

Coop, R.L. and Kyriazakis, I. (2001) Influence of host nutrition on the development and consequences of nematode parasitism in ruminants. *Trends in Parasitology* 17, 325–350.

Cooper, S.D.B., Kyriazakis, I. and Nolan, J.V. (1995) Diet selection in sheep: the role of the rumen environment in the selection of a diet from two feeds that differ in their energy density. *British Journal of Nutrition* 74, 39–54.

Coppock, C.E. (1970) Free choice mineral consumption by dairy cattle. *Proceedings of the Cornell Nutrition Conference for Feed Manufacturers*, pp. 29–35.

Coppock, C.E., Everett, R.W. and Belyea, R.L. (1976) Effect of low calcium or low phosphorous diets on free choice consumption of dicalcium phosphate by lactating dairy cows. *Journal of Dairy Science* 59, 571–580.

Damasio, A. (2003) *Looking for Spinoza: Joy, Sorrow, and the Feeling Brain*. Harcourt, New York.

Denton, D.A., Blair-West, J.R., McKinley, M.J. and Nelson, J.F. (1986) Physiological analysis of bone appetite (osteophagia). *BioEssays* 4, 40–42.

De Waal, F. (2001) *The Ape and the Sushi Master: Cultural Reflections of a Primatologist*. Basic Books, New York.

Distel, R.A. and Provenza, F.D. (1991) Experience early in life affects voluntary intake of blackbrush by goats. *Journal of Chemical Ecology* 17, 431–450.

Distel, R.A., Villalba, J.J. and Laborde, H.E. (1994) Effects of early experience on voluntary intake of low-quality roughage by sheep. *Journal of Animal Science* 72, 1191–1195.

Distel, R.A., Villalba, J.J., Laborde, H.E. and Burgos, M.A. (1996) Persistence of the effects of early experience on consumption of low-quality roughage by sheep. *Journal of Animal Science* 74, 965–968.

Dohi, H. and Yamada, A. (1997) Preference of sheep and goats for extracts from high-grain concentrate. *Journal of Animal Science* 75, 2073–2077.

Duncan, A.J. and Young, S.A. (2002) Can goats learn about foods through conditioned food aversions and preferences when multiple food options are simultaneously available? *Journal of Animal Science* 80, 2091–2098.

Dziba, L.E. and Provenza, F.D. (2006) Sagebrush monoterpenes influence feeding bouts and regulation of food intake by lambs. *Applied Animal Behaviour Science* (in press).

Dziba, L.E., Hall, J.O. and Provenza, F.D. (2006) Pharmacokinetics of 1,8-cineole administered intravenously and into the rumen influence feeding behavior of lambs. *Journal of Chemical Ecology* (in press).

Early, D. and Provenza, F.D. (1998) Food flavor and nutritional characteristics alter dynamics of food preference in lambs. *Journal of Animal Science* 76, 728–734.

Egan, A.R. (1980) Host animal–rumen relationships. *Proceedings of the Nutrition Society* 39, 79–87.

Egan, A.R. and Rogers, Q.R. (1978) Amino acid imbalance in ruminant lambs. *Australian Journal of Agricultural Research* 29, 1263–1279.

Engel, C. (2002) *Wild Health*. Houghton Mifflin, Boston, Massachusetts.

Foley, W.J. and McArthur, C. (1994) The effects and costs of allelochemicals for mammalian herbivores: an ecological perspective. In: Chivers, D.J. and Langer, P. (eds) *The Digestive System in Mammals: Food, Form and Function*. Cambridge University Press, Cambridge, UK, pp. 370–391.

Foley, W.J., McLean, S. and Cork, S.J. (1995) Consequences of biotransformation of plant secondary metabolites on acid-base metabolism in mammals – a final common pathway? *Journal of Chemical Ecology* 21, 721–743.

Foley, W.J., Iason, G.R. and McArthur, C. (1999) Role of plant secondary metabolites in the nutritional ecology of mammalian herbivores: how far have we come in 25 years? In: Jung, H.G. and Fahey, G.C. Jr (eds) *Nutritional Ecology of Herbivores. Proceedings of the 5th International Symposium on the Nutrition of Herbivores*. American Society of Animal Science, Illinois.

Francis, S.A. (2003) Investigating the role of carbohydrates in the dietary choices of ruminants with an emphasis on dairy cows. PhD thesis, University of Melbourne, Australia.

Freeland, W.J. and Choquenot, D. (1990) Determinants of herbivore carrying capacity: plants, nutrients, and *Equus asinus* in northern Australia. *Ecology* 71, 589–597.

Freeland, W.J. and Janzen, D.H. (1974) Strategies in herbivory by mammals: the role of plant secondary compounds. *American Naturalist* 108, 269–286.

Freeland, W.J., Calcott, P.H. and Anderson, L.R. (1985) Tannins and saponin: interaction in herbivore diets. *Biochemical Systematics and Ecology* 13, 189–193.

Fretwell, S.D. and Lucas, H.L. (1970) On territorial behavior and other factors influencing habitat distribution in birds. I. Theoretical development. *Acta Biotheoretica* 19, 16–36.

Gietzen, D.W. (1993) Neural mechanisms in the responses to amino acid deficiency. *Journal of Nutrition* 123, 610–625.

Ginane, C.A., Duncan, J., Young, S.A., Elston, D.A. and Gordon, I.J. (2005) Herbivore diet selection in response to simulated variation in nutrient rewards and plant secondary compounds. *Animal Behaviour* 69, 541–550.

Gladwell, M. (2005) *Blink: The Power of Thinking without Thinking*. Little, Brown, New York.

Glass, L. and MacKey, M.C. (1988) *From Clocks to Chaos: The Rhythms of Life*. Princeton University Press, Princeton, New Jersey.

Gordon, J.G., Tribe, D.E. and Graham, T.C. (1954) The feeding behaviour of phosphorous-deficient cattle and sheep. *British Journal of Animal Behaviour* 2, 72–74.

Green, G.C., Elwin, R.L., Mottershead, B.E. and Lynch, J.J. (1984) Long-term effects of early experience to supplementary feeding in sheep. *Proceedings of the Australian Society of Animal Production* 15, 373–375.

Greenhalgh, J.F.D. and Reid, G.W. (1971) Relative palatability to sheep of straw, hay and dried grass. *British Journal of Nutrition* 26, 107–116.

Grovum, W.L. (1988) Appetite, palatability and control of feed intake. In: Church, D.C. (ed.) *The Ruminant Animal*. Prentice-Hall, Englewood Cliffs, New Jersey, pp. 202–216.

Hao, S., Sharp, J.W., Ross-Inta, C.M., McDaniel, B.J., Anthony, T.G., Wek, R.C., Cavener, D.R., McGrath, B.C., Rudell, J.B., Koehnle, T.J. and Gietzen, D.W. (2005) Uncharged tRNA and sensing of amino acid deficiency in mammalian piriform cortex. *Science* 307, 1776–1778.

Hill, D.L. and Mistretta, C.M. (1990) Developmental neurobiology of salt taste sensation. *Trends in Neuroscience* 13, 188–195.

Hills, J., Kyriazakis, I., Nolan, J.V., Hinch, G.N. and Lynch, J.J. (1999) Conditioned feeding responses in sheep to flavoured foods associated with sulphur doses. *Animal Science* 69, 313–325.

Holliday, R.J. (2003) *Fundamentals of Animal Health: A Collection of Articles on Holistic Animal Health*. Waukon, Iowa.

Houston, D.C., Gilardi, J.D. and Hall, A.J. (2001) Soil consumption by elephants might help to minimize the toxic effects of plant secondary compounds in forest browse. *Mammal Review* 31, 249–254.

Howery, L.D., Provenza, F.D., Banner, R.E. and Scott, C.B. (1996) Differences in home range and habitat use among individuals in a cattle herd. *Applied Animal Behaviour Science* 49, 305–320.

Howery, L.D., Provenza, F.D., Banner, R.E. and Scott, C.B. (1998) Social and environmental factors influence cattle distribution on rangeland. *Applied Animal Behaviour Science* 55, 231–244.

Huffman, M.A. and Hirata, S. (2004) An experimental study of leaf swallowing in captive chimpanzees: insights into the origin of a self-medicative behavior and the role of social learning. *Primates* 45, 113–118.

Hutchings, M.R., Gordon, I.J., Kyriazakis, I., Robertson, E. and Jackson, F. (2002) Grazing in heterogeneous environments: infra- and supra-parasite distributions determine herbivore grazing decisions. *Oecologia* 132, 453–460.

Hutchings, M.R., Athanasiadou, S., Kyriazakis, I. and Gordon, I. (2003) Can animals use foraging behaviour to combat parasites? *Proceedings of the Nutrition Society* 62, 361–370.

Illius, A.W. and Jessop, N.S. (1995) Modeling metabolic costs of allelochemical ingestion by foraging herbivores. *Journal of Chemical Ecology* 21, 693–719.

Illius, A.W. and Jessop, N.S. (1996) Metabolic constraints on voluntary intake in ruminants. *Journal of Animal Science* 74, 3052–3062.

Katz, D. (1937) *Animals and Man. Studies in Comparative Psychology*. Longmans, Green & Co, London.

Kauffman, S. (1995) *At Home in the Universe*. Oxford University Press, New York.

Kauffman, S. (2000) *Investigations*. Oxford University Press, New York.

Keeler, R.F., Baker, D.C. and Evans, J.O. (1988) Individual animal susceptibility and its relationship to induced adaptation or tolerance in sheep to *Galea officinalis*. *Letters in Veterinary and Human Toxicology* 30, 420–423.

Konarzewski, M. and Diamond, J. (1994) Peak sustained metabolic rate and its individual variation in cold-stressed mice. *Physiological Zoology* 67, 1186–1212.

Kyriazakis, I. and Oldham, J.D. (1997) Food intake and diet selection of sheep: the effect of manipulating the rates of digestion of carbohydrates and protein of the foods offered as a choice. *British Journal of Nutrition* 77, 243–254.

Kyriazakis, I., Oldham, J.D., Coop, R.L.F. and Jackson, F. (1994) The effect of subclinical intestinal nematode infection on the diet selection of growing sheep. *British Journal of Nutrition* 72, 665–677.

Launchbaugh, K.L. and Provenza, F.D. (1993) Can plants practice mimicry to avoid grazing by mammalian herbivores? *Oikos* 66, 501–504.

Launchbaugh, K.L., Provenza, F.D. and Burritt, E.A. (1993) How herbivores track variable environments: response to variability of phytotoxins. *Journal of Chemical Ecology* 19, 1047–1056.

Launchbaugh, K.L., Provenza, F.D. and Werkmeister, M.J. (1997) Overcoming food neophobia. *Applied Animal Behaviour Science* 54, 327–334.

Lewontin, R. (2000) *The Triple Helix: Gene, Organism and Environment*. Harvard University Press, Cambridge, Massachusetts.

Lobley, G.E. and Milano, G.D. (1997) Regulation of hepatic nitrogen metabolism in ruminants. *Proceedings of the Nutrition Society* 57, 547–563.

Lozano, G.A. (1998) Parasitic stress and self-medication in wild animals. *Advances in the Study of Behavior* 27, 291–317.

Marsh, K.J., Wallis, I.R. and Foley, W.J. (2005) Detoxification rates constrain feeding in common brushtail possums (*Trichosurus vulpecula*). *Ecology* 86, 2946–2954.

Martinez, J.H., Ross, T.T., Becker, K.A. and Smith, G.S. (1993) Ingested dry snakeweed foliage did not impair reproduction in ewes and heifers during late gestation. *Proceedings of the Western Section of the American Society of Animal Science* 44, 32–35.

Maturana, H.R. and Varela, F.J. (1980) *Autopoiesis and Cognition*. Reidel, Dordrecht, The Netherlands, Boston, Massachusetts.

Mehiel, R. (1991) Hedonic-shift conditioning with calories. In: Bolles, R.C. (ed.) *The Hedonics of Taste*. Lawrence Erlbaum Associates, Hillsdale, New Jersey, pp. 107–126.

Mertens, D.R. and Loften, J.R. (1980) The effect of starch on forage fiber digestion kinetics *in vitro*. *Journal of Dairy Science* 63, 1437–1446.

Min, B.R. and Hart, S.P. (2003) Tannins for suppression of internal parasites. *Journal of Animal Science* 81, E102–E109.

Moore, D.S. (2002) *The Dependent Gene: The Fallacy of 'Nature vs. Nurture'*. Henry Holt and Company, New York.

Mosley, J.C. (1999) Influence of social dominance on habitat selection by free-ranging ungulates. In: Launchbaugh, K.L., Mosley, J.C. and Sanders, K.D. (eds) *Grazing Behavior of Livestock and Wildlife.* Idaho Forest, Wildlife and Range Experiment Station, Moscow, Germany, pp. 109–115.

Muller, L.D., Schaffer, L.V., Ham, L.C. and Owens, M.J. (1977) Cafeteria style free-choice mineral feeder for lactating dairy cows. *Journal of Dairy Science* 60, 1574–1582.

NRC (National Research Council) (1985) *Nutrient Requirements of Sheep,* 6th edn. National Academy Press, Washington, DC.

Osweiler, G.D., Carson, T.L., Buck, W.B. and Van Gelder, G.A. (1985) *Clinical and Diagnostic Veterinary Toxicology.* Kendall/Hunt, Dubuque, Iowa.

Pamp, D.E., Goodrich, R.D. and Meiske, J.C. (1976) A review of the practice of feeding minerals free choice. *World Review of Animal Production* 12, 13–18.

Pamp, D.E., Goodrich, R.D. and Meiske, J.C. (1977) Free choice minerals for lambs fed calcium- or sulfur-deficient rations. *Journal of Animal Science* 45, 1458–1466.

Pfister, J.A., Provenza, F.D., Manners, G.D., Gardner, D.R. and Ralphs, M.H. (1997) Tall larkspur ingestion: can cattle regulate intake below toxic levels? *Journal of Chemical Ecology* 23, 759–777.

Phy, T.S. and Provenza, F.D. (1998) Sheep fed grain prefer foods and solutions that attenuate acidosis. *Journal of Animal Science* 76, 954–960.

Provenza, F.D. (1994) Ontogeny and social transmission of food selection in domesticated ruminants. In: Galef, B.G. Jr, Mainardi, M. and Valsecchi, P. (eds) *Behavioral Aspects of Feeding: Basic and Applied Research in Mammals.* Harwood Academic Publishers, Singapore, pp. 147–164.

Provenza, F.D. (1995a) Postingestive feedback as an elementary determinant of food preference and intake in ruminants. *Journal of Range Management* 48, 2–17.

Provenza, F.D. (1995b) Tracking variable environments: there is more than one kind of memory. *Journal of Chemical Ecology* 21, 911–923.

Provenza, F.D. (1996) Acquired aversions as the basis for varied diets of ruminants foraging on rangelands. *Journal of Animal Science* 74, 2010–2020.

Provenza, F.D. (2003a) *Foraging Behavior: Managing to Survive in a World of Change.* Utah State University, Logan, Utah.

Provenza, F.D. (2003b) Twenty-five years of paradox in plant–herbivore interactions and 'sustainable' grazing management. *Rangelands* 25, 4–15.

Provenza, F.D. and Balph, D.F. (1988) The development of dietary choice in livestock on rangelands and its implications for management. *Journal of Animal Science* 66, 2356–2368.

Provenza, F.D. and Balph, D.F. (1990) Applicability of five diet-selection models to various foraging challenges ruminants encounters. In: Hughes, R.N. (ed.) *Behavioural Mechanisms of Food Selection. NATO ASI Series G: Ecological Sciences,* Vol. 20. Springer-Verlag, Berlin, Heidelberg, pp. 423–459.

Provenza, F.D. and Cincotta, R.P. (1993) Foraging as a self-organizational learning process: accepting adaptability at the expense of predictability. In: Hughes, R.N. (ed.) *Diet Selection.* Blackwell Scientific Publications, London, pp. 78–101.

Provenza, F.D., Burritt, E.A., Clausen, T.P., Bryant, J.P., Reichardt, P.B. and Distel, R.A. (1990) Conditioned flavor aversion: a mechanism for goats to avoid condensed tannins in blackbrush. *American Naturalist* 136, 810–828.

Provenza, F.D., Pfister, J.A. and Cheney, C.D. (1992) Mechanisms of learning in diet selection with reference to phytotoxicosis in herbivores. *Journal of Range Management* 45, 36–45.

Provenza, F.D., Lynch, J.J. and Nolan, J.V. (1993a) Temporal contiguity between food ingestion and toxicosis affects the acquisition of food aversions in sheep. *Applied Animal Behaviour Science* 38, 269–281.

Provenza, F.D., Lynch, J.J. and Nolan, J.V. (1993b) The relative importance of mother and toxicosis in the selection of foods by lambs. *Journal of Chemical Ecology* 19, 313–323.

Provenza, F.D., Lynch, J.J. and Nolan, J.V. (1994a) Food aversion conditioned in anesthetized sheep. *Physiology & Behavior* 55, 429–432.

Provenza, F.D., Lynch, J.J., Burritt, E.A. and Scott, C.B. (1994b) How goats learn to distinguish between novel foods that differ in postingestive consequences. *Journal of Chemical Ecology* 20, 609–624.

Provenza, F.D., Lynch, J.J. and Cheney, C.D. (1995) Effects of a flavor and food restriction on the intake of novel foods by sheep. *Applied Animal Behaviour Science* 43, 83–93.

Provenza, F.D., Villalba, J.J., Cheney, C.D. and Werner, S.J. (1998) Self-organization of foraging behavior: from simplicity to complexity without goals. *Nutrition Research and Review* 11, 199–222.

Provenza, F.D., Villalba, J.J. and Augner, M. (1999) The physics of foraging. In: Buchanan-Smith, J.G., Bailey, L.D. and McCaughey, P. (eds) *Proceedings of the XVIII International Grassland Congress. Extension Service, Saskatchewan Agriculture & Food*, Vol. III. Saskatoon, Saskatchewan, pp. 99–107.

Provenza, F.D., Burritt, E.A., Perevolotsky, A. and Silanikove, N. (2000) Self-regulation of intake of polyethylene glycol by sheep fed diets varying in tannin concentrations. *Journal of Animal Science* 78, 1206–1212.

Provenza, F.D., Villalba, J.J., Dziba, L.E., Atwood, S.B. and Banner, R.E. (2003a) Linking herbivore experience, varied diets, and plant biochemical diversity. *Small Ruminant Research* 49, 257–274.

Provenza, F.D., Villalba, J.J. and Bryant, J.P. (2003b) Foraging by herbivores: linking the biochemical diversity of plants with herbivore culture and landscape diversity. In: Bissonette, J.A. and Storch, I. (eds) *Landscape Ecology and Resource Management: Linking Theory with Practice*. Island Press, New York, pp. 387–421.

Reed, J.D. (1995) Nutritional toxicology of tannins and related polyphenols in forage legumes. *Journal of Animal Science* 73, 1516–1528.

Reynolds, E.B., Jones, J.M., Jones, J.H., Fudge, J.H. and Kleberg, J.F. (1953) Methods of supplying phosphorus to range cattle in south Texas. *Texas Agricultural Experiment Station Bulletin* 773, 3–16.

Richter, C.P. (1976) *The Psychobiology of Curt Richter*. York Press, Baltimore, Maryland.

Rogers, Q.R. and Egan, A.R. (1975) Amino acid imbalance in the liquid-fed lamb. *Australian Journal of Biological Science* 28, 169–181.

Rogosic, J., Pfister, J.A. and Provenza, F.D. (2006a) The effect of polyethylene glycol (PEG) and varied shrub selection on intake of Mediterranean shrubs by sheep and goats. *Rangeland Ecology & Management* (in press).

Rogosic, J., Pfister, J.A. and Provenza, F.D. (2006b) The effect of activated charcoal and varied shrub selection on intake of Mediterranean shrubs by sheep and goats. *Applied Animal Behaviour Science* (in press).

Rose, S. (1998) *Lifelines: Biology Beyond Determinism*. Oxford University Press, New York.

Russell, J.B., O'Connor, J.D., Fox, G.G., Van Soest, P.J. and Sniffen, C.J. (1992) A net carbohydrate and protein system for evaluating cattle diets. I. Ruminal fermentation. *Journal of Animal Science* 70, 3551–3561.

Schmidt-Nielsen, K. (1994) How are control systems controlled? *American Scientist* 82, 38–44.

Schulkin, J. (1991) *Sodium Hunger: The Search for a Salty Taste*. Cambridge University Press. New York.

Schulkin, J. (2001) *Calcium Hunger. Behavioral and Biological Regulation*. Cambridge University Press, Cambridge, UK.

Scott, L.L. and Provenza, F.D. (1999) Variation in food selection among lambs: effects of basal diet and foods offered in a meal. *Journal of Animal Science* 77, 2391–2397.

Scott, T.R. (1990) The effect of physiological need on taste. In: Capaldi, E.D. and Powley, T.L. (eds) *Taste, Experience, and Feeding*. American Psychological Association, Washington, DC, pp. 45–61.

Shaw, R.A., Villalba, J.J. and Provenza, F.D. (2006) Resource availability and quality influence patterns of diet mixing with foods containing toxins by sheep. *Journal of Chemical Ecology* (in press).

Simpson, S.J. and Raubenheimer, D. (2002) Herbivore foraging in chemically heterogeneous environments: nutrient and secondary metabolites. *Ecology* 83, 2489–2501.

Simpson, S.J., Sibly, R.M., Lee, K.P., Behmer, S.T. and Raubenheimer, D. (2004) Optimal foraging when regulating intake of multiple nutrients. *Animal Behaviour* 68, 1299–1311.

Sinclair, L.A., Garnsworthy, P.C., Newbold, J.R. and Buttery, P.J. (1993) Effect of synchronizing the rate of dietary energy and nitrogen release on rumen fermentation and microbial protein synthesis in sheep. *Journal of Agricultural Science (Cambridge)* 120, 251–263.

Skinner, B.F. (1981) Selection by consequences. *Science* 213, 501–504.

Skinner, B.F. (1984) The evolution of behavior. *Journal of Experiment and Analysis of Behavior* 41, 217–221.

Stephens, D.W. and Krebs, J.R. (1986) *Foraging Theory*. Princeton University Press, Princeton, New Jersey.

Stricker, E.M. and Verbalis, J.G. (1990) Sodium appetite. In: Stricker, E. (ed.) *Handbook of Behavioral Neurobiology*. Plenum, New York, pp. 387–418.

Taleb, N.N. (2001) *Fooled by Randomness: The Hidden Role of Chance in the Markets and in Life*. Texere, New York.

Ternouth, J.H. (1991) The kinetics and requirements of phosphorus in ruminants. In: Ho, Y.W., Wong, H.K., Abdullah, N. and Tajuddin, Z.A. (eds) *Recent Advances on the Nutrition of Herbivores*. Malaysian Society of Animal Production, University Pertanian Malaysia, Serdang, Selangor Darul Ehsan, Malaysia, pp. 143–151.

Tien, D.V., Lynch, J.J., Hinch, G.N. and Nolan, J.V. (1999) Grass odor and flavor overcome feed neophobia in sheep. *Small Ruminant Research* 32, 223–229.

Tordoff, M.G. (2001) Calcium: taste, intake, and appetite. *Physiological Review* 81, 1567–1597.

Van Soest, P.J. (1982) *Nutritional Ecology of the Ruminant*. O & B Books, Corvallis, Oregon.

Villalba, J.J. and Provenza, F.D. (1996) Preference for flavored wheat straw by lambs conditioned with intraruminal administrations of sodium propionate. *Journal of Animal Science* 74, 2362–2368.

Villalba, J.J. and Provenza, F.D. (1997a) Preference for wheat straw by lambs conditioned with intraruminal infusions of starch. *British Journal of Nutrition* 77, 287–297.

Villalba, J.J. and Provenza, F.D. (1997b) Preference for flavored foods by lambs conditioned with intraruminal administrations of nitrogen. *British Journal of Nutrition* 78, 545–561.

Villalba, J.J. and Provenza, F.D. (1997c) Preference for flavored wheat straw by lambs conditioned with intraruminal infusions of acetate and propionate. *Journal of Animal Science* 75, 2905–2914.

Villalba, J.J. and Provenza, F.D. (1999) Nutrient-specific preferences by lambs conditioned with intraruminal infusions of starch, casein, and water. *Journal of Animal Science* 77, 378–387.

Villalba, J.J. and Provenza, F.D. (2000a) Postingestive feedback from starch influences the ingestive behaviour of sheep consuming wheat straw. *Applied Animal Behaviour Science* 66, 49–63.

Villalba, J.J. and Provenza, F.D. (2000b) Discriminating among novel foods: effects of energy provision on preferences of lambs for poor-quality foods. *Applied Animal Behaviour Science* 66, 87–106.

Villalba, J.J. and Provenza, F.D. (2000c) Roles of novelty, generalization, and postingestive feedback in the recognition of foods by lambs. *Journal of Animal Science* 78, 3060–3069.

Villalba, J.J. and Provenza, F.D. (2000d) Roles of flavor and reward intensities in acquisition and generalization of food preferences: do strong plant signals always deter herbivory? *Journal of Chemical Ecology* 26, 1911–1922.

Villalba, J.J. and Provenza, F.D. (2001) Preference for polyethylene glycol by sheep fed a quebracho tannin diet. *Journal of Animal Science* 79, 2066–2074.

Villalba, J.J. and Provenza, F.D. (2002) Polyethylene glycol influences selection of foraging location by sheep consuming quebracho tannin. *Journal of Animal Science* 80, 1846–1851.

Villalba, J.J. and Provenza, F.D. (2005) Foraging in chemically diverse environments: energy, protein and alternative foods influence ingestion of plant secondary metabolites by lambs. *Journal of Chemical Ecology* 31, 123–138.

Villalba, J.J., Provenza, F.D. and Rogosic, J. (1999) Preference for flavored wheat straw by lambs conditioned with intraruminal infusions of starch administered at different times after straw ingestion. *Journal of Animal Science* 77, 3185–3190.

Villalba, J.J., Provenza, F.D. and Banner, R.E. (2002a) Influence of macronutrients and activated charcoal on utilization of sagebrush by sheep and goats. *Journal of Animal Science* 80, 2099–2109.

Villalba, J.J., Provenza, F.D. and Banner, R.E. (2002b) Influence of macronutrients and polyethylene glycol on intake of a quebracho tannin diet by sheep and goats. *Journal of Animal Science* 80, 3154–3164.

Villalba, J.J., Provenza, F.D. and Bryant, J.P. (2002c) Consequences of nutrient–toxin interactions for herbivore selectivity: benefits or detriments for plants? *Oikos* 97, 282–292.

Villalba, J.J., Provenza, F.D. and Han, G. (2004) Experience influences diet mixing by herbivores: implications for plant biochemical diversity. *Oikos* 107, 100–109.

Villalba, J.J., Provenza, F.D. and Shaw, R. (2006a) Sheep self-medicate with substances that ameliorate the negative effects of grain, tannins, and oxalates. *Animal Behaviour* (in press).

Villalba, J.J., Provenza, F.D. and Shaw, R. (2006b) Initial conditions and temporal delays influence preference for foods high in tannins and for foraging locations with and without foods high in tannins by sheep. *Applied Animal Behaviour Science* (in press).

Villalba, J.J., Provenza, F.D., Hall, J.O. and Peterson, C. (2006c) Phosphorus appetite in sheep: dissociating taste from postingestive effects. *Journal of Animal Science* (in press).

Wang, J. and Provenza, F.D. (1996) Food preference and acceptance of novel foods by lambs depend on the composition of the basal diet. *Journal of Animal Science* 74, 2349–2354.

Wang, J. and Provenza, F.D. (1997) Food deprivation affects preference of sheep for foods varying in nutrients and a toxin. *Journal of Chemical Ecology* 22, 2011–2021.

Wiedmeier, R.D., Provenza, E.A. and Burritt, E.A. (2002) Exposure to ammoniated wheat straw as suck-
ling calves improves performance of mature beef cows wintered on ammoniated wheat straw. *Jour-
nal of Animal Science* 80, 2340–2348.
Williams, J.L., Campos, D., Ross, T.T., Becker, K.A., Martinez, J.M., Oetting, B.C. and Smith, G.S. (1992)
Snakeweed (*Gutierrezia app.*) toxicosis in beef heifers. *Proceedings of the Western Section of the
American Society of Animal Science* 43, 67–69.

14 Feeding and Mastication Behaviour in Ruminants

R. Baumont,[1] M. Doreau,[1] S. Ingrand[2] and I. Veissier[1]

[1]INRA – Unité de Recherches sur les Herbivores; [2]INRA – Transformation des systèmes d'élevage, Centre de Clermont/Ferrand-Theix, 63122 St-Genès-Champanelle, France

Introduction

Feeding behaviour can be defined as the whole process by which the animal ingests the feeds that are able to satisfy organic needs and refuses non-alimentary or toxic compounds. Ruminants, like other species, try to adjust their food intake to match their nutritional requirements, especially those of energy. Although many animal species eat grass, few are as well adapted as the ruminant with its particular feeding behaviour combining ingestion and rumination. These activities are central to ruminant life and it is commonly said that consuming a diet of forage is a full-time job for an adult ruminant. Each day, the ruminant takes between 10 and 15 meals for a total duration of 5–9 h, and also spends 5–9 h ruminating (Jarrige et al., 1995). The total mastication time during ingestion and rumination generally exceeds more than 60% of daytime, and during this intense mastication effort the ruminant performs between 30,000 and 50,000 chewing movements (Welch and Hooper, 1988).

In this chapter, we will first review the specificities of feeding behaviour and mastication in ruminants and their implications for mastication measurement and analysis. We will then focus on the functional role of mastication during eating and ruminating, highlighting its major contribution to digestion and thus to

the control of intake. Finally we will review and discuss the variations of rhythm and duration of mastication activities in relation to feed and dietary characteristics on one hand and to animal and group characteristics on the other.

Studying feeding behaviour and mastication

The study of feeding behaviour and mastication in ruminants has to take account of two major constraints: (i) the necessity of recording both eating and ruminating activities; and (ii) the diversity of feeding situations – pasture or indoors, individual or group feeding – that influence behaviour. A detailed knowledge of behaviour patterns derived from direct observation is an essential prerequisite to developing methods and techniques for the collection and analysis of feeding behaviour and mastication data (Penning and Rutter, 2004). However, direct visual observation is laborious and often difficult, and indeed it does not allow a continuous detailed monitoring of jaw movements for several animals simultaneously. Thus, a significant effort has been directed towards developing systems to monitor feeding behaviour in ruminants fed indoors (Baumont et al., 2004) or at grazing (Penning and Rutter, 2004a).

Monitoring jaw movements

Monitoring jaw movements involves mounting a sensor on the head of the animal. A widespread technique is to place a small foam-filled balloon in the submandibular space and to connect it to a pressure transducer (Ruckebusch, 1963; Baumont *et al.*, 1998). Changes in air pressure when the jaws are opened can easily be recorded on paper charts, and enable easy distinction between eating and ruminating activities (Fig. 14.1). Another type of sensor introduced by Penning (1983) is a silicone rubber tube filled with graphite that is formed into a noseband, so that when the animal moves its jaws, the noseband is stretched and its electrical resistance increases.

Analysis of traces of jaw movements on paper charts is very tedious and time-consuming. Electronic devices enable automatic recording of jaw movements, and computer software has been developed to interpret the events and assist in data reduction (e.g. Brun *et al.*, 1984; Penning *et al.*, 1984; Luginbuhl *et al.*, 1987; Rutter *et al.*, 1997; Baumont *et al.*, 1998; Rutter, 2000). Most of these systems are mainly aimed at counting jaw movements in order to distinguish between, and estimate the time devoted to eating, ruminating and resting. Rumination can generally be easily identified on the basis of the regularity of both jaw movement frequency and time interval between the boluses (Fig. 14.1), whereas during eating, jaw movements are irregular due to small interruptions and alternation between prehension and mastication sequences. However, computerized recording and analysis can lead to some overestimation of eating time as it is very difficult to distinguish between a true eating period and other activities such as licking and self-grooming. Overestimation of eating time compared to visual observations has been reported by Beauchemin and Buchanan-Smith (1989) and by M. van Os and R. Baumont (unpublished data). A way to overcome this problem is to simultaneously record the variations in the weight of feed mangers for stall-fed animals (Baumont *et al.*, 1998).

From jaw movements to feeding behaviour recordings

Recording jaw movements is a useful tool for estimating the time spent eating and ruminating, but it is insufficient for comprehensive studies of feeding behaviour that require the nature and the amount of feeds ingested to be quantitatively characterized.

For animals fed indoors and individually penned, measuring feed intake by weighing the food offered and the food refused is easy to

Eating

| 10 min |

Recording at 15 cm/h: small pauses during the eating bouts can be distinguished.

Ruminating

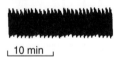

| 10 min |

Recording at 15 cm/h: the successive rumination cycles can be distinguished.

1   2

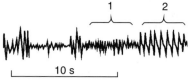

|_____ 10 s _____|

Recording at 30 cm/min: two types of jaw movements can be distinguished: (1) prehension with low amplitude and high frequency; (2) mastication with higher amplitude and lower frequency.

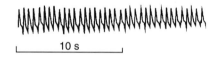

|_____ 10 s _____|

Recording at 30 cm/min: the regularity of jaw movements in amplitude and frequency can be clearly observed.

Fig. 14.1. Recordings of traces of jaw movements on paper charts using a small foam-filled balloon in the submandibular space connected to a pressure transducer. (From Baumont, in Jarrige *et al.*, 1995.)

achieve, but can prove disturbing for the animals when it is performed too frequently. Recording the daily kinetics of intake can be achieved by continuous recording of the weight of feed boxes connected to a weight sensor (Suzuki *et al.*, 1969; Baile, 1975; Forbes *et al.*, 1986; Heinrichs and Conrad, 1987; Baumont *et al.*, 1990a). Using a microcomputer to continuously monitor the weight of feeders makes it possible to record meaningful weights, i.e. stable weights corresponding to the small interruptions that occur during a meal to masticate and swallow the feeds. Together with monitoring jaw movements, this system allows the researcher to record the meal pattern in quantity and duration, as well as the rumination periods (Fig. 14.2). The system was then adapted to record intake in a choice situation (Baumont *et al.*, 1998). For group feeding, Forbes *et al.* (1986) and Ingrand *et al.* (1998) developed systems that combine equipment to identify which cows are eating and to weigh the food to which they have access. The system developed by Ingrand *et al.* (1998) provides free access to any feed manger for any cow. The access to the manger is fitted with a loop, which energizes the transponder of any cow eating in the

manger and thus identifies the cow (Fig. 14.3). This system aims to analyse the ability of cows to adapt their intake by modifying their feeding behaviour according to group composition and competition level for feed.

At pasture, intake measurements can only be achieved by indirect methods. Intake can be estimated on a day-to-day basis through marker techniques measuring faecal output and digestibility of the feed (see review by Penning, 2004). As the elementary event of the grazing process is biting, comprehensive study of ingestive behaviour at grazing requires bites to be analysed. By visual observations bites can be quantified in cattle using the sound made by severing the herbage, and in sheep using the jerk of the head. Penning *et al.* (1984) and Rutter *et al.* (1997) have proposed a way to count bites based on analysis of the signal trace shape of jaw movement recordings. This method requires a considerable amount of algorithm development and a careful choice of parameters used to analyse the data files. Delagarde *et al.* (1999) have developed and validated for dairy cows an automatic system that combines recording the sound made by severing the herbage and the position of the head.

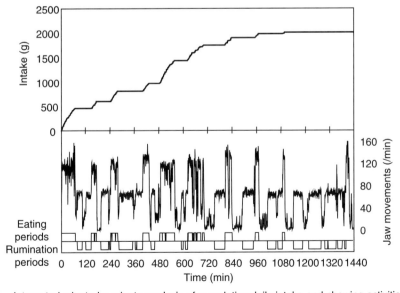

Fig. 14.2. Integrated minute-by-minute analysis of cumulative daily intake and chewing activities in sheep fed lucerne hay *ad libitum*. Eating periods are recognized by a number of chews per minute higher than 80 and a simultaneous increase of the cumulative intake curve. Rumination periods are recognized by a number of chews per minute between 40 and 80 without increase in the intake curve.

(a)   (b)

Fig. 14.3. A system for recording the kinetics of intake and eating behaviour in cattle fed in groups. (a) Cows inside the stall with their transponder around the head. (b) The six mangers with the loop antenna (feeding corridor side). (From Ingrand *et al.*, 1998.)

Analysis of the data

Feeding behaviour data are first analysed as the time spent eating and ruminating, usually per day, and as the number of eating and ruminating periods. When bites are counted or when the amount ingested is recorded, combined variables can be assessed, such as eating or biting rate, or the time spent masticating per unit ingested. However, as pointed out by Forbes (1995), it is first necessary to consider 'when is a meal not a meal?'. The short pauses devoted to the search and the selection of the food ingested during the meal must be differentiated from the longer gaps between separate meals. Analysis of the frequency distribution of the intervals between eating bouts shows that these intervals form two distinct categories that make it possible to separate intrameal and intermeal intervals. Using these types of analysis, Metz (1975) found that intrameal intervals are less than 4 min in cows and Forbes *et al.* (1986) found an intermeal interval close to 7 min for lactating cows fed grass silage. The mathematical method for determining these intervals remains a matter of debate. Recently, Tolkamp and Kyriazakis (1999) suggested using log normal models rather than log (cumulative) frequency models.

Feeding behaviour data are often summarized hourly for pattern analysis. Deswysen *et al.* (1989) attempted to extract rhythms from meal patterns recorded in heifers using Fourier analysis. Consistent cycles with phases of 24, 12 and 8 h were identified, but only the 24-h cycle could be explained in relation to the solar cycle and daily offering of fresh food. Continuous recording allows cumulative change in feeding behaviour parameters to be analysed, and this is of particular interest for intake. Simple exponential models fit with accuracy cumulative intake during meals in sheep and cows (Baumont *et al.*, 1989; Fig. 14.4).

Functional role of mastication during eating and ruminating

The pattern of eating and ruminating has been extensively studied and reviewed both at pasture (Arnold, 1981) and indoors (Dulphy *et al.*, 1980; Jarrige *et al.*, 1995). In natural conditions, eating pattern is characterized by two main periods at sunrise and sunset. When fed indoors, feed distributions, generally one or two, induce large eating periods followed by smaller ones (Fig. 14.2). Eating has an autonomous rhythm independent of the duration of rumination (Metz, 1975), the latter being inhibited when the eating rhythm is in the meal state. During non-eating periods, the rumination rhythm maintains a rather constant frequency.

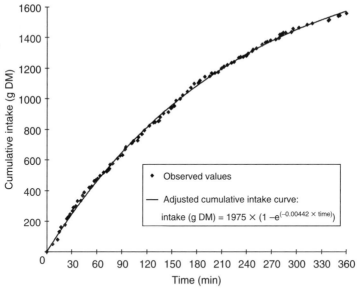

Fig. 14.4. Recording of cumulative intake and fitting of the data for sheep during 6-h access to *ad libitum* lucerne hay.

Collecting and chewing food

During eating, jaw movements allow the animal to collect the food into bites and to masticate it before swallowing. At pasture, ruminants gather and manipulate herbage in the sward with their tongues (cattle) and lips (sheep and goats), grip the herbage between their incisors and the upper dental pad and sever it from the sward, often, but not always, with a jerk of the head. The herbage severed by the bite is then manipulated, masticated and formed into a bolus before it is swallowed. Biting and masticating appear to be mutually exclusive in sheep (Penning *et al.*, 1991), but can be carried out simultaneously (Laca and WallisDeVries, 2000) in cattle. Much experimental effort has been devoted to analysing and quantifying relationships between sward and bite characteristics (Black and Kenney, 1984; Burlison *et al.*, 1991; Penning *et al.*, 1991; Prache, 1997; Garcia *et al.*, 2003 for sheep, and Ungar *et al.*, 1991; Laca *et al.*, 1992 for cattle). When sward height and biomass decrease, bite mass also decreases. In order to maintain intake rate, this is partly compensated by higher bite frequency and lower time per bite (Penning *et al.*, 1991).

With heterogeneous swards of high biomass, bite mass remains stable, but time per bite increases with the time spent by the animals in selecting the best parts of the sward (Garcia *et al.*, 2003). Time per bite is the sum of a prehension time, which depends primarily on the selection activity of the animal and on sward resistance (Griffiths and Gordon, 2003), and of a mastication time, which depends primarily on bite mass modulated by its resistance to chewing (Prache, 1997; Baumont *et al.*, 2004b).

When fed indoors, animals take bites from the manger that are larger but less frequent than at pasture because they do not have to search, collect and severe the forage (Jarrige *et al.*, 1995; Table 14.1). Thus, the eating time of animals fed indoors is 30–50% shorter than during grazing. During the meal, the rate of intake is highest at the beginning and then decreases continuously (satiation process) until satiety (Faverdin *et al.*, 1995). Intake rate at the beginning of the meal is 4–5 times higher than at the end for cows eating maize silage (Faverdin, 1985), and 3–6 times higher for sheep eating hays (Baumont *et al.*, 1989). As jaw movement frequency remains quite constant during meals, the decrease in intake rate results from: (i) a decrease in bite frequency and

Table 14.1. Eating (E) and rumination (R) chewing durations of ruminants depending on whether they graze or consume conserved forages. (From Jarrige et al., 1995.)

References	Forages	Animals	Grazing		Cut forage	
			Duration (h/day)			
			E	R	E	R
Lofgreen et al. (1957)	Lucerne or orchard grass grazed vs. cut	Cattle Sheep	6.8 8.7	5.3 3.9	4.6 4.3	7.0 6.0
Stricklin et al. (1976)	Grazed grass vs. grass silage	Cows	8.8	–	6.1	–
Journet et al. (unpublished data)	Grazed grass vs. maize silage (13 kg DM)	Dairy Cows	8.1	5.8	6.2	8.4
Dougherty et al. (1989)	Grazed lucerne vs. freshly cut in a swath	Cows	74 26	[chews/min] [bites/min]		83 5

longer mastication time per bite; and (ii) the increasing occurrence of small interruptions.

After a meal, rumination starts following a latency period that can vary from a few minutes to over 1 h (Dulphy and Faverdin, 1987). The duration of this latency period is inversely related to the size of the meal, larger meals inducing higher stimulation to ruminate than shorter ones (Jarrige et al., 1995). The first rumination periods after the meal are often relatively short (15–30 min) and the longest (up to 2 h) occur during the night (Welch and Hooper, 1988). Each of the 12–18 rumination periods that occur daily is a regular succession of cycles and small pauses lasting 4–8 s, during which the masticated bolus is reswallowed and the next one is regurgitated. Chewing frequency is lower and much more regular during rumination than during eating (Table 14.2). Daily time spent ruminating can reach a maximum of 10 h (Jarrige et al., 1995).

Particle size reduction and digesta outflow

Only small particles of feed residues (lower than 1–2 mm in sheep and 2–4 mm in cattle) are able to flow out of the reticulo-rumen. It is mainly mastication during eating and ruminating that ensures particle size reduction, microbial digestion and motility per se having a lesser impact (Ulyatt et al., 1986; McLeod and Minson, 1988).

Particle size is fairly constant in the duodenum. It decreases progressively from the offered feed to the swallowed ingesta and on to reticular digesta, the origin of ruminal digesta outflow (Ulyatt et al., 1986; Grenet, 1989; Bernard et al., 2000; Fig. 14.5). The effect of mastication during eating on ingesta particle size was modelled by Sauvant et al. (1996). For long forages (95% of the particles ≥1 mm) the proportion of large particles in the swallowed bolus ranges between 45% and 65% with a mean predicted value of 55% (Fig. 14.6). During rumination, the regurgitated digesta come from the cranial part of the reticulo-rumen. The size of the masticated bolus varies between 2.3 and 4.3 g dry matter (DM) in sheep and between 16.9 and 29.3 g in cattle (Ulyatt et al., 1986). Based on comparison of particle size in regurgitated and reswallowed bolus (Ulyatt et al., 1986; Grenet, 1989), it appears that 39–65% of the large particles are reduced to less than 1 mm (Table 14.3).

Table 14.2. Mastication movement frequency during ingestion and rumination in adult cattle and sheep eating forages. (From Jarrige *et al.*, 1995.)

	Cattle	Sheep
Ingestion		
Movements/min	65–85	90–140
Rumination		
Movements/min	55–65	85–110
Movements/cycle	50–60	60–75
Cycle duration (s)	45–65	40–60

Chewing during eating and chewing during rumination have different functions with respect to particle size reduction. Chewing during eating prepares feed for swallowing, releases soluble constituents, breaks down plant tissues for microbial digestion and reduces the shear force necessary to break the particles down (Boudon *et al.*, 2002a). The release of soluble constituents is mainly achieved by chewing during eating and digestion rather than by chewing during rumination (Ulyatt *et al.*, 1986; Boudon *et al.*, 2002b). The proportion of soluble constituents released during eating varies with their nature, sugar constituents being more easily released than protein constituents (Boudon and Peyraud, 2001). While feed is masticated only once during eating, the daily amount of DM

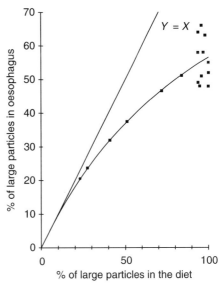

Fig. 14.6. Effect of mastication during eating on particle size reduction. (Filled boxes represent data from literature synthetised by Sauvant *et al.*, 1996.)

submitted to mastication during rumination represents 1.7–2.7 times the daily intake (Table 14.3) and 4.5–6 times the amount of DM in the rumen. The main function of chewing during rumination is to reduce the size of the refractory material. For fresh forages with low cell-wall and high soluble contents (Ulyatt *et al.*, 1986) and for mixtures of long and ground forages (Bernard *et al.*, 2000), the contribution of chewing during rumination to the whole process of particle size reduction is about 50%, while for hays with higher cell-wall and lower soluble contents (Ulyatt *et al.*, 1986) this contribution can reach 70%. In their study on cattle fed leaves or stems of grasses or lucerne, McLeod and Minson (1988) concluded that eating contributes to 25% and ruminating to 50% of particle size reduction.

Particle size reduction is the major, but not the only, contribution of feeding behaviour and mastication to digesta outflow. Particle size reduction is a prerequisite to outflow, but the material present in the rumen at any given time is predominantly below the threshold size (Poppi *et al.*, 1980; Baumont and Deswysen, 1991). Particles that could leave the rumen have a high density and remain in the bottom

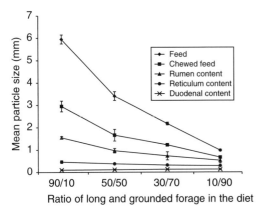

Fig. 14.5. Mean particle size of the feed offered, chewed feed, rumen content, reticulum content and duodenal content with different ratios of long and grounded forage in the diet. (Adapted from Bernard *et al.*, 2000.)

Table 14.3. Characteristics and effects of mastication during eating and ruminating in sheep fed fresh forages or chaffed hay. (From Ulyatt *et al.*, 1986.)

	Fresh forages		Hays		
	Rye grass	Red clover	Lucerne	Lucerne	Meadow
Digestibility (% of DM)	78	78	72	54	60
Intake (g DM/day)	861	918	952	946	943
Mastication during eating					
Duration (min/day)	210	83	70	123	88
Jaw movements/min	150	145	143	142	125
DM solubilized (% of intake)	37	38	32	23	20
% of large-particle DM reduced to less than 1 mm	49	52	45	37	35
Mastication during rumination					
Duration (min/day)	540	436	317	570	547
Jaw movements/min	108	107	108	86	86
Size of the bolus (g DM)	2.3	2.8	3.3	4.3	3.3
DM masticated (g/day)	1902	2004	1625	2449	2539
% of large-particle DM reduced to less than 1 mm	42	39	63	60	64

of the reticulo-rumen. Functional density of particles depends on the density of the plant structure of the particle, the liquid inside the particle and also the gas, i.e. on the degree of fermentation. During eating, mastication mixes particles with saliva and eliminates part of the air trapped inside, thereby increasing particle density from 0.05–0.1 g/l for dry forage to 0.8 g/l. Rumination contributes to sorting particles according to density. During regurgitation, a fluid rich in small particles of high density is immediately reswallowed, and the mastication of the bolus eliminates part of the fermentation gases.

Mastication, salivation and ruminal pH

A major function of mastication is to induce salivation, which is necessary for swallowing the feeds (Church, 1988). It has been clearly shown that salivary secretion in ruminants is positively correlated to feed intake (Silanikove and Tadmor, 1989; Maekawa *et al.*, 2002). Salivation is a constant process but it is increased by mastication. The causes of this increase have been discussed by Carter *et al.* (1990). A direct effect of chewing through mechanical stimulation of each parotid gland

has been demonstrated (Meot *et al.*, 1997). In addition, there is a probable effect of intake alone on salivary flow through the increase in blood flow. The rate of saliva production, measured by non-invasive methods, is higher by 20–50% during eating than during resting (e.g. Bowman *et al.*, 2003). Saliva production during rumination may be higher than during eating (Cirio *et al.*, 2000). As a consequence, salivation is higher with forage diets than with concentrate diets (see, e.g. Church, 1988).

Mastication through salivation indirectly affects ruminal pH. Due to the presence of HPO_4^- and especially HCO_3^-, saliva functions as a buffer, limiting the decrease in rumen pH caused by the production of volatile fatty acids. This major function is particularly necessary in the event of lactic acidosis. This disease occurs with diets rich in concentrates and is characterized by a rapid decrease in ruminal pH due to significant volatile fatty acid production. An insufficient mastication results in a lack of carbonate ions. A supply of highly fibrous feeds is thus recommended to alleviate acidosis. The prevalence of acidosis in early lactating cows could also be related, in addition, to a rapid shift from forage to concentrate in the diet, and to insufficient saliva production (Cassida and Stokes, 1986) due to a slow adaptation in chewing to the increase in DM intake (see later).

Influence of feed and diet characteristics

Feed characteristics affect mastication activity. A comprehensive analysis of these variations is important for diet formulation and feeding management, as mastication is a key factor of digestion and also an indicator of animal health and welfare.

Cell-wall content and physical presentation

In sheep fed with forages, Dulphy et al. (1980) analysed the relationship between voluntary intake and time spent eating and ruminating in relation to plant age during the first vegetation cycle. As voluntary intake decreases, eating time decreases slightly and rumination time increases. The total time spent chewing remains stable or increases slightly, but the chewing time per kilogram ingested increases sharply. Differences between botanical families exist. The decrease in intake and the increase in the time spent masticating in relation to

plant age are less pronounced with legumes than with grasses.

One factor to explain this relationship with plant age is cell-wall content. Rumination has been shown to be closely related to the intake of cell-wall content in both sheep and cattle (Dulphy et al., 1980; Welch and Hooper, 1988). Cell-wall content of the forage affects eating and rumination patterns by reducing the number of additional meals (Dulphy and Faverdin, 1987) and by stimulating the onset of rumination induced by forage distribution after a large meal (Baumont et al., 1997). Cell-wall content of forage also affects the rate at which it can be ingested (Fig. 14.7), as it affects the resistance to prehension and mastication during eating. Rye grass selected for a lower resistance to leaf shear is ingested at a faster rate (MacKinnon et al., 1988). Nevertheless, differences in intake rate are also related to sensory properties of forages, as is the case with grass silages (van Os et al., 1995).

The increase with forage cell-wall content in ruminating and chewing time needed per kilogram ingested can be quantified in sheep for forages of various types (Fig. 14.8a). About 1 min rumination and 1.5 min chewing time per kilogram ingested is induced by an

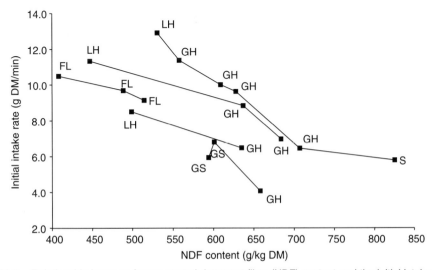

Fig. 14.7. Relationship between forage neutral detergent fibre (NDF) content and the initial intake rate measured in sheep fed *ad libitum*. (Data from Baumont *et al.*, 1989, 1997 and unpublished data.) S: straw; GH: grass hay; LH: lucerne hay; GS: grass silage; FL: fresh lucerne.

increase of 1 g neutral detergent fibre (NDF) per kg DM of the forage. Similar trends were found by Mertens (1997) summarizing data obtained in cows. Rumination and chewing time are also closely related to the retention time of DM in the rumen (Fig. 14.8b). An increase of the forage retention time of 1 h induces an increase of 15 min in the time spent ruminating per kg DM ingested, and of 21 min for the total time spent chewing.

The second major factor that affects mastication activity and intake is the physical presentation of forages. Chopping and primarily grinding dramatically reduce the time spent eating and ruminating (Fig. 14.9). The reduction in chewing time accompanying a decrease in mean particle size of dried herbage was followed by an increase of voluntary intake until mean particle size remained higher than 0.4 mm for grasses and 0.8 mm for legumes. In sheep, these particle sizes probably represent thresholds under which rumen function and thus intake is disturbed. A sufficient proportion of large particles in the rumen is necessary to ensure normal rumination behaviour (Deswysen and Ehrlein, 1981). In cows also, chopping and grinding decreases mastication times for various types of forages (Mertens, 1997). In sheep and cattle, the time spent masticating per kg DM

ingested can be related to the mean particle size of the forage fed to animals using quadratic equations (Sauvant, 2000). The time spent masticating decreases when mean particle size is below 2.5 mm in sheep and 5 mm in cattle.

Concentrate feeds are ground and their cell-wall content is generally lower than that of forages. For these reasons, they require little mastication. They are ingested at high intake rates – up to 10 times that of forages – and they do not induce rumination. In an extreme diet made of 100% concentrates, Bines and Davey (1970) observed the absence of rumination. Concentrate feeds composed of cereals require only 5–15 min mastication per kg DM, whereas forages require between 40 and 150 min (Sudweeks *et al.*, 1981).

Mastication and diet fibrosity

The mastication activity induced by a given diet depends primarily on the properties of the different feeds composing the diet, since the times spent eating and ruminating the different forages and concentrates can be considered additional factors (Sudweeks *et al.*, 1981). Thus, the mastication time induced by a diet will depend on its cell-wall content and on the

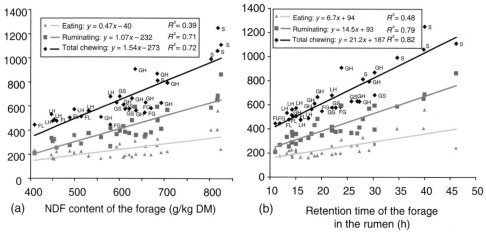

Fig. 14.8. Relationship between the time spent chewing during eating, ruminating and in total and (a) the neutral detergent fibre (NDF) content or (b) the retention time of the forage. Data obtained in sheep fed various types of forages *ad libitum*. (Data from Dulphy *et al.*, 1992; Chiofalo *et al.*, 1992; Baumont *et al.*, 1989, 1990a, 1997 and unpublished data) S: straw; GH: grass hay; LH: lucerne hay; GS: grass silage; FL: fresh lucerne; FG: fresh grass.

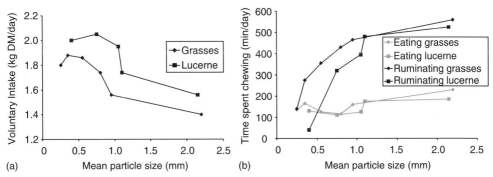

Fig. 14.9. Effect of decreasing the mean particle size by chopping and grounding forage (a) on voluntary intake and (b) time spent chewing during eating and ruminating in sheep fed *ad libitum*. (Adapted from Jarrige *et al.*, 1995.)

size of the feed particles. The combination of both factors characterizes feed and diet fibrosity, which is the resistance to particle size reduction. Accordingly, variations in diet mastication times are closely related to the percentage of concentrate when the forage of the diet is presented in long form (Jarrige *et al.*, 1995; Fig. 14.10).

Ruminant animals need a sufficient amount of long and fibrous particles in their diet. Indeed, in a free-choice situation, diet selection does not always maximize dietary energy density. Sheep eat some straw to prevent rumen disorders, even when a more concentrated feed is also on offer (Cooper *et al.*, 1995). Sheep fed with a 'long-fibre-free' diet will eat 10 mm polyethylene fibre to restore normal rumination activity (Campion and Leek, 1997). Inclusion of 20% straw in a con-

centrate diet is enough to restore a significant rumination activity (Bines and Davey, 1970). Diet fibrosity increases mastication time during eating and thus salivation, and stimulates rumination. Rumination time was increased by the presence in the rumen of polypropylene ribbon, which was regurgitated during the rumination process (Welch and Smith, 1974). It was also increased when the rumen contained inert compounds that were not regurgitated, such as polystyrene cubes (Baumont *et al.*, 1990b) or an artificial mechanical stimulating brush (Horiguchi and Takahashi, 2002), as a result of mechanical stimulation of the rumen dorsal wall.

Balch (1971) proposed a definition of the feed fibrosity based on mastication time per kg DM ingested. This concept was used by Sudweeks *et al.* (1981) and by Norgaard (1983)

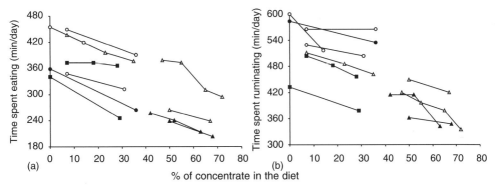

Fig. 14.10. Influence of the proportion of concentrate in the diet on the time spent (a) eating and (b) ruminating in dairy cows. Data connected by a same line stem from the same trial. (Adapted from Jarrige *et al.*, 1995.)

to characterize the feeds incorporated in dairy cow diets in order to formulate diets that ensure sufficient mastication time and thereby prevent acidosis. More recently, Mertens (1997) proposed a system to meet the fibre requirements of dairy cows based on the 'physically effective NDF content' of the diet, i.e. the cell-wall content of the diet particles larger than 8 mm. Increasing the intake of physically effective cell walls by increasing particle size of the diet increases mastication time and ruminal pH with diets based on lucerne silage (Krause *et al.*, 2002) or on a mixture of lucerne silage and hay (Beauchemin *et al.*, 2003), but is less effective on maize silage diets (Kononoff *et al.*, 2003; Fernandez *et al.*, 2004).

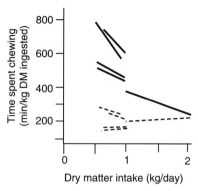

Fig. 14.11. Relationship between the level of intake and time spent chewing during eating (dotted line) and ruminating (solid line) in sheep. (From Grimaud, 1999.)

Mastication and level of intake

When a diet is not offered *ad libitum*, mastication activities largely depend on the level of intake. Time spent eating and to a lesser extent time spent ruminating logically increase with the level of intake, but the number of meals does not vary with intake. Variation in rumination time is due to variation in the number of rumination boli, but the duration of each rumination bolus and the number of chews per bolus remain constant (Luginbuhl *et al.*, 1989). Time spent eating 1 kg DM generally does not vary with intake level, whereas time spent ruminating 1 kg DM decreases when intake level increases, especially in sheep (Malbert and Baumont, 1989; review by Grimaud, 1999; Fig. 14.11).

When the level of intake is very low, the trends are similar. In cows, the decrease in intake level from maintenance to 30% maintenance requirements results either in an increase in time spent eating 1 kg DM and a stability in time spent ruminating 1 kg DM (Doreau *et al.*, 2004), or a stability in time spent eating 1 kg DM and an increase in time spent ruminating 1 kg DM (Grimaud *et al.*, 1999). In all cases, time spent chewing 1 kg DM is increased. A logical consequence could be a decrease in particle size in the rumen. This has been frequently observed, for example, by Luginbuhl *et al.* (1989) and Okine and Mathison (1991), but in some cases and for an unknown reason, particle size is not modified

by intake level (Grimaud *et al.*, 1999). When a diet is very digestible, a strong decrease in intake may result in disturbances in the rumination process. A 'pseudo-rumination', which corresponds to a rumination activity without food to be swallowed, has been observed by Doreau *et al.* (1986). Such a phenomenon is observed at fasting (Welch and Smith, 1968) and can be related to an insufficient load of long particles in the rumen.

Mastication, oral activities and welfare

Mastication activities contribute to animal welfare. This is of particular importance for young animals fed partly on liquid diets. In ruminants, feeding behaviour develops rapidly after birth. In natural conditions, at birth, the calf initially only suckles from its dam but after a few days it starts taking some solid foods in its mouth. A 10-day-old calf can spend 10% of the day eating grass, hay or concentrates and another 10% of the day ruminating them, and a 4-month-old calf can spend 25% of the day grazing or eating solid foods (Petit, 1972; Nicol and Sharafeldin, 1975; Webster *et al.*, 1985). It is likely that when feeding activities directed to solid foods cannot develop, cattle display non-nutritive oral activities. This is the case in veal calves that, until recently, were fed milk or milk replacer only (Webster *et al.*, 1985; Wiepkema *et al.*, 1987). The following non-nutritive oral activities have been described in cattle kept indoors (Redbo, 1990): nibbling or biting equipment, licking equipment, tongue rolling and tongue playing, i.e. the mouth is open and the tongue is rolled inside the mouth towards the pharynx (inner tongue rolling) or outside the mouth (tongue playing).

The origin of these activities has not been completely defined. Non-nutritive oral activities generally occur around feeding (Redbo, 1990; Veissier et al., 1998) and are linked to diet. In veal calves, the provision of solid foods that contain fibres (straw, hay) reduced the occurrence of non-nutritive nibbling, biting and tongue rolling or tongue playing (Veissier et al., 1998; Mattiello et al., 2002). This reduction seems to be linked to the increased time spent eating the solid foods rather than to the increased time spent ruminating (Veissier et al., 1998). This suggests that the lack of oral activities directed towards ingestion of solid foods is compensated for by non-nutritive oral activities, whereas the lack of chewing for rumination is not compensated for by any other activity, the animals spending more time inactive when not ruminating. This is further supported by the fact that tongue rolling and tongue playing resemble the movements cattle perform to pick up grass: the tongue is rolled around the grass to take it off the ground. Furthermore, non-nutritive oral activities are generally absent at pasture (Redbo, 1990). These observations suggest that, in cattle, non-nutritive oral activities derive from nutritive activities.

The frequency of tongue rolling or tongue playing is further increased when calves have no access to objects that can easily be bitten and when they are housed individually (Veissier et al., 1997, 1998). It appears that when solid foods are not provided in sufficient amounts (in terms of time needed to eat them), cattle develop non-nutritive nibbling, biting and licking, and the poverty of the environment, in addition to the lack of solid foods, triggers vacuum activities such as tongue rolling or biting (Veissier et al., 1999). According to Wiepkema (1983), the occurrence of non-nutritive activities at a rate of 1–5% of animals on a farm should be taken as an index of frustrated feeding activity, and thus poor welfare.

Influence of animal characteristics

Domestic ruminants comprise a large variety of species, breeds and type of animals linked with various production systems. Since ruminants are often managed in groups, it is important to account for animal characteristics that influence chewing behaviour and interactions between feeding and social behaviours.

Species and breeds

Differences between ruminant species in time spent chewing are moderate, despite the large size differences between sheep or goats and cattle. From 20 direct comparisons of sheep and goats receiving the same diets (Dulphy et al., 1995), it was shown that goats spent more time eating per g DM ingested, a higher number of meals, but much less time ruminating, and the total time spent chewing was slightly lower in goats than in sheep (Table 14.4). The comparison between sheep and cattle is more difficult due to differences in intake and in levels of requirements relative to maintenance. Data, therefore, have to be expressed per metabolic live weight. Compared to small ruminants, cattle eat for a longer time with more meals, and their total mastication time is generally longer. However, due to a higher level of intake per metabolic live weight, mastication time relative to intake is lower in cattle than in sheep (Dulphy et al., 1995).

Despite the large anatomical differences between camelids (camels and llamas) and other ruminants, chewing activities remain comparable (review by Dulphy et al., 1995). However, the number of rumination periods is lower in llamas than in sheep, and rumination occurs more often at night in llamas than in sheep (Lemosquet et al., 1996). These differences are too minor to explain the higher digestibility for a same diet in llamas, which may more likely be related to a higher retention rate of particles in the forestomach.

Differences between species are more pronounced at pasture due to interactions between sward structure and animal body mass (see review by Prache et al., 1998). Allometric relationships between bite area and sward structure show that bite mass increases faster

Table 14.4. Chewing activities in sheep and goats: summary of 20 comparisons. (Adapted from Dulphy *et al.*, 1995.)

	Sheep	Goats
Live weight (kg)	53	44
Intake		
(g DM/day)	1153	952
(g DM/kg LW$^{0.75}$)	56.5	54.1
Organic matter digestibility (%)	67.7	68.9
Time spent eating		
(min/day)	221	248
(min/g DM/kg LW$^{0.75}$)	3.91	4.58
Number of meals per day	6.4	8.1
Time spent ruminating		
(min/day)	500	407
(min/g DM/kg LW$^{0.75}$)	8.85	7.52
Time spent chewing		
(min/g DM/kg LW$^{0.75}$)	12.8	12.1

with sward height for large animals. Consequently, large animals are more handicapped on short swards, where intake per bite increases more slowly with body mass than energy requirements (Illius and Gordon, 1987).

Differences in chewing activities between breeds of the same species have not been studied in stall-fed ruminants under good environmental conditions. The differences may be minor, as only small differences in digestibility and associated digestive processes (retention time of particles, size of particles after mastication) have been observed between zebu and taurine species both adapted to hot climates (Grimaud *et al.*, 1998), between Holstein and Charolais cows (Doreau and Diawara, 2003) and between ovine meat breeds (Ranilla *et al.*, 2000) or dairy breeds (Molina *et al.*, 2001). However, digestive efficiency is better in desert goats than in European goats (Silanikove *et al.*, 1993).

Differences between breeds can be more pronounced at pasture as they interact with vegetation structure. For example, Salers heifers were found to be more efficient at grazing than Limousin heifers as their daily grazing time was shorter due to a higher biting rate (Petit *et al.*, 1995). In harsh environmental conditions, differences exist between local and imported breeds. Voluntary intake, which decreases when temperature and/or humidity increase, is often more limited in imported breeds due to an insufficient thermoregulation (see Morand-Fehr and Doreau, 2001), with the result that time spent

chewing decreases. In addition, the increase in solar radiation is less well tolerated by imported breeds, which look for shade and stop eating for longer times (Mualem *et al.*, 1990).

Age, live weight and nutritional needs

In growing cattle, the increase in live weight is related to a large increase in intake, but time spent eating and ruminating do not vary to a large extent, probably because the size of the organs involved in mastication and digestion (mouth, rumen) increases with live weight (Bae *et al.*, 1983). As a consequence, time spent chewing 1 kg DM decreases with animal age (and weight). However, there is no consequence on particle size reduction, because each bite is more efficient when age increases.

The same mechanisms occur in adult animals. Live weight differences result in differences in voluntary intake but not in time spent chewing. Time spent chewing 1 kg DM decreases when live weight increases (Dulphy *et al.*, 1980), but there is no difference in time spent chewing 1 kg DM/kg metabolic weight.

Ewes and sheep of the same weight, consuming the same diets with the same voluntary intake, demonstrate slight differences in chewing activities (Dulphy *et al.*, 1990). Ewes spend a longer time eating a higher daily number of meals, but ruminate less. Thus, total mastication time and the time spent masticating per gram ingested do not differ. The same authors compared activities in 550 kg cows and 350 kg young bulls. The cows had a higher level of intake and spent more time ruminating but not eating. Thus the time spent masticating per kilogram ingested was higher for cows, but as a consequence of higher intake.

Most chewing activity differences between physiological stages in females can be explained by differences in intake, which varies with nutritional requirements. Between pregnancy and lactation, the rapid increase in intake results either in an increase in time spent eating (Doreau and Rémond, 1982) or in an increase in eating rate. It can be suggested that both mechanisms occur, but that when chewing time has reached a maximal value, cows can only further increase intake by an increase in the rate of eating. However, when feed characteristics do not allow an increase in eating rate, cows increase time spent eating. In all cases, time spent ruminating increases moderately and the time spent masticating per kilogram ingested decreases (Coulon et al., 1987).

Social behaviour and group feeding

Social hierarchy and leadership are two major components of social behaviour that strongly influence the intake and feeding behaviour of cattle fed in groups (Albright, 1993). For example, high-ranking dairy cows spend more time eating and have a higher intake level than their submissive counterparts (Campling and Morgan, 1981). Social factors interact with the nutritional requirements of animals, because of the time required for daily feeding. For dairy cows or growing cattle, social impact is characterized by higher intake and lower feed conversion when fed in groups than when individually fed (Coppock et al., 1972; Phipps et al., 1983; Hasegawa et al., 1997). For animals with low requirement levels, an average decrease in intake of 1.1 kg DM/day was observed for loose-housed lactating Charolais cows, whereas there was no significant difference during pregnancy (Ingrand et al., 1999).

The characteristics and environment of the animal groups (competition for food, density, number of feeders, animal number and characteristics, stability in group composition)

modulate both feeding behaviour and intake. Increasing crowding and decreasing width of the trough or time of access to food induce a decrease in daily duration and in synchronization of eating activity, with an associated increase in eating rate that may help maintain intake levels, even with a high competition for food (Gonyou and Stricklin, 1981; Harb et al., 1985; Corkum et al., 1994). Group size has little effect on intake and feeding behaviour; it only plays a role in dictating the time required for the social hierarchy to be established (Penning et al., 1993; Lawrence, 1994).

The criteria used by farmers to assign animals to different groups do not concern only nutritional aspects, implying that each group can be non-homogeneous as regards nutrient requirements (Ingrand et al., 1999). Without competition for food, there was no incidence of group heterogeneity (dry and lactating beef cows in the same group) on daily intake levels, or on daily time spent eating (Ingrand et al., 2000). When the competition for food increased by reducing access to food with only one trough for two cows, the intake, eating duration, number of long meals and synchronization of eating activity decreased in homogeneous groups (Ingrand et al., 2001; Table 14.5). However, within heterogeneous groups dry cows were able to increase their intake with competition by 1.2 kg DM/day (Ingrand et al., 2001). Thus, interaction between social factors and nutritional demand has to be taken into account when constraints are high in relation to competition for food.

Conclusion

Domestic ruminants, especially sheep and cattle, belong to the most evolved species ingesting and digesting plant fibrous material. This relies on mastication activity and on the complex set of stomachs in which feeds are stored for 10–40 h and submitted to digestion by a symbiotic population of rumen microorganisms. Mastication during ingestion and rumination is essential to feed digestion. It ensures hydration and salivation of the consumed feeds as well as the particle size reduction of digesta which is a prerequisite for their outflow from the reticulo-rumen, reducing the residence time

Table 14.5. Effect of group composition (homogeneous or heterogeneous for nutritional demand) and of competition for food on intake and feeding behaviour in dry cows. (From Ingrand *et al.*, 2001.)

Competition (number of mangers)	Homogeneous			Heterogeneous		
	6	3	*F* test	6	3	*F* test
Intake (kg DM/cow/day)	14.1 ± 1.5	13.4 ± 1.6	***	11.4 ± 1.4	12.6 ± 1.9	*
Eating duration (min/cow/day)	290 ± 33	259 ± 35	***	287 ± 36	274 ± 19	
Rate of eating (g DM/min)	49 ± 6	52 ± 6		40 ± 8	46 ± 8	
Number of short meals[a]	9.1 ± 2.2	10.5 ± 1.9	**	9.2 ± 3.0	10.9 ± 2.9	*
Number of long meals[a]	1.1 ± 0.4	0.8 ± 0.3	**	1.0 ± 0.3	0.8 ± 0.2	
Synchronization of eating[b]	2.7 ± 0.1	2.2 ± 0.2	***	2.1 ± 0.2	1.7 ± 0.1	***

[a]Short meal: <60 min; long meal: ≥60 min.
[b]Number of cows eating at the same time.
* = $p < 0.05$
** = $p < 0.01$
*** = $p < 0.001$

of indigestible material in the rumen. However, the interactions between feed characteristics and mastication activity affecting the release of soluble constituents are less well understood. Modelling biting and mastication activities in relation to digestion enables comprehensive simulation of intake and feeding behaviour in ruminants fed with forages (Sauvant *et al.*, 1996; Baumont *et al.*, 2004b), and should be further developed in the future to improve the prediction of animal response to various types of diets.

Feeding behaviour and mastication have been widely studied in relation to animal, feed and diet characteristics. Feeding high-producing animals with high-energy-density diets, which have low fibre content and particle size, can dramatically reduce mastication time, leading to digestive disorders and health problems. Mastication activity must no longer be considered only a constraint to forage digestion, but as a necessity for the animal. It is part of animal behaviour and should also be taken into account from an ethological point of view in terms of health, social behaviour and welfare.

When given the choice, ruminants generally prefer feeds that can be ingested at high intake rate and that need less mastication to be digested. However, these choices are never absolute, probably partly in relation to the need to maintain sufficient fibre content in the diet. Therefore, when formulating appropriate diets, it is necessary to find the best trade-off between increasing the energy density to meet nutritional demand and production objectives and maintaining the health and welfare of the animals. Individual variations in intake and chewing behaviour are often reported without satisfactory explanation. A better understanding of these variations should contribute to animal selection and improve feeding management with respect to individual needs.

At pasture, ruminants are often faced with constraints on the time needed to sort and sever the best items of the sward. This can limit intake in particular for animals with high nutritional demand that need to spend longer time grazing. Selection of plants with lower shear resistance may contribute to better chewing efficiency at grazing.

References

Albright, J.L. (1993) Nutrition, feeding, and calves. Feeding behaviour of dairy cattle. *Journal of Dairy Science* 76, 485–498.

Arnold, G.W. (1981) Grazing behaviour. In: Morley, F.H.W. (ed.) *World Animal Science B1 Grazing Animals.* Elsevier, Amsterdam, The Netherlands, pp. 79–104.

Bae, D.H., Welch, J.G. and Gilman, B.E. (1983) Mastication and rumination in relation to body size of cattle. *Journal of Dairy Science* 66, 2137–2141.

Baile, C.A. (1975) Control of feed intake in ruminants. In: McDonald, I.W. and Warner, A.C.I. (eds.) *Digestion and metabolism in the ruminants*. University of New England Publishing Unit, pp. 333–350.

Balch, C.C. (1971) Proposal to use the time spent chewing as an index of the extent to which diets for ruminants possess the physical property of fibrousness characteristic of roughages. *British Journal of Nutrition* 26, 383–392.

Baumont, R. and Deswysen, A.G. (1991) Mélange et propulsion du contenu du réticulo-rumen. *Reproduction, Nutrition, Développement* 31, 335–359.

Baumont, R., Brun, J.P. and Dulphy, J.P. (1989) Influence of the nature of hay on its ingestibility and the kinetics of intake during large meals in sheep and cow. In: Jarrige, R. (ed.) *Proceedings of the XVI International Grassland Congress*, French Grassland Society, Versailles, France, pp. 787–788.

Baumont, R., Seguier, N. and Dulphy, J.P. (1990a) Rumen fill, forage palatability and alimentary behaviour in sheep. *Journal of Agricultural Science* 115, 277–284.

Baumont, R., Malbert, C.H. and Ruckebush, Y. (1990b) Mechanical stimulation of rumen fill and alimentary behaviour in sheep. *Animal Production* 50, 123–128.

Baumont, R., Dulphy, J.P. and Jailler, M. (1997) Dynamic of voluntary intake, feeding behaviour and rumen function in sheep fed three contrasting types of hay. *Annales de Zootechnie* 46, 231–244.

Baumont, R., Vimal, T. and Détour, A. (1998) An automatic system to record and analyse kinetics of intake in sheep fed indoor with one or two feeds offered at the same time. In: Gibb, M. (ed.) *Proceedings of the 9th European Intake Workshop*, IGER, North Wyke, Devon, UK, pp. 21–22.

Baumont, R., Chenost, M. and Demarquilly, C. (2004a) Measurement of herbage intake and ingestive behaviour by housed animals. In: Penning, P. (ed.) *Herbage Intake Handbook*. British Grassland Society, Reading, UK, pp. 121–150.

Baumont, R., Cohen-Salmon, D., Prache, S. and Sauvant, D. (2004b) A mechanistic model of intake and grazing behaviour integrating sward architecture and animal decisions. *Animal Feed Science and Technology* 112, 5–28.

Beauchemin, K.A. and Buchanan-Smith, J.G. (1989) Effects of dietary neutral detergent fiber concentration and supplement hay on chewing activities and milk production of dairy cows. *Journal of Dairy Science* 72, 2288–2300.

Beauchemin, K.A., Yang, W.Z. and Rode, L.M. (2003) Effects of particle size of alfalfa-based dairy cows diets on chewing activity, ruminal fermentation and milk production. *Journal of Dairy Science* 86, 630–643.

Bernard, L., Chaise, J.P., Baumont, R. and Poncet, C. (2000) The effect of physical form of orchardgrass hay on the passage of particulate matter through the rumen of sheep. *Journal of Animal Science* 78, 1338–1354.

Bines, J.A. and Davey, W.F. (1970) Voluntary intake, digestion, rate of passage, amount of material in the alimentary tract and behaviour in cows receiving complete diets containing straw and concentrates in different proportions. *British Journal of Nutrition* 24, 1013–1028.

Black, J.L. and Kenney, P.A. (1984) Factors affecting diet selection by sheep. II. Height and density of pasture. *Australian Journal of Agricultural Research* 35, 565–578.

Boudon, A. and Peyraud, J.L. (2001) The release of intracellular constituents from fresh ryegrass (*Lolium perenne* L.) during ingestive mastication in dairy cows: effects of intracellular constituents, season and stage of maturity. *Animal Feed Science and Technology* 93, 229–245.

Boudon, A., Mayne, S., Peyraud, J.L. and Laidlaw, A.S. (2002a) Effects of stage of maturity and chop length on the release of cell contents of fresh ryegrass (*Lolium perenne* L.) during ingestive mastication in steers fed indoors. *Animal Research* 52, 349–365.

Boudon, A., Peyraud, J.L. and Faverdin, P. (2002b) The release of cell contents from fresh ryegrass (*Lolium perenne* L.) during digestion in dairy cows: effects of intracellular constituents, season and stage of maturity. *Animal Feed Science and Technology* 97, 83–102.

Bowman, G.R., Beauchemin, K.A. and Shelford, J.A. (2003) Fibrolytic enzymes and parity effects on feeding behaviour, salivation and ruminal pH of lactating dairy cows. *Journal of Dairy Science* 86, 565–575.

Brun, J.P., Prache, S. and Béchet, G. (1984) A portable device for eating behaviour studies. In: *Proceedings of the 5th Meeting of European Grazing Workshop*, Edinburgh, UK.

Burlison, A.J., Hodgson, J. and Illius, A.W. (1991) Sward canopy structure and the bite dimensions and bite weight of grazing sheep. *Grass and Forage Science* 46, 29–38.

Campion, D.P. and Leek, B.F. (1997) Investigation of a fibre appetite in sheep fed a long fibre-free diet. *Applied Animal Behaviour Science* 52, 79–86.

Campling, R.C. and Morgan, C.A. (1981) Eating behaviour of housed dairy cows: a review. *Dairy Science Abstracts* 43(2), 57–63.

Carter, R.R., Grovum, L. and Greenberg, G.R. (1990) Parotid secretion patterns during meals and their relationships to the tonicity of body fluids and to gastrin and pancreatic polypeptide in sheep. *British Journal of Nutrition* 63, 319–327.

Cassida, K.A. and Stokes, M.R. (1986) Eating and resting salivation in early lactation dairy cows. *Journal of Dairy Science* 69, 1282–1292.

Church, D.C. (1988) *The Ruminant Animal: Digestive Physiology and Nutrition*. Prentice-Hall, Englewood Cliffs, New Jersey.

Cirio, A., Meot, F., Delignette-Muller, M.L. and Boivin, R. (2000) Determination of parotid urea secretion in sheep by means of ultrasonic flow probes and a multifactorial regression analysis. *Journal of Animal Science* 78, 471–476.

Cooper, S.D.B., Kyriazakis, I. and Nolan, J.V. (1995) Diet selection in sheep: the role of the rumen environment in the selection of a diet from two feeds that differ in their energy density. *British Journal of Nutrition* 74, 39–54.

Coppock, C.E., Noller, C.H., Crowl, B.W., McLellon, C.D. and Rhykerd, C.L. (1972) Effect of group versus individual feeding of complete rations on feed intake of lactating cows. *Journal of Dairy Science* 55(3), 325–327.

Corkum, M.J., Bate, L.A., Tennessen, T. and Lirette, A. (1994) Consequences of reduction of number of individual feeders on feeding behaviour and stress level of feedlot steers. *Applied Animal Behaviour Science* 41, 27–35.

Coulon, J.B., Doreau, M., Rémond, B. and Journet, M. (1987) Evolution des activités alimentaires des vaches laitières en début de lactation et liaison avec les quantités d'aliments ingérées. *Reproduction, Nutrition, Development* 27, 67–75.

Delagarde, R., Caudal, J.P. and Peyraud, J.L. (1999) Development of an automatic bitemeter for grazing cattle. *Annales de Zootechnie* 48, 329–340.

Deswysen, A.G. and Ehrlein, H.J. (1981) Silage intake, rumination and pseudo-rumination activity in sheep studied by radiography and jaw movements recordings. *British Journal of Nutrition* 46, 327–335.

Deswysen, A.G., Dutilleul, P.A. and Ellis, W.C. (1989) Quantitative analysis of nycterohemeral eating and ruminating patterns in heifers with different voluntary intakes and effects of monensin. *Journal of Animal Science* 67, 2751–2761.

Doreau, M. and Diawara, A. (2003) Effect of level of intake on digestion in cows: influence of animal genotype and nature of hay. *Livestock Production Sciences* 81, 35–45.

Doreau, M. and Rémond, B. (1982) Comportement alimentaire et utilisation digestive d'une ration de composition constante chez la vache laitière en fin de gestation et début de lactation. *Reproduction, Nutrition, Development* 22, 307–324.

Doreau, M., Lomri, A.I. and Adingra, K. (1986) Influence d'une faible niveau d'ingestion sur la digestion et le comportement alimentaire chez la vache recevant un régime très digestible. *Reproduction, Nutrition, Development* 26, 329–330.

Doreau, M., Michalet-Doreau, B. and Béchet, G. (2004) Effect of underfeeding on digestion in cows. Interaction with rumen degradable N supply. *Livestock Production Sciences* 88, 33–41.

Dougherty, C.T., Bradley, N.W., Cornelius, P.L. and Lauriault, L.M. (1989) Ingestive behaviour of beef cattle offered different forms of lucerne (*Medicago sativa* L.). *Grass and Forage Science* 44, 335–342.

Dulphy, J.P. and Faverdin, P. (1987) L'ingestion alimentaire chez le ruminants: modalités et phénomènes associés. *Reproduction, Nutrition, Development* 27, 129.

Dulphy, J.P., Rémond, B. and Thériez, M. (1980) Ingestive behaviour and related activities in ruminants. In: Ruckebusch, Y. and Thivend, P. (eds) *Digestive Physiology and Metabolism in Ruminants*. MTP Ltd Press, Lancaster, UK, pp. 103–112.

Dulphy, J.P., Carle, B. and Demarquilly, C. (1990) Quantités ingérées et activités alimentaires comparées des ovins, bovins et caprins recevant des fourrages conservés avec ou sans aliment concentré. I. Etude descriptive. *Annales de Zootechnie* 39, 95–111.

Dulphy, J.P., Balch, C.C. and Doreau, M. (1995) Adaptation des espèces domestiques à la digestion des aliments lignocellulosiques. In: Jarrige, R., Ruckebusch, Y., Demarquilly, C., Farce, M.H. and Journet, M. (eds) *Nutrition des Ruminants Domestiques*. INRA Editions, Paris, France, pp. 759–803.

Faverdin, P. (1985) Regulation de l'ingestion des vaches laitières en début de lactation: etude du rôle de l'insuline. PhD thesis, Institut National Agronomique, Paris, France.

Faverdin, P., Baumont, R. and Ingvartsen, K.L. (1995) Control and prediction of feed intake in ruminants. In: Journet, M., Grenet, E., Farce, M.-H., Thériez, M. and Demarquilly, C. (eds) *Recent Developments in the Nutrition of Herbivores*. INRA Editions, Paris, France, pp. 95–120.

Fernandez, I., Martin, C., Champion, M. and Michalet-Doreau, B. (2004) Effect of corn hybrid and chop length of whole-plant corn silage on digestion and intake by dairy cows. *Journal of Dairy Science* 87, 1298–1309.

Forbes, J.M. (1995) *Voluntary Food Intake and Diet Selection in Farm Animals*. CAB International, Wallingford, UK.

Forbes, J.M., Jackson, D.A., Johnson, C.L., Stockill, P. and Boyle, B.S. (1986) A method for the automatic monitoring of food intake and feeding behaviour of individual cattle kept in a group. *Research and Development in Agriculture* 3, 175–180.

Garcia, F., Carrère, P., Soussana, J.F. and Baumont, R. (2003) The ability of sheep at different stocking rates to maintain the quality and quantity of their diet during the grazing season. *Journal of Agricultural Science* 140, 113–124.

Gonyou, H.W. and Stricklin, W.R. (1981) Eating behaviour of beef cattle groups fed from a single stall or trough. *Applied Animal Ethology* 7, 123–133.

Grenet, E. (1989) A comparison of the digestion and reduction in particle size of lucerne hay and Italian ryegrass hay in the ovine digestive tract. *British Journal of Nutrition* 62, 493.

Griffiths, W.M. and Gordon, I.J. (2003) Sward structural resistance and biting effort in grazing ruminants. *Animal Research* 52, 145–160.

Grimaud, P. (1999) Effets de la sous-alimentation énergétique sur la digestion ruminale chez les bovins et les ovins. PhD thesis, Université de Rennes, Rennes, France.

Grimaud, P., Richard, D., Kanwé, A., Durier, C. and Doreau, M. (1998) Effect of undernutrition and refeeding on digestion in *Bos taurus* and *Bos indicus* in a tropical environment. *Animal Science* 67, 49–58.

Grimaud, P., Richard, D., Vergeron, M.P., Guilleret, J.R. and Doreau, M. (1999) Effect of drastic undernutrition on digestion in zebu cattle receiving a diet based on rice straw. *Journal of Dairy Science* 82, 974–981.

Harb, M.Y., Reynolds, V.S. and Campling, R.C. (1985) Eating behaviour, social dominance and voluntary intake of silage in group-fed milking cattle. *Grass and Forage Science* 40, 113–118.

Hasegawa, N., Nishiwaki, A., Sugawara, K. and Ito, I. (1997) The effects of social exchange between two groups of lactating primiparous heifers on milk production, dominance order, behavior and adrenocortical response. *Applied Animal Behaviour Science* 51, 15–27.

Heinrichs, A.J. and Conrad, H.R. (1987) Measuring feed intake patterns and meal size of lactating dairy cows. *Journal of Dairy Science* 70, 705–711.

Horiguchi, K.I. and Takahashi, T. (2002) Effect of ruminal dosing of mechanical stimulating brush on rumination time, ruminal passage rate and rumen fermentation status in Holstein steers fed a concentrate diet. *Animal Science* 73, 41–46.

Illius, A.W. and Gordon, I.J. (1987) The allometry of food intake in grazing ruminants. *Journal of Animal Ecology* 56, 989–999.

Ingrand, S., Vimal, T., Fléchet, J., Agabriel, J., Brun, J.P., Lassalas, J. and Dedieu, B. (1998) A free-access system for the long-term monitoring of individual intake of beef cows kept in group. In: Gibb, M. (ed.) *Proceedings of the 9th European Intake Workshop*, IGER, North Wyke, Devon, UK, pp. 17–20.

Ingrand, S., Agabriel, J., Lassalas, J. and Dedieu, B. (1999) How group feeding influences intake level of hay and feeding behaviour of beef cows. *Annales de Zootechnie* 48, 435–445.

Ingrand, S., Agabriel, J., Lassalas, J. and Dedieu, B. (2000) Effects of within-group homogeneity of physiological state on individual feeding behaviour of loose-housed Charolais cows. *Annales de Zootechnie* 49, 15–27.

Ingrand, S., Agabriel, J., Dedieu, B. and Lassalas, J. (2001) Effects of reducing access to food on intake and feeding behaviour of loose-housed dry Charolais cows. *Annales de Zootechnie* 50, 145–148.

Jarrige, R., Dulphy, J.P., Faverdin, P., Baumont, R. and Demarquilly, C. (1995) Activités d'ingestion et de rumination. In: Jarrige, R., Ruckebusch, Y., Demarquilly, C., Farce, M.H. and Journet, M. (eds) *Nutrition des Ruminants Domestiques*. INRA Editions, Paris, France, pp. 123–182.

Kononoff, P.J., Heinrichs, A.J. and Lehman, H.A. (2003) The effect of corn silage particle size on eating behaviour, chewing activities, and rumen fermentation in lactating dairy cows. *Journal of Dairy Science* 86, 3343–3353.

Krause, K.M., Combs, D.K. and Beauchemin, K.A. (2002) Effects of forage particle size and grain fermentability in midlactation cows. II. Ruminal pH and chewing activity. *Journal of Dairy Science* 85, 1947–1957.

Laca, E.A. and WallisDeVries, M.F. (2000) Acoustic measurement on intake and grazing behaviour of cattle. *Grass and Forage Science* 55, 97–104.

Laca, E.A., Ungar, E.D., Seligman, N. and Demment, M.W. (1992) Effects of sward height and bulk density on bite dimensions of cattle grazing homogeneous swards. *Grass and Forage Science* 47, 91–102.

Lawrence, N.G. (1994) Beef cattle housing. In: Wathes, C.M. and Charles, D.R. (eds) *Livestock Housing*. CAB International, Wallingford, UK, pp. 339–357.

Lemosquet, S., Dardillat, C., Jailler, M. and Dulphy, J.P. (1996) Voluntary intake and gastric digestion of two hays by llamas and sheep: influence of concentrate supplementation. *Journal of Agricultural Science* 127, 539–548.

Lofgreen, G.P., Meyer, J.H. and Hull, J.L. (1957) Behaviour patterns of sheep and cattle fed pasture or silage. *Journal of Animal Science* 16, 773–780.

Luginbuhl, J.M., Pond, K.R., Russ, J.C. and Burns, J.C. (1987) A simple electronic device and computer interface system for monitoring chewing behavior of stall-fed ruminant animals. *Journal of Dairy Science* 70, 1307–1312.

Luginbuhl, J.M., Pond, K.R., Burns, J.C. and Russ, J.C. (1989) Eating and ruminating behavior of steers fed coastal bermudagrass hay at four levels. *Journal of Animal Science* 67, 3410–3418.

MacKinnon, B.W., Easton, H.S., Barry, T.N. and Sedcole, J.R. (1988) The effect of reduced leaf shear strength on the nutritive value of perennial ryegrass. *Journal of Agricultural Science* 111, 469–474.

McLeod, M.N. and Minson, D.J. (1988) Large particle breakdown by cattle eating ryegrass and alfalfa. *Journal of Animal Science* 66, 992–999.

Maekawa, M., Beauchemin, K.A. and Christensen, D.A. (2002) Chewing activity, saliva production, and ruminal pH of primiparous and multiparous lactating dairy cows. *Journal of Dairy Science* 85, 1176–1182.

Malbert, C.H. and Baumont, R. (1989) The effects of intake of lucerne (*Medicago sativa* L.) and orchard grass (*Dactylis glomerata* L.) hay on the motility of the forestomach and digesta flow at the abomaso-duodenal junction of the sheep. *British Journal of Nutrition* 61, 699–714.

Mattiello, S., Canali, E., Ferrante, V., Caniatti, M., Gottardo, F., Cozzi, G., Andrighetto, I. and Verga, M. (2002) The provision of solid feeds to veal calves. II. Behaviour, physiology, and abomasal damage. *Journal of Animal Science* 80, 367–375.

Meot, F., Cirio, A. and Boivin, R. (1997) Parotid secretion daily patterns and measurement with ultrasonic flow probes in conscious sheep. *Experimental Physiology* 82, 905–923.

Mertens, D.R. (1997) Creating a system for meeting the fibre requirements of dairy cows. *Journal of Dairy Science* 80, 1463–1481.

Metz, J.H.M. (1975) Time patterns of feeding and rumination in domestic cattle. *Medelingen Landbouwhogeschool Wageningen Nederland* 75, 1–66.

Molina, E., Ferret, A., Caja, G., Calsamiglia, S., Such, X. and Gasa, J. (2001) Comparison of voluntary food intake, apparent digestibility, digesta kinetics and digestive tract content in Manchega and Lacaune dairy sheep in late pregnancy and early and midlactation. *Animal Science* 72, 209–221.

Morand-Fehr, P. and Doreau, M. (2001) Ingestion et digestion chez les ruminants soumis à un stress de chaleur. *INRA Productions Animales* 14, 15–27.

Mualem, R., Choshniak, I. and Shkolnik, A. (1990) Environmental heat load, bionergetics and water economy in two breeds of goats, the Mamber goat versus the desert Bedouin goat. *World Review of Animal Production* 25, 91–95.

Nicol, A.M. and Sharafeldin, M.A. (1975) Observations on the behaviour of single suckled calves from birth to 120 days. *Proceedings of the New Zealand Society of Animal Production* 35, 221–230.

Norgaard, P. (1983) Physical structure. In: Ostergaard, V. and Neimann-Sorensen, A. (eds) *Optimum Feeding of the Dairy Cow*. Statens Husdyrburgsforsog, Copenhagen, Denmark, Chapter 36.

Okine, E.K. and Mathison, G.W. (1991) Effects of feed intake on particle distribution, passage of digesta and extent of digestion in the gastrointestinal tract of cattle. *Journal of Animal Science* 69, 3435–3445.

Penning, P.D. (1983) A technique to record automatically some aspects of grazing and ruminating behaviour in sheep. *Grass and Forage Science* 38, 89–96.

Penning, P.D. (2004) Animal-based techniques for estimating herbage technique. In: Penning, P. (ed.) *Herbage Intake Handbook*. British Grassland Society, Reading, UK, pp. 53–94.

Penning, P.D. and Rutter, S.M. (2004) Ingestive behaviour. In: Penning, P. (ed.) *Herbage Intake Handbook*. British Grassland Society, Reading, UK, pp. 151–176.

Penning, P.D., Steel, G.L. and Johnson, R.H. (1984) Further development and use of an automatic recording system in sheep grazing studies. *Grass and Forage Science* 39, 345–351.

Penning, P.D., Parsons, A.J., Orr, R.J. and Treacher, T.T. (1991) Intake and behaviour responses by sheep to changes in sward characteristics under continuous stocking. *Grass and Forage Science* 46, 15–28.

Penning, P.D., Parsons, A.J., Newman, J.A., Orr, R.J. and Harvey, A. (1993) The effects of group size on grazing time in sheep. *Applied Animal Behaviour Science* 37, 101–109.

Petit, M. (1972) Emploi du temps des troupeaux de vaches-mères et de leurs veaux sur les pâturages d'altitude de l'Aubrac. *Annales de Zootechnie* 21, 5–27.

Petit, M., Garel, J.P., D'Hour, P. and Agabriel, J. (1995) The use of forages by the beef cows herd. In: Journet, M., Grenet, E., Farce, M.-H., Thériez, M. and Demarquilly, C. (eds) *Recent Developments in the Nutrition of Herbivores*. INRA Editions, Paris, France, pp. 473–496.

Phipps, R.H., Bines, J.A. and Cooper, A. (1983) A preliminary study to compare individual feeding through calan electronic feeding gates to group feeding. *Animal Production* 36, 544.

Poppi, D.P., Norton, B.W., Minson, D.J. and Hendricksen, R.E. (1980) The validity of the critical size theory for particles leaving the rumen. *Journal of Agricultural Science* 94, 275.

Prache, S. (1997) Intake rate, intake per bite and time per bite of lactating ewes on vegetative and reproductive swards. *Applied Animal Behaviour Science* 52, 53–54.

Prache, S., Gordon, I.J. and Rook, A.J. (1998) Foraging behaviour and diet selection in domestic herbivores. *Annales de Zootechnie* 47, 335–346.

Ranilla, M.J., Carro, M.D., Giraldez, F.J., Mantecon, A.R. and Gonzalez, J.S. (2000) Comparison of rumen fermentation patterns and *in situ* degradation of grazed herbage in Churra and Merino sheep. *Livestock Production Sciences* 62, 193–204.

Redbo, I. (1990) Changes in duration and frequency of stereotypies and their adjoining behaviours in heifers, before, during and after the grazing period. *Applied Animal Behaviour Science* 26, 57–67.

Ruckebusch, Y. (1963) Recherches sur la régulation centrale du comportement alimentaire chez les ruminants. Thèse de Doctorat en Sciences Naturelles, Université de Lyon, France.

Rutter, S.M. (2000) Graze: a program to analyse recordings of the jaw movements of ruminants. *Behaviour Research Methods, Instruments and Computers* 32, 86–92.

Rutter, S.M., Champion, R.A. and Penning, P.D. (1997) An automatic system to record foraging behaviour in free-ranging ruminants. *Applied Animal Behaviour Science* 54, 185–195.

Sauvant, D. (2000) Granulométrie des rations et nutrition du ruminant. *INRA Productions Animales* 13, 99–108.

Sauvant, D., Baumont, R. and Faverdin, P. (1996) Development of a mechanistic model of intake and chewing activities of sheep. *Journal of Animal Science* 74, 2785–2802.

Silanikove, N. and Tadmor, A. (1989) Rumen volume, saliva flow rate and systemic fluid homeostasis in dehydrated cattle. *The American Physiological Society* R809–R815.

Silanikove, N., Tagari, H. and Shkolnik, A. (1993) Comparison of rate of passage, fermentation rate and efficiency of digestion of high fiber diet in desert Bedouin goats compared to Swiss Saanen goats. *Small Ruminant Research* 12, 45–60.

Stricklin, W.R., Wilson, L.L. and Graves, H.B. (1976) Feeding behaviour of Angus and Charolais-Angus cows during summer and winter. *Journal of Animal Science* 43, 721–732.

Sudweeks, E.M., Ely, O., Mertens, D.R. and Sisk, L.R. (1981) Assessing minimum amount and form of roughages in ruminant diets: roughage value index system. *Journal of Animal Science* 53, 1406–1411.

Suzuki, S., Fujita, H. and Shinde, Y. (1969) Change in the rate of eating during a meal and the effect of the interval between meals on the rate at which cows eat roughages. *Animal Production* 11, 29–41.

Tolkamp, B.J. and Kyriazakis, I. (1999) To split behaviour into bouts, log-transform the intervals. *Animal Behaviour* 57, 807–817.

Ulyatt, M.J., Dellow, D.W., John, A., Reid, C.S.W. and Waghorn, G.C. (1986) Contribution of chewing during eating and rumination to the clearance of digesta from the reticulo-rumen. In: Milligan, L.P., Grovum, W.L. and Dobson, A. (eds) *Control of Digestion and Metabolism in Ruminants*. Prentice-Hall, Englewood Cliffs, New Jersey, pp. 498–515.

Ungar, E.D., Genezi, A. and Demment, M.W. (1991) Bite dimensions and herbage intake by cattle grazing short hand-constructed sward. *Agronomy Journal* 83, 973–978.

van Os, M., Dulphy, J.P. and Baumont, R. (1995) The effect of protein degradation products in grass silages on feed intake and intake behaviour in sheep. *British Journal of Nutrition* 73, 51–64.

Veissier, I., Chazal, P., Pradel, Ph. and Le Neindre, P. (1997) Providing social contacts and objects for nibbling moderates reactivity and oral behaviours in veal calves. *Journal of Animal Science* 75, 356–365.

Veissier, I., Ramirez, d.l.F.A.R. and Pradel, Ph. (1998) Non-nutritive oral activities and stress responses of veal calves in relation to feeding and housing conditions. *Applied Animal Behaviour Science* 57, 35–49.

Veissier, I., Sarignac, C. and Capdeville, J. (1999) Les méthodes d'appréciation du bien-être des animaux d'élevage. *INRA Productions Animales* 12, 113–121.

Webster, A.J.F., Saville, C., Church, B.M., Gnanasakthy, A. and Moss, R. (1985) The effect of different rearing systems on the development of calf behaviour. *British Veterinary Journal* 141, 249–264.

Welch, J.G. and Hooper, A.P. (1988) Ingestion of feed and water. In: Church, D.C. (ed.) *The Ruminant Animal, Digestive Physiology and Nutrition*. Prentice-Hall, Englewood Cliffs, New Jersey, pp. 108–116.

Welch, J.G. and Smith, A.M. (1968) Influence of fasting on rumination activity in sheep. *Journal of Animal Science* 27, 1734–1737.

Welch, J.G. and Smith, A.M. (1974) Physical stimulation of rumination activity. *Journal of Animal Science* 33, 1118–1123.

Wiepkema, P.R. (1983) On the significance of ethological criteria for the assessment of animal welfare. In: Smidt, D. (ed.) *Indicators Relevant to Farm Animal Welfare*. Martinus Nijhoff Publishers, Boston/The Hague/Dordrecht/Lancaster for the Commission of the European Communities, pp. 71–79.

Wiepkema, P.R., Van Hellemond, K.K., Roessingh, P. and Romberg, H. (1987) Behaviour and abomasal damage in individual veal calves. *Applied Animal Behaviour Science* 18, 257–268.

15 Food in 3D: How Ruminant Livestock Interact with Sown Sward Architecture at the Bite Scale

I.J. Gordon[1] and M. Benvenutti[2]

[1]CSIRO – Davies Laboratory, PMB PO Aitkenvale, Qld 4814, Australia; [2]INTA EEA Cerro Azul, C.C. 6 (3313) Cerro Azul, Misiones, Argentina

Introduction

Domestic ruminant livestock production occupies over 25% of the world's land area and provide the human population with food (meat, milk) and fibre (FAO, 2003). The majority of the food and fibre supplied to the human population comes from four domestic species (cattle: *Bos taurus* and *Bos indicus*; sheep: *Ovis aries*; goats: *Capra hircus*) although in certain regions other species have been domesticated (llama: *Llama glama* and alpaca: *Llama pacos* in South America; water buffalo: *Bubalus bubalus* in South-east Asia). In order to meet their demands for maintenance, production and reproduction domestic ruminants need to glean nutrients from the vegetation resources on offer. There are three main limitations to the ability of the animal to meet its demands: metabolic, digestive and ingestive. Metabolic constraints only operate where the forage provides a high rate of nutrient release in the rumen, and this can cause metabolic overload, leading to a reduction in intake (Illius and Jessop, 1996). Digestive constraints operate where the quality of the forage is low, resulting in a reduction in the rate at which undigested material can be passed from the rumen, to make way for newly ingested, more digestible material (Illius and Gordon, 1991; Allen, 1996). Finally, ingestive constraints operate where the availability of forage is such that the

animal cannot ingest sufficient material during its daily grazing cycle (Illius and Gordon, 1987). In general, grazing ruminants' nutrient intake is limited by either digestive (when there is a bulk of poor quality material) or ingestive (when there is low biomass of material of sufficient quality) constraints. Metabolic constraints only really apply to animals fed indoors or to high yielding genotypes, e.g. dairy cows, on sown pastures (Illius and Jessop, 1996).

The majority of domestic ruminants gain their nutrition from pastures that have been modified through improvement in pasture species, water availability and fertilizer inputs (Buringh and Dudal, 1987). In these circumstances the livestock manager attempts to change the availability of forage (in space and time) to meet the nutritional demands of ruminants by managing the interaction between the plant and the animal so as maximize the quantity and quality of the material the animal ingests (McCall and Sheath, 1993). This requires a detailed understanding of the ways in which the animal selects plant material from that which is on offer. Over the last 50 years, a great deal of research has been conducted to provide insights into the interactions between the herbivore and sown pastures, in both tropical and temperate pastures (see reviews by Demment and Laca, 1993; Gordon and Lascano, 1993; Demment *et al.*, 1995; Hodgson and Illius, 1998; Prache and

Peyraud, 2001; Sollenberger and Burns, 2001); however, to date, there have been no reviews of how both the vertical and horizontal distribution of plant material in the sward affects the intake and diet quality of the grazing ruminant. In this chapter, we will describe the small-scale, short-term behavioural responses (bite size and bite rate) of domestic ruminants to the vertical (e.g. height, leaf/stem ratios) and horizontal (e.g. sward density, spacing of leaves and stems, cover of a species and its degree of aggregation) availability of plant material in sown swards, since it is known that these behavioural responses are critical in determining the longer-term intake and diet selection of the grazing animal (Demment and Laca, 1993; Illius and Gordon, 1999). We do understand that behavioural responses can occur at a range of temporal and spatial scales (Senft et al., 1987); however, many of these are not relevant to the domestic ruminant and are beyond the scope of this review (but see review by Agreil, et al., Chapter 17, this volume). We will concentrate our review on research conducted on sown tropical and temperate pastures; however, where research has been conducted on extensive pastures, which highlights a hitherto unanswered question in sown swards, we will refer to that research. Firstly, we will describe the distribution of vegetation in sown swards. We will then consider how the ruminant interacts with that distribution of vegetation in selecting its diet. Following that, we will review models of intake and diet selection of ruminants in sown swards, highlighting issues raised in previous sections of the review that have not been incorporated into the models that are currently available. Finally, we will draw together the research in the light of provision of guidelines for managing sown swards so as best to meet the nutritional demands of domestic ruminants.

Sward Description

For many years farmers have attempted to manage sown swards so that swards present the best opportunity to meet the nutritional demands of livestock (Hodgson, 1990). Farm-

ers have used improved plant species (e.g. perennial ryegrass, *Lolium perenne* and white clover, *Trifolium repens*), fertilizer, irrigation and mowing to increase the amount of green leaf relative to stem and dead material, and use 'put and take' to maintain the sward at a height that most effectively maximizes the intake of the grazing animal. This suggests that sward structure is the primary means by which managers can influence animal production (Hodgson, 1990; Hodgson and Illius, 1998). In this section we will describe the distribution of plant material in sown swards, in the vertical and horizontal dimensions. This will form the platform for describing how the grazing animal interacts with plant material when selecting its diet and achieving its intake to satisfy its nutritional demands.

It goes without saying that a sward presents itself to the animal in three dimensions (3D). In the vertical dimension the sward has height and the distribution of sward components (lamina, stem and pseudostem) or species in different horizons of the sward, and the horizons themselves vary in density of plant components or species. In the horizontal dimension, there will be variation in tiller density, in plant component and species composition, in height and mass. If the vertical and horizontal dimensions are interlinked, the sward varies in bulk density and mass.

Hodgson (1985) was the first to provide a clear graphical indication of the structure of swards in the vertical dimension (Fig. 15.1).

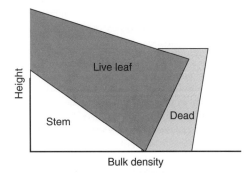

Fig. 15.1. Schematic diagram showing the relationship between bulk density and sward height for a sown sward. It also shows the distribution of different plant components in the sward. (After Hodgson, 1985.)

This demonstrates that, in a monoculture rye-grass (*Lolium perenne*) sward for instance, most of the biomass is in the lower part of the sward, which is primarily composed of stem and dead material. In the upper part of the sward the biomass is lower but it is composed primarily of live leaf material. Further analyses have demonstrated that sward management will affect the relative proportions of sward components and also the distribution of different species in the different horizons (Mitchley and Willems, 1995; Liira and Zobel, 2000).

Gibb and Ridout (1988) were the first to assess the spatial distribution of biomass in the horizontal dimension in mixed ryegrass–clover swards. They showed that cattle-grazed swards had a two-peaked distribution of sward height with parts of the sward being tall and others being short. This may be because of the cattle avoiding grazing near faecal deposits, thereby creating a relatively tall, ungrazed component of the sward and a shorter grazed component (Forbes and Hodgson, 1985), although it could also relate to differences in growth rate between areas that had received urine and those that had not (Matches, 1992). Similar results have been shown for herbage mass in a tropical, bahia grass (*Paspalum notatum*), sward-grazed by cattle, with the variance in sward mass across the sward increasing with the length of the grazing period and higher levels of sward utilization (Hirata, 2000). Edwards *et al.* (1996) measured the spatial distribution of clover in a mixed ryegrass–clover sward under cutting and grazing, and found that white clover occurred in patches of about 1 m in size finely intermixed with grass. These patches were separated by gaps of about 4 m. Individuals within plant species are generally aggregated, and within sown and natural swards there tend to be species that aggregate and those that do not (e.g. McNaughton, 1978; Purves and Law, 2002). In some instances this has been interpreted as more preferred species seeking refuge within less preferred species (plant defence guilds; McNaughton, 1978, see also Milchunas and Noy-Meir, 2002) as a way of avoiding being grazed.

To date there have been few attempts to describe tropical swards in the same way as for temperate swards (O'Reagain, 1993; Torales *et al.*, 2000). Tropical swards would be expected to differ from temperate sown swards in the vertical and horizontal distribution of material as a consequence of the differences in growth form between tropical and temperate grasses, and tropical and temperate legumes (O'Reagain, 1993).

Diet Selection and Intake

The fulcrum of interaction of the grazing ruminant with the vegetation is the mouth (Gordon and Lascano, 1993) (Fig. 15.2). The grazing process can be described as the bite and the rate at which the bites are taken. The biting process occurs when the animal severs the vegetation between its incisors, on the lower jaw, and the pad on the upper jaw (it should be noted that ruminants do not have incisors on the upper jaw, probably because these would hinder the rumination process). When the animal takes a bite the amount of material removed from the sward will be a function of the volume of material gathered into the mouth. This will be limited by the bite area, and the depth to which the animal inserts its mouth into the sward (bite depth). The rate at which the animal bites is limited by three processes: one based on the interaction between the sward and the animal (prehension), and the other two are animal-based. The prehension process is the rate at which the animal selects the preferred components from the sward, which is reliant upon its selection criteria and the ease with which the preferred components of the sward can be prehended (Gordon and Illius, 1988; Gordon and Lascano, 1993). The solely animal-based processes are the rate at which the animal can process the material contained within the mouth (i.e. chew), and swallow that material to make way for new material to be bitten (Spalinger and Hobbs, 1992). We will not address the chewing and swallowing processes as this is beyond the scope of this review (interested readers should refer to the review by Perez-Barberia and Gordon, 1998).

In the next three subsections we will summarize research that has been conducted to assess the effects of the vertical and horizontal distribution of vegetation in the sward and their interaction with the diet selection and intake of

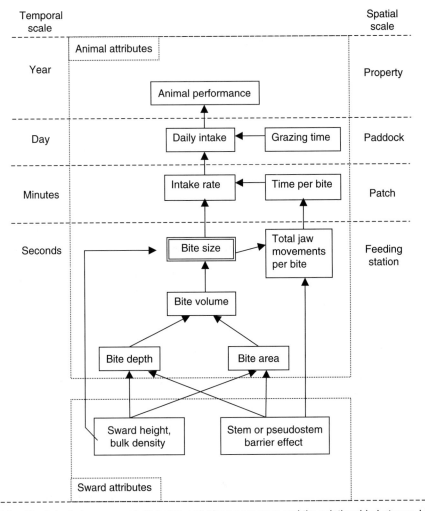

Fig. 15.2. The links between sward attributes and bite parameters and the relationship between bite parameters, intake and animal performance.

the grazing ruminant. We concentrate our review at the scale of the bite taken within a feeding station (Senft *et al.*, 1987). Since this is an extremely difficult component of grazing behaviour to study in free-ranging ruminants, the majority of the studies referred to have been conducted in highly controlled conditions where the swards are presented within confined conditions (see Gordon, 1995 for a review of these methodologies). Whilst there are issues concerning the applicability of the results to field conditions, these studies have given researchers a great deal of insight into the interaction between the grazing animal and the sward.

Vertical dimension

Whilst a number of experimental studies have addressed the effects of vertical variation (Black and Kenney, 1984; Ungar *et al.*, 1991; Laca *et al.*, 1992; Flores *et al.*, 1993; Ginnett *et al.*, 1999; Bergman *et al.*, 2000) in the sward structure on ingestive behaviour and intake rate in temperate pastures, very few have addressed this variation in tropical pastures (Cangiano *et al.*, 2002). However, it is likely that the plant–herbivore interaction principles generated in temperate swards can be applied to tropical pastures and vice versa;

therefore, we will start by describing the findings in temperate swards and then discuss findings from tropical swards and highlight the possible differences between the two. In addition, we will divide this subsection into research findings from vegetative and reproductive swards because they pose different challenges to the animal.

Vegetative swards

In the vertical dimension a vegetative sward can vary in height, density and relative lengths of pseudostem and lamina. Various experimental studies have addressed the effects of sward height (Black and Kenney, 1984; Ungar et al., 1991; Laca et al., 1992) on ingestive behaviour and intake rate in vegetative temperate swards. Generally, bite mass and intake rate increase with sward height. Sward height affects bite size, and thus intake rate, mainly through its effect on bite depth; the taller the sward, the deeper the bite (Fig. 15.3). In studies with artificial microswards, the bite depth is almost a constant proportion, near to 50%, of the pasture height for both sheep and cattle (Burlison et al., 1991; Ungar et al., 1991; Laca et al., 1992; Flores et al., 1993; Ginnett et al., 1999). This was

later confirmed by studies with seedling microswards (Ungar and Ravid, 1999; Cangiano et al., 2002; Griffiths et al., 2003). Given the importance of the effect of bite depth on bite size and consequently intake rate, several hypotheses have been proposed to explain the constancy of the ratio of bite depth to sward height (see Griffith and Gordon, 2003 for a review). As for temperate pastures, in tropical swards the bite depth of cattle seems to be almost a constant fraction (≈ 50%) of the sward height (Fig. 15.3), although it is likely that the relationship between bite depth and sward height will vary as a consequence of differences in the force required to sever swards of different densities and proportion of pseudostem relative to vegetative components in the sward strata (Griffiths et al., 2003).

Pseudostems do not appear to limit bite depth in tropical species (Cangiano et al., 2002); however, in temperate species, the pseudostem has been suggested to form a physical barrier to bite depth (Barthram, 1981; Arias et al., 1990; Dougherty et al., 1992; but see Flores et al., 1993). This possible barrier effect depends upon pseudostem characteristics and the species' ruminant foraging on the sward. Wright and Illius (1995) found a marked difference in the shear and tensile strength of

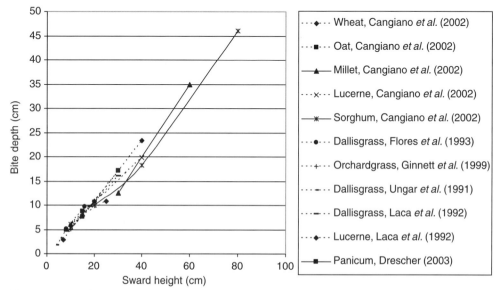

Fig. 15.3. Relationship between bite depth and sward height for cattle grazing vegetative tropical (——) and temperate (------) swards.

pseudostems in different forage species. Pseudostems of lower tensile strength are less likely to form a barrier to grazing than those of higher tensile strength. Also, because of their larger size, cattle are less likely to find pseudostems a barrier to bite depth than do sheep, goats and deer (Barthram, 1981; Arias et al., 1990; Mitchell et al., 1991). However, experimental quantification of the relationship between the barrier effect of pseudostems and their physical properties is still required (Prache and Peyraud, 2001).

Sward height affects not only bite depth but also bite area; the taller the pasture, the bigger the bite area (Ungar et al., 1991; Laca et al., 1992; Flores et al., 1993). In tall swards animals such as cattle can fully exploit the benefits of the tongue sweep movements to gather the forage, and thus they can achieve the maximum bite area. In contrast, on short swards the tongue fails to gather the forage, and thus bite area is reduced (Illius, 1997). This limiting effect of sward height on bite area is likely to be less severe for sheep and goats that do not use their tongues to gather herbage into the mouth before severing it with the incisors.

Since bite size increases with sward height, the time required for chewing each bite increases also and there is, therefore, increased competition between prehending a new bite and chewing the bite just taken before swallowing. Consequently, bite rate falls with increasing sward height (Laca et al., 1994). When the oral processing capacity of the animal is saturated, intake rate becomes asymptotic. In contrast, as sward height decreases, bite rate increases in attempt to compensate the decreasing bite size, but this compensating capacity is usually only partial. Therefore, in vegetative swards, bite rate is negatively related to bite size in a curvilinear fashion (Laca et al., 1994).

Reproductive swards

In the vertical dimension, a reproductive sward varies in the proportion and relative length of the plant's different parts. Therefore, the herbivore faces a more complex situation in selecting its diet from reproductive swards where leaves are in close contact with stems, decreasing the accessibility of leaves, as compared to the situation in vegetative swards. Leaves and stems are usually unevenly distributed within a grass plant (Orians and Jones, 2001), and within-plant distribution of these plant parts varies between species with different growth forms (O'Reagain, 1993) and between plants that have suffered different grazing histories (Ruyle et al., 1987). Stems may represent a physical barrier to bite penetration into the sward. Flores et al. (1993) and Ginnett et al. (1999) studied the effect of the presence of stems in the lower strata of the sward on ingestive behaviour and found that the stem stratum stopped the animal penetrating deeper into the canopy. Therefore, bite depth was not related to overall sward height but to the length of the lamina. Intuitively, it would be expected that the degree to which stems form a barrier to bite penetration into the sward would depend upon the physical properties of the stem; tender stems should not represent a limitation to bite depth as compared to mature stems. However, the effect of the physical properties of stems on grazing behaviour has not yet been studied and we will discuss this aspect of bite selection later. In the reproductive stage, tropical pastures present more extreme contrasts in plant morphology and maturity than do temperate swards (Hodgson et al., 1994); therefore, it would be expected that when grazing, tropical swards herbivores would be under greater pressure to select leaves as opposed to stems than in temperate swards, if their diet quality is not to be compromised.

Horizontal dimension

As for the vertical distribution of plant components, a number of experimental studies have addressed the effects of horizontal (Black and Kenney, 1984; Ungar et al., 1991; Laca et al., 1992) variation in the sward structure on ingestive behaviour and intake rate in temperate pastures, and only a few have addressed this interaction for tropical pastures (Drescher, 2003). As for temperate pastures, in the reproductive stage, tropical species present an extreme contrast in terms of plant morphology and level of maturity (Hodgson et al., 1994); therefore, tropical pastures would be expected to represent a greater challenge for herbivores in terms of avoiding stems in order to graze leaves than do temperate swards.

Reproductive swards

The main sward factors that vary in the horizontal dimension in reproductive swards are leaf bulk density, leaf, tiller and stem density and their interaction. Leaf and tiller density can markedly vary among species and defoliation regimes both in tropical and temperate pastures; however, tropical species often have a lower leaf bulk density or tiller density than do temperate species (Stobbs, 1973; Sollenberger and Burns, 2001). Therefore, sward height is more likely to limit bite size in temperate swards, whereas leaf bulk density is likely to limit bite size in tropical swards (Cosgrove, 1997). The intimate mixing of leaves and stems decreases the ability of the animal to access high-quality leaves from within the sward (Prache et al., 1998), which results in a decrease in forage intake for animals that are being selective, or a decrease in diet quality when selectivity is low. Therefore, forage intake rates and diet quality depend on the density and accessibility of preferred forage parts (Fryxell, 1991; Drescher, 2003). Forage intake rate in *Panicum maximum* swards increased with increasing leaf bulk density and decreased with increasing stem density (Drescher, 2003). However, the negative effect of stem density on intake rate was greater at high leaf densities. The reduction in intake rate with increasing stem density was the consequence of the reduction both in bite size and bite rate (Drescher, 2003). For a constant sward height, bite depth was not affected by stem density; therefore, the reduction in bite size with increasing stem density was a consequence of a decreased bite area. This reduction in bite area can be explained mechanically as resulting from the physical interference of the stems with the cropping process; in leafy swards the cattle apply circular tongue movements to gather groups of leaves in each bite; however, in swards with a high stem density, the stems physically interfere with the tongue sweep. Consequently, the animal reduces the reach of its tongue sweep to avoid including

stems in the bite. In the more extreme cases the animal changes its cropping style, abandoning the tongue sweep and gathering individual groups of leaves, clamping them between its lips (Ruyle et al., 1987; Drescher, 2003). This change in cropping style leads to an increased investment in handling time and a decrease in bite rate (Stobbs, 1973; Ruyle et al., 1987; Drescher, 2003).

The use of the muzzle and lips allows the animal to select leaves between stems even when the distance between stems is smaller that the animal's mouth. However, under these conditions bite size, bite rate and intake rate fall dramatically. When the distance between stems is larger than the width of the animal's mouth, bite size, bite rate and intake rate are expected to be greater than where the distance between stems is narrower than the animal's mouth; however, the foraging behaviour parameters may still not reach the values achieved on vegetative swards. For example, Drescher (2003) found that with leaf clusters twice as large as the maximum potential bite area of the animal, intake rate never exceeded 50% of the intake rate from vegetative swards. It appears as though this is because of an increase in the overlap of bites as the leaf clusters became depleted; attempts to crop residual material near a leaf cluster edge presumably resulted in smaller bite area, and thus bite size, as compared to leaf-only swards.

In swards with a high density of stems, bite area is the single most important variable controlling the rate of forage intake (Drescher, 2003); therefore, there should be an interaction between stem density and the animal mouth dimension, with a small mouth being better able to take a bite between stems in comparison with a bigger mouth. This means that small animals should be more capable to select leaves between stems and thus achieve a higher intake rate than large ones, although this hypothesis has not been tested yet.

The ability of animals to discriminate between leaf and mature stem, by penetrating through the upper strata of the sward canopy to graze preferred material at the base of the sward is well established both in cattle (Gardener, 1980) and sheep (L'Huillier et al., 1984). However, the degree of avoidance of stems in preference for leaves is likely to reflect the differences in structural strength and shear strength between these two plant components, and thus their resistance to grazing (Hodgson *et al.*, 1994). Therefore, there may also be effects of stem structural characteristics on diet selection and intake rate. Evidence from field studies demonstrates the influence of both spatial distribution and maturity of plant components on the degree of selectivity whilst grazing. For example, Forbes (1988) quotes examples of intake per bite being depressed in cattle following culm development in species like bromegrass (*Bromus inermis*), but not in species such as bermudagrass (*Cynodon dactylon*). Avoiding poor-quality stems seems to be logical, where the animal can select a better diet quality than the average vegetation quality on offer. However, in swards consisting of high-quality stems avoidance of stems may not be necessary because this behaviour may reduce bite size, bite rate and intake rate without substantially increasing diet quality. Again, this hypothesis has not been tested.

In many improved tropical pastures, the less preferred plant parts, such as stems and dead material, remain intimately attached to the preferred leaf fraction; therefore, leaves are not easily prehended and separated from the stem and dead fraction, resulting in low diet quality (Burns *et al.*, 1991). Local aggregation and density of stem, leaves and dead material can vary on small spatial scales throughout the grass sward. The distribution of preferred and avoided plant fraction is usually uneven within and between grass species due to their different growth forms (O'Reagain, 1993) and different grazing history between plants (Ruyle *et al.*, 1987).

Species distribution patterns

Over the last 50 years a great deal of research has been conducted on the factors (animal and plant) that affect the selection of different plant species from sown swards (Ungar, 1996). The main effects that have been studied include the proportion of different species in the sward, and the physiological state of the animal selecting the diet (Prache *et al.*, 1998).

Much evidence shows that the proportion of plant species affects the proportion in the diet, and how this interacts with the preferences that animals show for different species (Parsons *et al.*, 1994). One comparatively poorly studied area is the effect that species dispersion in the sward has on diet selection, and how this is affected by the preferences for the different species. Recent studies have shown that the size of patches at the relatively large scale affects the diet composition of foraging herbivores (Clarke *et al.*, 1995; Dumont *et al.*, 2002), with animals tending to increase the proportion of the preferred species in the diet when it is in large patches (m²); however, at the scale of the bite very little research has been conducted to assess the effects of dispersion of plant species on diet selection. Plant ecologists have long known that there are associations between species that affect the degree to which different species are defoliated (McNaughton, 1978, see also Milchunas and Noy-Meir, 2002); however, little is known about how, for example, the interaction between the size of an animal's mouth (Gordon and Illius, 1988) and the size of a plant species patch affects diet selection. Here we would expect that the animal would exert greater selection for species for which they have relatively greater preference, and that there will be an interaction between the degree of relative preference and the degree of aggregation of the species on the diet choice of the foraging herbivore (Fig. 15.4). We hypothesized that the proportion in the diet of a preferred species will decrease when the pattern of distribution of this species is intimately mixed with a less preferred species; accordingly, the proportion of the less preferred species will increase. Several studies have focused on the issue of preferences among different foods in order to estimate the parameters that lead the preferences (Berteaux *et al.*, 1998; Illius *et al.*, 1999), and several studies have shown that clover planted in strips (Clark and Harris, 1985; although see Carrere *et al.*, 2001) leads to higher selection for clover than where it is sown in an intimate mixture with ryegrass; however, little is understood as to how the preferences among plant species change within feeding stations when dispersion varies.

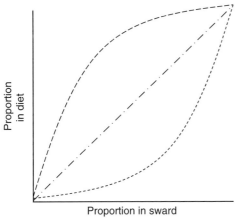

Fig. 15.4. The relationship between the proportion of a plant component in the sward and the proportion in the diet. Intimate mixture of preferred and less preferred components (- - - - -), preferred components aggregated (–·–·–·–·–), and less preferred components aggregated (·····).

3 Dimensions

In 3D, a vegetative pasture varies only in bulk density. In contrast to temperate species, tropical pastures vary greatly in plant component composition throughout the vertical plane. As a consequence of these changes, bulk density markedly varies from top to bottom of the canopy (Stobbs, 1975; Sollenberger and Burns, 2001). Bulk density variation in the vertical plane is associated with changes in the proportion of plant fractions, and thus in nutritive value. In tropical species, while leaf percentage and leaf bulk density markedly decrease from top to bottom of the canopy, stem bulk density increases. Therefore, tropical swards are often more heterogeneous than temperate swards in regard to plant component proportion and nutritive value in the vertical plane (Sollenberger and Burns, 2001).

Bulk density

Various experimental studies have addressed the effects of bulk density (Black and Kenney, 1984; Ungar et al., 1991; Laca et al., 1992) on ingestive behaviour and intake rate in temperate pastures. Bite dimension seems to be affected by bulk density: as bulk density increases, bite area and bite depth decrease but bite size increases because the increasing bulk density is enough to offset the decreasing bite dimensions. It was suggested that the animal reduces bite area when bulk density increases to keep constant the force required to sever the bite (Ungar, 1996). As in temperate pastures, forage intake rate in tropical pastures, such as *Panicum maximum*, increases with increasing leaf bulk density (Drescher, 2003).

There is an interaction between sward height and bulk density in terms of their effect on bite dimensions (Laca et al., 1992). In tall swards bulk density has a greater effect on bite area than it does on short swards. This is due to the animal's capacity to adjust bite area in tall pastures by modifying the reach of the tongue sweep, for example. On short swards the amount of vegetation that can be captured in the mouth is limited and bite area is restricted to the dimension of the arcade breadth (Illius and Gordon, 1987).

Modelling Plant–ruminant Interactions

Many attempts have been made to model intake and diet selection of grazing ruminants (Mertens, 1994; Allen, 1996; Illius et al., 1999; Baumont et al., 2000). The majority of models are based upon diet quality (see review by Illius et al., 1999); however, a few estimate intake from the relationship between the animal and the sward characteristics (Illius and Gordon, 1987; Baumont et al., 2004). Swards, however, are generally described in terms of sward height, as this is a primary driver of intake rate in vegetative swards, which means that many of the sward structural characteristics described in this review are not taken into account. One of the first attempts to include sward structural characteristics in modelling plant–herbivore interactions was by Illius and Gordon (1987), who described sward structure in relation to bulk density, live/dead ratios and sward height. The bite was described in terms of the area encompassed by the opening of the mouth or the tongue sweep, the quantity of material incised by the action of the incisors against the pad on the upper jaw and the depth to which the animal penetrated into the sward. Therefore, the intake per bite was predicted from the bite area and the bite depth interaction with the sward density and the vertical distribution of material in the sward (bulk density), and the diet quality was estimated from the bite depth and the distribution of live and dead material in the grazed horizon. It was suggested that the interspecific allometric relationship between body weight (BW) and bite area changes with sward height (vertical plane) (Gordon et al., 1996). On short swards, where only a narrow band of tillers can be prehended, bite area is determined by incisor arcade breadth (scaled to $BW^{0.36}$) (Illius and Gordon, 1987), whereas on tall swards, where the maximum bite area can be achieved, bite area is related to the square of the incisor arcade breadth (scaled to $BW^{0.66}$) (Gordon et al., 1996). Consequently, when grazing on short swards, intake by large species is expected to be more severely restricted than that of small species, due to a greater limitation in bite dimensions (Illius and Gordon, 1987). From this mechanistic description of the grazing process and how the animal decisions affect bite quantity and quality, a number of more detailed approaches to modelling plant–herbivore interactions in sown swards have been adopted. Several provide detailed descriptions of the sward structure and how grazing affects the growth and dynamics of the plant material (Hutchings and Gordon, 2001), whilst others partition the vertical dimension of the sward into grazed horizons that are characterized by plant biomass and quality (e.g. Baumont et al., 2004). However, little attempt has been made to incorporate this detailed mechanistic understanding of the biting behaviour of ruminants in relation to sward structure into models that can be used to support management decisions (but see Armstrong et al., 1997) particularly for tropical swards.

Management

The manager has a range of methods (e.g. fertilizer application and cutting) available to alter the sward structure and composition (e.g. increasing the quality of new leaf growth, minimizing the proportion of leaf to stem in the sward and increasing the proportion of preferred species in the sward) in a way that gives more opportunity for the grazing animal to increase the intake of nutrients and thereby raise animal performance. Fertilizer accelerates the rate of production of new leaf, but also hastens the rate at which swards senesce. Cutting can be used as a method to rejuvenate a sward, reducing the proportion of dead to live material in the sward and also increasing the length of the lamina relative to the pseudostem. The proportion of preferred species within the sward can be regulated during sowing or by cutting frequency and height.

One area that requires further research and development is the formation of practical management guidelines from ruminant grazing behaviour principles. A typical example is the use of sward height thresholds as a means to achieve higher daily intake in temperate grazing systems (Parsons and Johnson, 1985; Hodgson, 1990). As sward height and bite size are positively correlated in temperate sown swards, intake rate also increases until the forage-processing capacity of the animal is saturated and intake rate achieves an asymptote (Laca et al., 1994). Based on this plant–herbivore interaction principle, practical recommendations can be derived. For example, for temperate swards based on ryegrass and white clover a sward surface height of about 6 cm results in the best animal performance for sheep (Hodgson, 1990), whereas for cattle it is recommended that swards be maintained at a surface height of approximately 8–10 cm (Wright and Russel, 1987; Wright and Whyte, 1989).

In either vegetative or reproductive swards, pseudostems or stems may, respectively, represent a barrier to bite depth (Arias et al., 1990; Dougherty et al., 1992; Griffiths and Gordon, 2003). In this situation, bite depth and bite size are no longer related to sward height but to lamina or regrowth length (Flores et al., 1993; Griffiths and Gordon, 2003); therefore, sward height is of limited practical use in these circumstances. Where stems act as a barrier to bite depth, lamina length may be used for management purposes; however, given the fact that pseudostems or stems only act as a partial regulator of bite depth (Griffiths and Gordon, 2003), a mechanistic understanding of the relationship between the physical properties of pseudostems or stems and their effect as limiters to bite depth is urgently required to predict when these vegetation components will act as a barrier (Prache and Peyraud, 2001).

Managers will sow mixtures of species of fodder plants for a range of reasons: some related to animal nutrition, others related to the maintenance of sward nutritional quality. Generally, when sward managers sow mixed species swards, they include all of the species in the seeding machine simultaneously. However, research has shown that sowing species in monoculture strips can increase the proportion of the preferred species in the diet relative to the less preferred species (Clark and Harris, 1985) and thus increase daily intake and animal performance (Marotti *et al.*, 2002). Whilst monoculture strips may be difficult to achieve in practice, it would be a method for improving animal performance, and it is worth pursuing further research in this area to assist managers to improve animal performance. Recently, it has been proposed that further improvement of the animal production in grazing systems will need to identify and manipulate the behavioural responses of animals to the characteristics of the pasture (Marotti *et al.*, 2002), allowing animals to express their grazing behaviour rather than inhibit their behaviour. For example, the fact that growing grass and clover in separate strips facilitates selection of the preferred diet (a mixture of the two species), thereby increasing daily intake and animal performance, demonstrates that allowing animals to express their preference and behaviour can have a positive impact on the animal productivity. Another behavioural response that may be used for management proposes is the typical pattern that grazing time follows in rotational grazing systems. During depletion of a paddock, grazing time initially increases as the animals attempt to

compensate for a decrease in bite size and intake rate (Illius, 1997). However, during the latter stages of sward depletion, grazing time decreases despite continued reduction in intake rate, leading to a reduction in animal performance (Prache and Peyraud, 2001). Therefore, changes in grazing time in a rotational grazing system could be used for grazing management purposes to provide the grazier with information as to when to move the animals to a new paddock. The biomass of green leaf has been proposed as the indicator variable for pasture management under rotational grazing to ensure maximum intake where leaf/stem ratio changes rapidly (Penning *et al.*, 1994; Orr *et al.*, 2004). However, monitoring grazing time may be a better or complementary method particularly in heterogeneous pastures where forage characteristics are difficult to assess. Today, the technology to monitor grazing behaviour remotely, without the necessity of capturing the animal to retrieve data and make management decisions, is available (Gordon, 1995).

As we have shown, a great deal is now known about the effect of sward structure on bite dimensions and intake of grazing ruminants in temperate swards. Some of this understanding is being taken up within management systems. However, little is known about plant–herbivore interactions in tropical pastures, where the influence of plant structural characteristics is likely to be large and where management interventions could have dramatic effects on improving sward structure and longevity, and ultimately animal performance. Given the expanding demand for ruminant livestock products in the tropics, we would urge researchers to fill the gap in knowledge urgently.

Acknowledgements

This review was supported through funding from CSIRO and INTA. We would like to thank an anonymous referee for valuable comments on a previous version of the chapter.

References

Allen, M.S. (1996) Relationship between forage quality and dairy cattle production. *Animal Feed Science and Technology* 59, 51–60.

Arias, J.E., Dougherty, C.T., Bradley, N.W., Cornelius, P.L. and Lauriault, L.M. (1990) Structure of tall fescue swards and intake of grazing cattle. *Agronomy Journal* 82, 545–548.

Armstrong, H.M., Gordon, I.J., Sibbald, A.R., Hutchings, N.J., Illius, A.W. and Milne, J.A. (1997) A model of grazing by sheep on hill systems in the UK. II. The prediction of offtake by sheep. *Journal of Applied Ecology* 34, 186–207.

Barthram, G.T. (1981) Sward structure and the depth of the grazed horizon. *Grass and Forage Science* 36, 130–131.

Baumont, R., Prache, S., Meuret, M. and Morand-Fehr, P. (2000) How forage characteristitcs influence behaviour and intake in small ruminants: a review. *Livestock Production Sciences* 64, 15–28.

Baumont, R., Cohen-Salmon, D., Prache, S. and Sauvant, D. (2004) A mechanistic model of intake and grazing behaviour in sheep integrating sward architecture and animal decisions. *Animal Feed Science and Technology* 112, 5–28.

Bergman, C.M., Fryxell, J.M. and Gates, C.C. (2000) The effects of tissue complexity and sward height on the functional response of Wood Bison. *Functional Ecology* 14, 61–69.

Berteaux, D., Crete, M., Huot, J., Maltais, J. and Ouellet, J.P. (1998) Food choice by white-tailed deer in relation to protein and energy content of the diet: a field experiment. *Oecologia* 115, 84–92.

Black, J.L. and Kenney, P.A. (1984) Factors affecting diet selection by sheep. II. Height and density of pastures. *Australian Journal of Agriculture Research* 35, 551–563.

Buringh, P. and Dudal, R. (1987) Agricultural land use in space and time. In: Wolman, W.G. and Fournier, F.G.A. (eds) *Land Transformation in Agriculture, SCOPE 32*. John Wiley & Sons, Chichester, UK, pp. 9–43.

Burlison, A.J., Hodgson, J. and Illius, A.W. (1991) Sward canopy structure and the bite dimensions and bite weight of grazing sheep. *Grass and Forage Science* 46, 29–38.

Burns, J.C., Pond, K.R. and Fisher, D.S. (1991) Effects of grass species on grazing steers. II. Dry matter intake and digesta kinetics. *Journal of Animal Science* 69, 1199–1204.

Cangiano, C.A., Galli, J.R., Pece, M.A., Dichio, L. and Rozsyplalek, S.H. (2002) Effect of liveweight and pasture height on cattle bite dimensions during progressive defoliation. *Australian Journal of Agriculture Research* 53, 541–549.

Carrere, P., Louault, F., Carvalho, P.C.D., Lafarge, M. and Soussana, J.F. (2001) How does the vertical and horizontal structure of a perennial ryegrass and white clover sward influence grazing? *Grass and Forage Science* 56, 118–130.

Clark, D.A. and Harris, P.S. (1985) Composition of the diet of sheep grazing swards differing in white clover content and spatial distribution. *New Zealand Journal of Agricultural Research* 28, 233–240.

Clarke, J.L., Welch, D. and Gordon, I.J. (1995) The influence of vegetation pattern on the grazing of heather moorland by red deer and sheep. 1. The location of animals on grass heather mosaics. *Journal of Applied Ecology* 32, 166–176.

Cosgrove, G.P. (1997) Grazing behaviour and forage intake. In: Gomide, J.A. (ed.) *International Symposium on Animal Production under Grazing.* Universidade Federal de Vicosa, Vicosa, Brazil, pp. 59–80.

Demment, M.W. and Laca, E.A. (1993) The grazing ruminant: models and experimental techniques to relate sward structure and intake. *Proceedings of the VII World Conference on Animal Production, Invited Papers*, Vol. 1. Canadian Society of Animal Science, Edmonton, Canada, pp. 439–460.

Demment, M.W., Peyraud, J.-L. and Laca, E.A. (1995) Herbage intake at grazing: a modelling approach. In: Journet, M., Grenet, E., Farce, M.-H., Thériez, M. and Demarquilly, C. (eds) *Proceedings of the IV International Symposium on the Nutrition of Herbivores*, INRA Editions, Paris, France, pp. 121–141.

Dougherty, C.T., Bradley, N.W., Lauriault, L.M., Arias, J.E. and Cornelius, P.L. (1992) Allowance-intake relations of cattle grazing vegetative tall fescue. *Grass and Forage Science* 47, 211–219.

Drescher, M. (2003) Grasping complex matter. Large herbivore foraging in patches of heterogeneous resources. PhD thesis, Wageningen University, Wageningen, The Netherlands.

Dumont, B., Carrere, P. and D'Hour, P. (2002) Foraging in patchy grasslands: diet selection by sheep and cattle is affected by the abundance and spatial distribution of preferred species. *Animal Research* 51, 367–381.

Edwards, G.R., Parsons, A.J., Newman, J.A. and Wright, I.A. (1996) The spatial pattern of vegetation in cut and grazed grass white clover pastures. *Grass and Forage Science* 51, 219–231.

FAO (Food and Agriculture Organization of the United Nations) (2003) Food and Agriculture Organization of the United Nations statistical databases. Available at: www.fao.org/ag/aga/glipha/index.jsp

Flores, E.R., Laca, E.A., Griggs, T.C. and Demment, M.W. (1993) Sward height and vertical morphological differentiation determine cattle bite dimensions. *Agronomy Journal* 85, 527–532.

Forbes, T.D.A. (1988) Researching the plant–animal interface: the investigation of ingestive behaviour in grazing animals. *Journal of Animal Science* 66, 2369–2379.

Forbes, T.D.A. and Hodgson, J. (1985) The reaction of grazing sheep and cattle to the presence of dung from the same, or the opposite species. *Grass and Forage Science* 40, 177–182.

Fryxell, J.M. (1991) Forage quality and aggregation by large herbivores. *American Naturalist* 138, 478–498.

Gardener, C.J. (1980) Diet selection and liveweight performance of steers on *Stylosanthes hamata* – native grass pastures. *Australian Journal of Agicultural Research* 31, 379–392.

Gibb, M.J. and Ridout, M.S. (1988) Application of double normal frequency distributions fitted to measurements of sward height. *Grass and Forage Science* 43, 131–136.

Ginnett, T.F., Dankosky, J.A., Deo, G. and Demment, M.W. (1999) Patch depression in grazers: the roles of biomass distribution and residual stems. *Functional Ecology* 13, 37–44.

Gordon, I.J. (1995) Animal-based techniques for grazing ecology research. *Small Ruminant Research* 16, 203–214.

Gordon, I.J. and Illius, A.W. (1988) Incisor arcade structure and diet selection in ruminants. *Functional Ecology* 2, 15–22.

Gordon, I.J. and Lascano, C. (1993) Foraging strategies of ruminant livestock on intensively managed grasslands: potential and constraints. In: Baker, M.J. (ed.) *Proceedings of the 17th International Grassland Congress.* SIR Publishing, Wellington, New Zealand, pp. 681–689.

Gordon, I.J., Illius, A.W. and Milne, J.D. (1996) Sources of variation in the foraging efficiency of grazing ruminants. *Functional Ecology* 10, 219–226.

Griffiths, W.M. and Gordon, I.J. (2003) Sward structural resistance and biting effort in grazing ruminants. *Animal Research* 52, 145–160.

Griffiths, W.M., Hodgson, J. and Arnold, G.C. (2003) The influence of sward canopy structure on foraging decisions by grazing cattle. II. Regulation of bite depth. *Grass and Forage Science* 58, 125–137.

Hirata, L. (2000) Quantifying spatial heterogeneity in herbage mass and consumption in pastures. *Journal of Range Management* 53, 315–321.

Hodgson, J. (1985) The significance of sward characteristics in the management of temperate sown pastures. In: Nishi-Nasuno and Tochigi-Ken (eds) *Proceedings of the XV International Grassland Congress.* Science Council of Japan and Japanese Society of Grassland Science, Kyoto, Japan, pp. 63–67.

Hodgson, J. (1990) *Grazing Management. Science into Practice.* Longman, Hong Kong, China.

Hodgson, J. and Illius, A.W. (1998) *The Ecology and Management of Grazed Ecosystems.* CAB International, Wallingford, UK.

Hodgson, J., Clark, D.A. and Mitchell, R.J. (1994) Foraging behaviour in grazing animals and its impact on plant communities. In: Fahey, G.C. Jr (ed.) *Forage Quality, Evaluation, and Utilization.* American Society of Agronomy, Crop Science Society of America & Soil Science Society of America, Madison, Wisconsin, pp. 796–827.

Hutchings, N.J. and Gordon, I.J. (2001) A dynamic model of herbivore–plant interactions on sown swards. *Ecological Modelling* 136, 209–223.

Illius, A.W. (1997) Advances and retreats in specifying the constraints on intake in grazing ruminants. In: Buchanan-Smith, J.G., Bailey, L.D. and McCaughey, P. (eds) *Proceedings of the XVIII International Grassland Congress.* United States Department of Agriculture, Manitoba, Canada, pp. 109–118.

Illius, A.W. and Gordon, I.J. (1987) The allometry of food intake in grazing ruminants. *Journal of Animal Ecology* 56, 989–999.

Illius, A.W. and Gordon, I.J. (1991) Prediction of intake and digestion in ruminants by a model of rumen kinetics integrating animal size and plant characteristics. *Journal of Agricultural Science* 116, 145–157.

Illius, A.W. and Gordon, I.J. (1999) Scaling-up from daily food intakes to numerical responses of vertebrate herbivores. In: Olf, H., Brown, V.K. and Brent, R. (eds) *Herbivores: Between Plants and Predators. Symposium of the British Ecological Society.* Blackwell Scientific Publications, Oxford, UK, pp. 397–425.

Illius, A.W. and Jessop, N.S. (1996) Metabolic constraints on voluntary intake in ruminants. *Journal of Animal Science* 74, 3052–3062.

Illius, A.W., Gordon, I.J., Elston, D.A. and Milne, J.D. (1999) Diet selection in goats: a test of intake rate maximization. *Ecology* 80, 1008–1018.

Laca, E.A., Ungar, E.D., Silegman, N.G., Ramey, M.R. and Demment, M.W. (1992) Effects of sward height and bulk density on bite dimensions of cattle grazing homogeneous swards. *Grass and Forage Science* 47, 91–102.

Laca, E.A., Ungar, E.D. and Demment, M.W. (1994) Mechanisms of handling time and intake rate of a large mammalian grazer. *Applied Animal Behaviour Science* 39, 3–19.

L'Huillier, P.J., Poppi, D.P. and Fraser, T.J. (1984) Influence of green leaf distribution on diet selection by sheep and the implications for animal performance. *Proceedings of the New Zealand Society of Animal Production* 44, 105–107.

Liira, J. and Zobel, K. (2000) Vertical structure of a species-rich grassland canopy, treated with additional illumination, fertilization and mowing. *Plant Ecology* 146, 185–195.

Marotti, D.M., Cosgrove, G.P., Chapman, D.F., Parsons, A.J. and Egan, A.R. (2001) Novel methods of forage presentation to boost nutrition and performance of grazing dairy cows. *Australian Journal of Dairy Technology* 56, 159.

Matches, A.G. (1992) Plant-response to grazing – a review. *Journal of Production Agriculture* 5, 1–7.

McCall, D.G. and Sheath, G.W. (1993) Development of intensive grassland systems: from science to practice. In: *Proceedings of the XVII Grassland Congress*. New Zealand Grassland Association, Tropical Grassland Society of Australia, New Zealand Society of Animal Production, Australian Society of Animal Production – Queensland Branch, New Zealand Institute of Agricultural Science, Palmerston North, New Zealand, pp. 1257–1265.

McNaughton, S.J. (1978) Serengeti ungulates: feeding selectivity influences the effectiveness of plant defense guilds. *Science* 199, 806–807.

Mertens, D.R. (1994) Regulation of forage intake. In: Fahey, G.C. Jr (ed.) *Forage Quality, Evaluation, and Utilization*. American Society of Agronomy, Crop Science Society of America and Soil Science Society of America, Madison, Wisconsin, pp. 450–493.

Milchunas, D.G. and Noy-Meir, I. (2002) Grazing refuges, external avoidance of herbivory and plant diversity. *Oikos* 99, 113–130.

Mitchell, R.J., Hodgson, J. and Clark, D.A. (1991) The effect of varying leafy sward height and bulk density on the ingestive behaviour of young deer and sheep. *Proceedings of the New Zealand Society of Animal Production* 51, 159–165.

Mitchley, J. and Willems, J.H. (1995) Vertical canopy structure of Dutch chalk grasslands in relation to their management. *Vegetatio* 117, 17–27.

O'Reagain, P.J. (1993) Plant structure and the acceptability of different grasses to sheep. *Journal of Range Management* 46, 232–236.

Orians, C.M. and Jones, C.G. (2001) Plants as resource mosaics: a functional model for predicting patterns of within-plant resource heterogeneity to consumers based on vascular architecture and local environmental variability. *Oikos* 94, 493–504.

Orr, R.J., Rutter, S.M., Yarrow, N.H., Champion, R.A. and Rook, A.J. (2004) Changes in ingestive behaviour of yearling dairy heifers due to changes in sward state during grazing down of rotationally stocked ryegrass or white clover pastures. *Applied Animal Behaviour Science* 87, 205–222.

Parsons, A.J. and Johnson, I.R. (1985) The physiology of grass growth under grazing. In: Frame, J. (ed.) *Occasional Symposium No. 19*. British Grassland Society, Hurley, UK, pp. 3–13.

Parsons, A.J., Newman, J.A., Penning, P.D., Harvey, A. and Orr, R.J. (1994) Diet preference of sheep: effects of recent diet, physiological state and species abundance. *Journal of Animal Ecology* 63, 465–478.

Penning, P.D., Parsons, A.J., Orr, R.J. and Hooper, G.E. (1994) Intake and behaviour response by sheep to changes in sward characteristics under rotational grazing. *Grass and Forage Science* 49, 476–486.

Perez-Barberia, F.J. and Gordon, I.J. (1998) Factors affecting food comminution during chewing in ruminants: a review. *Biological Journal of the Linnaean Society* 63, 233–256.

Prache, S. and Peyraud, J.L. (2001) Foraging behaviour and intake in temperate cultivated grasslands. In: *Proceeding of the XIX International Grassland Congress*. Fundacao de Estudos Agrarios Luiz de Queiroz, Piracicaba, Brazil, pp. 309–319.

Prache, S., Roguet, C. and Petit, M. (1998) How degree of selectivity modifies foraging behaviour of dry ewes on reproductive compared to vegetative sward structure. *Applied Animal Behaviour Science* 57, 91–108.

Purves, D.W. and Law, R. (2002) Fine-scale spatial structure in a grassland community: quantifying the plant's-eye view. *Journal of Ecology* 90, 121–129.

Ruyle, G.B., Hasson, O. and Rice, R.W. (1987) The influence of residual stems on biting rates of cattle grazing *Eragrostis lehmanniana* Nees. *Applied Animal Behaviour Science* 19, 11–17.

Senft, R.L., Coughenour, M.B., Bailey, D.W., Rittenhouse, L.R., Sala, O.E. and Swift, D.M. (1987) Large herbivore foraging and ecological herbivores. *Bioscience* 37, 789–799.

Sollenberger, L.E. and Burns, J.C. (2001) Canopy characteristics, ingestive behaviour and herbage intake in cultivated tropical grasslands. *Proceedings of the XIX International Grassland Congress*. Fundacao de Estudos Agrarios Luiz de Queiroz, Piracicaba, Brazil, pp. 321–327.

Spalinger, D.E. and Hobbs, N.T. (1992) Mechanisms of foraging in mammalian herbivores: new models of functional response. *American Naturalist* 140, 325–348.

Stobbs, T.H. (1973) The effect of plant structure on the intake of tropical pastures. I. Variation in the bite size of grazing cattle. *Australian Journal of Agricultural Research* 24, 809–819.

Stobbs, T.H. (1975) Factors limiting the nutritional value of grazed tropical pastures for beef and milk production. *Tropical Grasslands* 9, 141–149.

Torales, A.T.A., Acosta, G.L., Deregibus, V.A. and Moauro, P.M. (2000) Effects of grazing frequency on the production, nutritive value, herbage utilisation, and structure of a *Paspalum dilatatum* sward. *New Zealand Journal of Agricultural Research* 43, 467–472.

Ungar, E.D. (1996) Ingestive behaviour. In: Hodgson, J. and Illius, A.W. (eds) *The Ecology and Management of Grazing Systems*. CAB International, Wallingford, UK, pp. 185–218.

Ungar, E.D. and Ravid, N. (1999) Bite horizons and dimensions for cattle grazing herbage to high levels of depletion. *Grass and Forage Science* 54, 357–364.

Ungar, E.D., Genizi, A. and Demment, M.W. (1991) Bite dimensions and herbage intake by cattle grazing short hand-constructed swards. *Agronomy Journal* 83, 973–978.

Wright, I.A. and Russel, A.J.F. (1987) The effect of sward height on beef cow performance and on the relationship between calf milk and herbage intakes. *Animal Production* 44, 363–370.

Wright, I.A. and Whyte, I. (1989) Effect of sward surface height on the performance of continuously stocked spring-calving beef cows and their calves. *Grass and Forage Science* 44, 259–266.

Wright, W. and Illius, A.W. (1995) A comparative study of the fracture properties of five grasses. *Functional Ecology* 9, 269–278.

16 Physiology and Models of Feeding Behaviour and Intake Regulation in Ruminants

W. Pittroff[1] and P. Soca[2]

[1]Department of Animal Science, University of California, 1 Shields Ave, Davis, CA 95616, USA; [2]Facultad de Agronomía, Universidad de la Republica, Ruta 3, km 363, CP 60000, Paysandú, Uruguay

Introduction

Considerable research funds are being spent on maps of the genome of ruminant livestock species, and predictions of complete understanding of functional genomics abound. At the same time, ruminant livestock producers around the world are faced with a seemingly simple yet intractable problem: the prediction of feed intake. The voluntary consumption of feed and its regulation constitutes the first and arguably most important interface between the farm animal and its environment. It is of pervasive economic significance to provide feed to farm animals commensurate with their requirements, and manage grazing animals with minimal negative impact on their resource base. This requires the ability to predict intake; however, this ability has not been acquired yet for ruminants. Ruminants constitute the economically most significant sector of global livestock production.

This chapter intends to accomplish the following:

1. Present and discuss conceptual models of intake regulation and feeding behaviour of ruminants.
2. Discuss quantitative modelling of intake regulation and feeding behaviour.

3. Summarize and discuss current understanding of neurohormonal information pathways implicated in regulation of intake and feeding behaviour.
4. Propose priorities for future research.

Conceptual Models of Feeding Behaviour and Intake Regulation

Attempts at formulating conceptual approaches to the understanding of feed intake regulation date back to the 19th century, when Settegast (1868) introduced the term 'ballast' for indigestible feed components. Since then, a truly vast literature has developed, reflecting both the significance of the problem, and the difficulties encountered in devising solutions. Previous reviews of intake modelling concepts were presented by Laca and Demment (1996), Pittroff and Kothmann (1999) and Illius et al. (2000).

Ruminants are managed in two distinct worlds: (i) pastures, where they select their own feed; and (ii) housed facilities, where they are provided with feedstuffs. These differences in feed on offer lead to important differences in feeding behaviour and complexity of theoretical and quantitative approaches to intake prediction. Collier and Johnson (2004) presented an interesting discussion of these differences. Further, conceptual models of feed intake regulation for ruminants can be classified into two categories: (i) models considering nutrient requirements of the ruminant and various types of intake constraints; and (ii) models considering only metabolic regulation, i.e. postulating regulation being driven by nutrient requirements of the animal only. These principles both apply to models that explicitly consider grazing, and to those that do not. Accordingly, in order to structure our discussion, topics 1 and 2 will be subdivided, as needed, into addressing problems specific to housed ruminants on the one hand, and to pastured ruminants on the other hand.

Conceptual models without consideration of grazing

Intake constraints models for ruminants

INTAKE REGULATION IN TWO PHASES.   Nutrient requirement prediction systems currently in use (e.g. NRC, 2000, 2001) resort to empirical equations developed from regression meta-analyses of published data. The resulting equations link intake to body weight and production level, with or without multiplicative adjustments for environmental conditions and animal properties. National Research Council (NRC) (e.g. NRC, 2000) further assumes a quadratic relationship between intake and energy concentration in the diet. Despite these systems abstaining from explicit commitment to a conceptual model of feed intake regulation, the currently prevailing view holds that intake is regulated by ruminants in two distinct phases (along a gradient of digestibility or energy concentration): the physical constraint phase during which feed bulk determines clearance from the rumen and hence intake, and the physiological phase when physical restrictions are not operating and intake follows metabolic requirements. Although the formulation of the bi-phasic regulation concept is commonly attributed to Conrad et al. (1964), it was Svoboda (1937) who first presented this conceptual model. If correct, this model requires that the following propositions must hold:

1. Postfeeding rumen fill of low-quality forages is higher than postfeeding rumen fill of high-quality forages.
2. Fill is physical in nature, i.e. effects of fill depend on mass only (in the absence of toxic or adverse properties of fill).
3. Intake varies quadratically with digestibility.
4. Energy supplements that do not add fill cannot decrease intake of low-quality forages in the same proportion as high-quality forages.
5. Dietary interventions that do not interfere with digestion in the rumen but change absorbed nutrient supply cannot change intake of low-quality forages.

It was shown in reviews by Grovum (1987), Ketelaars and Tolkamp (1992), Weston (1996), Pittroff and Kothmann (1999), Poppi et al. (2000) and Weston (2002) that these propositions are contradicted by a wealth of published data, and consequently, this model must be rejected. This does not necessarily imply that physical properties of forages do not play a role in intake regulation: physical properties of feedstuff could also condition metabolic responses. Hence, a considerably more differentiated research hypothesis is required. It is somewhat alarming that research exploring such alternatives is lacking, as remarked by Illius et al. (2000). Despite being rejected repeatedly, this theory still dominates the literature and underlies major ruminant production system models and forage classification systems currently in use.

DIFFERENTIATED INTAKE CONSTRAINTS MODEL. Weston (1996) presented a concept that addresses the shortcomings of the bi-phasic model. His model, called here 'Differentiated Intake Constraints Model', combines considerations of energy transactions in the ruminant, aspects of rumen function and their interactions with chemical and physical properties of feedstuffs. The central postulates of the model (Fig. 16.1, reprinted with permission) are:

1. The sum of energy demands of the ruminant (maintenance, growth, lactation, fibre, reproduction) constitutes its capacity to use energy.
2. This capacity for energy use is reduced when:

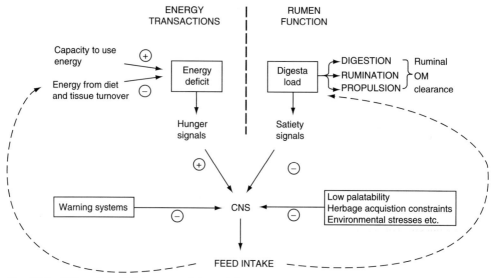

Fig. 16.1.   Differentiated intake constraints model.

(a) oxidation or synthesis is impaired by inadequate nutrients (composition, deficiency, presence of toxins);

(b) dissipation of heat is constrained by environmental temperatures.

3. Forage diets generally fail to meet potential energy requirements in domestic livestock, creating an effective energy deficit, due to:

(a) physical capacity of the rumen not allowing intake commensurate with potential requirements;

(b) low palatability;

(c) harvesting efficiency not being optimum (spatial distribution, density of forage on offer);

(d) environmental stress.

4. The energy deficit generates hunger signals related to the magnitude of the deficit.

5. Clearance rate of digesta in the rumen and the quantity of organic matter (OM) in ruminal digesta load are directly related to an upper, as of yet undefined, limit.

6. Digesta load in the rumen generates a satiety signal.

7. The feeding drive (i.e. hunger is the basic state) is overcome when the sum of constraints more than balances the signals from the energy deficit.

8. Mechano-, osmo- and chemoreceptors and associated neurohormonal pathways, sub-

servient to the energy-regulating system, maintain functional homeostasis in the ingestive and digestive system.

This concept allows generating testable research hypotheses, and indeed resolves the most problematic aspects of the bi-phasic model, such as the postulates of fixed limits for rumen digesta load and energy concentration of the diet. Weston's model further proposes a definitive solution to an issue that seems to be improperly addressed in many if not most experimental and review reports on intake regulation in ruminants: the question of short-term vs. long-term intake regulation, or in other words, the role of functional digestive homeostasis in intake regulation. Once it is agreed that hunger is the basic state, as proposed by Weston (1996), the understanding that meal-patterning mechanisms cannot be basic to intake regulation is both straightforward and, indeed, groundbreaking. This will be discussed further while discussing current issues in neurohormonal regulation of intake.

Nevertheless, open questions remain with Weston's concept. Perhaps the most important issue awaiting elucidation is the interplay between meal patterning and intake regulation. One could argue, for example, that although meal patterning is not functionally basic to intake regulation, its metabolic consequences, integrated over time, are. This issue is probably most critical in understanding the role of digesta load. Sprinkle et al. (2000) presented data showing that lactating cows on pasture had higher digesta loads than non-lactating ones, but identical residence times and rates of passage. Taweel et al. (2004) observed very large differences in rumen digesta loads after the three major feeding bouts of grazing, lactating dairy cattle. As discussed by Weston (1996), upper limits to intake capacity as determined by clearance rate of OM from the reticulo-rumen have not been identified. The report of Taweel et al. (2004), for example, concluded, from measuring the highest digesta load at the 23:30 h rumen evacuation, that intake at that time was limited by rumen fill, while it was not limited during the other two daily grazing bouts as rumen fill was substantially smaller. This does not appear to be a solid basis for the claim that the fill observed at night actually constitutes the absolute physical limit. If indeed fill comes into play as an intake-limiting constraint (meal terminating), it would require a functional explanation why in the same animal, in the same day, it is sometimes a constraint, and sometimes not.

Leek's review (1986) of sensory receptors in the ruminant digestive tract suggests that the principal role of mechano- and chemoreceptors is to maintain homeostasis of digestion processes. Accordingly, a relationship between meal patterning and intake regulation must be defined if a physical constraint to intake, whatever its nature and extent, is postulated. Moreover, it seems necessary to consider change of state (size and specific gravity, i.e. two properties) of rumen digesta in order to understand limits to passage. For feed particles to leave the rumen, they must overcome buoyancy and acquire a specific weight that allows them to move towards the ostium reticulum-omasicum. Note, however, Ørskov's view (1994) of specific gravity being an elusive parameter due to entrapment of gases. The process of change of specific gravity is determined by both feed and animal properties: mastication, remastication (rumination), hydration and microbial colonization. Once they are located at the ostium,

however, the flow constraint supersedes particle properties because, regardless of particle size of OM in the rumen, there is a constraint to its ready passage (Pittroff and Kothmann, 1999; Poppi et al., 2000). Being independent of particle size, this constraint can only be animal related, i.e. dependent upon ruminal motility. Indeed, eating and ruminating provide the major motor inputs for rumen motility, and the early work of Freer and Campling (1965) already showed that the number of rumen contractions increases almost exponentially with the amount of food eaten per unit of time. These observations have several ramifications. First, if Weston's concept (1996) of ruminants being able to regulate rumen load according to energy deficit is correct, as suggested by abundant published data, it follows that prediction of feed intake is neither possible from knowledge of animal properties nor from knowledge of feed properties alone. Second, compartmental modelling of the passage of digesta through ruminants can provide no insight useful for prediction of intake because animal and feed factors remain intractably confounded. Third, experimental designs that can resolve the confounding influences of feed and animal properties are required to test the Differentiated Intake Constraints Model. This will likely prove a formidable challenge.

SATIETY–MALAISE CONTINUUM.   In recent years the work of Provenza (see summary article Provenza, 1996; Forbes and Provenza, 2000) has attracted considerable interest. His postulate states: 'The same mechanisms decrease preference for foods deficient in nutrients, foods containing excess toxins, and foods containing too large a portion of rapidly digestible nutrients.' This mechanism is considered to be aversion, i.e. learned preference (or aversion) based on post-ingestive feedback. In this concept, excess nutrients cause 'malaise' and activate behavioural adjustment akin to that responding to intake of toxins. While considerable literature data illustrate the capability of ruminants to make structured food selection decisions that can be related to nutrient and toxin contents (see Forbes and Kyriazakis, 1995), the conclusion of a continuum between satiety and malaise remains controversial. The substance used by Provenza to induce aversion, LiCl, interferes with a rather unspecific second messenger system involved in the reception of many peptide hormones and neurotransmitters, the phosphatidylinositol cycle. Consequently, it appears to be impossible to attribute the effects of LiCl in the creation of aversion to any specific system involved in neurohormonal intake (or behaviour) regulation. Foley *et al.* (1999) further emphasized that emesis and nausea (malaise) remain ill-defined concepts, precluding generalized conclusions about plant toxin emesis 'with so little evidence'. It appears logical to extend this caution to the interpretation of effects of nutrients on nausea or malaise. The hypothesis of a satiety–malaise continuum has been suggested before (see, e.g. van der Kooy, 1984), but many studies produced equivocal data. For example, Moore and Deutsch (1985) concluded that exogenous cholecystokinin (CCK) cannot serve as a model for studying satiety, and further, that exogenous CCK causes malaise and, therefore, malaise is not a factor causing normal satiety. Mele *et al.* (1992) reported that emesis can be suppressed by a specific serotonin receptor blocker, but not a concurrently established conditioned taste aversion. Ervin *et al.* (1995) presented a comprehensive study assessing anorectic and aversive properties of a variety of substances known to cause both effects (satiety and malaise) in monogastrics. The authors concluded that there is no evident relationship between satiety and malaise effects. Thus, it appears that current data do not provide firm support for the Satiety–Malaise Continuum Model. However, this area of research is important as ruminants are increasingly used to reduce biomass of invasive plants and fire fuel matrices in brush communities. These plants are commonly characterized by the presence of plant secondary metabolites (PSM). The effects of PSM on ingestion or digestion of ruminants are still weakly understood, and functional models suggesting definitive answers are not available (see the comprehensive review by Foley *et al.*, 1999). Foley *et al.* (1999) have described numerous ways in which detoxification and energy metabolism might be linked. These interrelationships may be very complex, involving, for example, acid–base balance and its effects on protein metabolism. Thus, the need of energy metabolism data for the interpretation of feed intake data is further substantiated. Further, there are well-published cases of herbivores consuming large amounts of forages containing toxins that cause fatal diseases without any apparent aversive feedback, such as nigropallidal encephaloamalacia in horses (Cordy, 1954), which is linked to the ingestion of *Centaurea solstitialis*. The need for much more detailed research in this area is obvious. Such research should not be constrained by the early adoption of a paradigm at odds with current data.

Metabolic regulation models for ruminants

COST–BENEFIT BALANCE MODEL.   Ketelaars and Tolkamp (1992, 1993) and Tolkamp and Ketelaars (1992) presented a conceptual model, which linked intake regulation of the ruminant to 'costs and benefits' of feed consumption. The central postulate of their model is that feed consumption benefits the animal by providing nutrients but presents costs in the form of damaging by-products of metabolism (free radicals). Feed intake regulation is understood as a process of optimizing a balance between costs and benefits. Forbes (1995) observed the absence of identified mechanisms by which this proposed optimization could operate. A more general objection to this concept would be the consideration of the evolutionary background of herbivores. Although systematic studies are lacking, one might argue that the co-evolution of ruminants and plants eaten by them mandates maximum intake at times when plant quality is high, because the plant growth cycle always imposes times of nutritional deficiency on the ruminant, for which the animal has to prepare by accumulating body reserves. It is difficult to see how these conflicting goals could be reconciled. Certainly, experimental challenge of the Cost–Benefit Balance Model would require real-time monitoring of metabolizable energy intake, and some quantification of free radicals. It would further require the identification of information pathways linking cellular events (free radicals) to behaviour – integrating centres – an intimidating challenge.

The Cost–Benefit Balance Model predicts maximum feed intake to occur at the level of maximum efficiency of oxygen consumption expressed as net energy intake per unit of oxygen consumed. This is equivalent to predicting maximum intake at the point of zero marginal increase in balanced (useful) nutrient harvest from additional intake. This observation will be discussed further below.

EFFECTIVE ENERGY REQUIREMENTS MODEL. The major nutrient requirement systems predict feed intake as a function of requirements, approximated by calculated demands for maintenance and products. Emmans (1997) remarked that such systems are static, and that food intake should be an explicit function of time, an approach especially applicable to growing animals. As a solution, Emmans (1997) proposed a model that linked growth rate, maintenance and composition of growth to food intake. His system of differential equations drew heavily on the genetic scaling rules proposed by Taylor (1980). However, close inspection of the resulting equations reveals no principal difference to any existing nutrient requirement system approach, for two reasons: (i) there is no explicit consideration of conditions where nutrients would be limiting; and (ii) intake is still simply an implicit function of requirements. Perhaps the only conceptual difference between Emmans' model (1997) and major requirement systems (e.g. NRC, 2000) is the fact that the former makes intake a dynamic function of requirements, whereas the latter models intake as only dependent upon weight of the animal and energy concentration of the diet. However, major nutrient requirement systems (e.g. NRC, 2000, 2001) also model the change in energy requirements due to change in growth rate and dynamic change of composition of growth. Hence, what Emmans (1997) perceived to be a limitation (intake a static function of weight) could be easily modified in such systems. However, the real problem is that feed intake is not a simple function of requirements and, moreover, that energy requirements and feed properties are interdependent, as discussed by Weston (1996).

REALIZED ENERGY DEMAND MODEL. A concept that focuses on a dynamic interaction between animal requirements and feed properties in determining energy demand and intake regulation was presented by Pittroff and Kothmann (1999). The postulates of this model are as follows:

1. Intake is a function of energy requirements.
2. An animal has a genetically determined potential energy demand (PED).
3. The interaction between animal requirements and feed properties determines realized energy demand (RED).
4. RED is the controlling element in intake regulation.

This model can function without the assumption of a physical capacity limit per se. Whether or not the evolutionary history of ruminants would have allowed for a physical capacity constraint is a question beyond this discussion, but it might be considered that the survival of a species requires optimum function of ingestive and digestive systems. Therefore, the proposal of physical capacity as the principal intake constraint for domestic ruminants as mostly grass and roughage eaters (Hofmann, 1988) is not necessarily self-evident. On the other hand, nutrient requirements substantially altered by selection for extremely high performance levels may have outgrown what was previously an 'optimum' ingestive and digestive capacity. The fundamental aspect of this model is the requirement of an iterative solution. This can be explained by the following example. Suppose a growing ruminant is fed a diet that is highly concentrated in energy, but deficient in protein. In order to satisfy its potential demand for energy, this animal, according to the model, would be expected to maximize intake up to the point where further increase in intake will not lead to a marginal increase in protein supply, or in other words, an increase in useful energy supply. Accordingly, an iterative calculation of intake is required in order to quantify this level.

The PED–RED model (Fig. 16.2) adopted key aspects of the fuel partitioning–intake regulation concept proposed by Friedman (1998). In this model, interference with the energy generation from ingested energy substrates elicits a feeding response. Friedman and his colleagues have demonstrated this effect clearly with agents interfering with ATP generation from long-chain fatty acid and glucose oxidation. Houpt (1974) showed that the same observation can be made in ruminants when he demonstrated stimulation of feed intake in goats following the administration of 2-deoxy-D-glucose, a selective inhibitor of glucose oxidation. The PED–RED model can accommodate variable loads at constant passage rates, or significant load differences at meal termination. With more detailed information on diurnal patterns of energy utilization available (e.g. conditioned by diurnal patterns of neurohormonal regulation of energy metabolism), realistic meal patterns should be a testable outcome of this model. The experimental challenge of this concept requires research into the metabolic consequences of diets differing in energy concentration, protein concentration, energy/protein ratio and rate of digestion. In order to be able to interpret feed and digesta composition data, a method for monitoring metabolizable energy intake, e.g. by measuring energy expenditure by heart rate (Brosh *et al.*, 2002), is required. For such an approach to achieve the precision called for by Poppi *et al.* (2000), the interaction between animal properties and feed properties determining nutrient release and absorption in the gut must be researched. It is pertinent to point out that the PED–RED approach would predict maximum intake at the point of zero marginal increase in balanced (useful) nutrient harvest from additional intake, just as the Cost–Benefit Balance Model would.

Summary

A consensus about the functional basis of intake regulation has not been achieved. At least six distinct theoretical approaches have been proposed, and while there is partial convergence between some, considerable differences remain. This is not discouraging, but actually presents an opportunity for formulating research hypotheses and conducting experiments to challenge them. However, we fail to see current research that specifically attempts to challenge these hypotheses. This may be explained in part by the difficulty of obtaining research funds for whole-animal oriented work.

It is important to point out the fact that most intake prediction models in practical use implement the 'ballast' and/or bi-phasic regulation paradigms one way or another. Often, they work very well, and available field data usually do not allow the parameterization of more complex approaches. The correspondence between observed intake and intake

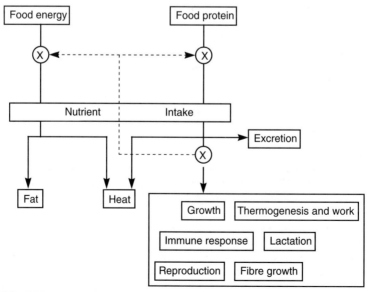

Fig. 16.2. PED–RED model.

predicted by these models is leading many to disregard the fact that the theoretical concepts underlying them are not tenable. Indeed, a conceptually wrong model that makes good predictions is a dangerous thing: it requires substantial expert knowledge to identify those situations where it should not be used. Most users of practical predictions models cannot be expected to possess this knowledge. More implications will be discussed in the final section of this chapter.

Conceptual models with consideration of grazing

Theory

When considering intake of grazing animals, the number of factors involved in intake regulation increases considerably. These include: spatial distribution of forage on offer; composition and density of sward; landscape properties such as topography, shade, protection from wind, distance to water; and animal properties such as breed-specific adaptation characteristics, which usually lack proper description. An example for the latter would be the alleged exceptional foraging capacity of the Texas Longhorn breed. Given the bewildering number of possible factors determining intake of grazing animals, it is understandable that considerable research has been directed towards exploring simplified hypotheses. The dominating hypothesis, formulated as the optimal diet or optimal foraging theory (OFT) states that animals forage to optimize retained nutrient intake. As pointed out by Belovsky *et al.* (1999), OFT is an extension of the theory of evolution. It is beyond the scope of this discussion to critically review the literature of OFT related to ruminants. However, it should be noted that key tenets of OFT are by no means universally accepted as sound ecological theory. Schluter (1981) presented an oft-cited summary of criticism based on experimental and empirical evidence, and Bergman *et al.* (2001) showed that the question of nutrient intake maximization is not resolved for large ungulates. Alm *et al.* (2002) demonstrated how food quality and relative abundance influenced food choice in deer, but their data did not suggest optimizing behaviour. Mårell *et al.* (2002)

demonstrated that free-ranging reindeer selected feeding sites characterized by higher biomass of two plant species, but choice was not explained by N content, a commonly used proxy for forage quality.

In the following sections we discuss some methodological issues that should be considered by quantitative models of feed intake of grazing animals.

General principles

Should optimizing models be used if we do not know whether animals optimize? This question seems to be pertinent as there is not even an agreement as to what constitutes the appropriate objective function. Models assuming that animals maximize for nutrient intake have been proposed, as well as those in which animals minimize harvesting time, or models in which they optimize thermal load, or yet others, in which they minimize (starvation) risk. Laca and Demment (1996) presented a summary of grazing models that illustrates the many different concepts. Given the diversity in approaches, there should be (but is usually lacking) an answer to the following questions: (i) Do animals optimize? (ii) What is the objective function of optimizing behaviour? (iii) What are the relevant constraints animals are dealing with? Possible objective functions may include:

- Maximize energy intake.
- Maximize useful energy intake.
- Minimize time spent eating.
- Minimize energy spent on harvesting.
- Minimize general energy expenditure.
- Balance nutrient intake with intake of anti-nutritious compounds.
- Minimize exposure to predators.

The formulation of constraints appears to be much more complex. Some of the objectives may also be employed as constraints. A problem of concern is the fact that modelling of intake of grazing animals seems to lag behind progress achieved in ruminant nutrition physiology in key issues. For example, physical intake constraint is still discussed without consideration of the insights of Weston (1996) and others, for example, as they relate to variable rumen loads. This is evident from the extensive review by

Belovsky *et al.* (1999) and the model of Baumont *et al.* (2004).

Diet selection

Newman *et al.* (1995), in their presentation of an intake model for grazing ruminants, emphasized the need to consider diet selection of grazers. In order to accommodate selectivity, Provenza (1996) proposed that herbivores forage selectively to balance nutrient intake with intake of potentially deleterious PSM. However, considering selectivity (based on whatever criterion) introduces another complexity: distribution of plant species. Belovsky *et al.* (1999) pointed out that the implementation of a selectivity algorithm based on quality ranking (the classical model) has to assume random distribution of plant species in the environment, an assumption that usually does not hold. They contended further that quantitative prediction based on a preference index tends to fail when implemented in a quantitative model. It is clear that a desirability or preference index approach cannot take into account distribution and abundance properties of plant species in the pasture without additional factors modifying their preference rank. This was indirectly achieved, for example, by Newman *et al.* (1995), by varying the encounter probabilities for the two species in the sward they modelled. Since $\rho_{grass} = 1-\rho_{clover}$ in that model (ρ denoting encounter probabilities), obviously complete and uniformly random species mixture must be assumed. Any other scenario would introduce the need for some function linking encounter probability to animal behaviour, requiring explicit statements about how grazing ruminants search. We are not aware of appropriate hypotheses guiding experimental work required to derive such a function.

The rather complex analyses of Mårell *et al.* (2002) of foraging search paths of reindeer, for example, definitely did not reveal any optimizing behaviour in diet selection related to search.

Independent and dependent variables

Theoretically, intake on pasture could be measured by the following equation:

$$I = MBM \times MBR \times AGT$$

where I represents intake per day, MBM denotes mean bite mass, MBR is the mean bite rate and AGT is the total active grazing time per day. The problem is that the measurement of right-hand variables does not produce an equation that could be generalized. The situation is not helped much by further decomposing bite rate into components related to jaw movements and their time budget, as described by Ungar (1996). The reason is that all right-hand variables are highly plastic and themselves depend on a multitude of factors. Some may be related to physical and nutritional properties of the sward, as is exemplified in the models of Baumont *et al.* (2004) and Woodward (1997). Others are clearly dependent upon animal characteristics. For example, intake rate and grazing time have been shown to depend on animal metabolic state (see below), and thermoregulation requirements (see, e.g. Sprinkle *et al.*, 2000). Table 16.1 summarizes experiments in which intake rate, bite mass and bite rate were measured in sheep and cattle after imposed fasting of variable length.

The variability observed impressively demonstrates the effect of metabolic status on these variables. Further, AGT not only depends on animal properties but also on distribution of forage on offer. Distributional properties include certainly more dimensions than sward height in all but homogeneous, sown pastures.

Summary

The preceding suggests:

1. There is no unequivocal evidence supporting the view that grazing ruminants expose optimizing behaviour.
2. While there is evidence for goal-seeking behaviour in grazing ruminants, no consensus exists as to the nature of the goal(s).
3. There is no consensus as to what are the relevant constraints.
4. There is no agreement as to what constitutes an appropriate time horizon in models of diet selection and grazing.

Table 16.1. Effect of fasting time in the forage intake rate of beef cattle and sheep grazing at small time and spatial scale.

Author	Pasture and animals	Treatment	Response		
			IIR	BW	BR
Dougherty et al. (1989a)	*Festuca arundinacea* (Height = 28 cm) beef cattle	Fasting time before grazing meal (1, 2 and 3 h)			
		F_1	370	0.67	44
		F_2	390	0.72	44
		F_3	360	0.59	48
Dougherty et al. (1989b)	*Medicago sativa* (Height = 42 cm) beef cattle	Fasting time before grazing meal (1, 2 and 3 h)			
		F_1	440	1.30	26
		F_2	510	1.49	27
		F_3	590	1.56	30
Greenwood and Demment (1988)	*Lolium* sp. beef cattle	Fasting 36 h before turnout			
		F_0	174	0.44	31
		F_{36}	221	0.48	31
Newman et al. (1995)	*Lolium* sp. and *Trifolium repens* sheep	24 h fasting			
		F_0	206	0.82	66
		F_{24}	315	0.4	83
Chilibroste (1999)	*Lolium perenne* dairy cows	Fasting (F_1 = 16.5 h; F_2 = 2, 5 h) and artificial fill of rumen (L)			
		F_1L	430	0.75	59
		F_1	470	0.72	61
		F_2L	350	0.63	59
		F_2	490	0.84	62
Patterson et al. (1998)	*Lolium perenne* and *Trifolium repens* dairy cows	Fasting time (hours)			
		F_1	483	1.1	45
		F_3	550	1.1	50
		F_6	733	1.4	53
		F_{13}	735	1.3	55

IIR: intake rate (g DM/100 kg live weight/h).

BW = Bite weight (g DM/100 kg live weight).

BR = Bite rate (bite/min).

5. There is no consensus as to how diet selection could be modelled in a spatially heterogeneous environment.

6. Many studies suggest non-random, but not optimizing grazing and foraging behaviour of ruminants.

We conclude that models attempting to incorporate grazing behaviour must specify the limitations of its intended application. Those that assume optimizing behaviour of the grazing ruminant cannot claim to be grounded in solid ecological theory. If variables such as grazing time, bite size, bite rate or search time are to be supplied as model constants or user inputs, the authors should clearly describe the situations where such simplified assumptions (i.e. external supply of these key parameters) would not be applicable.

Quantitative Models of Feeding Behaviour and Intake Regulation

Several comparative analyses of quantitative intake prediction models are available (e.g. Elsen *et al.*, 1988; Ingvartsen, 1994; Roseler *et al.*, 1997; Keady *et al.*, 2004). Often regression-type models such as those included in the nutrient requirement prediction systems constitute the major number of models reviewed; however, by design these do not offer explanatory insight and are therefore not of interest in a discussion of concepts and implementations. Most model comparisons do not entail a systematic analysis of conceptual, mathematical and implementation properties. This was attempted by Pittroff and Kothmann (2001a–c).

Models without consideration of grazing

The majority of 23 models for sheep and cattle analysed by Pittroff and Kothmann (2001a–c) were designed without the explicit consideration of grazing. It is noteworthy that only ten models analysed were based explicitly on the concept of bi-phasic intake regulation. The other models did not make any statements about an underlying conceptual model of intake regulation, but implemented empirical relationships relating intake to differentiated energy requirements and feed properties based on implicit cause–effect relationships.

Example calculations for the comparison of those models that could be parameterized with typically available forage quality data produced substantial variability in results. Pittroff and Kothmann (2001a–c), unlike Keady *et al.* (2004), did not compare published intake data with model calculations, and thus could not offer a quantitative assessment of the predictive capability of the models analysed. How-

ever, such comparisons suffer from the fact that actual and predicted intakes are both variables measured with error.

Many of the models analysed by Pittroff and Kothmann (2001a–c) suffered from mathematical and/or implementation problems. One has to then wonder about the exact significance of deviations of predicted from actual intake, whether directly observed or inferred from the literature. Very frequently, intake prediction models contain multiplicative correction factors addressing animal properties, feed properties or climatic conditions. Multiplicative correction factors introduce implicit interaction between independent variables, and present non-trivial requirements (usually not met) for model testing. How far such equations or equation systems can be extrapolated thus remains an open question, creating potentially serious pitfalls for their users.

Models explicitly considering grazing

Several models for sheep and cattle analysed by Pittroff and Kothmann (2001a–c) contain a component addressing the effects of grazing in intake regulation. Three models for sheep (Arnold *et al.*, 1977; Christian *et al.*, 1978; Finlayson *et al.*, 1995) but none for cattle implement diet selection as a factor in intake regulation. All models use the classical concept of a preference index based on quality of sward components. None of these models provides any experimental data for model development or model validation. The schemes proposed do not consider spatial distribution of forage on offer, yet clearly illustrate the complexities involved in modelling effective forage availability and resultant quality for grazing ruminants. All three models further contain mathematical inconsistencies. Two of the models (Arnold *et al.*, 1977; Christian *et al.*, 1978), however, produced similar intake prediction data for the forage data sets used in the comparison.

Although not amenable to parameterization for a comparative analysis, the models published by Thornley *et al.* (1994), Parsons *et al.* (1994) and Newman *et al.* (1995, all from the same research group) warrant discussion because they collectively illustrate key unresolved issues of intake prediction for grazing ruminants. Thornley *et al.* (1994) call their model teleonomic as opposed to optimizing, which is confusing because intake is obtained in a maximization algorithm. Maximized is diet quality in terms of benefits (net energy retention) subject to costs imposed by sward composition. Composition of the diet (clover vs. grass) depends on sward properties. The algorithm must assume (but does not state) random distribution of clover and grass in the sward. The model does not explain degree of preference for clover, but requires it as an input. Cost of selectivity depends strongly on physical sward properties, adjusted by a scaling factor provided as input. The model appears to be most sensitive to this scaling factor.

Newman *et al.* (1995) presented a model that optimizes diet selection. The objective function is based on the amount of retained feed energy at the end of the grazing day. The algorithm is stochastic programming, employing as terminal reward function the probability of survival at the beginning of the next grazing day. Three (physiological) states are evaluated (stored energy, digestible gut fill, indigestible gut fill). Five behaviours (rest, ruminate, graze grass, grass clover, graze either grass or clover) can be exercised during each model stage. The stage of the model is a period of 8 min, and the algorithm is implemented as a Markov process, with transition probabilities of stage $t + 1$ depending only on the state of the animal at stage t. A combination of behaviours during subsequent stages that maximizes expectation for survival at the beginning of the next grazing period is derived by backward recursive solution. This combination is then simulated to obtain specific species' combinations of diets. The model seems to have an infinite planning horizon, which is problematic because it introduces the question of time constancy of the attributes that describe the states. As has to be expected, the equations determining states are rather simplistic, and we note in passing that the rate of passage in the model depends upon particle breakdown by rumination, which is clearly not correct. A result of this model should be meal patterning; however, this is not presented. Some results of the model are counterintuitive: e.g. proportion of clover in the diet increases when bite rate for grass surpasses bite rate for clover. This is counter to Thornley *et al.*'s model (1994), where cost of selectivity would increase (and proportion of clover decrease) if bite rate for grass would exceed bite rate for clover. Finally, Parsons *et al.* (1994) developed a 'mechanistic model of diet selection' based on purely physical properties of the sward, principally leaf area index and bulk density. Intake rate is derived from these parameters based on bite mass and handling time. We note that Parsons *et al.* (1994) stated the assumption of random species distribution. The value of this model lies in providing a basis for estimating energy cost of foraging.

All three models consider only a homogeneous two-species sward, the second simplest scenario for grazing animals. Yet, as acknowledged by the authors, many equations are based on assumptions, and some assumptions are at least in their implications not acknowledged (such as random distribution). However, this group has also demonstrated that there is likely not a single most appropriate approach to modelling intake of grazing ruminants.

Summary

The models that could be parameterized for a comparison with easily available forage quality and animal property data (15 out of 23 in Kothmann and Pittroff, 2001a–c) must be considered relatively simplistic. There are of course other intake regulation models. However, the number of models demanding input data that are not available in any but research situations seems to be growing. A good example is the model of Baumont *et al.* (2004), which links an intake model (Sauvant *et al.*, 1996) to a diet selection model. The paucity of available data for such exercises is well illustrated by the fact that a key parameter of the intake rate equation derived by Sauvant *et al.* (1996) apparently was based on one of the data sets (Baumont, 1989, cited following Sauvant *et al.*, 1996) on which the full model was tested.

Keady *et al.* (2004) reported a regression-type intake model (Vadiveloo and Holmes,

1979) as the most accurate model for predicting food intake of lactating cattle. This is not surprising, given that that model was developed for lactating cattle of the same type and for similar feedstuffs. Nobody would seriously attempt to claim generality for such a model. Generality, however, becomes an issue once intake prediction is part of an animal performance or even herd-level production system model. Such a task requires a more mechanistic approach, and these models do not fare well in terms of predictive power. The assertion of Illius *et al.* (2000) that intake models, if not predicting well, should at least motivate research, is of course correct. However, we fail to see a response in practical research. New hypotheses, in particular those pointing to the metabolic background of intake regulation in ruminants, currently do not appear to be challenged by experimental work.

Probably the most desolate situation is seen in intake prediction of freely grazing animals. Metabolic intake regulation and grazing behaviour must be tightly coupled processes. Ruminants clearly make structured choices on pasture – in terms of selection of feeding patch, time spent grazing, intake rate, etc. Most authors agree that the resulting behaviour is a response to metabolic needs and the constraints imposed by the environment to fulfilling them; however, no agreement exists about the appropriate currency (nutrient uptake optimization, harvesting time minimization, risk aversion, adaptive behaviour). Thus, the choice of model algorithm appears arbitrary.

As long as variables such as bite rate, bite mass and time spent eating are not viewed as responses, no functional understanding can be gained from measuring them. Being functionally identical to intake per se, their measurement must be accompanied by concurrent recording of energy intake. Although other goals or objective functions have been proposed, ultimately energy is the currency that matters.

Perhaps the most difficult problem to solve is the issue of diet selection. Preference indices, no matter how derived, collapse under spatial variability, and it appears to be

extremely difficult to design research providing data suitable for mechanistic relationships. In this context, given the proliferation of experiments in which position data of grazing animals are collected as time series of Global Positioning System (GPS) coordinates, we feel compelled to ask what hypotheses underlie such work, and how they may be tested. Suppose one positions 20 animals in a pasture that provides realistic spatial variability in all parameters potentially relevant (topography, distance to water, sward heterogeneity, etc.); would we expect a meaningful overlap of the spatial and temporal coordinates of their feeding stations? How would we interpret significant differences, however measured, in site preference? For example, it has been suggested that there is between-animal variability of preference of riparian habitat, opening the possibility for selection (culling and genetic) against riparian habitat users in a herd of cows. As Belovsky *et al.* (1999) have pointed out, this scenario would contradict teleonomic behaviour, which we assume to be based on retained energy acquisition. As long as no studies are conducted that monitor the energy intake of animals in real time while providing them with structured choices in a grazing environment, such hypotheses necessarily remain speculative.

Thornley *et al.* (1994) stated that a purely mechanistic approach cannot consider aspects of behaviour about which insufficient information for mechanistic description is available because they are determined by the brain. The current state of that field, not yet part of intake models but ultimately of decisive importance in understanding intake regulation, will be summarized next.

Neurohormonal Mechanisms Involved in Intake Regulation

Introduction

While evidence for the dominance of a metabolic background in intake regulation is convincing, open questions remain. For example, the issue of the role of physical properties of feedstuffs beyond effects on nutrient supply (in the true sense of a physical constraint or fill effect) is indeed relevant for both assessment of feed quality and for the design of supplementation (or total mixed ration) regimes. Further, factors affecting meal patterning must be considered of intrinsic relevance to the understanding of grazing behaviour. Hence, the functional understanding of the neurohormonal pathways of intake regulation is critical to develop quantitative models based on a thorough understanding of the relevant biology.

A literature of staggering proportions on neurohormonal intake regulation pathways has been developed. Most of the work is being conducted with monogastric animal models with the clear objective to identify drug targets for human applications. Obesity is a global epidemic, likely to be followed by a global diabetes epidemic, and pharmaceutical approaches seem to have considerably more appeal to administrators of research than efforts directed at behavioural change. As a consequence, it is indeed a challenge to filter the resulting material for work specifically relevant to ruminants. Because only comparatively few experiments on neurohormonal intake regulation pathways have been conducted with ruminants, the major benefit derived from perusing this literature may be the distillation of mechanisms that should at least in their basic properties also be applicable to ruminants. As shall be seen, the major open question emerging from such a synthesis attempt is the applicability to ruminants of functional understanding of neurohormonal effects on meal patterning.

It is convenient and customary to organize an overview of neurohormonal intake regulation pathways into separate discussions of major hormones and neurotransmitters. We consider here: serotonin, CCK, leptin, neuropeptide Y (NPY) and ghrelin. There are numerous others implicated, and some will be discussed within the context of our overview. For each one of these messengers, exhaustive reviews are available. We are not aware of recent work, however, that would discuss this literature in relation to the specific aspects of ruminant biology.

Key neurohormonal agents involved in intake regulation

Serotonin

Serotonin (5-hydroxytryptamine, 5-HT) is a neurotransmitter in the central nervous system and in the gut. It is found in high concentrations in enterochromaffin cells in the alimentary canal and blood platelets. Anderson et al. (1992) suggested that serotonergic regulatory mechanisms lead animals to select a relatively constant proportion of their diet as protein. Consequently, if the diet does not provide for sufficient protein quantity or quality, intake will be adjusted accordingly. This view is apparently supported by a large number of reports on various species, including pigs and sheep, consistently selecting diets according to protein content (Kyriazakis et al., 1990, 1993, 1994; Kyriazakis and Emmans, 1991; Kyriazakis and Oldham, 1993; Cooper et al., 1995). Further, anorectic responses can be linked to diets with amino acid imbalances. Gietzen (1993) presented a comprehensive review of neural mechanisms involved in the responses to amino acid deficiency in the form of experimentally imbalanced diets. Amino acid–imbalanced diets (IMB) lead to rapid, predictable and (depending on the degree of imbalance) transient decreases in intake. Strong anorectic responses to these diets have also been observed in suckling lambs (Rogers and Egan, 1975) and developed ruminants (Egan and Rogers, 1978). In the experiments conducted by Rogers and Egan (1975), lambs did not show an adaptation; intake was permanently reduced.

The anorectic effects of this paradigm involve serotonergic pathways. This was shown by Washburn et al. (1994), who found that tropisetron, a potent serotonin antagonist at the $5-HT_3$ receptor, abolishes the anorectic effects of IMB in rats. Hoebel et al. (1992) speculated that the observed inhibition of feed intake exerted by serotonin might involve the inhibition of reward functions of dopamine in the mesolimbic system. Potent antidepressants are drugs inhibiting 5-HT reuptake. These serotonergic antidepressant drugs also decrease feed intake. Fernstrom's review (1992) of the role of serotonin in food intake regulation focused on the site of action (peripheral vs. central). The author proposed that peripheral serotonin may produce central behavioural adjustments by the direct effect of serotonin on reduction of gastric emptying. Reduced gastric emptying might be expected to produce a physiological inhibition of feed intake. This explanation relates to CCK and its function. We conclude that serotonin is a neurotransmitter with important functions in intake regulation. Its activity may be strongly influenced by composition of absorbed nutrients and is particularly closely linked with amino acid metabolism. Therefore, alternative hypotheses about the link between protein metabolism and serotonergic activity as related to intake control should be explored. The preceding discussion indicates that two possible alternative hypotheses exist for the explanation of serotonergic action in intake control: (i) serotonin is involved in intake control under physiological conditions; and (ii) serotonin is involved in the generation of an anorectic response to diets that generate a metabolic disturbance. Currently available experimental evidence from ruminant livestock indicates the second alternative, but cannot exclude the first. The role of serotonergic transmission in meal patterning, however, seems to be unclear.

Cholestokinin

At least two active forms of CCK have been related to intake regulation. The two corresponding receptor types are CCK receptor type A (pancreas) and CCK receptor type B (central nervous system – CNS). High concentrations of CCK-8 are found in the hippocampus, cerebral cortex and amygdala. Kruk and Pycock (1991) described the primary functional role of CCK as co-transmission. CCK distribution overlaps with dopamine- and serotonin-containing parts in the brain (substantia nigra, raphe nuclei). A close interaction between CCK and dopamine has been suggested (Kruk and Pycock, 1991). CCK has been shown to either inhibit or potentiate dopamine and serotonin action in the mesolimbic system. The administration of serotonin at 5-HT$_1$ receptors produced a significant attenuation of CCK and bombesin-induced satiety, leading to the conclusion that the

effects of these peptides must depend on the activity of serotonin (Gibbs and Smith, 1992).

Farningham et al. (1993) found a synergistic decreasing effect of hepatic infusions of propionate and CCK-8 on feed intake of sheep. These authors identified migrating binding sites in similar quantities on both sides of a ligated vagus fibre, indicating bi-directional axonal transport. However, the receptors were predominantly of type B, as identified by incubation with the specific type B receptor antagonist L-365,260. In both studies, feed intake effects were short-lived; 24 h intake was unchanged.

The intake effects of CCK are complicated by its interaction with gastric distension and gut motility. Forbes (1995) suggested that the major effect of exogenous CCK is to cause contractions of the digestive tract that activate mechanoreceptors. This information may be vagally relayed to the CNS and consequently followed by behavioural intake adjustments. This viewpoint is supported by data reported by Cottrell and Iggo (1984). They found that close-arterial bolus injections of CCK produced an increase in impulse activity of tension receptors in the duodenum of sheep. The increased activity coincided with phasic contractions of the muscularis externa of the duodenum.

Smooth muscle contractions excite tension receptors more than passive activation caused by, for example, distension of the lumen; therefore, substantial indirect effects of CCK seem likely (Leek, 1986). However, the effects of CCK on gut motility are by no means straightforward. Increased duodenal contractions, caused by CCK injections in sheep (Cottrell and Iggo, 1984), contrast with marked reductions of reticular motility, caused by continuous intravenous infusions of CCK that were not effective in causing a reduction in feed intake in sheep (Grovum, 1981; Onaga et al., 1995). The difficulty of dissociating the motility and intake effects makes it hard to attribute a direct satiating property to CCK (Grovum, 1981). Geary (2004) further noted that the question of physiological vs. pharmacological dose effectiveness in satiation remains unresolved. If distension of the stomach and CCK release both act as satiating agents, it would be logical to assume synergistic depressing effects on intake, even if CCK works rather indirectly by eliciting smooth muscle contractions. This synergy could be based on two different mechanisms: (i) distension and CCK; or (ii) presence of nutrients and CCK. However, Cox (1996) showed that gastric mechanoreceptors do not contribute to satiety independently of nutrient flow. Geary (2004) pointed out that CCK reduces meal size but either does not change meal frequency or shortens between-meal intervals.

Oxytocin antagonists abolish the satiety effects of exogenous CCK (Uvnäs-Moberg, 1994), indicating that the role of neurotransmitters and hormones in information processing is very complicated. Oxytocin is elevated in pregnancy, and the interaction between oxytocin and CCK may indicate the mechanism by which lactating animals abolish milk production under food deprivation. Thus, vagotomized rats, while feeding normally, gradually abolished lactation (Uvnäs-Moberg, 1994), apparently because information about the status of nutrient acquisition did not reach the CNS. In lactating dairy cows, longer periods of food deprivation were required to cause relatively small decreases in plasma levels of CCK (Samuelsson et al., 1996), while changes in oxytocin levels remained insignificant. The slower change in plasma levels of gut hormones in dairy cows compared to laboratory animals may be explained by the gradual reduction of digesta flow from the forestomach of dairy cows. The primary function of CCK seems to be participation in the regulation of gut motility. Neurons that process mechanical stimuli respond to CCK, either directly, or indirectly, by muscular contractions. Although CCK injections, even at physiological levels, may cause satiety, the primary function of CCK does not appear to be that of a satiety agent. CCK response is very specific to nutrient type and is influenced by metabolic state of the animal. The strong substrate specificity of CCK release suggests a primary function of co-transmission, facilitating primary aspects of digestion such as motility regulation and secretion of digestive enzymes. Weston and Poppi (1987) concluded that CCK as well as other peptides, hormones and neurotransmitters influence meal patterns but not feed intake when integrated over time. The current literature suggests that mechanisms involved in meal termination do not entirely depend on, or may even not acutely require, a direct metabolic function, e.g. energy transactions in the liver.

Leptin

Leptin, a 167-amino acid protein secreted from adipocytes, is a product of the *ob* gene. Leptin receptors are highly concentrated in the hypothalamus but are also expressed in non-neuronal tissues. The discovery of leptin occurred in ob/ob mice that show early and massive obesity, hyperphagia, non-insulin-dependent diabetes mellitus, defective thermoregulation and infertility. These animals are deficient in leptin, and exogenous leptin largely eliminates the pathological symptoms (Hamann and Matthaei, 1996). Similar symptoms are presented by animals genetically lacking a functional leptin receptor (fa/fa (Zucker) rats and db/db (diabetes) mice).

Most research has focused on leptin's effects on feed intake (see, e.g. voluminous reviews by Friedman, 2001 and Ingvartsen and Boisclair, 2001), but the literature is not conclusive. First, as Geary (2004) pointed out, the evidence that physiological doses of leptin affect eating is not convincing. Second, testing for the effects of removal and replacement of leptin is currently only possible in genetic disorders where either leptin or its receptors are missing. However, such models are not ideal because the null mutation has: (i) pervasive effects throughout all tissues; and (ii) is effective throughout all stages of development and may provoke syndromes involving compensatory effects (Geary, 2004). Hence, the development of inducible and ideally tissue-specific knockouts is required to test the removal/replacement criterion.

While direct intake regulation effects are controversial, evidence for a strong involvement of leptin in energy metabolism is mounting. For example, Wang et al. (1998) reported that a nutrient-sensing pathway regulates leptin gene expression in muscle and fat. These findings point out the interesting possibility that leptin levels may react rather acutely to changes in nutrient or, more specifically, in energy supply. Further, Rosenbaum et al. (2002) reported that physiological exogenous leptin restored thyroid hormone activity in human subjects maintained at 90% of normal body weight. Their report was confirmed by Legradi et al. (1997), who showed that the starvation-induced fall in plasma levels of thyroid hormones is largely prevented by exogenous leptin in rats. Ob/ob mice have reduced oxygen consumption compared with normal controls (Hwa et al., 1997). Further, leptin injected into ob/ob mice reduced carbohydrate oxidation and increased fat oxidation (Hwa et al., 1997). This observation is consistent with the theory that satiety-contributing effects of metabolic fuels rest on their primary use in ATP-generating oxidation (Friedman, 1998). In cultured pancreatic islet cells, leptin prevented triglyceride formation from free fatty acids and increased fatty acid oxidation (Shimabukuro et al., 1997). In normal rats transfected with an adenovirus-carried gene causing hyperleptinemia, triglyceride content in the liver, skeletal muscle and pancreas was depleted without concomitant increases in plasma free fatty acids or ketones. This finding was interpreted to indicate increased intracellular oxidation of fatty acids. Numerous interactions of leptin with other components of regulation of energy metabolism are known. There are frequent reports of glucocorticoids upregulating leptin (Berneis et al., 1996; Kiess et al., 1996; Larsson and Ahren, 1996; Miell et al., 1996; Considine et al., 1997; Papaspyrou-Rao et al., 1997). Miell et al. (1996) pointed out that while glucocorticoids stimulate appetite in humans, they also seem to upregulate leptin. That apparent contradiction strongly suggests the need to include metabolic background, i.e. energy metabolism, in the assessment of regulatory effects of hormones and neurotransmitters.

Even if it were shown that leptin at physiological doses affects eating and it passed the removal/replacement test, the effects of leptin on energy metabolism could not be dissociated. Thus, as a key agent in energy metabolism leptin must be expected to affect intake regulation. However, its effects on intake may be indirect.

Neuropeptide

Neuropeptide Y is the strongest known stimulant of feed intake when administered centrally

(Stanley and Leibovitz, 1985). It is also the most abundant peptide in mammalian brain and a multitude of functions are known. NPY has stimulating effects on corticosterone, aldosterone, insulin and vasopressin (Hoebel et al., 1992). It is conceivable that in periods of fasting, NPY is increased to enhance a drive for food search. This speculation is supported by the fact that centrally administered NPY remains active if feed access is delayed for up to 4 h after injection, whereas levels of NPY in injected animals return to normal after 30 min if feed is available (Miner, 1992). Likewise, central administration of NPY lowers thyroid hormones, similarly to what is observed in fasted animals (Fekete et al., 2001). However, NPY is not required for thyroid hormone levels to fall in starvation.

Considerable literature discusses the interactions between NPY and leptin (see the comprehensive review by Pedrazzini et al., 2003). Leptin receptors and NPY neurons are co-localized in the hypothalamus, and adipocytes possess NPY receptors regulating leptin secretion. Thus, a regulatory pathway balancing anorexigenic and orexigenic agents, with insulin, leptin and NPY as key elements, has been proposed (Pedrazzini et al., 2003). All agents in this proposed pathway are connected to energy metabolism, reinforcing the notion of the dominant nature of energy metabolism in regulation of feed intake. However, this picture is probably grossly oversimplified. To date there is no conclusive evidence about the specific receptor(s) involved in the orexigenic effects of NPY. While knockout mice for two different NPY receptor subtypes all became obese, it was shown that the elimination of one receptor type caused obesity due to decreased energy expenditure, while the other receptor elimination caused obesity due to hyperphagia (Marsh et al., 1998; Kanatani et al., 2000). Williams et al. (2001, 2004) pointed out the existence of several parallel systems involved in orexigenic action. Parallel, redundant systems explain why NPY knockout mice eat and grow normally (Erickson et al., 1996). It is concluded that while strong evidence for a role of NPY in intake regulation exists, the effects of this neuropeptide appear to be tightly linked to energy metabolism. Similar to leptin, NPY demonstrates the need to understand energy metabolism in developing functional models of intake regulation.

Ghrelin

Ghrelin is yet another potent orexigenic peptide hormone, released primarily in the stomach. The discussion by Geary (2004) suggests an important role of ghrelin in meal patterning. Yet, it must be pointed out that ghrelin is also closely tied to energy metabolism: it is the ligand for the pituitary growth hormone secretogogue receptor (Kojima *et al.*, 2001). Tschöp *et al.* (2000) reported ghrelin-inducing obesity by the reduction of oxidation of fat, and notably not by hyperphagia. Thus, the interesting fact is established that the exogenous administration of a growth hormone (GH) secretogogue does not lead to the expected, i.e. a lipolytic effect, but to the exact opposite. However, it promotes a positive energy balance and thus creates conditions maximizing the anabolic action of GH. Tschöp *et al.* (2000) further noted that sugar intake, but not stomach expansion, decreases circulating ghrelin levels.

As was the case with leptin and NPY, ghrelin seems to affect intake, but its effects on energy metabolism cannot be separated and may be more important than any acute intake regulation effects.

Summary

The preceding discussion can only provide a glimpse of the truly vast area of neurohormonal regulation of feed intake. However, several salient points emerge.

1. The study of singular agents in isolation will probably produce in-depth knowledge about their expression, regulation, specific metabolism and receptor physiology, but it will not specifically aid in the systematic, holistic understanding of feed intake. The first and simplest reason for this is the presence of many redundant systems. The second reason is the presence of multiple interactions between the known agents involved. Considering how relatively recently the discoveries of leptin and ghrelin occurred, it would be preposterous to assume that we possess complete knowledge of all major neurohormonal players involved in the regulation of feed intake and energy metabolism.

2. All agents involved in neurohormonal regulation of intake are indirectly or directly linked to energy metabolism. This even applies to CCK, whose secretion depends (rather specifically) on the presence of nutrients in the upper gut.

3. The disciplines generating the bulk of the knowledge on neurohormonal regulation of food intake insist on the meal as the basic functional behavioural unit of eating (Eckel, 2004). This is clearly problematic for the evaluation of the relevance of this knowledge to ruminants. Ruminants do eat meals, but meal eating and nutrient absorption are less sharply defined than in monogastrics. Consequently, the possibility exists that metabolic messages influencing meal patterns in monogastrics are quantitatively, or even qualitatively, different. The monogastric literature, for example, speaks of 'gastric satiation' – how would that apply to a ruminant? It seems intuitively unlikely that gastric satiation that is considered to be volumetric (Powley and Phillips, 2004) would similarly apply to ruminants. In particular CCK, a key messenger implied in gastric satiation, has clear functions related to the processing of nutrients in the duodenum by regulating flow of pancreatic enzymes (and stimulating intestinal motility). Thus, it cannot be related to distension of the rumen. This would be one indication that gastric distension plays different roles in ruminants and monogastrics.

4. Many studies have shown that interference with the messengers will impact meal size and meal frequency in monogastrics, but when integrated, energy intake remains the same. Exceptions are experiments where the persistent administration of exogenous quantities of orexigenic or anorexigenic agents perpetuates states serving as transient adaptations to changes in energy supply.

5. It has often been shown that the interactions between messengers involved in meal patterning can be functionally explained by the need of dynamic adaptation of energy consumption to changing physiologic requirements. A good example for short-term effects is the role of estradiol and its interaction with CCK in meal patterning (Eckel, 2004). An example for more long-term regulation adjustment is provided by Zammit (1990), who pointed out that hypergluconemia in late pregnancy and hypoinsulinemia in early lactation in ruminants lead to drastic changes in fatty acid metabolism (depression of malonyl-CoA levels, a key intermediate in acylglyceride synthesis, and elevation of carnitine palmitoyl-transferase I, the key enzyme for transport of long-chain fatty acids into the mitochondrium for oxidation). Thus, intake depression under these conditions comes secondary to fundamental shifts towards fatty acid oxidation (Zammit, 1990). The often confusing reports about the effects of neurohormonal messengers in feed intake regulation begin to make sense when adopting a systemic view, recognizing that feed intake is subservient to energy metabolism, and energy metabolism is subservient to vital functions.

6. It appears that the study of neurohormonal regulation pathways alone cannot provide functional insight. It is time to realize that separating energy metabolism and regulatory mechanisms cannot provide conclusive answers about the regulation of feed intake.

7. If energy requirements change dynamically, and indeed cyclically in ruminants, it has to be expected that the exact same basic adjustments in energy metabolism and energy partitioning occur as in monogastrics. Consequently, corresponding effects on intake regulation must occur that are intrinsic in nature and may affect the way the interaction of feeds and energy requirements determines realized energy demand. Thus, effects of feedstuffs on intake (they definitely include purported 'fill values' of ruminant feeds) cannot be fixed quantities. This was authoritatively explained by Weston (1996). However, it seems that this insight has not permeated to the extent where it is guiding research.

Conclusions, Implications and Future Research

Progress in the understanding of intake regulation in ruminants is slow, due to the unique challenges of a symbiont-dependent digestion system, and the complexities of free-range grazing. Several well-described conceptual models of intake regulation have been proposed. Research should systematically challenge the propositions of these models. The evidence for intake regulation dominated by metabolic processes is so strong that experimental research without concurrent monitoring of metabolizable energy intake cannot be expected to contribute unequivocal information. Since appropriate technology is available, there is no reason not to expect faster progress with such designs. However, gaining understanding of intake regulation of ruminants on pasture is much more complex. The intrinsic difficulties of deficient understanding of metabolic intake regulation compound with lack of knowledge about how grazing ruminants perceive and react to their environment at different scales. Rather than attempting to find evidence in support of conceptual models, such as the OFT, research should employ systematic designs challenging specific hypotheses. It is obvious that the evaluation of teleonomic or optimizing concepts must measure the 'currency' of what is optimized. Although most models claim to be based on such measurements, the many assumptions underlying, for example, estimated metabolizable or net energy intake render such claims not credible.

The proposal of PSM playing a role in regulating intake and motivating grazing behaviour, especially selectivity, further requires demonstrating that their effects are indeed deleterious, i.e. reduce effective or useful nutrient intake. It is not helpful to postulate ill-defined states that cannot be measured, such as malaise.

Collier and Johnson (2004) emphasized that meal frequency, meal size, rate of eating and distribution of meals are much more variable in the same individual than overall intake, clearly illustrating the plasticity of these components of feed intake. It is rather obvious, therefore, that simple, meal-oriented satiation models cannot work, and that further the question of time horizon across which an animal integrates information relevant for intake regulation is critical. Again, without monitoring of energy intake (over long enough experimental periods), experimental intake data can only lead to speculative conclusions. The highest priority for future research should be the evaluation of the role of energy demand in intake regulation, with systematic research of the question of time horizons across which animals adjust to changes in feed on offer and requirements.

Our brief summary of conceptual and quantitative models of intake regulation of grazing ruminants thus demonstrates that, although individual aspects of foraging behaviour of ruminants are well researched, integration into an intake prediction model remains an unresolved issue. Given the enormous economic significance of the problem, it appears that thorough questioning of research priorities in the animal sciences should no longer be postponed. Formidably complex interactions between metabolism, environment and behaviour suggest that without meaningful whole-animal research, the most important interface between animal and environment will never be understood.

Acknowledgements

The authors sincerely thank Robert Weston, (Sidney, Australia) and Andrew Illius, (Edinburgh, Scotland) for their collegial, constructive and helpful discussion and review.

References

Alm, U., Birgersson, B. and Leimar, O. (2002) The effect of food quality and relative abundance on food choice in fallow deer. *Animal Behaviour* 64, 439–445.

Anderson, G.H., Black, R.M. and Li, E.T.S. (1992) Physiological determinants of food selection: association with protein and carbohydrate. In: Anderson, G.H. and Kennedy, S.H. (eds) *The Biology of Feast and Famine*. Academic Press, San Diego, California, p. 73.

Arnold, G.W., Campbell, N.A. and Galbraith, K.A. (1977) Mathematical relationships and computer routines for a model of food intake, liveweight change and wool production in grazing sheep. *Agricultural Systems* 2, 209–226.

Baumont, R. (1989) Etat de réplétion du réticulo-rumen et ingestion des fourrages. PhD dissertation, Institut National Agronomique Paris Grignon, Paris, France.

Baumont, R., Cohen-Salmon, D., Prache, S. and Sauvant, D. (2004) A mechanistic model of grazing behaviour in sheep integrating sward architecture and animal decisions. *Animal Feed Science and Technology* 112, 5–28.

Belovsky, G.E., Fryxell, J.M. and Schmitz, O.J. (1999) Natural selection and herbivore nutrition: optimal foraging theory and what it tells us about the structure of ecological communities. In: Jung, H.G. and Fahey, G.C. (eds) *Nutritional Ecology of Herbivores. Proceedings of the V International Symposium on the Nutrition of Herbivores*. American Association of Animal Science, Savoy, Illinois, pp. 1–700.

Bergman, C.M., Fryxell, J.M., Cormack Gates, C. and Fortin, D. (2001) Ungulate foraging strategies: energy maximizing or time minimizing? *Journal of Animal Ecology* 70, 289–300.

Berneis, K., Vosmeer, S. and Keller, U. (1996) Effects of glucocorticoids and of growth hormone on serum leptin concentrations in man. *European Journal of Endocrinology* 135, 663.

Brosh, A., Aharoni, Y. and Holzer, Z. (2002) Energy expenditure estimation from heart rate, validation by long-term energy balance measurement in cows. *Livestock Production Science* 77, 287–299.

Chilibroste, P. (1999) Grazing time: the missing link. A study of the plant–animal interface by integration of experimental and modeling approaches. PhD thesis, Wageningen Agricultural University, Wageningen, The Netherlands, 190 pp.

Christian, K.R., Freer, M., Donnelly, J.R., Davidson, J.L. and Armstrong, J.S. (1978) *Simulation of Grazing Systems*. Centre for Agricultural Publishing Documentation, Wageningen, The Netherlands.

Collier, G. and Johnson, D.F. (2004) The paradox of satiation. *Physiology & Behavior* 82, 149–153.

Conrad, H.R., Pratt, A.D. and Hibbs, J.W. (1964) Regulation of feed intake in dairy cows. I. Change in importance of physical and physiological factors with increasing digestibility. *Journal of Dairy Science* 47, 54–62.

Considine, R.V., Nyce, M.R., Kolaczynski, J.W., Zhang, P.L., Ohannesian, J.P., Moore, J.H. Jr, Fox, J.W. and Caro, J.F. (1997) Dexamethasone stimulates leptin release from human adipocytes: unexpected inhibition by insulin. *Journal of Cellular Biochemistry* 65, 254.

Cooper, S.D.B., Kyriazakis, I. and Nolan, J.V. (1995) Diet selection in sheep: the role of the rumen environment in the selection of a diet from two feeds that differ in their energy density. *British Journal of Nutrition* 74, 39–54.

Cordy, D.R. (1954) Nigropallidal encephalomalacia in horses associated with ingestion of yellow star this-
   tle. *Journal of Neuropathology and Experimental Neurology* 13, 330–342.
Cottrell, D.F. and Iggo, A. (1984) The responses of duodenal tension receptors in sheep to pentagastrin,
   cholecystokinin and some other drugs. *Journal of Physiology* 354, 477–496.
Cox, J.E. (1996) Effect of pyloric cuffs on cholecystokinin satiety. *Physiology & Behavior* 60, 1023–1026.
Dougherty, C.T., Bradley, N.W., Cornelius, P.L. and Lauriault, L.M. (1989a) Short-term fasts and the ingestive behaviour of grazing cattle. *Grass and Forage Science* 44, 295–302.
Dougherty, C.T., Cornelius, P.L., Bradley, N.W. and Lauriault, L.M. (1989b) Ingestive behaviour of beef
   heifers within grazing sessions. *Applied Animal Behaviour Science* 23, 341–351.
Eckel, L.A. (2004) Estradiol: a rhythmic, inhibitory, indirect control of meal size. *Physiology & Behavior* 82,
   35–41.
Egan, A.R. and Rogers, Q.R. (1978) Amino acid imbalance in ruminant lambs. *Australian Journal of Agricultural Research* 29, 1263–1279.
Elsen, J.M., Wallach, D. and Charpenteau, J.L. (1988) The calculation of herbage intake of grazing sheep:
   a detailed comparison between models. *Agricultural Systems* 26, 123–160.
Emmans, G.C. (1997) A method to predict the food intake of domestic animals from birth to maturity as a
   function of time. *Journal of Theoretical Biology* 186, 189–199.
Erickson, J.C., Clegg, K.E. and Palmiter, R.D. (1996) Sensitivity to leptin and susceptibility to seizures of
   mice lacking neuropeptide Y. *Nature* 381, 415–421.
Ervin, G.N., Birkemo, L.S., Johnson, M.F., Conger, L.K., Mosher, J.T. and Menius, J.A. Jr (1995) The effects
   of anorectic and aversive agents on deprivation-induced feeding and taste aversion conditioning in
   rats. *Journal of Pharmacology and Experimental Therapy* 273, 1203.
Farningham, D.A.H., Mercer, J.G. and Lawrence, C.B. (1993) Satiety signals in sheep: involvement of CCK,
   propionate, and vagal binding sites. *Physiology & Behavior* 54, 437–442.
Fekete, C., Kelly, J., Mihaly, E., Sarkar, S., Rand, W.M., Legradi, G., Emerson, C.H and Lechan, R.M.
   (2001) Neuropeptide Y has a central inhibitory action on the hypothalamic–pituitary–thyroid axis.
   Endocrinology 142, 2606–2613.
Fernstrom, J.D. (1992) Diet, food intake regulation, and brain serotonin: an overview. In: Bray, G.A. and
   Ryan, D.H. (eds) *The Science of Food Regulation: Food Intake, Taste Nutrient Partitioning, and
   Energy Expenditure*. Louisiana State University Press, Baton Rouge, Louisiana, pp. 195–209.
Finlayson, J.D., Cacho, O.J. and Bywater, A.C. (1995) A simulation model of grazing sheep. I. Animal
   growth and intake. *Agricultural Systems* 48, 1–25.
Foley, W.J., Iason, G.R. and MacArthur, C. (1999) Role of plant secondary metabolites in the nutritional
   ecology of mammalian herbivores: how far have we come in 25 years? In: Jung, H.G. and Fahey, G.C.
   (eds) *Nutritional Ecology of Herbivores. Proceedings of the V International Symposium on the Nutri-
   tion of Herbivores*. American Association of Animal Science, Savoy, Illinois, pp. 130–209.
Forbes, J.M. (1995) Physical limitation of feed intake in ruminants and its interactions with other factors
   affecting intake. In: von Engelhardt, W., Breves, G., Leonhard-Marek, G.S. and Giesecke, D. (eds)
   *Ruminant Physiology: Digestion, Metabolism, Growth and Reproduction. Proceedings of the 8th
   International Symposium in Ruminant Physiology*. Ferdinand Enke Verkag, Stuttgart, Germany, p. 217.
Forbes, J.M. and Kyriazakis, I. (1995) Food preferences in farm animals: why don't they always choose
   wisely? *Proceedings of the Nutrition Society* 54, 429–440.
Forbes, J.M. and Provenza, F.D. (2000) Integration of learning and metabolic signals into a theory of
   dietary choice and food intake. In: Cronjé, P.B. (ed.) *Ruminant Physiology: Digestion, Metabolism,
   Growth and Reproduction*. CAB International, Wallingford, UK, pp. 3–19.
Freer, M. and Campling, R.C. (1965) Factors affecting the voluntary intake of food by cows. *British Jour-
   nal of Nutrition* 19, 195–207.
Friedman, J.M. (2001) Leptin and the regulation of body weight. *Harvey Lectures* 95, 107–136.
Friedman, M.I. (1998) Fuel partitioning and food intake. *American Journal of Clinical Nutrition* 67(suppl.),
   513S–518S.
Geary, N. (2004) Endocrine controls of eating: CCK, leptin and ghrelin. *Physiology & Behavior* 81, 719–733.
Gibbs, J. and Smith, G.P. (1992) Cholecystokinin and bombesin: peripheral signals for satiety? In: Bray,
   G.A. and Ryan, D.H. (eds) *The Science of Food Regulation: Food Intake, Taste Nutrient Partitioning,
   and Energy Expenditure*. Louisiana State University Press, Baton Rouge, Louisiana, p. 224.
Gietzen, D.W. (1993) Neural mechanisms in the responses to amino acid deficiency. *Journal of Nutrition*
   123, 610–625.
Greenwood, G.B. and Demment, M.W. (1988) The effects of fasting on short-term grazing behaviour.
   Grass and Forage Science 46, 29–38.

Grovum, W.L. (1981) Factors affecting the voluntary intake of food by sheep. *British Journal of Nutrition* 45, 183–201.

Grovum, W.L. (1987) A new look at what is controlling intake. In: Owens, F.N. (ed.) *Feed Intake by Beef Cattle*. Symposium Proceedings, Oklahoma State University, Oklahoma, pp. 1–40.

Hamann, A. and Matthaei, S. (1996) Regulation of energy balance by leptin. *Experimental Clinical Endocrinology and Diabetology* 104, 293.

Hoebel, B.G., Leibowitz, S.F. and Hernandez, L. (1992) Neurochemistry of anorexia and bulimia. In: Anderson, G.H. and Kennedy, S.H. (eds) *The Biology of Feast and Famine*. Academic Press, San Diego, California, p. 21.

Hofmann, R.R. (1988) Anatomy of the gastrointestinal tract. In: Church, D.C. (ed.) *The Ruminant Animal*. Prentice-Hall, Englewood Cliffs, New Jersey, pp. 14–43.

Houpt, T.R. (1974) Stimulation of food intake in ruminants by 2-deoxy-D-glucose and insulin. *American Journal of Physiology* 22, 161–167.

Hwa, J.J., Fawzi, A.B., Graziano, M.P., Ghibaudi, L., Williams, P., Van Heek, M., Davis, H., Rudinski, M., Sybertz, E. and Strader, C.D. (1997) Leptin increases energy expenditure and selectively promotes fat metabolism in ob/ob mice. *American Journal of Physiology* 272, R1204.

Illius, A.W., Jessop, N.S. and Gill, M. (2000) Mathematical models of food intake and metabolism in ruminants. In: Theodorou, M.K. and France, J. (eds) *Feeding Systems and Feed Evaluation Models*. CAB International, Wallingford, UK, pp. 21–39.

Ingvartsen, K.L. (1994) Models of voluntary food intake in cattle. *Livestock Production Science* 39, 19–38.

Ingvartsen, K.L. and Boisclair, Y.R. (2001) Leptin and the regulation of food intake, energy homeostasis and immunity with special focus on periparturient ruminants. *Domestic Animal Endocrinology* 21, 215–250.

Kanatani, A., Mashiko, S., Murai, N., Sugimoto, N., Ito, J., Fukuroda, T., Fukami, T., Morin, N., MacNeil, D.J., Van der Ploeg, L.H.T., Saga, Y., Nishimura, S. and Ihara, M. (2000) Role of the Y1 receptor in the regulation of neuropeptide Y-mediated feeding: comparison of wild-type, Y1 receptor-deficient, and Y5 receptor-deficient mice. *Endocrinology* 141, 1011–1016.

Keady, T.W.J., Mayne, C.S. and Kilpatrick, D.J. (2004) An evaluation of five models commonly used to predict food intake of lactating cattle. *Livestock Production Science* 89, 129–138.

Ketelaars, J.J.M.H. and Tolkamp, B.J. (1992) Toward a new theory of feed intake regulation in ruminants. 1. Causes of differences in voluntary feed intake: critique of current views. *Livestock Production Science* 30, 269–296.

Ketelaars, J.J.M.H. and Tolkamp, B.J. (1993) Toward a new theory of feed intake regulation in ruminants. 3. Optimum feed intake: in search of a physiological background. *Livestock Production Science* 31, 235–258.

Kiess, W., Englaro, P., Hanitsch, S., Rascher, W., Attanasio, A. and Blum, W.F. (1996) High leptin concentrations in serum of very obese children are further stimulated by dexamethasone. *Hormone and Metabolic Research* 28, 708.

Kojima, M., Hosoda, H., Matsuo, H. and Kangawa, K. (2001) Ghrelin: discovery of the natural endogenous ligand for the growth hormone secretogogue receptor. *Trends in Endocrinology and Metabolism* 12, 118–126.

Kruk, Z.L. and Pycock, C.J. (1991) *Neurotransmitters and Drugs*. Chapman & Hall, London, 204 pp.

Kyriazakis, I. and Emmans, G.C. (1991) Diet selection in pigs: dietary choices made by growing pigs following a period of underfeeding with protein. *Animal Production* 52, 337–346.

Kyriazakis, I. and Oldham, J.D. (1993) Diet selection in sheep: the ability of growing lambs to select a diet that meets their crude protein (nitrogen × 6.25) requirements. *British Journal of Nutrition* 69, 617–629.

Kyriazakis, I., Emmans, G.C. and Whittemore, C.T. (1990) Diet selection in pigs: choices made by growing pigs given foods of different protein concentrations. *Animal Production* 51, 189–199.

Kyriazakis, I., Emmans, G.C. and Taylor, A.J. (1993) A note on the diets selected by goats given a choice between two foods of different protein concentrations from 44 to 103 kg live weight. *Animal Production* 56, 151–154.

Kyriazakis, I., Oldham, J.D., Coop, R.L. and Jackson, F. (1994) The effect of subclinical intestinal nematode infection on the diet selection of growing sheep. *British Journal of Nutrition* 72, 665–677.

Laca, E.A. and Demment, M.W. (1996) Foraging strategies of grazing animals. In: Hodgson, J. and Illius, A.W. (eds) *The Ecology and Management of Grazing Systems*. CAB International, Wallingford, UK, pp. 137–158.

Larsson, H. and Ahren, B. (1996) Short-term dexamethasone treatment increases plasma leptin independently of changes in insulin sensitivity in healthy women. *Journal of Clinical Endocrinology and Metabolism* 81, 4428.

Leek, B.F. (1986) Sensory receptors in the ruminant alimentary tract. In: Milligan, L.P., Grovum, W.L. and Dobson, A. (eds) *Control of Digestion and Metabolism in Ruminants*. Prentice-Hall, Englewood Cliffs, New Jersey, pp. 3–17.

Legradi, G., Emerson, C.H., Ahima, R.S., Flier, J.S. and Lechan, R.M. (1997) Leptin prevents fasting-induced suppression of prothyrotropin-releasing hormone messenger ribonucleic acid in neurons of the hypothalamic paraventricular nucleus. *Endocrinology* 138, 2569.

Mårell, A., Ball, J.P. and Hofgaard, A. (2002) Foraging and movement paths of female reindeer: insights from fractal analysis, correlated random walks, and Levy flights. *Canadian Journal of Zoology* 80, 854–865.

Marsh, D.J., Hollopeter, G., Kafer, K.E. and Palmiter, R.D. (1998) Role of the Y5 neuropeptide Y receptor in feeding and obesity. *Nature Medicine* 4, 718–721.

Mele, P.C., McDonough, J.R., McLean, D.B. and O'Halloran, K.P. (1992) Cisplatin-induced conditioned taste aversion: attenuation by dexamethasone but not zacopride or GR38032F. *European Journal of Pharmacology* 218, 229–236.

Miell, J.P., Englaro, P. and Blum, W.F. (1996) Dexamethasone induces an acute and sustained rise in circulating leptin levels in normal human subjects. *Hormone and Metabolic Research* 28, 704.

Miner, J.L. (1992) Recent advances in the central control of intake in ruminants. *Journal of Animal Science* 70, 1283–1289.

Moore, B.O. and Deutsch, J.A. (1985) An antiemetic is antidotal to the satiety effects of cholecystokinin. *Nature* 315, 321.

Newman, J.A., Parsons, A.J., Thornley, J.H.M., Penning, P.D. and Krebs, J.R. (1995) Optimal diet selection by a generalist herbivore. *Functional Ecology* 9, 255–268.

NRC (National Research Council) (2000) *Nutrient Requirements of Beef Cattle*. National Academy Press, Washington, DC, 248 pp.

NRC (National Research Council) (2001) *Nutrient Requirements of Dairy Cattle*. National Academy Press, Washington, DC, 408 pp.

Onaga, T., Onodera, T., Mineo, H. and Kato, S. (1995) Cholecystokinin does not act on the efferent pathway of cholinergic and adrenergic nerves to inhibit ruminal contractions in sheep (*Ovis aries*). *Comparative Biochemistry and Physiology* 111A, 51.

Ørskov, E.R. (1994) Plant factors limiting roughage intake in ruminants. In: Thacker, P.A. (ed.) *Livestock Production in the 21st Century*. University of Saskatchewan, Canada, pp. 1–9.

Papaspyrou-Rao, S., Schneider, S.H., Petersen, R.N. and Fried, S.K. (1997) Dexamethasone increases leptin expression in humans *in vivo*. *Journal of Clinical Endocrinology and Metabolism* 82, 1635.

Parsons, A.J., Thornley, J.H.M., Newman, J. and Penning, P.D. (1994) A mechanistic model of some physical determinants of intake rate and diet selection in a two-species temperate grassland sward. *Functional Ecology* 8, 187–204.

Patterson, D.M., Mc Gilloway, D.A., Cushnahan, A., Mayne, C.S. and Laidaw, A.S. (1998) Effect of duration of fasting period on short-term intake rates of lactating dairy cows. *Animal Science* 66, 299–305.

Pedrazzini, T., Pralong, F. and Grouzmann, E. (2003) Neuropeptide Y: the universal soldier. *Cell and Molecular Life Sciences* 60, 350–377.

Pittroff, W. and Kothmann, M.M. (1999) Intake regulation and diet selection in herbivores. In: Jung, H.G. and Fahey, G.C. (eds) *Nutritional Ecology of Herbivores. Proceedings of the V International Symposium on the Nutrition of Herbivores*. American Association of Animal Science, Savoy, Illinois, pp. 366–422.

Pittroff, W. and Kothmann, M.M. (2001a) Quantitative prediction of feed intake in ruminants. 1. Conceptual and mathematical analysis of models for sheep. *Livestock Production Science* 71, 131–150.

Pittroff, W. and Kothmann, M.M. (2001b) Quantitative prediction of feed intake in ruminants. 2. Conceptual and mathematical analysis of models for cattle. *Livestock Production Science* 71, 151–169.

Pittroff, W. and Kothmann, M.M. (2001c) Quantitative prediction of feed intake in ruminants. 3. Comparative example calculations and discussion. *Livestock Production Science* 71, 171–181.

Poppi, D.P., France, J. and McLennan, S.R. (2000) Intake, passage and digestibility. In: Theodorou, M.K. and France, J. (eds) *Feeding Systems and Feed Evaluation Models*. CAB International, Wallingford, UK, pp. 35–52.

Powley, T.L. and Phillips, R.J. (2004) Gastric satiation is volumetric, intestinal satiation is nutritive. *Physiology & Behavior* 81, 69–74.

Provenza, F.D. (1996) Acquired aversions as the basis for varied diets of ruminants foraging on rangelands. *Journal of Animal Science* 74, 2010–2020.

Rogers, Q.R. and Egan, A.R. (1975) Amino acid imbalance in the liquid-fed lamb. *Australian Journal of Biological Science* 28, 169–181.

Roseler, D.K., Fox, D.G., Pell, A.N. and Chase, L.E. (1997) Evaluation of alternative equations for prediction of intake for Holstein dairy cows. *Journal of Dairy Science* 80, 864.

Rosenbaum, M., Murphy, E.M., Heymsfeld, S.B., Matthews, D.E. and Leibel, R.L. (2002) Low dose leptin administration reverses effects of sustained weight-reduction on energy expenditure and circulating concentrations of thyroid hormones. *Journal of Clinical Endocrinology and Metabolism* 87, 2391–2394.

Samuelsson, B., Uvnäs-Moberg, K., Gorewit, R.C. and Svennersten-Sjaunja, K. (1996) Profiles of the hormones somatostatin, gastrin, CCK, prolactin, growth hormone and cortisol. II. In dairy cows that are milked during food deprivation. *Livestock Production Science* 46, 57.

Sauvant, D., Baumont, R. and Faverdin, P. (1996) Development of a mechanistic model of intake and chewing activities of sheep. *Journal of Animal Science* 74, 2785–2802.

Schluter, D. (1981) Does the theory of optimal diets apply in complex environments? *American Naturalist* 118, 139–147.

Settegast, H. (1868) *Die Tierzucht*. Verlag Korn, Breslau, Germany.

Shimabukuro, M., Koyama, K., Chen, G., Wang, M.Y., Trieu, F., Lee, Y., Newgard, C.B. and Unger, R.H. (1997) Direct antidiabetic effect of leptin through triglyceride depletion of tissues. *Proceedings of the National Academy of Sciences USA* 94, 4637.

Sprinkle, J.E., Holloway, J.W., Warrington, B.G., Ellis, W.C., Stuth, J.W., Forbes, T.D.A. and Greene, L.W. (2000) Digesta kinetics, energy intake, grazing behaviour, and body temperature of grazing beef cattle differing in adaptation to heat. *Journal of Animal Science* 78, 1608–1624.

Stanley, B.G. and Leibowitz, S.F (1985) Neuropeptide Y injected in the paraventricular hypothalamus: a powerful stimulant of feeding behavior. *Proceedings of the National Academy of Sciences* 82, 3940–3943.

Svoboda, F. (1937) Das Sättigungsproblem bei der Milchviehfütterung. *Ber. d. XI. Milchw. Weltkongr. Berlin* 1, 168.

Taweel, H.Z., Tas, B.M., Dijkstra, J. and Tamminga, S. (2004) Intake regulation and grazing behaviour of dairy cows under continuous stocking. *Journal of Dairy Science* 87, 3417–3427.

Taylor, St. C.S. (1980) Genetic size-scaling rules in animal growth. *Animal Production* 30, 161–165.

Thornley, J.H.M., Parsons, A.J., Newman, J., Penning, P.D. (1994) A cost–benefit of grazing intake and diet selection in a two-species temperate grassland sward. *Functional Ecology* 8, 5–16.

Tolkamp, B.J. and Ketelaars, J.J.M.H. (1992) Toward a new theory of feed intake regulation in ruminants. 2. Costs and benefits of feed consumption: an optimization approach. *Livestock Production Science* 30, 297–317.

Tschöp, M., Smiley, D.L. and Heiman, M.L. (2000) Ghrelin induces adiposity in rodents. *Nature* 407, 908–913.

Ungar, E.D. (1996) Ingestive behaviour. In: Hodgson, J. and Illius, A.W. (eds) *The Ecology and Management of Grazing Systems*. CAB International, Wallingford, UK, pp. 185–218.

Uvnäs-Moberg, K. (1994) Role of efferent and afferent vagal nerve activity during reproduction: integrating function of oxytocin on metabolism and behaviour. *Psychoneuroendocrinology* 19, 687.

Vadiveloo, J. and Holmes, G. (1979) The prediction of the voluntary feed intake of dairy cows. *Journal of agricultural Science* 93, 553–562.

van der Kooy, D. (1984) Area postrema: site where cholecystokinin acts to decrease food intake. *Brain Research* 295, 345.

Wang, J., Liu, R., Hawkins, M., Barzilai, N. and Rossetti, L. (1998) A nutrient sensing pathway regulates leptin gene expression in muscle and fat. *Nature* 393, 684–687.

Washburn, B.S., Jiang, J.C., Cummings, S.L., Dixon, K. and Gietzen, D.W. (1994) Anorectic responses to dietary amino acid imbalance: effects of vagotomy and tropisetron. *American Journal of Physiology* 266(B part 2), R1922–R1927.

Weston, R.H. (1996) Some aspects of constraint to forage consumption by ruminants. *Australian Journal of Agricultural Research* 47, 175–197.

Weston, R.H. (2002) Constraints on feed intake by grazing sheep. In: Freer, M. and Dove, H. (eds) *Sheep Nutrition*. CAB International, Wallingford, UK, pp. 27–50.

Weston, R.H. and Poppi, D.P. (1987) Comparative aspects of food intake. In: Hacker, J.B. and Ternouth, J. (eds) *Nutrition of Herbivores*. Academic Press, Sydney, Australia, p. 138.

Williams, G., Bing, C., Cai, X.J., Harrold, A., King, P.J. and Liu, X.H. (2001) The hypothalamus and the control of energy homeostasis. Different circuits, different purposes. *Physiology & Behavior* 74, 883–701.

Williams, G., Cai, X.J., Elliot, J.C. and Harrold, J.A. (2004) Anabolic neuropeptides. *Physiology & Behaviour* 81, 211–222.

Woodward, S.R. (1997) Formulae for predicting animals' daily intake of pasture and grazing time from bite weight and composition. *Livestock Production Science* 52, 1–10.

Zammit, V.A. (1990) Ketogenesis in the liver of ruminants – adaptations to a challenge. *Journal of Animal Science* 115, 155–162.

17 Adjustment of Feeding Choices and Intake by a Ruminant Foraging in Varied and Variable Environments: New Insights from Continuous Bite Monitoring

C. Agreil,[1] M. Meuret[1] and H. Fritz[2]

[1]INRA – Unité d'Ecodéveloppement, Site Agroparc – Domaine Saint Paul, 84914 Avignon Cedex 9, France; [2]CNRS – Centre d'Etudes Biologiques de Chizé UPR 1934 du CNRS, 79360 Beauvoir-sur-Niort, France

Introduction

European natural grasslands and rangelands are attracting new attention for their environmental values as habitats for threatened fauna and flora species. They also contribute to the diversity of landscapes, greatly appreciated by urban people seeking the calm and pleasure of nature. This explains why the European agri-environmental policy (Buller *et al.*, 2000) is now encouraging livestock farmers to adopt grazing practices that contribute to biodiversity conservation. Consequently, nature and land-use managers, as well as livestock technicians, have to propose technical specifications to be included in grazing contracts of livestock farmers. This leads them to questioning the scientific community about the relationship between domestic herbivores and the environment.

However, there is a serious lack of knowledge on how to graze flocks on highly diversified plant communities. This is the case in particular when plant communities are subject to shrubby dynamics, as a consequence of sudden land abandonment after several centuries of multipurpose uses, as in most of continental European rangelands (Deverre *et al.*, 1995; Hubert *et al.*, 1995). Livestock farmers are not familiar with grazing management on 'plant mosaics' that support biodiversity (Wallis de Vries *et al.*, 1998), as they were strongly encouraged by agricultural modernization to focus on cultivated and homogeneous herbaceous pastures. As a consequence, technical specifications in most agri-environmental contracts favour mechanized land clearance (Agreil, 2003), despite the cost, thus undermining the expected impact by the flocks. Such specifications are supported by the paradigm that considers that domestic flocks are able to forage only on 'herbaceous' structures: either grasses and forbs within the herbaceous strata or the current-year growth of shoots from shrubs and lianas. This contradicts the fact that, notably for habitat restoration, grazing impact is expected on coarse and large grass tussocks, together with tree and shrub foliages. Knowledge should hence be produced about the feeding value of natural grasslands and rangelands and the appropriate grazing rules to be included in agri-environmental contracts.

How to Cope with a Varied and Variable Environment?

Domestic flocks have to cope with varied and variable vegetation

The feed offer for domestic flocks grazed on natural grasslands and rangelands is highly varied and variable. When the aim is to have grazing impact on shrub dynamics that endanger biodiversity, flocks are grazed on successive, relatively small, paddocks or herding sectors, in order to apply high grazing pressure. Such management creates emerging characteristics of the vegetation that have been poorly investigated in relation to their effect on grazing behaviour. On the one hand, it is highly heterogeneous in terms of plant part nature, size, structure and nutrient content, a situation that has been widely studied in ecology, but generally over large spatio-temporal scales. On the other hand, it varies very quickly during a paddocking sequence as a result of grazing, which is a situation usually studied in domestic herbivores, but on rather homogeneous vegetation such as cultivated grasslands.

In these varied and variable grazing situations, recent advances in herbivory have pointed out that the evaluation of feeding value will benefit from first considering the animal's point of view (Illius and Hodgson, 1996; Provenza and Launchbaugh, 1999; Dumont *et al.*, 2001; Provenza, 2003). From a research perspective, this means a better understanding of the herbivore's feeding strategy at various scales: the set of behavioural rules and adjustments allowing them to adapt their feeding behaviour so as to satisfy their nutritional requirements within an environment that varies both in space and in time (Laca and Demment, 1996).

Underlying knowledge on herbivore feeding strategies

Many studies of the behavioural and mechanistic adjustment of grazing have been carried out. Results obtained on the foraging behaviour in rather simplified grazing contexts, i.e. homogeneous vegetation or artificial swards in which heterogeneity remains controlled (e.g. Black and Kenney, 1984; Lundberg and Danell, 1990; Spalinger and Hobbs, 1992; Ginnett and Demment, 1995), led to a detailed knowledge of both digestive physiology and the harvesting and chewing processes.

Three plant characteristics are generally used to describe, explain and predict the foraging behaviour of ruminants (Laca and Demment, 1996): plant structure, digestible matter content and toxin content. First, plant structure determines the accessibility and size of plant organs, hence influencing the two central components of the instantaneous intake rate (IR, g/min): bite mass (BM, g/bite) and bite frequency (BF, bite/min) (Spalinger and Hobbs, 1992). Consequently, plant structure may have a major impact on the daily intake as it is the product of IR and daily feeding time. Second, the digestible matter content of plant organs also plays a role in diet choices. It determines the turnover rate in the gut and can become a constraining factor when fibre content is too high, and hence digestibility too low (Westoby, 1974; Belovsky, 1986; Illius and Gordon, 1991). Herbivores thus face a trade-off between food items that ensure adequate intake quantity and those that ensure adequate intake quality (Van der Wal *et al.*, 2000). Finally, the toxin content in plant parts can also strongly influence the diet choices. In heterogeneous vegetation, the very wide spectrum of plant species and plant parts observed in the ruminant's diet has been interpreted as an adaptive response that can dilute the effects of plant toxins (Launchbaugh, 1996; Hägele and Rowell-Rahier, 1999).

These results have been considered in relation to a global effort for a theoretical formalization of the foraging behaviour, which gave birth to many controversies and debates in particular because of a great diversity of grazing contexts. One of the first theoretical frameworks is the 'functional response', a behavioural adjustment that consists of a compensation for the decrease in BM by an increase in BF (Allden and Whittaker, 1970; Spalinger and Hobbs, 1992). Because ungulates are more usually faced with heterogeneous vegetation, nutritional ecologists have focused heavily on the quality–quantity trade-off. This trade-off has been especially developed for grazers within the 'forage maturation hypothe-sis' (Fryxell, 1991), which states that availabil-ity (biomass or sward height) is the main constraint when animals graze on short swards, while quality (digestibility or nutrient content) is the major constraint on taller swards, i.e. when forage is mature. As a consequence, herbivores are expected to select swards of intermediate height as a trade-off between constraints (Fryx-ell, 1991; Wilmshurst *et al.*, 1999).

Despite the strong influence of plant characteristics on the foraging behaviour, experimental results and their integration into a theoretical framework, if taken alone, are often poorly predictive of behavioural pat-terns, when ruminants graze natural grasslands and rangelands. This may be attributed to the fact that many animal characteristics also impact the expected behavioural responses. Animal species determines mouth size (Pérez-Barberia and Gordon, 1998), prehensive apti-tude (Owen-Smith and Cooper, 1987a) and digestive ability (Van Soest, 1994). Models have also been built for taking into account their nutritional requirements, as a function of their physiological stage (Mertens, 1996).

It has been more difficult to take account of the various temporal dependencies on past events: the long-term influence of learning on preferences (Provenza, 1995; Forbes and Provenza, 2000), the influence of basal diet on ruminal flora composition (Van Soest, 1994), the effect of recent diet on foraging choices (Newman *et al.*, 1992), the effects of the dynamics of rumen filling and nutrient kinetics on satiety (Forbes, 1995), the appearance of a specific satiety for each feed when consumed during a certain lapse of time (Rolls, 1986). The grazing of domestic ruminants on hetero-geneous vegetation in paddocks results in dynamic behavioural patterns that not only depend on plant and animal characteristics but are also time-dependent. Hence, when design-ing experimental plans, it is important to con-sider the management practices to which the animals have become accustomed since they determine the animal's feeding habits.

In this chapter, we focus on the dynamic behavioural patterns that occur at the short timescales: from the paddocking sequence to the intraday organization as described by instantaneous behaviour at the bite scale. We

start by emphasizing the importance of the research design phase. We then review the recording methods that make it possible to acquire multiscale and multivariable data-sets on foraging behaviour in heterogeneous vege-tation. We also present some recent advances in understanding the behavioural response to both diversity and variability of food offer. And finally, we conclude this chapter with a pro-posal for rather simple herd management in the paddock, based on the identification of feeds that animals use in implementing their feeding strategy.

Principles for On-farm Behavioural Studies

Implementation of environmental policies that involve grazing cannot be postponed until knowledge has been accumulated on each process that finally leads to the production of synthetic, efficient action models. Moreover, the need for action, regardless of the inevitable uncertainties that will subsist in some of its determining factors, seems to call for systemic research produced from on-site observation (Checkland, 1981; Legay, 1997).

This requires special attention to ensure that new research mechanisms meet the needs for both theoretical and practical knowledge as realistically as possible. What phenomena should be observed, where, when and how, in order to produce models with overall values based on site-specific and timebound observa-tions that can be translated into proposals for rules of action? These are the thoughts that underlie the agronomic work on practices and techniques (Brossier *et al.*, 1993) and the eco-logical work on environments and species (Legay, 1993). It has already led to the pro-duction of original ways to managing herds on the rangelands (Meuret, 1993; Hubert and Bonnemaire, 2000; Agreil, 2003). What are the main lessons that we can use for the sub-ject of this chapter?

The relevance of research depends on the relevance of the objects being studied. The full complexity of these objects must be recog-nized because of their mixture of ecological and biological processes, human practices, which affect and are affected by these processes, and even the scientist's knowledge, which guides him in selecting the conditions and methods of observation. We are indeed no longer working with a 'laboratory model' (Hatchuel, 2000) that tries to isolate the world under observation, reconstitute it experimentally in order to manipulate certain selected processes or test the effect of factors duly identified beforehand. We are now dealing with a type of 'field model' (Hatchuel, 2000), underpinned by the 'naturalization' of the objects under study. We therefore have to be careful that the weight of prior knowledge does not mark the new scientific design by imposing too many conceptual constraints.

In our case, we would like to propose the following approach, on the basis of our experience:

1. The object studied is the 'feeding motivation' of the domestic flocks, their strategy on diverse, variable vegetation for constituting satisfactory diets with regards to the production objectives. This is different from the traditional approach that seeks to construct feed value tables for the different feeds. In tables, an intrinsic value is assigned to each plant, depending on its phenological state, conditions of growth and, if applicable, technological processing. Animal expression is considered to be the animal's ingestive capacity and, more importantly, its digestive capacity and nutrient assimilation. In this exercise we are somehow returning the domestic animals to nature, 'renaturalizing' them, by giving them total freedom to choose and compose their diet, which is close to the observation approach used for wild animals.

2. We have to make observations at temporal and spatial scales defined by grazing practices set out in biodiversity conservation contracts:

• Space is one of the portions of the territory given to grazing: fenced paddocks for farmed animals or grazing sectors for herded flocks.
• Time is the duration of the paddocking sequence or of the successive passages by a shepherd in a given sector; this duration has to be long enough to obtain a satisfactory impact on the vegetation and its multiannual

dynamics ('modes of plot exploitation', according to Hubert, 1993).

These two dimensions are constitutive of the object of research: what is the feeding motivation of flocks when driven to paddocks during the days required to fulfil the grazing contract? The animals whose behavioural response interests us must be made accustomed to these grazing conditions, and any simplification can be detrimental to the quality of the results.

3. The observation technique is applied to the ingestive behaviour, which traditionally is investigated at the individual scale. The individuals being observed are part of the herd or at least part of a group of farmed animals. These are the usual groups composed of young animals, nursing animals, a mixture of the two, etc. and not groups composed exclusively for the experiment. The purpose is to avoid disturbances caused by herd compositions that do not depict the individual's hierarchical habits and affinities.

4. One prerequisite is that the animals have to be knowledgeable about the grazing conditions. There is no sense in using naive animals that are accustomed to other farming practices or animals that come from an experimental station that does not use the selected grazing area regularly.

5. Preference should be given to 'non-invasive' measuring techniques. In animal farming, direct observation can be adapted to changing and even unexpected conditions, which cannot be done by automated data collection machines. But to avoid distortive errors, when using this technique, special attention should be given to ensure mutual familiarization between the group of animals, the monitored individuals and the observers. Furthermore, the observers can never replace the herder or shepherd, even temporarily, because the animals must remain totally indifferent to the observer's constant presence.

6. The animals should be very familiar with the observation seasons and consequently with the range of vegetation available, both regarding the proportions and the phenological states. To help the expression of a feeding strategy, the vegetation should be diversified, not only regarding its strata (grasses, shrubs, trees, liana, etc.) and specific composition but also regarding the diversity of the plant organs, considering how they can be used by the herbivores under observation (morphology, format, nutritional value, etc.). Concerning vegetation variability over days, it is important for the animals to be able to anticipate the length of time the paddock or grazing area can be used in relation to the changes they perceive, over the days, in the nature and structure of the available vegetation. The length of an observation period in a given place should be established according to the herder's management rules, especially when he is carrying out a grazing contract, and not according to the scientists' usual practices.

Direct Observation, a Fitted Method for Multiscale Analysis of Foraging Behaviour

Recording the diversity of feed items and facilitating multiscale analyses

The nature of data recorded during observation strongly affects future possibilities for making analyses and for modelling. Observation method should be chosen or designed carefully, bearing in mind the objectives of the current research and methods available for analysing and modelling the recorded data. To understand foraging behaviour in spatially varied and temporally variable vegetation, particular attention should be given to both the duration and the resolution of the observations.

Behavioural data have been produced at different timescales by observation methods based on sampling and designed for vegetation assumed to be relatively homogeneous (Forbes, 1988). These methods involve average measurements for a given time interval during which the characteristics of the ingested material remain relatively homogeneous. Using this line of reasoning on diversified grazing environments, the time period being measured has to be shortened because of the frequent changes in food item choices; experimental tests are often carried out during time periods of less than 1 h. But these snapshots of ingestive behaviour do not provide sufficient data to explain the dynamics of the grazing

process, in particular on composite vegetation where the rules adopted by the animals to govern their decisions are complex. More generally, from the viewpoint of ecology, Schmitz (2000) observed: 'Field studies that do generate long-term time series data often do not provide the kind of resolution needed to understand which underlying mechanisms are driving the dynamics.' The same observation was also made (Bergman *et al.*, 2001) for shorter time periods when studying ingestive behaviour, which requires information on short-term IRs (scale of a few minutes) and the longer-term nutritional effects, i.e. from a few hours to a whole day, depending on the author. This implies that observations should not only be maintained during long periods of time but also that observations should be made on the same individuals throughout the whole observation period.

Further, in an effort to take into account plant heterogeneity, many authors have coupled the usual recording methods with specific internal markers (Mayes and Dove, 2000). However, marker-based techniques only distinguish a limited number of species and are not able to identify specific features of the foraging behaviour in response to the characteristics of the plant consumed. More importantly, they do not distinguish the different plant parts of a single species, which may be of very different size, structure and nutritive quality. For this reason, we advocate high-resoluted methods, based on observations at the bite scale.

Direct observation methods thus seem to be an appropriate tool for recording data on diversified vegetation and at different timescales, from the bite level (a few seconds) to the paddocking sequence (several days or weeks). Literature reports that these methods have been widely used during the last 30 years for a variety of ruminant species and plant formations, mostly for recording the amount of material eaten on a daily scale (Neff, 1974; Stobbs, 1975; Currie *et al.*, 1977; Bourbouze, 1980; Meuret *et al.*, 1985; Parker *et al.*, 1993; Genin *et al.*, 1994; Dumont *et al.*, 1995; Gillingham *et al.*, 1997; Mofareh *et al.*, 1997; Kohlmann *et al.*, 1999; Agreil and Meuret, 2004). Since these methods are non-invasive, they are particularly suitable for *in situ* studies of animals on private farms or tractable wild animals (Litvaitis, 2000). These methods are also used as a reference for validating data obtained by automatic bite recorders (Abijaoudé et al., 1999; Delagarde et al., 1999). The main problem with these methods is their BM estimation accuracy (Gordon, 1995). In an effort to improve the accuracy, some authors stratified bites taken off each one of the eaten plant species into different categories of mass and structural composition (Meuret, 1989; Parker et al., 1993; Dumont et al., 1995). These developments were proposed for shrubby rangelands where animals graze each species in successive sequences of monospecific bites. Studying black-tailed deer browsing in deciduous forest, Parker et al. (1993) designed their method to be specific for forbs, shrubs and deciduous trees for which ruminants generally bite individual leaves. In a more diversified grazing environment, where the number of eaten species can be as high as 50/day, many bite categories (BCs) have to be defined.

Description of a global direct observation method

A complete method has recently been proposed for direct and continuous observation of bites during an on-farm experiment (Agreil and Meuret, 2004). It includes some improvements on the earlier methods for recording foraging choices and intake in heterogeneous environments. The principal recent improvements are: (i) continuous recording instead of 1 min grouped observations (Meuret et al., 1994); and (ii) refinement of BCs to be able to record bites in a highly heterogeneous and variable environment. As it is well suited for analysing foraging behaviour adjustments in heterogeneous environments, we are presenting the three main steps involved: a familiarization procedure to ensure that the flock and the observers are accustomed to each other; the designing of a bite coding grid for an accurate description of feed items harvested; and the recording of flock activities to validate the consistency of the monitored individual's behaviour.

The familiarization procedure

The purpose of the familiarization procedure is to accustom the animals to having an observer very close to them. It is obvious that both the familiarization procedure and the observations should be carried out by the same observers. First the observer spends several whole days in the paddock in order to accustom the whole flock to his presence. The success of the exercise is measured in terms of the distance the animals keep from the observer (Fig. 17.1). This first step ends when the observer can move near and within the flock without any visible effect on the animals' movements and attitudes. Second, while alternating movement within the flock and close monitoring, the observer selects one animal within the flock for close and continuous focal observation during full daytime periods. He looks for individuals that seem indifferent to his uninterrupted presence. The full-time presence of an observer automatically changes the social status of an individual. This is why, at the beginning of this step, the individual being identified for close monitoring should not be a leader or a potential leader. Here again, it is important to listen to the herder's advice if he knows the social hierarchy of his flock well. At the end of this step, about 15–20% of the individuals are considered to meet the prerequisites for full-time close monitoring. The monitored individual is subsequently tested to ensure that its behaviour is consistent with that of the flock. The familiarization procedure ends when the observer can move around and talk for the whole day while accompanying the monitored individual at a distance of ~0.5–1.5 m.

The bite coding grid

The proposed bite coding grid was designed for small ruminants (sheep and goats). The bites were grouped into 40 BCs based on the shape and size of the plant organs selected and the way they were cropped. To ensure that the BCs were as broadly applicable as possible to small ruminants grazing in various conditions (but not too numerous), botanical distinctions were carefully avoided. Information on plant

Fig. 17.1. During direct observation of the continuously monitored animal, the observer remained alongside the animal so as not to obstruct its spontaneous line of movement. By following the animal closely, the observer can easily identify bites and dictate the corresponding codes on an audio-recorder.

species was only recorded during the experiments and added as a modifier to BC.

The bite coding grid is illustrated in Fig. 17.2. It includes the 40 BCs and allows for the categorization of bites taken off several hundred plant species. Each little pictogram is roughly sketched and represents the 'ideal type' of a BC, in other words, the median case of all possible observations, grouped into one single category. For each category, the physiognomy of the plant matter in the bite is described by symbols (see Fig. 17.2). For 19 BCs, the bitten matter, not stretched out, is described on the left-hand side, in centimetres. The U-shape icon with a thicker line stands for the animal's mouth. It is drawn to most accurately reflect the feeding technique for a given BC. Codes on this grid are selected to form a 'language', which the observer can easily learn by heart. Since the codes are monosyllabic, they are easy to dictate immediately after the name of the selected botanical species.

During the experiment itself, the observer remains alongside the animal so as not to obstruct its spontaneous line of movement (Fig. 17.1). He follows the monitored animal in order to observe its mouth continuously and to identify BC and plucked plant species. Back

at the laboratory, audio recordings of the chronology of bites can easily be converted into tables using behaviour-specific software (e.g. *The Observer* software; Noldus Information Technology, 1995).

After each daily observation sequence, the observer removes plant samples by manually simulating each BC recorded. These samples are then used to estimate BM and nutritive quality, using near infrared reflectance spectrometry (Schenk *et al.*, 1979; Dardenne, 1990).

Validation of consistency in the individual's behaviour

In order to test the consistency in the individual's behaviour, common measurements are then carried out on the selected animal and on the rest of the animals in the group. A flock observer records, by scan sampling (Altman, 1974), the time devoted by the flocks to ingestion, movement and rest. Every 10 min and for each major collective interruption in activities, the proportion of individuals allocated to each of the three activities is recorded. This makes it possible to compare the daily duration of the

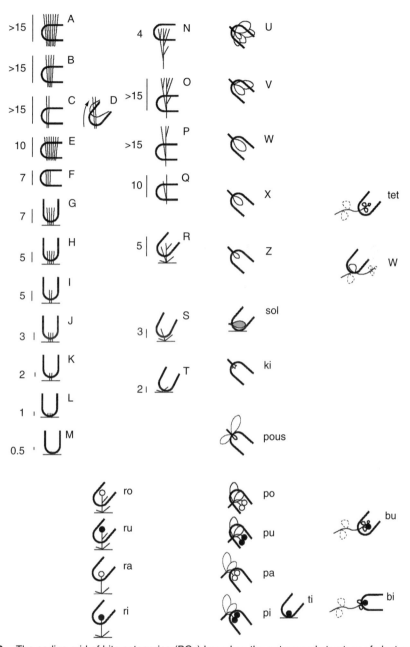

Fig. 17.2. The coding grid of bite categories (BCs) based on the nature and structure of plant portions clipped. (From Agreil and Meuret, 2004.) The U-shaped icon stands for the jaws of the small ruminant. The plant portions are indicated by small icons that symbolize their physiognomy: fine lines (chlorophyllic stems of shrubs or leaf blades and stems of grasses), ovals (tree and shrub leaves), open circles (flowers) and filled circles (fruits). The length of leaves, laid out but not stretched, is indicated in centimetres to the left of the icons for 19 BCs. The mono- and bi-syllabic codes dictated during the observation are given to the right of the icons.

feeding activity, and its distribution throughout each day, for the flock and for the individual.

An individual's behavioural consistency is also tested by comparing the faecal matter reflection in near infrared spectroscopy (NIRS). This measurement is considered to be a reliable indicator of differences in the quality and composition of diets between the individuals (Lyons and Stuth, 1992).

Advantages, shortcomings and future prospects

Intakes estimated by direct observation methods have been validated by making comparisons with intakes estimated by other methods: (i) values obtained with an external marker (Meuret *et al.*, 1985); (ii) BMs estimated using different techniques (Parker *et al.*, 1993); and (iii) estimated intake of oesophageally fistulated animals (Wallis de Vries, 1995). The strongest point in direct observation is that variables known to be indicators of intake in ruminants (nitrogen or fibre content, intake at preceding meal, etc.), expressed for long time intervals, can thus be related to instantaneous bite choices observed at the bite scale.

The observation and recording method can be used to observe feed intake continuously for days on end, and hence allows for dynamic exploration of ingestive behaviour. This should pave the way to an original approach to a series of research issues. First, the data-sets should be useful in exploring variations in feed preferences within a day or from one day to the next, provided the feed availability could be accurately estimated. Studies of feeding behaviour should not ignore the animals' habits and diets before and after an experimental test: preferences are governed by the animals' usual diet (Newman *et al.*, 1992), by their physiological state (Parsons *et al.*, 1994) and by the items most recently consumed during recent meals (Newman *et al.*, 1995). Second, continuous recording makes it possible to characterize the temporal organization of feeding choices during these meals, and thus highlight effects related to feeding synergy between plants or effects resulting from the way the farmer grazes his animals (Meuret *et al.*, 1994). Third, direct observation can also generate data of use in the study of the

mechanics of biting. How ruminants harvest feed is a process with various successive, sometimes overlapping, steps: identification of plant organ(s), prehension, clipping, chewing, swallowing. Functional relations have been established between these behavioural steps and the characteristics of the plant parts removed (size and composition) (Spalinger and Hobbs, 1992; Ginnett and Demment, 1995; Woodward, 1998). Recent improvements of direct observation methods should lead to adapted data-sets for carrying out such studies in diversified vegetation.

The originality of this method results from its capacity to couple detailed and short-term measurements at bite scale with the long-term intake balance (Agreil *et al.*, 2005). The method is designed to consider first and foremost the animal's 'point of view' on the available feeding resources, as recommended by Illius and Hodgson (1996).

As described above, the direct recording of bites also has some shortcomings. Its practical utilization is restricted to animals that can be followed at a few metres. Hence, it seems to be inappropriate for wild animals, except in the particular cases of tamed animals (Litvaitis, 2000).

Behavioural Patterns of Sheep Grazing in Heterogeneous Environments

To understand the ruminants' feeding strategy in spatially varied and temporally variable environments, the choice rules in response to diversity and variability need to be explored at relevant timescales. Indeed, choice rules determine the nature of selected plant organs and hence, intake levels and diet composition. But on heterogeneous vegetation, the usual concepts related to choice, such as preference and selection, should be considered with caution. The livestock farmers' usual grazing practices and the great heterogeneity of plant parts on offer give dynamic properties to the foraging behaviour. Hence, particular emphasis should be given to the identification and the choice of the appropriate temporal scales. Some recent studies on ingestive behaviour and choice rules pointed out some limits and possible improvements to the usual theoretical frameworks.

Very high daily intake in small ruminants grazing on varied environments

Observations we made in southern France revealed very high daily intakes in sheep and goats grazing in various conditions and environments (Meuret et al., 1985; Meuret, 1989; Meuret and Bruchou, 1994; Agreil, 2003; Agreil and Meuret, 2004). Data are presented in Fig. 17.3. They are compared to the reference models published by Morley (1981) and Van Soest (1994) (dotted line, bottom right), which were confirmed by data (open circles) obtained with sheep fed indoors on green fodder (Dulphy et al., 1999). Daily dry matter (DM) intake is plotted against the daily average

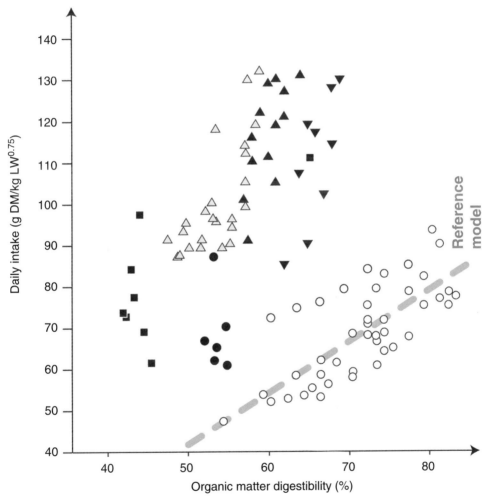

Fig. 17.3. Relation between daily intake of dry matter (DM) and organic matter digestibility (OMd) of the daily diet. Each filled symbol represents a single observation of daily intake for an individual: lactating goats herded during the summer in oak coppice (black triangles on their base; Meuret et al., 1994; Meuret, 1997b; Baumont et al., 2000); lactating goats fed fresh leafy oak branches frequently renewed in digestibility crates (open triangles; Meuret, 1988; Meuret and Giger-Reverdin, 1990); dry ewes grazed in early spring on natural swards encroached by edible shrubs (black circles; Agreil et al., 2005) and dry ewes grazed in summer on natural swards encroached by edible shrubs (black squares; Agreil et al., 2005). The open circles represent daily intakes for dry ewes fed fresh grass at the trough (Baumont et al., 1999). The dotted line symbolizes the reference models published by Morley (1981) and Van Soest (1994).

of the diet's organic matter digestibility (OMd). To accommodate differences in live weight (LW), all daily intakes were expressed in g/kg of metabolic weight (MW = $LW^{0.75}$).

Some of our observations come from lactating goat flocks that were herded during the summer in oak coppice (black triangles on their base), and only received a small supplement twice daily. In that situation, it was clearly the herder's interventions during the grazing routes that essentially boosted the IR during each meal and consequently increased the intake from pasture (Meuret *et al.*, 1994; Meuret, 1997a; Baumont *et al.*, 2000). These surprisingly high intake levels, ranging from 90 to 130 g DM/kg MW for an OMd limited to about 60%, were similar to those obtained from goats grazing in spring within paddocks composed of cultivated swards (black triangles on top). These intake levels are also consistent with those obtained from goats fed fresh leafy oak branches frequently renewed in digestibility crates (Meuret, 1988; Meuret and Giger-Reverdin, 1990; open triangles).

With dry ewes grazed in early spring on natural swards encroached by edible shrubs (Agreil *et al.*, 2005; black circles on Fig. 17.3), daily intakes were partly similar to those with sheep fed indoors on excellent forage, i.e. fresh lucerne (Baumont *et al.*, 1999). But, for grazing in heterogeneous environment in summer (black squares), intakes by sheep were considerably higher than those estimated through the reference model for about the same OMd values. In most of the cases we observed, daily intake levels were twice as high as the ones predicted from the reference model.

These intake values seem to be driven by the animals' dietary diversity. A wide variety of ingestive bites were observed during the meals. Each meal included not only several vegetation species but also a large number of BCs (see the third section in this chapter) for each species. As an example, the meal observed during the summer, when sheep entered a new paddock composed of natural swards encroached by shrubs, is shown in Fig. 17.4 (Agreil and Meuret, 2004). This meal lasted about

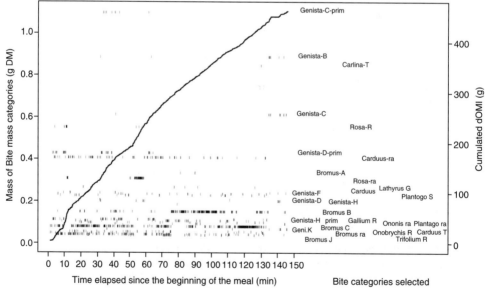

Fig. 17.4. Profile of a meal observed for dry ewes grazed in summer on natural swards encroached by edible shrubs. (From Agreil and Meuret, 2004.) The figure gives a chronicle of bite category (BC) selection: each little vertical bar represents a bite. Its position in relation to the vertical axis on the left gives information on its mass (DM, g). For the 28 BCs most recorded, the name of the plant followed by the code derived from the BC grid (see Fig. 17.1) are given on the right. The effects of these choices on cumulated digestible organic matter ingested (dOMI, g: vertical axis on the right) are shown on the curve.

150 min with almost continuous intake. It was composed of more than 3700 bites taken from 13 botanical species and grouped into 34 BCs. The mass of BCs ranges from 0.03 to 1.09 g DM/bite. The profile of this meal (expressed as cumulated digestible OM intake) resembles the profile of the meals observed for sheep eating indoors a single very palatable fodder (R. Baumont, personal communication, Paris, 2004). Does such a diverse feeding offer work as a 'complex' that boosts the appetite? In the case of goats foraging in oak coppice undergrowth, a specific diversity optimum has been identified as the main determinant factor for intake levels at the meal scale (Meuret and Bruchou, 1994). This raises the question of how to identify the nature of the diversity ruminants perceive and use when achieving their feeding strategies.

Studying the choice rules in varied and variable environments

It is not easy to study the choice rules of ruminants faced with a diversity of feed items. To apply the concept of feeding preferences requires not only the recording of foraging behaviour by distinguishing the selected plant organs but also the description of available plant organs. The particular features of the feeding contexts we deal with in this chapter should also be considered, as the temporal variation of feed items on offer strongly affects the choice rules.

The question of defining feeding preferences in a heterogeneous environment

To analyse the ruminants' choice rules, literature often refers to the common concept of feeding preferences (Newman et al., 1992; Parsons et al., 1994; Provenza, 1995; Dumont, 1996; Ginane et al., 2002). Ecology has a precise definition for feeding preference, i.e. the probability that a feed will be selected when it is offered in the same conditions as other feed (Johnson, 1980). The differential between the selection rates for two feeds has

to be studied, by considering the 'selection rate' as the offtake/availability ratio (Manly et al., 1993).

But there are two problems that arise in studying preferences. First, certain authors explicitly point out the difficulty in deciding on the relevant variable to record to quantify them (Owen-Smith and Cooper, 1987b). The result is that work on feeding preferences is often based on the measurement of different variables that are evaluated and sometimes compared on the basis of measurements of quantities ingested (Newman et al., 1992; Parsons et al., 1994; Wang and Provenza, 1996), number of bites (Newman et al., 1992) or a recording of the order of consumption of the feed items that have been identified during the experimental trial (O'Reagain and Grau, 1995; Prache et al., 1996). Second, defining available feed items on offer in order to calculate the offtake/availability ratio is not a straightforward exercise in a heterogeneous environment. Some authors have defined available feed items as those encountered by the animal along its path (Etzenhouser et al., 1998). But defining a feed encounter is somewhat speculative. Little is known about the size of the relevant environment, as animals simultaneously use their memory of places, and visual and smell cues to identify available feed items. The question of defining the feed 'on offer' is not specific to animals that are free to move at pasture. It also concerns controlled feeding conditions during laboratory experiments, e.g. when evaluating the feeding value of tree foliages for goats (Meuret et al., 1991). Some authors chose to pre-cut leaves and young shoots before offering. Others preferred to let the animals make their own choices, but used the potential edible matter (PEM) that had been defined previously through observation of the foraging behaviour of very hungry animals (Meuret, 1988).

Progress still needs to be made in understanding the interaction between ruminants and feed on offer. Which cues do animals use for identifying, locating and remembering the feed items on offer? What are the spatio-temporal scales to consider when defining availability? With this in mind, we argue in favour of experimental plans that do not include an a priori decision on the single variable and the single temporal scale to be analysed.

Varied vs. variable: defining the feeding context

When studying the feeding choice rules of ruminants in a heterogeneous environment, a distinction should be made between two feeding contexts. It would be easier to understand foraging adjustments if there were explicit information on whether ruminants were faced with a food on offer that was constant in time or whether the food on offer varied during the time of the study.

- A first approach could be to characterize the feeding choices in relation to a resource that is varied but whose variations in time are not described in the study, which therefore has to be rather short. In this scenario, no reference is made to temporal evolution, and the aim is to quantify the feeding choices as reflected in the assessment made for the selected time period. Such an approach would mean identifying the choice criteria in a constant context.
- A second approach consists of characterizing behavioural responses to a varied offer whose characteristics change during the time period of the study. These temporal variations can be caused by changes in the vegetation (growth, senescence, phenological change) or can be the consequence of grazing (shrinking size, growing scarcity of feed after consumption). In this case, changes in the composition of daily diets are the direct consequence of the consumption order of the feed items. Thus, the phenomenon that this second approach seeks to characterize is changes in the choice criteria in a constant context (defined above), in response to changes in the characteristics of the offer. This approach could lead to the identification of prioritization criteria in a variable context.

Identifying relevant timescales for describing feeding adjustments

After being identified, choice rules should not be considered to have an absolute value. Many studies have shown that foraging behaviour is dynamic. Selection, for example, is a dynamic process. Some authors argued that considering

the encountered feed items is not adequate, regardless of the variable being recorded (total biomass, selected biomass, potential bite size, etc.) (Gillingham *et al.*, 1997). Our observations were consistent with these reports, since the small ruminants we observed selected particular plant parts as they passed through a vegetation unit and later in the bout, when passing through the same vegetation unit again, selected other plant parts and ignored the first ones (C. Agreil and M. Meuret, Drôme, France, 2002; Savoie, France, 2004). As a consequence, the temporal scale adopted in a study is of utmost importance.

Usual timescales in feeding science and nutritional ecology

Available studies cover a broad range of temporal scales: from small timescales at the bite sequence level to intermediate timescales at the feeding bout level, and even to long timescales at the day level or the level of seasonal intakes, including nutritional requirement coverage.

Small-scale studies have led to both mechanistic description of ingestive behaviour (Hanley, 1997) and clarification of choice rules that lead to a mixed diet (as reviewed by Provenza *et al.*, 2003). First, small-scale studies have been useful in investigating ruminants' choice rules. At this scale and in rather controlled experiments, many papers confirm that the plant organs' structure, nutritive quality and toxin content affect choice rules (Launchbaugh, 1996; Ungar, 1996). Second, the functional response of ruminants has been widely modelled on the basis of the initial work of Spalinger and Hobbs (1992), linking animal characteristics (body mass, mouth size, physiological state, etc.) with plant characteristics (nutritional quality, fibrousness, plant organ morphology, etc.).

Despite the high relevance of ingestive bouts for ruminants, few studies have investigated ingestive behaviour dynamics at such an intermediate timescale. On the one hand, when studying choice rules, authors largely considered the ingestive bout as a discrete time interval: use and selection of the feeds or species on offer were quantified and interpreted as a result of choice behaviour (Meuret et al., 1985; Parsons et al., 1994). On the other hand, ingestive behaviour during bouts has been seen as a dynamic process, but attention has only been given to its quantitative contribution to intake. For domestic cows and sheep fed indoors or grazed on homogeneous sward, the cumulative intake curves were modelled as a simple saturation process (Sauvant et al., 1996). In this case, two main factors were considered to affect appetite and satiety, and hence the shape of the cumulative curves: ruminal storage of feeds and nutrients liberation kinetics (Pittroff and Kothmann, 1999). For black-tailed deer browsing on heterogeneous vegetation, Gillingham et al. (1997) observed more diverse shapes of the cumulative curves. The same observation was made with domestic goats grazed on rangelands (Meuret, 1989; Dumont and Meuret, 1993).

For understanding the feeding strategy on highly heterogeneous and variable vegetation, the scale of the bout as a whole is of little relevance. By construction, cumulative signals cannot be affected by short-term variations in feeding choices during bouts, which are of primary importance here. The bout in Fig. 17.4 has a shape that is typical of a saturation process with an exponentially decreasing slope, despite a huge diversity of selected plant parts and its variation during this bout. In order to determine if a feeding choice dynamic is hidden behind this highly regular cumulative curve, we have to consider and model the first derivative as a time series, i.e. IR kinetics, with a short-term resolution (about 1 min). Derivative-based analyses have already been developed for use in analysing the way shepherds stimulate intake during grazing trips by an ordered offer of vegetation diversity (Meuret et al., 1994).

Large-scale studies have been less used for studying the ruminant's ingestive responses when faced with a diversity of feeds. Studies on the ecology of wild ungulates have mainly focused on the matching of nutritional requirements with variations of abundance, morphology and the nutritional quality of plant parts throughout the year (Maizeret, 1988), while studies on domestic ruminants have essentially considered the critical management indicators for animal growth

and performance. Mertens (1996) pointed out that when dealing with the formulation of feed rations for long-term performances, e.g. in lactating cows, nutritionists are more interested in average daily intake than in ingestive response at the meal scale. When forage availability changes rapidly as a result of repeated grazing within a paddock, some descriptive studies have identified sequences of plant species selection by cattle and sheep during the paddocking sequence (Stoltsz and Danckwerts, 1990; O'Reagain and Grau, 1995), but with no clear repeatable pattern.

In varied and variable environments, efforts to understand foraging strategy should combine relevant timescales, since long-term processes cannot be deduced from direct integration of short-term results (Schmitz, 2000). When ruminants are grazed on such vegetation, a major challenge is to obtain data-sets fitted for linking small-scale mechanisms of biting with the dynamics of intake at intermediate and large timescales. By maintaining bite-scaled observations (as described in the third section of this chapter) during full days or weeks, transient instantaneous behaviours could be situated within the intermediate and long-term dynamics of the daily and interday ingestive response to feed diversity and variability.

The case of sheep flocks grazing on shrublands within paddocks

The interest of analysing behavioural data at several temporal scales has been recently illustrated for domestic sheep grazed on heterogeneous environment (Agreil et al., 2005). Experiments were carried out on private farms that were engaged in biodiversity conservation grazing contracts. Flocks of dry ewes were grazed for approximately 2-week periods within paddocks consisting of different kinds of broom shrublands. In this section we present one of these experiments during which a flock of 214 ewes was grazed for 16 days in a 4.5 ha paddock in the Drôme region (southern France, see Fig. 17.5).

With such a high grazing pressure, ewes were able to maintain high daily intake levels, although the feed offer dwindled as the days went by (77.9 ± 10.2 g DMI/kg MW). Since the

Fig. 17.5.   A flock of dry ewes grazing within a 4.5 ha paddock consisting of a calcareous sward invaded by shrubs (*Genista cinerea* DC., *Juniperus communis* L. and *Buxus sempervirens* L.) and trees (*Pinus sylvestris* L. and *Pinus nigra* Arn. var *austriaca* Loud.) in the Drôme region (southern France).

daily foraging duration was kept unchanged, the steadily high daily intake levels result from the stability of their average IR.

When foraging behaviour is expressed in daily average values, we confirm that the decline of BM is compensated by an increase in BF (Arnold and Dudzinski, 1978). But the range of BMs we observed on broom shrubland was extremely narrow (0.15–0.08 g DM/bite), compared to the range observed for sheep on monospecific sward (Allden and Whittaker, 1970). As sheep and goats on rangelands are known to be able to select a huge variety of BMs (2–0.03 g DM/bite; Meuret, 1997b), this raises the question of choice patterns within and between days that result in this slight variation in BM daily average, and in the maintenance of daily intake levels.

Linking the daily average values with the elementary instantaneous ingestive behaviour gives insight of the behavioural adjustments small ruminants achieve when facing a varied and variable food offer. Interday variation in instantaneous BMs (calculated over 20 s periods of time) is presented in Fig. 17.6. BMs are expressed on a log scale and horizontal lines are proportionate to: BF (left box), contribu-

tion to daily intake (central box), OMd (right box).

The decrease of average BM (thick line in the left box) is small in comparison to the range of BMs used each day. As the days went by, the size of the remaining plant parts dwindled and the ewes selected numerous small BMs at high BFs (left box). These small bites, however, do not contribute significantly to daily intake (central box). From day 7, ewes added to their diet bites of high masses that contributed most to daily intake. This late selection of large BMs may be attributed to their poor nutritive quality. However, this was not the case, as large BMs selected during the last days had an OMd (top right in the right box) that was at least equivalent to the bites selected during the first days.

In conclusion, sheep facing spatially varied and temporally variable environments do not systematically prioritize large BMs during the first days in a paddock, even though they are of high OMd. Ewes observed in this study chose an increasingly broad range of BMs and the larger ones gradually made up the greatest part of their daily intake during the last days in the paddock, as the size and structure of plant parts decreased.

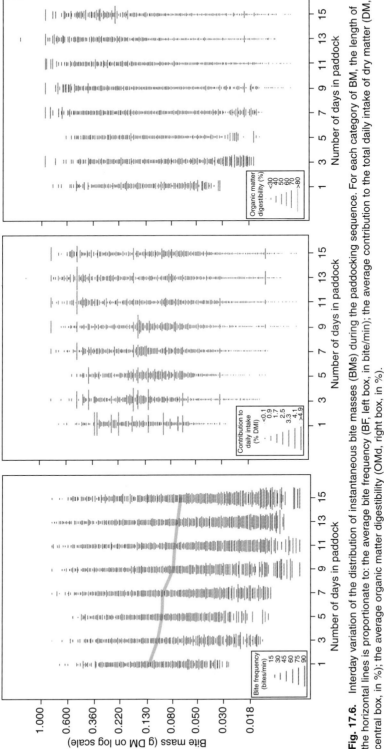

Fig. 17.6. Interday variation of the distribution of instantaneous bite masses (BMs) during the paddocking sequence. For each category of BM, the length of the horizontal lines is proportionate to: the average bite frequency (BF, left box, in bite/min); the average contribution to the total daily intake of dry matter (DM, central box, in %); the average organic matter digestibility (OMd, right box, in %).

The quantity–quality trade-off revisited under varied and variable grazing conditions

Traditional theoretical frameworks are not supported by results obtained with sheep on broom shrubland. They assume that ruminants facing a diversified availability of forage prioritize the feed items that are most satisfactory in a 'quantity–quality trade-off'. The formalization of this trade-off has been based mainly on the 'forage maturation hypothesis' (Westoby, 1974; Belovsky, 1986; Fryxell, 1991), according to which the feeding choices are predicated on two main constraints: fodder availability and fodder quality. But the relative importance of these two constraints is still being debated, and depends upon the pastoral conditions and the timescale (Hixon, 1982; Newman *et al.*, 1995; Wilmshurst *et al.*, 1999; Bergman *et al.*, 2001). Further, the forage maturation hypothesis has mainly been used to analyse the behavioural response when a fibre content gradient can be observed from the distal to the proximal parts of the plant. This is the case for grasses (McNaughton, 1979; Bergman *et al.*, 2001) and for the foliage of deciduous trees (Lundberg and Danell, 1990; Shipley *et al.*, 1999). Faced with such a gradient, ruminants have to choose either small plant parts, which means small BM and highly nutritive bites, or large plant parts, which means bigger BM but poorer quality (Vivas *et al.*, 1991; Shipley *et al.*, 1999). In the case of heterogeneous vegetation comprising several plant life forms and architecture (from forbs, shrubs, tree foliages and lianas), BM and bite quality are not well correlated (Van Soest, 1982).

Stabilizing vs. maximizing: how to cope with variety and variability?

Rather than prioritizing bites that are both heavy and OMd-rich, ruminants may achieve adjustments of day-to-day feeding behaviour that lead to the stabilization of daily average digestibility and BM. This may contribute to the current debates in domestic ruminant feeding science and in nutritional ecology, concerning the choice of variables and 'objective functions' to be included in models. In the optimization models, maximization is the function

used most, by far (Stephens and Krebs, 1986, p. 7), together with the variables that describe ingestion efficiency (IR, nutritive value, etc.). The debate on the validity of these models and the way to depict the ruminants' behaviour remains open because of the continued inflow of new field data (see review by Sih and Christensen, 2001).

A schematic representation of the consequences of two different feeding strategies is given in Fig. 17.7. The volumes shown in grey represent the range of IRs and digestibility of the DM ingested each day: the darker the grey, the greater the contribution. In the case of maximization functions (Fig. 17.7, left), absolute maximization is never really observed (Black and Kenney, 1984; Illius *et al.*, 1999), and thus daily values cover a rather broad range and the grey colour is darker on the top right side. When the ruminants are kept in the closed paddock, the ingestion maximization hypothesis would lead to gradual depletion of the high-quality feed offer and a shift to a range of feed with lower IR and digestibility values.

Sheep we observed have adopted a different feeding strategy (Fig. 17.7, right). Their daily ingestive behaviour included each day a large range of IR and OMd, without giving more importance to the high values (Agreil, 2003). As a consequence, the centre of gravity of the daily volumes should be placed at the centre (Fig. 17.7, right). This represents a sheep that would adopt a stabilizing feeding strategy, modelled by stabilization functions rather than maximization functions.

Such stabilizing behavioural adjustments may be of particular interest for grazing conditions in which the individual animal can never obtain full information on the range of available feed or the pace of their depletion because the animals in the group eat all together, at the same time and in the same enclosure.

Managing a Paddock by Identifying Time-saving Feeds

Heterogeneous environments have advantages that are not found in the herbaceous pastures. Considering the diversity of plant structures and communities, adjustments can be made in all three spatial dimensions. But the fourth

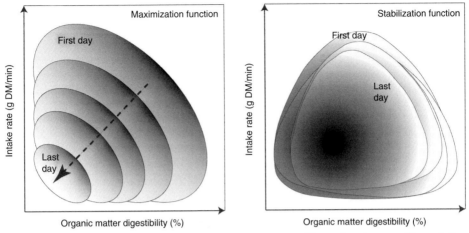

Fig. 17.7. Schematic representation of the consequences of two different functions used in modelling feeding strategies: maximization function (left) and stabilization function (right). The grey volumes represent the range of intake rates (IRs) and digestibility of the organic matter (OM) ingested each day: the darker the grey, the greater the contribution.

dimension, i.e. temporal adjustments during the days that the flock stays on a given grazing area, has been neglected all too often. The adjustment here is not only a continued increase in BF to maintain the IR. Direct benefit is derived from the plant structure, especially when it allows for large bites at the end of the grazing sequence. The study of the fourth dimension of the grazing process contributes to knowledge of the qualities of the three spatial dimensions.

An understanding of the feeding strategies also makes it possible to propose some rather simple rules for grazing management. Figure 17.8 gives a schematic representation of the rationale to be used in anticipating behavioural adjustments before deciding whether to leave a flock of small ruminants in a paddock or withdraw it. After entering the paddock (left in Fig. 17.8), each individual prioritizes intermediary BMs and BFs (masses of around ~0.1, in other words, very similar on the grasslands and the rangelands). On the grasslands the only possible adjustment would be to increase BFs to compensate for the gradual decrease in BMs (see arrow A). The decision to remove the animals from the paddock (Out-1) is taken when the structure of the grasses is such that the animals can no longer maintain a BF that allows them to maintain an average IR of over 3 g

MS/min. In heterogeneous environments, the animals find feed to maintain their IR by grazing on plant structures that allow them to consume bites of over 0.5 g MS (arrow B), and thus, temporarily but frequently, to reduce their BF during meals. This gives them the possibility to explore and to devote time to consuming small plant parts with very small BMs (arrow A') without jeopardizing average IR. In this scenario, the decision to leave can be postponed (Out-2). In certain cases, especially when the paddock contains edible vegetation that can be consumed in very large BMs, i.e. over 2 g MS (arrow C), the departure can be organized even later (Out-3). The decision to withdraw the flock from the paddock should be based on the observation of the animals' stimulation and be related to what we call 'time-saver' feeds (B and C), which are often not even of least nutritional value.

Our knowledge of the ruminants' feeding strategy confirms the assertion that several types of vegetation have been unjustly denigrated in livestock farming. More specifically, we are referring to vegetation that includes time-saver feeds, in other words, feed that can be consumed in large bites. In the situation we studied, this included grass tussocks and shrubs with chlorophylian stems, which could be offered in 10 cm-long bunches (e.g. medium

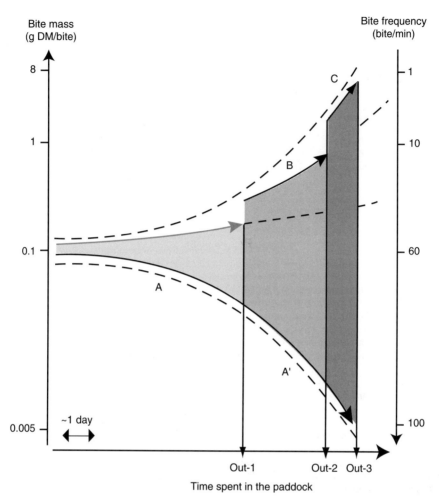

Fig. 17.8. Schematic representation on the rationale to be used before deciding whether to leave a flock of small ruminants in a paddock or to withdraw it. In a heterogeneous environment that varies from day to day (*x*-axis), behavioural adjustments lead to a gradual increase of the range of bite masses (BMs) used each day (left *y*-axis), and of the associated bite frequency (right *y*-axis). When plant structure allows the animals to take bigger BMs, the decision to withdraw them out of the paddock can be postponed (from Out-1 to Out-2 or even Out-3), without jeopardizing the daily average intake rate.

and large size brooms); ramified twigs from the small shrubs such as calluna; foliage from trees that could be cut into bunches of leaves or needles (e.g. oak, willow, pine).

Conclusion

The recent advances presented in this chapter should contribute to a better understanding of specific feeding strategies developed by domestic ruminants when grazing in natural and heterogeneous environments. It is often pointed out that coping with plant diversity and variability is one of the major challenges ruminants are faced with (Provenza *et al.*, 2003). But little is known about the functionality of plant diversity when grazing in variable conditions.

In this chapter, we demonstrated the high interest for scientific studies in small ruminants to describe long-term behavioural trends (within day or interday adjustments) by the elementary decisions (the bites), and vice versa. The coupling of the instantaneous-scaled analysis of ingestive behaviour with the large-scale expression of intake levels and diet quality gave a more realistic overview of behavioural adjustment achieved by small ruminants. One of the major challenges for scientists is probably to model such behavioural adjustments under the hypothesis of an incomplete knowledge by the animals of the available feeds. This leads us to argue for a better consideration of stabilization functions when modelling feeding strategies. An interesting framework is the 'satisficing rule' (Ward, 1992), which postulates that animals only modify their feeding choices when the consequences of previous ones are below a particular threshold. A better exploration of such 'reference thresholds' would be beneficial for the understanding of ruminants' feeding strategies, provided they are defined as dynamic reference values.

References

Abijaoudé, J.A., Morand-Fehr, P., Béchet, G., Brun, J.P., Tessier, J. and Sauvant, D. (1999) A method to record the feeding behaviour of goats. *Small Ruminant Research* 33, 213–221.

Agreil, C. (2003) Pâturage et conservation des milieux naturels: une approche fonctionnelle visant à qualifier les aliments à partir de l'analyse du comportement alimentaire chez la brebis (in French). PhD dissertation, Institut National Agronomique, Paris, France, 351 pp.

Agreil, C. and Meuret, M. (2004) An improved method for quantifying intake rate and ingestive behaviour of ruminants in diverse and variable habitats using direct observation. *Small Ruminant Research* 54, 99–113.

Agreil, C., Fritz, H. and Meuret, M. (2005) Maintenance of daily intake through bite mass diversity adjustment in sheep grazing on heterogeneous and variable vegetation. *Applied Animal Behaviour Science* 91, 35–56.

Allden, W.G. and Whittaker, I.A. (1970) The determinants of herbage intake by grazing sheep: the interrelationship of factors influencing herbage intake and availability. *Australian Journal of Agriculture Research* 21, 755–766.

Altman, J. (1974) Observational study of behaviour: sampling methods. *Behaviour* 9, 227–265.

Arnold, G.W. and Dudzinski, M.L. (1978) *Ethology of Free-Ranging Domestic Animals. Development in Animal and Veterinary Sciences*, Vol. 2. Elsevier Scientific Publishing Company, Amsterdam, The Netherlands, 198 pp.

Baumont, R., Champciaux, P., Agabriel, J., Andrieu, J., Aufrère, J., Michalet-Doreau, B. and Demarquilly, C. (1999) Une démarche intégrée pour prévoir la valeur des aliments pour les ruminants: prévAlim pour INRAtion. *INRA Productions Animales* 12, 183–194.

Baumont, R., Prache, S., Meuret, M. and Morand-Fehr, P. (2000) How forage characteristics influence behaviour and intake in small ruminants: a review. *Livestock Production Sciences* 64, 15–28.

Belovsky, G.E. (1986) Optimal foraging and community structure: implications for a guild of generalist grassland herbivores. *Oecologia* 70, 35–52.

Bergman, C.M., Fryxell, J.M., Gates, C.C. and Fortin, D. (2001) Ungulate foraging strategies: energy maximizing or time minimizing? *Journal of Animal Ecology* 70, 289–300.

Black, J.L. and Kenney, P.A. (1984) Factors affecting diet selection by sheep. II. Height and density of pasture. *Australian Journal of Agricultural Research* 35, 551–563.

Bourbouze, A. (1980) Goats grazing on woody rangeland (in French, with English abstract). *Fourrages* 82, 121–144.

Brossier, J., de Bonneval, L. and Landais, E. (eds) (1993) *Systems Studies in Agriculture and Rural Development*. INRA Editions, Paris, France, 415 pp.

Buller, H., Wilson, G.A. and Höll, A. (eds) (2000) *European Agri-environmental Policy in the European Union*. Ashgate Publishing, Basingstoke, UK.

Checkland, P. (1981) *System Thinking: Systems Practices*. John Wiley & Sons, Chichester, UK.

Currie, P.O., Reichert, D.W., Malecheck, J.C. and Wallmo, O.C. (1977) Forage selection comparisons for mule deer and cattle on the managed Ponderosa Pine. *Journal of Range Management* 30, 352–356.

Dardenne, P. (1990) Contribution à l'utilisation de la spectrométrie dans le proche infrarouge pour l'étude de critères de qualité des céréales et des fourrages [in French]. PhD dissertation, University of Gembloux, Gembloux, Belgium.

Delagarde, R., Caudal, J.-P. and Peyraud, J.-L. (1999) Development of an automatic bitemeter for grazing cattle (in French, with English abstract). *Annales de Zootechnie* 48, 329–339.

Deverre, C., Hubert, B. and Meuret, M. (1995) The know-how of livestock farmers challenged by new objectives for european farming. I. Rangelands usages, greatness and decay. In: West, N.E. (ed.) *Fifth International Rangeland Congress*. Salt Lake City, Utah, pp. 115–116.

Dulphy, J.-P., Baumont, R., L'Hotelier, L. and Demarquilly, C. (1999) Improvement of the measurement and prediction of forage ingestibility in sheep by taking into account the variations of their intake capacity using a standard forage (in French, with English abstract). *Annales de Zootechnie* 48, 469–476.

Dumont, B. (1996) Préférence et sélection alimentaire au pâturage [in French, with English abstract]. *INRA Productions Animales* 9, 359–366.

Dumont, B. and Meuret, M. (1993) Intake dynamics of goats and llamas feeding on garrigue grazing lands (in French, with English abstract). *Annales de Zootechnie* 42, 193.

Dumont, B., Meuret, M. and Prud'Hon, M. (1995) Direct observation of biting for studying grazing behaviour of goats and llamas on garrigue rangelands. *Small Ruminant Research* 16, 27–35.

Dumont, B., Meuret, M., Boissy, A. and Petit, M. (2001) Grazing from the animals' point of view: behavioural mechanisms and applications to animal husbandry (in French, with English abstract). *Fourrages* 166, 213–238.

Etzenhouser, M.J., Owens, M.K., Spalinger, D.E. and Blake Murden, S. (1998) Foraging behaviour of browsing ruminants in a heterogeneous landscape. *Landscape Ecology* 13, 55–64.

Forbes, J.M. (1995) *Voluntary Food Intake and Diet Selection in Farm Animals*. CAB International, Wallingford, UK, 532 pp.

Forbes, J.M. and Provenza, F.D. (2000) Integration of learning and metabolic signals into a theory of dietary choice and food intake. In: Cronje, P. (ed.) *Ruminant Physiology: Digestion, Metabolism, Growth and Reproduction*. CAB International, Wallingford, UK, pp. 3–19.

Forbes, T.D.A. (1988) Researching the plant–animal interface: the investigation of ingestive behaviour in grazing animals. *Journal of Animal Research* 66, 2369–2379.

Fryxell, J.M. (1991) Forage quality and aggregation by large herbivores. *American Naturalist* 138, 478–498.

Genin, D., Villca, Z. and Abasto, P. (1994) Diet selection and utilization by llama and sheep in a high altitude-arid rangeland of Bolivia. *Journal of Range Management* 47, 245–248.

Gillingham, M.P., Parker, K.L. and Hanley, T.A. (1997) Forage intake by black-tailed deer in a natural environment: bout dynamics. *Canadian Journal of Zoology* 75, 1118–1128.

Ginane, C., Petit, M. and D'Hour, P. (2002) How do grazing heifers choose between maturing reproductive and tall or short vegetative swards? *Applied Animal Behaviour Science* 83, 15–27.

Ginnett, T.F. and Demment, M.W. (1995) The functional response of herbivores: analysis and test of a simple mechanistic model. *Functional Ecology* 9, 376–384.

Gordon, I.J. (1995) Animal-based techniques for grazing ecology research. *Small Ruminant Research* 16, 203–214.

Hägele, B.F. and Rowell-Rahier, M. (1999) Dietary mixing in three generalist herbivores: nutrient complementation or toxin dilution? *Oecologia* 119, 521–533.

Hanley, T.A. (1997) A nutritional view of understanding and complexity in the problem of diet selection by deer (Cervidae). *Oikos* 73, 537–541.

Hatchuel, A. (2000) Intervention research and the production of knowledge. In: Cerf, M., Gibbon, D., Hubert, B., Ison, R., Jiggins, J., Paine, M., Proost, J. and Röling, N. (eds) *Cow up a Tree: Knowing and Learning in Agriculture*. INRA Editions, Paris, France, pp. 55–68.

Hixon, M.A. (1982) Energy maximizers and time minimizers: theory and reality. *American Naturalist* 119, 596–599.

Hubert, B. (1993) Modelling pastoral land-use practices. In: Brossier, J., de Bonneval, L. and Landais, E. (eds) *Systems Studies in Agriculture and Rural Development*. INRA Editions, Paris, France, pp. 235–258.

Hubert, B. and Bonnemaire, J. (2000) Building research 'objects' in targeted interdisciplinary studies: the new demands set to evaluation procedures (in French, with English abstract). *Nature Sciences et Sociétés* 8(3), 5–19.

Hubert, B., Deverre, C. and Meuret, M. (1995) The know-how of livestock farmers challenged by new objectives for european farming. II. Reassigning rangelands to new, environment-related usages. In: West, N.E. (ed.) *Fifth International Rangeland Congress*. Salt Lake City, Utah, pp. 251–252.

Illius, A.W. and Gordon, I.J. (1991) Prediction of intake and digestion in ruminants by a model of rumen kinetics integrating animal size and plant characteristics. *Journal of Agricultural Science* 116, 145–157.

Illius, A.W. and Hodgson, J. (1996) Progress in understanding the ecology and management of grazing systems. In: Hodgson, J. and Illius, A.W. (eds) *The Ecology and Management of Grazing Systems*. CAB International, Wallingford, UK, pp. 429–458.

Illius, A.W., Gordon, I.J., Elston, D.A. and Milne, J.D. (1999) Diet selection in goats: a test of intake-rate maximization. *Ecology* 80, 1008–1018.

Johnson, D.H. (1980) The comparison of usage and availability measurements for evaluating resource preference. *Ecology* 61, 65–71.

Kohlmann, S.G., Matis, J.H. and Risenhoover, K.L. (1999) Estimating handling times for herbivore diets: a statistical method using the gamma distribution. *Journal of Animal Ecology* 68, 638–645.

Laca, E.A and Demment, M.W. (1996) Foraging strategies of grazing animals. In: Hodgson, J. and Illius, A.W. (eds) *The Ecology and Management of Grazing Systems*. CAB International, Wallingford, UK, pp. 137–158.

Launchbaugh, K.L. (1996) Biochemical aspects of grazing behaviour. In: Hodgson, J. and Illius, A.W. (eds) *The Ecology and Management of Grazing Systems*. CAB International, Wallingford, UK, pp. 159–184.

Legay, J.-M. (1993) Une expérience est-elle possible? In: Lebreton, J.D. and Asselain, B. (eds) *Biométrie et Environnement*. Masson, Paris, France, pp. 1–14.

Legay, J.-M. (1997) L'expérience et le modèle: un discours sur la méthode. In: *Sciences en Question*. INRA Editions, Paris, France, 110 pp.

Litvaitis, J.A. (2000) Investigating food habits of terestrial vertebrates. In: Boitani, L. and Fuller, T. (eds) *Research Techniques in Animal Ecology. Controversies and Consequences*. Columbia University Press, New York, pp. 160–180.

Lundberg, P. and Danell, K. (1990) Functional response of browsers: tree exploitation by moose. *Oikos* 58, 378–384.

Lyons, R.K. and Stuth, J.W. (1992) Faecal NIRS equations for predicting diet quality of free-ranging cattle. *Journal of Range Management* 45, 238–244.

Maizeret, C. (1988) Stratégies alimentaires des chevreuils: les fondements écologiques d'une diversification du regime. *Acta Oecologica Oecologia Applicata* 9, 191–211.

Manly, B., McDonald, L. and Thomas, D. (1993) *Resource Selection by Animals: Statistical Design and Analysis for Field Studies*. Chapman & Hall, London, 177 pp.

Mayes, R.W. and Dove, H. (2000) Measurement of dietary nutrient intake in free-ranging mammalian herbivores. *Nutrition Research Review* 13, 107–138.

McNaughton, S.J. (1979) Grazing as an optimization process: grass–ungulate relationships in the Serengeti. *American Naturalist* 113, 691–703.

Mertens, D.R. (1996) Methods in modelling feeding behaviour and intake in herbivores. *Annales de Zootechnie* 45, 153–164.

Meuret, M. (1988) Feasibility of *in vivo* digestibility trials with lactating goats browsing fresh leafy branches. *Small Ruminant Research* 1, 273–290.

Meuret, M. (1989) Valorisation par des caprins laitiers de rations ligneuses prélevées sur parcours [in French]. PhD dissertation, University of Gembloux, Gembloux, Belgium, 248 pp.

Meuret, M. (1993) Feeding management on rangelands. An analytical investigation including a system approach for action. In: Brossier, J., De Bonneval, L. and Landais, E. (eds) *System Studies in Agriculture and Rural Development*. INRA Editions, Paris, France, pp. 185–190.

Meuret, M. (1997a) How do I cope with that bush? Optimizing on less palatable feeds at pasture using the MENU model. Recent advances in small ruminant nutrition. In: Lindberg, J.E., Gonda, H.L. and Ledin, I. (eds) *Options Méditerranéennes* A-34, 53–57.

Meuret, M. (1997b) Préhensibilité des aliments chez les petits ruminants sur parcours en landes et sous-bois [in French, with English abstract]. *INRA Productions Animales* 10, 391–401.

Meuret, M. and Bruchou, C. (1994) Modelling voluntary intake related to dietary choices diversity in goat grazing on rangelands (in French, with English abstract). *Rencontres Recherches Ruminants* 1, 225–228.

Meuret, M. and Giger-Reverdin, S. (1990) A comparison of two ways of expressing the voluntary intake of oak foliage-based diets in goats raised on rangelands. *Reproduction, Nutrition, Développement* 2(suppl.), 205.

Meuret, M., Bartiaux-Thill, N. and Bourbouze, A. (1985) Feed intake of dairy goats on woody rangelands: -direct observation of biting method; -chromic oxide method (in French, with English abstract). *Annales de Zootehnie* 34, 159–180.

Meuret, M., Boza, J., Narjisse, H. and Nastis, A. (1991) Evaluation and utilisation of rangeland feeds by goats. In: Morand-Fehr, P. (ed.) *Goat Nutrition*. IEAAP Publications 46, Pudoc, Wageningen, The Netherlands, pp. 160–171.

Meuret, M., Viaux, C. and Chadoeuf, J. (1994) Land heterogeneity stimulates intake during grazing trips. *Annales de Zootechnie* 43, 296.

Mofareh, M.M., Beck, R.F. and Schneberger, A.G. (1997) Comparing techniques for determining steer diets in northern Chihuahuan Desert. *Journal of Range Management* 50, 27–32.

Morley, F.H.W. (1981) Grazing animals. In: Neimann-Sorensen, A. and Tribe, D.E. (eds) *World Animal Science B1*. Elsevier Publishing Company, London, 411 pp.

Neff, D.J. (1974) Forage preference of trained mule deer on the Beavercreek watersheds. *Arizona Game and Fish Department Special Report* 4, 1–61.

Newman, J.A., Parsons, A.J. and Harvey, A. (1992) Not all sheep prefer clover; diet selection revisited. *Journal of Agricultural Science* 119, 275–283.

Newman, J.A., Parsons, A.J., Thornley, J.H.M., Penning, P.D. and Krebs, J.R. (1995) Optimal diet selection by a generalist grazing herbivore. *Functional Ecology* 9, 255–268.

Noldus Information Technology (1995) The observer, base package for the Macintosh. Reference Manual, Version 3.0.

O'Reagain, P.J. and Grau, E.A. (1995) Sequence of species selection by cattle and sheep on South African sourveld. *Journal of Range Management* 48, 314–321.

Owen-Smith, N. and Cooper, S.M. (1987a) Palatability of woody plants to browsing ruminants in a South African savanna. *Ecology* 68, 319–331.

Owen-Smith, N. and Cooper, S.M. (1987b) Assessing food preferences of ungulates by acceptability indices. *Journal of Wildlife Management* 51, 372–378.

Parker, K.L., Gillingham, M.P. and Hanley, T.A. (1993) An accurate technique for estimating forage intake of tractable animals. *Canadian Journal of Zoology* 71, 1462–1465.

Parsons, A.J., Newman, J.A., Penning, P.D. and Harvey, A. (1994) Diet preference of sheep: effects of recent diet, physiological state and species abundance. *Journal of Animal Ecology* 63, 465–478.

Pérez-Barberia, F.J. and Gordon, I.J. (1998) Factors affecting food comminution during mastication in herbivorous mammals: a review. *Biological Journal of the Linnaean Society* 63, 233–256.

Pittroff, W. and Kothmann, M.M. (1999) Regulation of intake and diet selection by herbivores. In: Jung, H.J.G. and Fahey, G.C. (eds) *Nutritional Ecology of Herbivores. Proceedings of the V International Symposium on the Nutrition of Herbivores in San Antonio, Texas*. American Society of Animal Science, Savoy, Illinois, pp. 366–422.

Prache, S., Roguet, C., Louault, F. and Petit, M. (1996) Evolution des choix alimentaires d'ovins entre talles végétatives et épiées au cours de l'exploitation d'un couvert épié de Dactyle. In: *Proceedings of the 3rd Rencontres Recherches Ruminants*, Paris, France, pp. 89–92.

Provenza, F.D. (1995) Postingestive feedback as an elementary determinant of food preference and intake in ruminants. *Journal of Range Management* 48, 2–17.

Provenza, F.D. (2003) Foraging behavior: managing to survive in a world of change. *Behavioral Principles for Human, Animal, Vegetation and Ecosystems Management*. USDA-NRCS, Maryland, 63 pp.

Provenza, F.D. and Launchbaugh, K.L. (1999) Foraging on the edge of chaos. In: Launchbaugh, K.L., Sanders, K.D. and Mosley, J.C. (eds) *Grazing Behavior of Livestock and Wildlife*. Idaho Forest, Wildlife and Range Experiment station, University of Idaho, Moscow, Idaho, 1–12.

Provenza, F., Villalba, J., Dziba, L., Atwood, S. and Banner, R. (2003) Linking herbivore experience, varied diet and plant biochemical diversity. *Small Ruminant Research* 49, 257–274.

Rolls, B.J. (1986) Sensory-specific satiety. *Nutrition Review* 44, 93–101.

Sauvant, D., Baumont, R. and Faverdin, P. (1996) Development of a mechanistic model of intake and chewing activities of sheep. *Journal of Animal Science* 74, 2785–2802.

Schenk, J.S., Westerhaus, M.O. and Hoover, M.R. (1979) Analysis of forage by infrared reflectance. *Journal of Dairy Science* 62, 807–812.

Schmitz, O.J. (2000) Combining field experiments and individual-based modeling to identify the dynamically relevant organizational scale in a field system. *Oikos* 89, 471–484.

Shipley, L.A., Illius, A.W., Danell, K., Hobbs, N.T. and Spalinger, D.E. (1999) Predicting bite size selection of mammalian herbivores: a test of a general model of diet optimization. *Oikos* 84, 55–68.

Sih, A. and Christensen, B. (2001) Optimal diet theory: when does it work, and when and why does it fail? *Animal Behaviour* 61, 379–390.

Spalinger, D.E. and Hobbs, N. (1992) Mechanisms of foraging in mammalian herbivores: new models of functional response. *American Naturalist* 140, 325–348.

Stephens, D.W. and Krebs, J.R. (1986) *Foraging Theory.* Princeton University Press, Princeton, New Jersey, 247 pp.

Stobbs, T.H. (1975) The effect of plant structure on the intake of tropical pasture. III. Influence of fertilizer nitrogen on the size of bite harvested by Jersey cows grazing *Setaria anceps* c.v. *Kazungula* swards. *Australian Journal of Agricultural Research* 26, 997–1007.

Stoltsz, C.W. and Danckwerts, J.E. (1990) Grass species selection patterns on rotationally-grazed Dohne Sourveld during autumn and early winter. *Journal of Grassland Society of South Africa* 7, 92–96.

Ungar, E.D. (1996) Ingestive behaviour. In: Hodgson, J. and Illius, A.W. (eds) *The Ecology and Management of Grazing Systems.* CAB International, Wallingford, UK, pp. 185–218.

Van der Wal, R., Madan, N., van Lieshout, S., Dormann, C., Langvatn, R. and Albon, S.D. (2000) Trading forage quality for quantity? Plant phenology and patch choice by Svalbard reindeer. *Oecologia* 123, 108–115.

Van Soest, P.J. (1982) *Nutritional Ecology of the Ruminant.* Cornell University Press, London, 476 pp.

Van Soest, P.J. (1994) *Nutritional Ecology of the Ruminant,* 2nd edn. Cornell University Press, New York, 476 pp.

Vivas, H.J., Saether, B.E. and Andersen, R. (1991) Optimal twig-size selection of a generalist herbivore, the moose *Alces alces*: implications for plant–herbivore interactions. *Journal of Animal Ecology* 60, 395–408.

Wallis de Vries, M.F. (1995) Estimating forage intake and quality in grazing cattle: a reconsideration of the hand-plucking method. *Journal of Range Management* 48, 370–375.

Wallis de Vries, M.F., Bakker, J.P. and Van Wieren, S.E. (1998) *Grazing and Conservation Management.* Kluwer Academic Publishers, Dordrecht, The Netherlands, 374 pp.

Wang, J. and Provenza, F.D. (1996) Food preference and acceptance of novel foods by lambs depend on the composition of the basal diet. *Journal of Animal Science* 74, 2349–2354.

Ward, D. (1992) The role of satisficing in foraging theory. *Oikos* 63, 312–317.

Westoby, M. (1974) An analysis of diet selection by large generalist herbivores. *American Naturalist* 108, 290–304.

Wilmshurst, J.F., Fryxell, J.M. and Colucci, P.E. (1999) What constrains daily intake in Thomson's gazelles? *Ecology* 80, 2338–2347.

Woodward, S.J.R. (1998) Bite mechanics of cattle and sheep grazing grass-dominant swards. *Applied Animal Behaviour Science* 56, 203–222.

18 Feeding Free-range Poultry and Pigs

P.C. Glatz,[1] Z.H. Miao[1] and Y.J. Ru[1,2]

[1]Livestock Systems, South Australian Research and Development Institute, Roseworthy Campus, Roseworthy, South Australia 5371, Australia; [2]Danisco Animal Nutrition, 61 Science Park Road, The Galen #06-16 East Wing, Singapore Science Park III, Singapore 117525

Introduction

Free-range systems allow poultry and pigs freedom of movement to forage and explore outdoors, increasing their choice of environments and food sources and thus improving their welfare (Stoll and Hilfiker, 1995; Nielsen et al., 2003). Under free-range conditions poultry spend 7–25% of their time foraging (Appleby et al., 1989), and they are capable of selecting a diet that is adequate for all their requirements (Hughes, 1984). Allowing pigs to forage on pasture is particularly suited to the pigs' natural behaviour (Stoll and Hilfiker, 1995). This chapter evaluates feed sources from temperate and tropical regions for free-range pigs and poultry, feeding strategies and the role of integrated farming in free-range farming systems.

Feed Sources for Free-range Poultry and Pigs

Poultry

Under natural conditions poultry consume seeds (dicotyledons and grass), fruit, herbage (roots and stems) (Savory et al., 1978), invertebrates (McBride et al., 1969) and crop pests such as beetles, shield-backer, bugs, muscoid

flies and ants (Clark and Gage, 1996). The diet of free-range poultry also changes considerably with the season. Young poultry require a high-protein diet mainly consisting of invertebrates, while adults consume cereals in autumn and winter and grass and herbage in spring and summer (Savory et al., 1978; Savory, 1989).

Currently, free-range poultry are supplemented with commercial feed. Over 60% of the supplementary feed (for both meat and egg-laying types) comprises cereals (maize, wheat, sorghum and barley). The most common sources of energy in commercial poultry feeds are soybean oil and cereals. The primary protein source is soybean meal, either whole or dehulled, while fish meal is used where available and cost-effective. Minerals and vitamins are provided as supplements to ensure poultry receive adequate amounts of trace elements (Walker and Gordon, 2003).

Pigs

Under natural conditions pigs scavenge and forage in forests, woods, pasture and orchards (Thornton, 1988). Wild pigs show a large seasonal variation (Dardaillon, 1989) in grasses foraged and are able to select plants or parts of plants to meet their nutritional requirements (Gustafson and Stern, 2003).

Free-range pigs are fed commercial feed. Feed is the major cost of free-range pig production. Alternative sources of feed with high fibre (grass and silage) are being used as pigs can digest a certain amount of fibre. For example, when lucerne leaf meal is included in a barley-based diet for grower pigs, absorption of essential amino acids is increased (Reverter and Lindberg, 1998). Kass *et al.* (1980) reported that volatile fatty acids are produced in the large intestine of pigs and can provide up to 12.5% of the energy required for maintenance in grower pigs. Integration of free-range pig farming with an arable farming system is a major ingredient for success in most situations (Thornton, 1988).

Potential feed ingredients for free-range poultry and pigs

Table 18.1 lists the major feed resources for poultry and pigs commonly used in temperate and tropical climates. The performances of poultry and pigs have been largely determined by studies with intensively housed animals but have relevance when preparing supplementary diets for free-range poultry and pigs.

Feeding Free-range Poultry and Pigs

Poultry

Considerable information on nutrient requirements is available for various strains of poultry housed under intensive systems (NRC, 1994). The nutrient requirement is higher for free-range poultry than for those housed intensively. However, the amount and type of nutrients foraged and variable environmental conditions of free-range poultry makes it difficult to apply the nutritional management guidelines recommended for intensive poultry. The amount of feed offered to foraging poultry should be the amount of feed required minus the intake from foraging in order to reduce feed cost. Tadelle and Ogle (2000) determined seeds, plant materials, worms, insects and unidentified materials in the crop contents of

scavenging hens (Table 18.2) and found seasonal variations in the percentage of seeds, worms and insects. The energy and protein supplied from the forage resources were 11.97 MJ/kg and 8.8%, respectively (Table 18.3). The protein content was lower during the short rainy and dry seasons, while the energy supply was more critical in the drier months (Tadelle and Ogle, 2002). These values were below the protein requirement of free-range hens in the tropics, estimated at about 11 g/bird/day. The metabolizable energy (ME) supply could only meet the requirement of a non-laying hen (Scott *et al.*, 1982), indicating the limitations of foraging feed resources in supplying nutrients for production.

Feeding poultry outside the shelters in the morning encourages poultry to forage further, while feeding poultry in the afternoon in the shelter assists in getting poultry back from the paddock. However, there is a biosecurity concern as it also provides food for wild poultry that could be a source of disease (Thear, 1997). Bogdanov (1997) fed 150 g of cereals (wheat, maize and barley) plus 30 g/hen fresh sliced nettle in the late afternoon and obtained a mean laying rate of 57% over 164 days.

A choice feeding system can be used to feed free-range laying hens because they are able to select from the various feed ingredients on offer to satisfy their needs and production capacity. Choices of feed for laying hens should be protein, energy and mineral source feeds. Choice feeding free-range poultry is beneficial to small poultry producers because this feeding system provides a more effective way to use home-produced grains and by-products (Henuk and Dingle, 2002). Restrict-feeding free-range poultry such as giving half their daily allocation (55 g/hen) in the morning and the other half in the evening will encourage poultry to forage outside.

Feeding chicks

The utilization of fat is poor for chicks in the first week of life. Use of vegetable fat such as soybean or canola oil has limited value. The inclusion of palm oil and animal fats in the diet

Table 18.1. Some potential feed ingredients for free-range poultry and pigs.

Ingredient	General description	Feeding value		Inclusion rate		References
		Pig	Poultry	Pig	Poultry	
Maize	Contains high energy and low protein, fibre and minerals. Irradiated maize improves the starch availability. Fat in maize is highly unsaturated. Yellow maize is rich in beta-carotene and poor in vitamin B12 and D. Xanthic pigment improves yolk and skin colour. White maize decolours egg yolks.	Most common cereal in piglet diets with low fibre and high energy. Used in creep or weaner diets. DE 17.1 MJ/kg.	60% and 70% utilization rate. ME 16.0 MJ/kg.	Useful as sole grain in diets for young pigs and porkers. Finisher diets limited to 30%. Minimum rate of 25% for sows.	Chick 30%. Broiler 50%. Breeder 50%. Layer 50%.	Eusebio (1980); Ralph Say (1987); Van der Poel *et al.* (1989); Ewing (1997); ALFID (2002); Medel *et al.* (2004).
Wheat	Three types: hard, soft and durum. New high-yield varieties for livestock have high energy, low fibre and high palatability. Wheat can be fed at higher levels with inclusion of enzymes.	CP 13%, starch 64% and fibre 3% as a naked grain. Wheat should not be ground too finely. DE 16.0 MJ/kg.	Limit soft wheat in non-pelleted feed to 30% to avoid pasting of beak. ME 15.1 MJ/kg.	Can be fed as sole cereal in pig diets. In lactating and dry sow diets high starch causes enterotoxaemia or bowel tympany. Creep 60%. Weaner 55%. Grower, finisher, sow 50%.	Chick 50%. Broiler 60%. Breeder 65%. Layer 60%.	Eusebio (1980); Ralph Say (1987); Ewing (1997); ALFID (2002).
Sorghum	Classified into two types according to tannin level. High tannin varieties used as forage. Seeds are low in fibre, high in energy. Dark brown or purple seeds contain high tannin and affect digestibility of other nutrients. White seeds have little tannin and are an ideal feed.	Type of sorghum should be known. DE 16.0 MJ/kg.	High tannin varieties can be used in broiler starter diets. ME 15.6 MJ/kg.	Limits are related to tannin content. Low tannin varieties can be fed as sole grain to pigs. Rolled sorghum is easier to digest. The maximal utilization rates are 20–70%. Tannin content varies from 0.2–3%.	Breeder 5%. Layer 5%.	Eusebio (1980); Ralph Say (1987); Jacob *et al.* (1996); Ewing (1997); ALFID (2002).

Feed	Description	Composition	ME	Pigs	Poultry	Reference
Triticale	Naked grain – excellent energy source. Quality can be variable.	CP 60–180 g/kg DM. DE 13.5 MJ/kg.	ME 14.5 MJ/kg.	Creep 20%. Weaner 30%. Grower 35%. Finisher 40%. Sow 25%.	Chick 10%. Broiler 20%. Breeder 30%. Layer 35%.	Ewing (1997); ALFID (2002).
Barley	Common feed in Africa, Europe and Australia. Should be ground before use. Reasonable energy source for pigs and poultry.	CP 11–12%. DE 14.5%.	High fibre and energy. ME 13.6 MJ/kg.	Limit inclusion to 20% in weaners due to effects of fibre on digestibility.	Soluble NSP content of barley limits compatibility with legumes (lupins).	Eusebio (1980); Ewing (1997); ALFID (2002).
Oats	Poor in energy compared with wheat and barley, but high in oil and fibre content. Less suitable for young livestock.	CP 11.8%. DE 13.2 MJ/kg.	ME 12.5 MJ/kg.	High-fibre content restricts use to 30% of the ration for grower or finisher pigs and for brood sows.	Breeder 15%. Layer 15%.	Eusebio (1980); Ewing (1997).
Australian sweet lupins	Economic alternative to soybean meal or canola meal. Medium protein quality, rich source of energy. Low NE yield due to beta-galactan (rather than starch), which is fermented to VFAs rather than sugars.	CP 32%. DE 13.2 MJ/kg.	Dehulled lupins have low fibre content. ME 11.0 MJ/kg.	Used for all ages of pigs. Maximum 10% for weaners, 30% for growers and 20% for sows.	Breeder 7.5%. Layer 7.5%.	Ewing (1997); ALFID (2002).
Chick pea	In some countries, used as energy and protein feed in pig diets.	CP 19.6%. DE 13.5 MJ/kg.	ME 12.23 MJ/kg.	80% inclusion does not improve live weight gain, but slight improvement in feed efficiency.	–	Eusebio (1980); Evans (1985).
Soybean	Major source of vegetable oil and protein meal. Good source of essential amino acids and essential fatty acids. Reasonable energy content. Full fat soybean meal contains about 38–40% of CP.	DE 19.3 MJ/kg.	ME 16.9 MJ/kg.	Creep 20%. Weaner 20%. Grower 15%. Finisher 10%. Sow 10%.	Chick 20%. Broiler 25%. Breeder 20%. Layer 20%.	Ewing (1997); ALFID (2002).

Continued

Table 18.1. Some potential feed ingredients for free-range poultry and pigs. – *Continued*

Ingredient	General description	Feeding value		Inclusion rate		References
		Pig	Poultry	Pig	Poultry	
Vetch	Limitation is cyanoalanine content, toxic non-protein amino acids and other anti-nutritional factors.	CP 24.2–32.6%.	–	–	–	Petterson and Mackintosh (1994).
Wheat middlings and bran	Common by-product of flour milling, comprising coarse outer coating of wheat. Composition varies with variety and milling technique. High protein content and medium energy. Valuable for pigs, but its high-fibre content limits its inclusion.	CP 17.5%. DE 11.5 MJ/kg.	ME 7.5 MJ/kg.	High bran levels increase fibre and lower DE value for pigs. Wheat bran only used in diets for grower or finisher pig and brood sow. In some countries wheat bran is used up to 40% in diets for brood sows.	In broiler feeds, middling use is avoided. Bran included up to 25%. Layers and pullets fed up to 25% of middlings and 40% for bran.	Eusebio (1980); Ralph Say (1987); Arosemena *et al.* (1995); Ewing (1997); ALFID (2002).
Rice bran	Germ and bran layers separate from endosperm during rice milling to produce bran (3–8% of the grain). Composition depends on milling process, heating, additives and fat extracted. Contains high fibre, rich in thiamine and nicotinamide.	CP 15%. DE 7.7 MJ/kg.	ME 6.5 MJ/kg.	Rice bran is used up to 50% in rations for grower and finisher pigs.	Weight gain decreases and feed conversion increases for broilers at 20% inclusion.	Eusebio (1980); Zombade and Ichhponani (1983); Ralph Say (1987); Arosemena *et al.* (1995); Ewing (1997); Amissah *et al.* (2003).

Feedstuff	Comments	Composition	Energy	Pig inclusion	Poultry inclusion	References
Brewer's grains	Residue of malted and unmalted cereals and other starchy products. High in digestible fibre, good quality protein and source of phosphorus.	DM 23%. CP 25%.	–	–	40% inclusion in layers reduces laying rate, but increases feed consumption. Pullets tolerate up to 20% inclusion, but increases FCR.	Ralph Say (1987); Ewing (1997).
Molasses	Has high feeding value. Used in small amounts to prevent diuresis. High potassium may cause wet droppings.	CP 4–7%. DE 13.9 MJ/kg for beet molasses. DE 14.1 MJ/kg for cane molasses.	ME 11.3 MJ/kg for beet molasses. 11.5 MJ/kg for cane molasses.	Less than 30% recommended for fattening pigs and up to 5% for young pigs.	Chick 1%. Broilers 3%. Breeder 3%. Layer 3%.	Eusebio (1980); Ralph Say (1987); Ewing (1997).
Soybean meal	High in protein and energy, good amino acid profile with high lysine and low methionine. Solvent extracted soybean meal product has better quality than expeller-processed. In poultry-feeding, soybeans may be generously used.	Fat 17% and CP 36%. DE 14.8–15.5 MJ/kg.	ME 10.7–12.0 MJ/kg.	Common protein supplement for young pigs and brood sows.	Chick 25%. Broiler 30%. Breeder 35%. Layer 35%.	Eusebio (1980); Ralph Say (1987); Ewing (1997).
Canola meal	Valuable source of protein and energy, but contains high fibre.	CP 34.2–44.7%, energy 19.3–21.1 MJ/kg and fibre 11%. DE 12.0 MJ/kg.	ME 10.5 MJ/kg. TME range 9.1–11.2 MJ/kg.	Grower 2.5%. Finisher 5%. Sow 2.5%.	Broiler 2.5%. Layer 5.0%.	Eusebio (1980); Ralph Say (1987); Ewing (1997); ALFID (2002).
Groundnut cake	Use limited by aflatoxin content. Fungi on the groundnut produces the toxin. The latter are poor in methionine and lysine. Groundnut meal can become rancid when stored.	CP 45% for expellers and 50% for extraction cakes. Cellulose 5–7%, fat 4–8% for expellers. Fat 1% for the extraction cakes. DE 17.8 MJ/kg.	ME 12.9 MJ/kg.	Up to 5% in grower pig diets.	Broiler 2.5%. Breeder 4.0%. Layer 4.0%.	Eusebio (1980); Ralph Say (1987); Ewing (1997).

Continued

Table 18.1. Some potential feed ingredients for free-range poultry and pigs. – *Continued*

Ingredient	General description	Feeding value		Inclusion rate		References
		Pig	Poultry	Pig	Poultry	
Linseed oil meal	High in energy, oil and available carbohydrates, low in essential amino acids. Good source of digestible fibre.	CP 36–38%. DE 14.9 MJ/kg.	ME 11.5 MJ/kg.	Maximum use is 8%. 5% inclusion in diets for grower pigs and brood sows.	Breeder 2.5%. Layer 2.5%.	Eusebio (1980); Ewing (1997).
Safflower meal (*Carthamus tinctorius*)	Feeding value is 65% of soybean oil meal. High digestible protein, methionine and cystine but low lysine.	CP 19.0%.	–	12.5% inclusion has no beneficial effect on live weight gain and feed efficiency. 25% inclusion causes a growth depression.	Breeder 5%. Layer 5%.	Eusebio (1980); Ewing (1997).
Cotton seed cakes	Amino acid composition is well balanced, but deficient in methionine and lysine. Contains gossypol (toxin) in the kernel. Growth reduced at 0.04 ppm. Death occurs at higher doses. Decorticated cake (8% cellulose content) is usable by poultry. Useful source of digestible fibre.	CP 40–50%. DE 11.1–13.3 MJ/kg.	Laying performance reduced by gossypol values >0.2 ppm, abnormal yolk coloration with 0.5 ppm. ME 9.1 MJ/kg.	Up to 10% growers. Finisher 2.5%. Sow 5%.	–	Eusebio (1980); Ralph Say (1987); Ewing (1997).
Coconut by-products	Two by-products: (i) coconut oil meal, coconut cake (copra meal). Extracted oil from the coconut is 34% to 42%. (ii) broken kernel (raw copra). Oil residue of products ranges from 1% to 22%. Coconut cake or meal must be fresh to prevent diarrhoea. Copra meal is an excellent source of	CP 19–23% cellulose 9–24% for copra cakes. DE 12.5–13.2 MJ/kg.	5.8–6.2 MJ/kg.	Copra meal can comprise 25% of the diet. 10% copra meal gives poor growth rate. No further deterioration up to 30% inclusion for pigs housed at a constant temperature (25°C).	Up to 20% inclusion. 40% copra meal promotes growth and efficiency. Can include 42% solvent extracted	Thomas and Scott (1962); Nagura (1964); Eusebio (1980); Hutagalung (1981); Panigrahi *et al.* (1987); Ralph Say (1987); Thorne *et al.* (1992); Ewing (1997); O'Doherty and

	energy for pigs. Copra cakes contain medium CP, low in lysine and energy, limiting use in grower poultry, but not layers.				coconut if diets have balanced protein and ME. Chick growth depressed when 50% copra cake included.	McKeon (2000).
Palm kernel cakes or meal	Produced by extracting oil from the kernel. Palm oil (African) contains 80% of unsaturated fatty acids, 10% of linoleic acid (essential fatty acid requirement of 0.1% for pigs). The residue from heat extraction is similar to ash and unusable for poultry. Water-extracted residue is nutritious and palatable. High fibre promotes gizzard development. Aflatoxin problem if not stored correctly.	CP 12–23%. DE 12.5–12.9 MJ/kg.	ME 7.9–8.1 MJ/kg.	Used up to 5% in dry diets to improve palatability, reduce dustiness, supply vitamins and improve texture of rations before pelleting.	Low energy excludes use in grower poultry. Should not exceed 15% in layer feeds. Up to 30% of palm kernel cake could be used in poultry diets.	Devendra (1977); Nwokolo *et al.* (1977); Eusebio (1980); Hutagalung and Mahyudin (1981); Onwudike (1986); Ralph Say (1987); NRC (1988); Perez (1997).
Fish meal	Good source of protein and minerals. Quality depends on the manufacturing process.	CP 72.0–77.0%. DE 15.8–17.0 MJ/kg.	ME 16.3–17.9 MJ/kg.	Inclusion levels range from 2% to 10%. Never use in finisher rations.	Up to 10% for broilers at 0–4 weeks and 3–5% for layers.	Eusebio (1980); Ralph Say (1987); Ewing (1997).
Skim milk	Feeding value is similar to that of soybean meal. Useful for pig starter diets.	—	—	10–30% inclusion for pig starter diets, but not for brood sow and finisher hogs due to high cost.	—	Eusebio (1980).

Continued

Table 18.1. Some potential feed ingredients for free-range poultry and pigs. – *Continued*

Ingredient	General description	Feeding value		Inclusion rate		References
		Pig	Poultry	Pig	Poultry	
Sweet potato	High in energy, low in protein, fat and fibre. A source of vitamins (vitamin A, ascorbic acid, thiamine, riboflavin and niacin). Dustiness and fungal growth during sun-drying can be overcome by boiling the tubers, commonly practised by smallholders. Peels are useful for scavenging. Can be fed raw, cooked or as silage for pigs. Uncooked starch is resistant to hydrolysis. Cooking increases hydrolysable starch fraction from 4% to 55%. Can be chopped, sun-dried and used as an energy source for pigs.	Fresh vines provide 27% of the DM and 40% of dietary protein for grower or finisher pigs. Contains 80–90% DM of carbohydrate. DE 16.3 MJ/kg.	ME 14.8 MJ/kg.	33% inclusion for grower pigs (25–60 kg). Replaces 30–50% of grain in pig diets. In Taiwan, sweet potato and cassava are used to replace 40% of maize in grower and grower or finisher diets.	Broiler 5%. Breeder 10%. Layers 35%.	Cerning-Beroard and Le Dividich (1976); Wu (1980); Fashina-Bombata and Fanimo (1994); Ewing (1997); Perez (1997).
Cassava	Widely grown in tropics. Root cortex or peel is used by itself or as a component of whole-root chips. Leaves also used. Fresh cassava roots contain 65% water and are dried or processed to extend shelf-life. Major concern is cyanide content (140–890 ppm DM basis), in form of cyanogenic glucoside (liamarin). Cyanide	CP 1–2%, ether extract 0.2–0.5%, crude fibre 0.8–1.0% and ash 1–2%. Methionine supplementation (0.2%) improves weight gain and feed efficiency. Palm oil and glucose addition improves the weight gain and feed efficiency of pigs. DE	ME 14.9 MJ/kg.	Dried whole cassava plant (40% flour, 10% peels and 10% leaves plus tender stems). Up to 60% for grower pigs has no adverse effect on performance. Over 50% decreases live weight gain and feed conversion efficiency.	15% cassava leaf meal (oven-dried 60°C) in broiler diets. Sun-dried whole cassava plant meal included at 12.5–25% for broilers'	Eusebio (1980). Luteladio and Egumah (1981); Gomez *et al.* (1984) Ravindran *et al.* (1986) Ofuya and Obilor (1993); Ewing (1997); Akinfala and Tewe (2001); Akinfala *et al.* (2002).

	Description	Composition		Notes	Reference
	content reduced by sun-drying, oven-drying, ensiling and boiling. Leaf protein has low sulphur amino acids. Harvesting leaves every 2 months does not affect root yield and increases leaf yield. Leaf meal is protein feed for livestock. Dried cassava or cassava flour has high feed value. Fresh or dried roots are good energy source for pigs. Fermented cassava peel used for young poultry.	15.15 MJ/kg.		starter diets. Supplementation with methionine improves growth. Growth and feed conversion impaired on 12.5% cassava plant meal diets. 20–45% cassava peel meal can be fed to chickens.	Calles *et al.* (1970).
Banana	Rich source of carbohydrate. High level of tannins in green bananas and low presence in fresh ripe bananas depresses protein digestibility. Bananas can be fed to pigs fresh, ensiled or as dry meal. Ripe bananas are palatable and degree of ripeness affects performance. Palatability of green bananas affects intake and reduces performance. Sliced drying can improve consumption.	CP 1.0%, DM 20% for whole unpeeled, ripe bananas. 20–22% of DM, mainly starch, for green bananas.	–	Banana leaf meal replaces 15% of dietary DM for grower pigs. 30% waste banana meal can be included as supplement for growers.	

Continued

Table 18.1. Some potential feed ingredients for free-range poultry and pigs. – *Continued*

Ingredient	General description	Feeding value		Inclusion rate		References
		Pig	Poultry	Pig	Poultry	
Legume leaf meal	Leaf meal from leguminous shrubs and trees contains high CP, high fibre and low energy. Leaf meals (moisture content 120 g/kg) can be stored without deterioration in quality. Level of carotenoids in leaf meals depends on duration and method of drying. Has low protein digestibility and inferior ME compared to legume grains.	High beta-carotene (227–248 mg/kg DM) and xanthophylls (741–766 total xanthophylls/kg).	–	*L. leucocephala* leaf meal at 100 g/kg gives good growth and at 500 g/kg depresses growth.	Depends on level of anti-nutritional factors. Young broiler chicks grow well at 150 g/kg diet. *L. leucocephala* leaf meal included at 100 g/kg diet for older broiler chicks reduces the growth.	Malynicz (1974); D'Mello and Taplin (1978); D'Mello *et al.* (1987); D'Mello (1992,1995);
Grass (fresh)	Nutritional value of grass depends on the species, location, season, growth stage and fertilizer used. Early in growing season, it has high water, organic acid and protein content, low carbohydrates and lignin.	CP 4–30%, crude fibre 20–45%. DE 8.2 MJ/kg.	ME 4.8MJ/kg.	–	–	Ewing (1997).
Lucerne (dried)	A good protein and digestible fibre source for livestock, mainly ruminants. Good source of beta-carotene. Nutritional value depends on the stage of harvest and process conditions.	CP 18%, crude fibre 23.0%. DE 7.5 MJ/kg.	ME 6.0 MJ/kg.	–	–	Ewing (1997).

Table 18.2. Effect of season on crop contents (visual observation) of scavenging local hens. (From Tadelle and Ogle, 2000.)

Season	Number of poultry	Physical components (% fresh basis)				
		Seeds	Plants	Worms	Insects	Others
Short Rainy	90	37.5	22.5	2.6	14.6	22.7
Rainy	90	25.8	31.8	11.2	7.7	23.4
Dry	90	29.5	27.7	6.2	11.1	25.6
Means ± SD	270	30.9 ± 7.9	23.3 ± 6.0	6.7 ± 4.5	11.1 ± 4.5	23.9 ± 4.6

can limit the uptake of essential elements (e.g. Ca, P) and many of the trace elements due to the formation of insoluble soaps with minerals. The diet in the first few weeks should be palatable and rich in digestible carbohydrates. Maize, wheat and barley are a good source of energy (see Table 18.1). Grit should be available in the paddock to stimulate early gizzard development (Portsmouth, 2000). The diet specification recommended by Portsmouth (2000) for free-range poultry is listed in Table 18.4.

The growing stage

The energy level in the diet is critical during the post-chick feeding stage, particularly in the

period from 10 weeks to the start of lay. Maintaining optimum stocking density is important to ensure that all poultry have access to feeders and drinkers to avoid uneven growth. Low-energy diets should be avoided between 6 and 15 weeks. The inclusion of enzymes in a free-range poultry ration could improve the energy utilization efficiency, especially if large amounts of fibre are consumed by birds foraging on pastures.

PRE-LAY TO EARLY LAY. This is a very critical period and many free-range flocks are held back by poor pre-lay nutrition. The requirements for pre-lay are listed in Table 18.5 and the types of feed sources that could be used are shown in Table 18.1. Ca is important for the development

Table 18.3. Chemical composition of the crop contents of scavenging hens, overall means, SD and range from three of the seasons and study sites. (From Tadelle and Ogle, 2000.)

	Chemical composition (%)	
	Means ± SD (270)	Range (270)
Dry matter	50.7 ± 12.5	26.4–85.8
As % of dry matter		
Crude protein	8.8 ± 2.3	4.3–15.4
Crude fibre	10.2 ± 1.6	6.5–14.0
Ether extract	1.9 ± 0.9	0.3–4.7
Ash	7.8 ± 2.7	1.6–15.7
Calcium	0.9 ± 0.4	0.2–1.9
Phosphorus	0.6 ± 0.3	0.1–2.4
Energy (ME, Kcal/kg calculated)	2864.3 ± 247	2245.1–3528.1

Table 18.4. Chick starter and pre-lay feed specifications. (From Portsmouth, 2000.)

Nutrient	Chick starter specification	Pre-lay specification
Protein %	20–21	16–17
ME (MJ/kg)	12.0	11.5
Lysine (total) %	1.15	0.75
Lysine (available) %	0.99	0.65
Methionine (total) %	0.50	0.34
Methionine (available) %	0.45	0.31
Methionine + cystine %	0.83	0.60
Tryptophan %	0.20	0.17
Threonine %	0.73	0.50
Calcium %	0.90	2.0
Available phosphorus %	0.45	0.40
Sodium %	0.20	0.16
Linoleic acid %	1.25	1.25
Added fat/oil %	Optional	2–3

Table 18.5. Composition of crop contents of free-range chicks. (From Tadelle and Ogle, 2003; Lomu *et al.*, 2004 and unpublished data.)

| | Crop contents (%) | | |
| | | Lomu *et al.* (2004) | |
	Tadelle and Ogle (2003)	Exclusion pen (without supplementation)	Free-range (with supplementation)
Seeds	30.9	17.8	55.3
Plant materials	23.3	51.2	22.1
Worms	6.7	–	–
Stones	–	20.8	7.9
Insects	11.1	4.5	11.4
Unidentified materials	23.9	5.8	3.4

of medullary bone, but only 2% is recommended for the pre-lay diet (Portsmouth, 2000). Oyster shell is a better Ca source than limestone granules because the rate of solubilization of limestone is too rapid to maximize blood Ca levels over long periods (Bogdanov, 1997; Portsmouth, 2000). An excessive amount of Ca can have a negative effect on feed intake.

Pig

Nutrient requirements of various breeds and different growth stages of pigs are available housed under an intensive system. Outdoor sows are normally fed on a conventional concentrated diet once daily (Petherick and Blackshaw, 1989). The suitable feed forms for free-range pigs are a cob, roll or biscuit, which reduces waste from treading into the mud and is less likely to be carried away by birds. Shy feeders can obtain a share and move off to eat in relative peace (Thornton, 1988). A common practice for free-range pigs is to allow grazing on fresh strip pastures and rotate pigs between paddocks. The huts, feeding troughs and water are moved every 2–3 weeks (Gustafson and Stern, 2003). Water can be provided through alkathene pipes, either on the surface or mole-ploughed in (Thornton, 1988).

Feeding weaners

The digestive system of weaners (3–4 weeks of age) is immature. Non-milk nutrients are not easily digested by weaners. A two-stage feeding of weaners is recommended, with a starter diet followed by a grower ration. The starter diet must be fresh and palatable, and should contain some milk products and cooked cereals. The starter diet can be fed to 4–12 kg live weight weaners, and then gradually changed to a grower ration, which can be fed *ad libitum* to 10–12 kg live weight weaners (Thornton, 1988).

Feeding growers

Ad libitum feeding of grower pigs on a high-energy diet (DE 13.8 MJ/kg) results in maximum growth rate and efficiency of feed utilization. Limited feeding produces a leaner carcass, but a lower growth rate (Pond *et al.*, 1995). Pigs can be fed *ad libitum* on a conventional grower feed until 60 kg and thereafter provided a restricted diet of 2.8 kg/pig/day (Hogberg *et al.*, 2001). It is common to feed finisher pigs complete commercial diets in the form of pellets or paste. Limiting the feed allowance of finisher pigs to 70–80% of *ad libitum* consumption reduces carcass fatness, but also reduces rate and feed efficiency of gain because of the greater proportion of daily intake needed for maintenance (Pond *et al.*, 1995).

Feeding sows

Under an intensive production system, commercial dry sows are currently housed in individual stalls and subject to restricted feeding. However, under restricted feeding systems, group-housed dry sows are also aggressive. (See Nielsen *et al.*, Chapter 10, this volume, for details of feeding behaviour of pigs.) The feed intake of these sows is variable, depending on their social rank, which results in different milk production and variable piglet performance. These problems can be solved by *ad libitum* feeding systems, but the large capacity of intake by dry sows will not allow this feeding system to be practical, as high feeding level during pregnancy can reduce reproduction performance of sows. Under free-range systems, it is generally agreed that pigs should be restrictively fed during gestation at a level of about 1.5× maintenance, which is about 0.6× *ad libitum* intake (Barnett *et al.*, 2001), and a small amount of pasture intake by foraging. Keeping primiparous pregnant sows on a plane of 26 MJ DE/day ensures optimum back fat and live weight gain. Feeding sows 19 MJ DE/day causes mobilization of fat reserves. Feeding pregnant sows 33 MJ DE/day does not have a positive effect on milk composition and litter performance (Santos Ricalde and Lean, 2002). Because pregnant and lactating sows are more active, they need 10–15% more dietary energy, especially in winter. Suitable supplementary lactation feeds contain 17.4% crude protein (CP), 1.01% lysine, 0.62% phosphorus (P) and an ME of 14.5 MJ/kg (Jost, 1995).

Sow feeding should be done as quickly as possible to avoid fighting and stress, by spreading the feed out in a long line on the ground and providing 2 m of feeding space per sow. If the ground is wet, temporary troughs can be used in the gestation paddocks. Extra feed may be needed in cold, wet periods. A small allowance should be made if sufficient pasture or stubble is available. In most situations, Thornton (1988) recommends the feeding allowance for sow and gilt as shown in Table 18.6.

Factors Affecting the Feed Allowance of Free-range Poultry and Pigs

One of the difficulties facing nutritionists developing rations for free-range poultry and

Table 18.6. Sow and gilt feeding allowance, diet with 13.0 MJ/kg DE and 16% CP. (From Thornton, 1988.)

Period	Daily intake (kg)
Sows	
Weaning to re-mating	4.0–4.5
First 3 weeks after mating	2.5–3.5
3 weeks after mating until	
3 weeks before farrowing	3.0
Last 3 weeks of gestation	3.5–4.2
Lactation	First week build up to appetite level, then *ad libitum*, or 7.0–8.0
Gilts	
On delivery and during	
acclimatization	2.2–2.4
1 week before and after	
mating	3.5
Remainder of gestation	2.5
Last 2–3 weeks of gestation	3.5–4.0
Lactation	First week build up to appetite level, or *ad libitum*

pigs is to determine the quantity and nutritional value of forage consumed. It is difficult to measure the intake of foraging pigs and poultry due to the lack of an appropriate method. The visual separation of crop contents can give a guideline on the diet composition, but cannot be used to further quantify the pasture species ingested by poultry. Measuring the availability and botanic composition of pastures pre- and post-foraging might indicate the preference of foraging poultry for pasture species, but the result is influenced by the sampling method, regrowth of pastures and patchy foraging. Currently a method using plant alkanes as a marker has been developed to measure forage intake of grazing sheep (Dove and Mayes, 1991) and deer (Ru *et al.*, 2002b).

The contribution of foraging to the nutrition of poultry and pigs has generally been considered negligible. The nutrient content of forage depends on the species, stage of growth (young grass CP = 31.2 g/kg, mature grass 28.2 g/kg), soil type, grazing systems (rotation frequency, grazing pressure, stocking rate), cultivation conditions and fertilizer application (McDonald *et al.*, 1995). If poultry consume grass largely, the nutritional value of the diet is relatively poor. In spring, grass has high sugar and protein and a relatively low fibre content. In summer the sugar and protein content falls before resurgence in early autumn (Walker and Gordon, 2003). The net energy value of autumn grass (4.3 MJ/kg) is often lower than that of spring grass (5.2 MJ/kg), even when the two are cut at the same stage of growth and are equal in digestibility or ME (McDonald *et al.*, 1995). The botanical composition of permanent grassland and older sown pastures is dependent on management factors, such as drainage, compaction, acidity, fertility, weed population, stocking rate and grazing control, as well as factors beyond the farmer's control such as altitude, aspect, soil type, geology and climate. These factors affecting the nutritional quality of pasture have been documented for ruminants, but not for poultry (Walker and Gordon, 2003).

When properly managed, short good-quality pasture is beneficial in spring and early summer and reduces the consumption of supplementary feed by 5% for free-range poultry (Thompson, 1952 cited by Walker and Gordon, 2003). Some herbs (rosemary: *Rosmarinus officinalis*; sage: *Salvia officinalis*; oregano: *Origanum vulgare* and thyme: *Thymus vulgaris*) have powerful antioxidant capabilities and moderate antimicrobial activities (Adams, 1999) and could be included in the sward to reduce the need to feed synthetic antioxidants.

Poultry

The amount of feed required changes with season to meet the changing energy costs for maintaining body temperature. High temperature in summer reduces feed intake and performance. Increasing the nutrient density (amino acids and essential fatty acids) in supplementary feed can increase the nutrient intake (Portsmouth, 2000).

Small amounts of pasture were ingested by free-range poultry reported by Lomu *et al.* (2004) and Tadelle and Ogle (2003) (Table 18.5). Hughes and Dun (1983) also found that free-range hens consumed at least 50 g DM of pasture/day with access to a mash feed.

In broiler diets 10% inclusion of medic, vetch and oaten hay did not significantly affect the feed conversion ratio (FCR), which ranged from 1.81 (control) to 1.92 (medic hay diet) (B.O. Hughes, 2004, South Australia, Australia, unpublished data). In addition, 25% inclusion of rice bran, palm kernel cake and wheat mill run and 10% inclusion of rice hulls did not significantly affect the FCR (J. Pandi, 2004, South Australia, Australia, unpublished data). Grass only constitutes a source of energy and fibre, and would make little protein contribution. Nutritionists therefore need to know the full nutritional value of pasture and the intake by the poultry, so that adjustments can be made to feed formulations to minimize the risk of impairing performance.

Chickens also feed on a wide range of macro invertebrates living in the soil surface, including ground beetles (*Carabidae*), rove beetles (*Staphlinidae*), spiders (*Araneae*) and earthworms (*Lumbricidae*; Clark and Gage, 1996). The ME value of earthworms, grasshoppers (*Orthoptera*) and housefly (*Musca domestica*) pupae range from 12.4 to 12.8 MJ/kg DM and the CP and lysine contents are high (Bassler *et al.*, 1999). Walker and Gordon (2003) recommended a CP content of 170 g/kg and energy content of 11.7 MJ/kg for free-range hen rations. Free-range hens consume about 130 g feed/day and 22 g CP/day at 16°C. The equivalent of this intake would be provided by the consumption of 36 g of earthworms (DM basis). The feed formulations for free-range table chickens are listed in Table 18.7 (Walker and Gordon, 2003).

Pig

The feed intake is affected by the quantity and quality of forage intake of free-range pigs. Thornton (1988) recommended that pasture at best quality could save 0.25–0.5 kg feed/sow/day. Stern and Andresen (2003) suggested that successive rotation could reduce the feed allowance by 20% and increase the nutrient intake from the herbage by ~5%. Ru *et al.* (2001) reported that grower pigs could be fed half their daily allowance and the remainder could be obtained from forage. Andresen and Redbo (1999) found that continuous allocation

Table 18.7. Feed formulation for organic table chickens. (From Walker and Gordon, 2003.)

Ingredient	Quantity (kg/t)		
	Starter (0–10 days)	Grower (11–28 days)	Finisher (29–81 days)
Wheat	550	700	710
Wheat feed	105	50	50
Full-fat soybean	260	198	192
Peas (*Pisum sativum*)	50	20	17
Starter supplement[a]	35	–	–
Grower supplement	–	32	–
Finisher supplement	–	–	30

[a]A supplementary mineral and vitamin premix.

of new land might be an efficient tool to stimulate a high level of foraging and explorative activity in outdoor pigs on grassland.

Temperature plays an important role on free-range pig production. Verstegen *et al.* (1982) reported that during the growth period (25–60 kg), pigs need about 25 g/day of feed to compensate for a 1°C fall in temperature. During the fattening period (60–100 kg), the requirement is 39 g/day for a diet containing 0.012 MJ ME/g and 13.4% digestible CP. In sows there is an increased maintenance requirement of 250 g feed/day for each 5°C decrease in environmental temperature below the lower critical temperature (King 1990, 1991).

The depression in feed intake and growth rate in heat-stressed animals can be alleviated by lowering dietary protein level and reducing the levels of essential amino acids (Waldroup *et al.*, 1976). Increasing the energy density of the diet in a hot environment improves milk yield at all stages of lactation. Conversely, the addition of fibre in a hot environment depresses milk yield and the weight of piglets at weaning (Schoenherr *et al.*, 1989a,b). It is difficult to control the fibre intake of free-range pigs although a high-energy, low-fibre supplementary diet can be used.

Under humid tropical climatic conditions, pig diets with high-energy feeds can result in a protein deficiency. Energy in diets should be slightly lower than diets fed to pigs in a temperate climate. Vitamin A and some of the B-complex vitamins (pantothenic acid and pyridoxine) are important for pigs raised under conditions where sudden changes in environmental temperature occur frequently (Eusebio, 1980).

Integration of Poultry and Pigs into a Farming System Production, Agronomic and Sustainability Aspects

Poultry

Worldwide there is a great threat from the spread of weeds and diseases and a reduction in soil fertility in cropping areas, costing the grain industry billions per year. To control weeds, pests and diseases and improve soil fertility, farmers are using chemicals heavily, resulting in the resistance of weeds, pests and diseases, and the chemical contamination of farm products, which is a genuine concern for consumers. The incorporation of free-range poultry into a cropping system can assist in weed, pest and disease control in the crop phase, reduce chemical input, improve soil fertility and crop yield, and change consumer perceptions. Poultry can reduce the abundance of Colorado potato beetle on potato (Casagrande, 1987) and flies in cattle pasture (Salatin, 1991). Incorporation of free-range poultry into a cropping system housing poultry in ecoshelters has public appeal. It allows the poultry to express their natural behaviours and provides a wide variety of food resources for poultry to forage.

Glatz and Ru (2002) assessed the potential of using free-range poultry in a crop–pasture rotation system (Fig. 18.1). Free-range chickens were compared to sheep. The herbage availability was greater in the chicken paddocks than in the sheep paddocks after 3 months of foraging (Table 18.8). Sheep grazed the medic pods heavily, leaving only

Fig. 18.1. Free-range poultry.

30 g/m^2 of pods, while poultry left 965 g/m^2, reducing the need to resow for the next pasture season. Sheep were very effective in grazing the wire weed that contaminated the paddocks, whereas poultry avoided this weed. In contrast, the number of unidentified weeds in the sheep paddock were greater than those in the poultry paddock. This raises the possibility that sheep and poultry could be grazed together, to provide a method for reducing weed build-up. Soil fertility was not different between the sheep and poultry paddocks.

Poultry were very active in the paddock during overcast conditions and also when there was drizzly rain. Similar results were reported by Dawkins *et al.* (2003).

Pigs

It has been recognized that grazing and rooting behaviour of free-range pigs reduces external inputs for soil tillage and weed control in the cropping system (Andresen and Redbo, 1999). However, one of the key public concerns for

Table 18.8. Comparison of the agronomic, snail, weed and soil fertility in paddocks grazed by sheep and poultry. (From Glatz and Ru, 2002.)

Variable	Poultry (110 poultry/ha)	Sheep (12 sheep/ha)	P value
Biomass (g/m^2)	491	132	***
Dry matter (g/m^2)	417	109	**
Crude protein (g/m^2)	50	6	**
Organic matter (g/m^2)	374	91	**
Snails (number/m^2)	4	2	NS
Medic pods (number/m^2)	418	69	**
Wire weed (number/m^2)	23	0	**
Unidentified weeds (number/m^2)	5	16	***
Nitrate N (mg/kg soil)	18	24	NS
Ammonia N (mg/kg soil)	0	0.1	NS

***Significantly different at $P < 0.01$; **different at $P < 0.05$; NS: not significant.

the free-range pig production systems is the impact on the environment. In some systems, pigs are held in the same paddock for a long period at a high stocking rate, which damages vegetation, and increases nutrient load in the soil, nitrate leaching and gas emission (Worthington and Danks, 1992; Rachuonyo et al., 2002). One way to avoid this is to integrate outdoor pigs in the cropping pasture system, with the stocking rate related to the amount of feed given to the animals.

Glatz and Ru (2002) assessed the potential of using free-range pigs in a crop–pasture rotation system (Fig. 18.2). Free-range pigs were compared to sheep. The herbage availability was greater in the pig paddocks than in the sheep paddocks after 4½ months of foraging on medic pasture and a barley crop (Table 18.9). Sheep grazed the barley crop and pasture heavily, leaving only 152 and 15.0 g/m² of the crop and pasture, respectively. There was no significant difference in wire weed population in the paddocks after grazing by pigs or sheep. In contrast to the medic pasture, the barley paddocks grazed by sheep had higher nitrate N and ammonia N compared to the paddocks grazed by pigs.

Regular herd movement to clean ground also is considered a major contribution to the maintenance of herd health. Movement of the herd to clean ground helps to avoid build-up of parasites and limits the problems that arise from overgrazing and the build-up of mud (Thornton, 1988).

The effect of feeding level on foraging behaviour of outdoor pigs

The activities of pigs, especially those associated with foraging behaviour, are affected by feeding levels. Pigs on a high feeding level spend most of their time on rooting (Stern and Andresen, 2003). Feeding high-fibre diets to outdoor sows, at either a restricted level or *ad libitum*, reduces foraging behaviour but does not reduce pasture damage (Braund et al., 1998). Sugarbeet pulp has proved to be particularly effective in reducing foraging, and reduces the appetite for other fibrous materials (Brouns et al., 1995). Time spent grazing, grazing activity and distances walked are significantly reduced when dietary energy increases from 19 to 33 MJ DE/day (Santos Ricalde and Lean, 2002). Studies under indoor conditions suggest that feed with inadequate CP content can induce rooting behaviour (Jensen et al., 1993) and pigs may select a

Fig. 18.2. Free-range pigs.

Table 18.9. Herbage availability, snail numbers and the number of weeds in the paddocks grazed by sheep and pigs during the season. (From Ru *et al.*, 2002a.)

Variable	Barley			Pasture		
	Pig (12 pigs/ha)	Sheep (12 sheep/ha)	*P* value	Pig (12 pigs/ha)	Sheep (12 sheep/ha)	*P* value
Mass (air dry g/m²)	426.0	152.0	***	174.5	15.0	***
Dry matter (g/m²)	369.5	132.1	***	139.9	12.4	**
Organic matter (g/m²)	317.9	118.9	***	109.9	9.0	**
Crude protein (g/m²)	28.7	6.3	**	15.0	1.3	***
Snails (number/m²)	0	0	NS	0	2.5	NS
Barley grass (g/m²)	0.4	0	**	514.6	150.8	NS
Rye grass (g/m²)	9.2	0.4	NS	0.8	1.3	NS
Barley (g/m²)	239.2	45.8	**	–	–	–
Other grass (g/m²)	0	0.8	NS	0.4	1.3	NS
Wire weed (number/m²)	0.8	0.4	NS	4.2	0	NS
Caltrops (number/m²)	10.0	9.2	NS	0.4	15.0	NS
Medic/clover (number/m²)	15.0	13.8	NS	41.3	10.0	NS
Soursobs (number/m²)	0	0.4	NS	15.0	31.7	NS
Other weeds (number/m²)	4.2	4.2	NS	26.7	60.0	NS
Nitrate N (mg/kg soil)	20.1	20.9	**	40.6	30.6	NS
Ammonia N (mg/kg soil)	0	0.4	***	6.4	3.0	NS

***Significantly different at $P < 0.01$; **different at $P < 0.05$; NS: not significant.

diet suitable for their needs if given the choice (Kyriazakis and Emmans, 1991). However, high ambient temperature has a greater effect on grazing behaviour and energy intake in pregnant sows kept outdoors under tropical conditions (Santos Ricalde and Lean, 2002).

Conclusion

The diet composition that free-range poultry and pigs consume is complex. They can forage soil, pebbles, grass, weeds, crop seeds and insects in the paddock. Pigs can scavenge and forage in forests, woods and pasture. This makes it difficult to develop a supplementary feeding strategy to meet their nutritional requirements. Ideally, the amount of supplement required should be based on the amount of nutrient foraged and the total requirement to reduce the feed cost. However, there is limited information on the forage intake of free-range poultry and pigs during the season. A better understanding of foraging behaviour and forage intake of free-range poultry and pigs will enable producers to develop an economic feeding system. Given that free-range poultry and

pigs consume a significant amount of forage, the nutritive value of forages for free-range poultry and pigs will be crucial for the development of a supplementary feeding system.

Free-range poultry and pigs can be incorporated into a crop–pasture rotation system to maintain soil fertility and control weeds. The sustainability of these systems requires further assessment, especially the optimum stocking rate and the most suitable pastures under different soil types.

It is well known that there is seasonal fluctuation in the production of free-range poultry and pigs. High temperature in summer is one of the key factors causing low animal production under free-range conditions. While a number of nutritional strategies are already applied by free-range farmers, there are new products available on the market for reducing heat stress and should be assessed for free-range poultry. Pigs exposed to temperatures above the evaporative critical temperature have a low feed intake and milk yield and poor reproductive performance and growth rate. Water drippers, sprays and wallows are effective in reducing the impact of ambient temperature and improve the production of free-range pigs.

References

Adams, C. (1999) *Nutricines. Food Components in Health and Nutritions.* Nottingham University Press, Nottingham, UK.
Akinfala, E.O. and Tewe, O.O. (2001) Utilization of whole cassava plant in the diets of growing pigs in the tropics. *Livestock Research for Rural Development* 13(5), 1–6.
Akinfala, E.O., Aderibigbe, A.O. and Matanmi, O. (2002) Evaluation of the nutritive value of whole cassava plant as replacement for maize in the starter diets for broiler chicken. *Livestock Research for Rural Development* 14(6), 1–6.
ALFID (2002) *Australian Livestock Feed Ingredient Database.* SARDI, Roseworthy, South Australia.
Amissah, J.G.N., Ellis, W.O., Oduro, I. and Manful, J.T. (2003) Nutrient composition of bran from new rice varieties under study in Ghana. *Food Control* 14, 21–24.
Andresen, N. and Redbo, I. (1999) Foraging behaviour of growing pigs on grassland in relation to stocking rate and feed crude protein level. *Applied Animal Behaviour Science* 62, 183–197.
Appleby, M.C., Hughes, B.O. and Hogarth, G.S. (1989) Behaviour of laying hens in a deep litter house. *British Poultry Science* 30, 545–553.
Arosemena, A., DePeters, E.J. and Fadel, J.G. (1995) Extent of variability in nutrient composition within selected by-product feedstuffs. *Animal Feed Science and Technology* 54, 103–120.
Barnett, J.L., Hemsworth, P.H., Cronin, G.M., Jongman, E.C. and Hutson, G.D. (2001) A review of the welfare issues for sows and piglets in relation to housing. *Australian Journal of Agricultural Research* 52, 1–28.
Bassler, A., Ciszuk, P. and Sjelin, K. (1999) Management of laying hens in mobile houses – a review of experiences. In: Hermansen, J.E., Lund, V.V. and Thuen, E. (eds) *Ecological Animal Husbandry in the Nordic Countries. Proceedings of the NJF Seminar,* Danish Research Centre for Organic Farming, No. 303, Horsens, Denmark, pp. 45–50.
Bogdanov, I.A. (1997) Seasonal effects on free-range egg production. *World Poultry-Misset* 13, 47–49.
Braund, J.P., Edwards, S.A., Riddoch, I. and Buckner, L.J. (1998) Modification of foraging behaviour and pasture damage by dietary manipulation in outdoor sows. *Applied Animal Behaviour Science* 56, 173–186.
Brouns, F., Edwards, S.A. and English, P.R. (1995) Influence of fibrous feed ingredients on voluntary intake of dry sows. *Animal Feed Science and Technology* 54, 301–313.
Calles, A., Clavijo, H., Hervas, E. and Maner, J.H. (1970) Ripe bananas (*Musa* sp.) as energy source for growing-finishing pigs. *Journal of Animal Science* 31, 197 (Abst.).
Casagrande, R.A. (1987) The Colorado potato beetle: 125 years of mismanagement. *Bulletin of the Entomological Society of America* 33, 142–150.
Cerning-Beroard, J. and Le Dividich, J. (1976) Valeur alimentaire de quelques produits amylaces d'origine tropicale. *Annual Zootechnology* 25(2), 155–168.
Clark, M.S. and Gage, S.H. (1996) Effects of free-range chickens and geese on insect pests and weeds in an agroecosystem. *American Journal of Alternative Agriculture* 11, 39–47.
Dardaillon, M. (1989) Age–class influences on feeding choices of free-ranging wild boars *Sus scrofa. Canadian Journal of Zoology* 67, 2792–2796.
Dawkins, M.S., Cook, P.A., Whittingham, M.J., Mansell, K.A. and Harper, A.E. (2003) What makes free-range broiler chickens range? *In situ* measurement of habitat preference. *Animal Behaviour* 66, 151–160.
Devendra, C. (1977) Utilization of feedingstuffs from the oil palm. In: *Proceedings of the Symposium on Feedingstuffs for Livestock in South East Asia,* pp. 116–131.
D'Mello, J.P.F. (1992) Chemical constraints to the use of tropical legumes in animal nutrition. *Animal Feed Science and Technology* 38, 237–261.
D'Mello, J.P.F. (1995) Leguminous leaf meals in non-ruminant nutrition. In: D'Mello, J.P.F. and Devendra, C. (eds) *Tropical Legumes in Animal Nutrition.* CAB International, Wallingford, UK, pp. 247–282.
D'Mello, J.P.F. and Taplin, D.E. (1978) *Leucaena leucocephala* in poultry diets for the tropics. *World Review of Animal Production* 14, 41–47.
D'Mello, J.P.F., Acamovic, T. and Walker, A.G. (1987) Evaluation of *Leucaena* leaf meal for broiler growth and pigmentation. *Tropical Agriculture (Trinidad)* 64, 33–35.
Dove, H. and Mayes, R.W. (1991) The use of plant wax alkanes as marker substances in studies of the nutrition of herbivores: a review. *Australian Journal of Agricultural Research* 42, 913–952.
Eusebio, J.A. ((1980) Pigbiology (nutrition and feeding). In: Payne, W.J.A. ((ed.) *Pig Production in the Tropics.* Intermediate Tropical Agriculture Servies, Longman Group, London.

Evans, M. (1985) *Nutrition Composition of Feedstuffs for Pigs and Poultry.* Queensland Department of Primary Industries, Queensland, Australia.

Ewing, W.N. (1997) Wheat. In: Ewing, W.N. (ed.) *The Feeds Directory*, Vol. 1. British Library Cataloguing in Publication Data, 112 pp.

Fashina-Bombata, H.A. and Fanimo, O.A. (1994) The effects of dietary replacement of maize with sun dried sweet potato meal on performance, carcass characteristics and serum metabolites of weaner-grower pigs. *Animal Feed Science and Technology* 47, 165–170.

Glatz, P.C. and Ru, Y.J. (2002) Free-range poultry in a pasture/crop rotation system. In: *Proceedings of the 2002 Poultry Information Exchange*, 14–16 April 2002. Poultry Information Exchange Association, Caboolture, Queensland, Australia, pp. 7–10.

Gomez, G., Santos, J. and Valdivieso, M. (1984) Evaluation of methionine supplementation to diets containing cassava meal for swine. *Journal of Animal Science* 58, 812–820.

Gustafson, G.M. and Stern, S. (2003) Two strategies for meeting energy demands of growing pigs at pasture. *Livestock Production Sciences* 80, 167–174.

Henuk, Y.L. and Dingle, J.G. (2002) Practical and economic advantages of choice feeding systems for laying poultry. *World's Poultry Science Journal* 58, 199–208.

Hogberg, A., Pickova, J., Dutta, P.C., Babol, J. and Bylund, A.C. (2001) Effect of rearing system on muscle lipids of gilts and castrated male pigs. *Meat Science* 58, 223–229.

Hughes, B.O. (1984) The principles underlying choice feeding behaviour in fowls – with special reference to production experiments. *World's Poultry Science Journal* 40, 141–150.

Hughes, B.O. and Dun, P. (1983) *A comparison of laying stock: housed intensively in cages and outside on range.* Research and Development Publication No. 18, The West of Scotland Agricultural College, Edinburgh, Scotland, 17 pp.

Hutagalung, R.I. (1981) The use of tree crops and their by-products for intensive animal production. In: Smith, A.J. and Gunn, R.G. (eds) *Intensive Animal Production in Developing Countries.* British Society of Animal Production, Occasional Publication No. 4, London, pp. 151–188.

Hutagalung, R.I. and Mahyudin, M. (1981) Feeds for animals from the oil palm. *Proceedings of the International Conference on Oil Palm in Agriculture in the Eighties*, pp. 609–622.

Jacob, J.P., Mitaru, B.N., Mbugua, P.N. and Blair, R. (1996) The effect of substituting Kenyan Serena sorghum for maize in broiler starter diets with different dietary crude protein and methionine levels. *Animal Feed Science and Technology* 61, 27–39.

Jensen, M.B., Kyriazakis, I. and Lawrence, A.B. (1993) The activity and straw directed behaviour of pigs offered foods with different crude protein content. *Applied Animal Behaviour Science* 37, 211–221.

Jost, M. (1995) Feeding pigs on pasture. *Agrarforschung* 2(2), 68–69.

Kass, M.L., Van Soest, P.J., Pond, W.G., Lewis, B. and McDowell, R.E. (1980) Utilization of dietary fibre from alfalfa by growing swine. 2. Volatile fatty acid concentrations in and disappearance from the gastrointestinal tract. *Journal of Animal Science* 50, 192–197.

King, R. (1990) Requirements for sows. In: *Pig Rations: Assessment and Formulation 1990. Proceedings of the Refresher Course for Veterinarians*, Postgraduate Committee in Veterinary Science, University of Sydney, Sydney, pp. 25–44.

King, R. (1991) The basic of sow feeding and management. In: *Proceedings of the Saskatchewan Pork Industry Symposium*, Saskatoon, Saskatchewan, Canada.

Kyriazakis, I. and Emmans, G.C. (1991) Diet selection in pigs: dietary choices made by growing pigs following a period of underfeeding with protein. *Animal Production* 52, 337–346.

Lomu, M.L., Glatz, P.C. and Ru, Y.J. (2004) Metabolizable energy of crop contents in free range hens. *International Journal of Poultry Science* 3(11), 728–732.

Luteladio, N.B. and Egumah, H.C. (1981) Cassava leaf harvesting in Zaire. In: Terry, K.A. and Oduro, F. (eds) *Proceedings of the First Triennial Root Crops Symposium of the International Society for Tropical Root Crops.* Africa Branch, Caveness IDRC–163.

Malynicz, G. (1974) The effect of adding *Leucaena leucocephala* meal to commercial rations for growing pigs. *Papua New Guinea Agriculture Journal* 25, 12–14.

McBride, G., Parer, I.P. and Foenander, F. (1969) The social organization and behaviour of feral domestic fowl. *Animal Behaviour Monographs* 2, 125–181.

McDonald, P., Edwards, R.A., Greenhalgh, J.F.D. and Morgan, C.A. (1995) *Grass and Forage Crops. Animal Nutrition*, 5th edn. Longman Scientific and Technical, London, pp. 434–450.

Medel, P., Latorre, M.A., de Blas, C., Lazaro, R. and Mateos, G.G. (2004) Heat processing of cereals in mash or pellet diets for young pigs. *Animal Feed Science and Technology* 113, 127–140.

Nagura, D. (1964) The use of coconut oil meal in chick diets. *Ceylon Veterinary Journal* 12(3), 40–42.

Nielsen, B.L., Thomsen, M.G., Sørensen, P. and Young, J.F. (2003) Feed and strain effects on the use of outdoor areas by broilers. *British Poultry Science* 44(2), 161–169.

NRC (National Research Council) (1988) *Nutrient Requirement of Domestic Animals. Nutrient Requirement of Swine*, 8th edn. National Academy Press, Washington, DC.

NRC (National Research Council) (1994) *Nutrient Requirements of Poultry*. National Academy Press, Washington, DC.

Nwokolo, E.N., Bragg, D.B. and Saben, H.S. (1977) A nutritive evaluation of palm kernel meal for use in poultry rations. *Tropical Science* 19, 147–154.

O'Doherty, J.V. and McKeon, M.P. (2000) The use of expeller copra meal in grower and finisher pig diets. *Livestock Production Sciences* 67, 55–65.

Ofuya, C.O. and Obilor, S.N. (1993) The suitability of fermented cassava peel as a poultry feedstuff. *Bioresource Technology* 44, 101–104.

Onwudike, O.C. (1986) Palm kernel meal as a feed for poultry. 1. Composition of palm kernel meal and availability of its amino acids to chicks. *Animal Feed Science and Technology* 16, 179–186.

Panigrahi, S., Machin, D.H., Parr, W.H. and Bainton, J. (1987) Responses of broiler chicks to dietary copra cake of high lipid content. *British Poultry Science* 28, 589–600.

Perez, R. (1997) *African oil palm*. Food and Agriculture Organization of the United Nations, Rome.

Petherick, J.C. and Blackshaw, J.K. (1989) A note on the effect of feeding regime on the performance of sows housed in a novel group-housing system. *Animal Production* 49, 523–526.

Petterson, D.S. and Mackintosh, J.B. (1994) *The Chemical Composition and Nutritive Value of Australian Grain Legumes*. Grains Research and Development Corporation, Canberra, Australia.

Pond, W.G., Church, D.C. and Pond, K.R. (1995) Swine. *Basic Animal Nutrition and Feeding*, 4th edn. John Wiley & Sons, New York, 461 pp.

Portsmouth, J. (2000) The nutrition of free-range layers. *World Poultry* 16, 16–18.

Rachuonyo, H.A., Pond, W.G. and McGlone, J.J. (2002) Effects of stocking rate and crude protein intake during gestation on ground cover, soil-nitrate concentration, and sow and litter performance in an outdoor swine production system. *Journal of Animal Science* 80, 1451–1461.

Ralph Say, R. (1987) Feeding. *Manual of Poultry Production in the Tropics*. Translated by Ralph Say, Technical Centre for Agricultural and Rural Co-operation, CAB International, Wallingford, UK.

Ravindran, V., Kornegay, E.T., Rajaguru, A.S.B., Potter, L.M. and Cherry, J.A. (1986) Cassava leaf meal as a replacement for coconut oil meal in broiler diets. *Poultry Science* 65, 1720–1727.

Reverter, M. and Lindberg, J.E. (1998) Ileal digestibility of amino acids in pigs given a barley-based diet with increasing inclusion of lucerne leaf meal. *Animal Science* 67, 131–138.

Ru, Y.J., Glatz, P.C., Reimers, H., Fischer, M. and Wyatt, S. (2001) Performance of grazing pigs in summer in southern Australia. *Manipulating Pig Production*, Vol. VIII, 44 pp.

Ru, Y.J., Glatz, P.C., Reimers, H., Fischer, M., Wyatt, S. and Rodda, B. (2002a) Grazing pigs and sheep have similar influence on soils. *Pig and Poultry Fair 2002, Seminar Notes and Research Summaries* pp.71.

Ru, Y.J., Miao, Z.H., Glatz, P.C. and Choct, M. (2002b) Predicting feed intake of fallow deer (*Dama dama*) using alkanes as a marker. *Asian-Australasian Journal of Animal Science* 15, 209–212.

Salatin, J. (1991) Profit by appointment only. *The New Farm* 13(6), 8–12.

Santos Ricalde, R.H. and Lean, I.J. (2002) Effect of feed intake during pregnancy on productive performance and grazing behaviour of primiparous sows kept in an outdoor system under tropical conditions. *Livestock Production Sciences* 77, 13–21.

Savory, C.J. (1989) The importance of invertebrate food to chicks of gallinaceous species. *Proceedings of the Nutrition Society* 48, 113–133.

Savory, C.J., Wood-Gush, D.G.M. and Duncan, I.J.H. (1978) Feeding behaviour in a poultry population of domestic fowls in the wild. *Applied Animal Ethology* 4, 13–27.

Schoenherr, W.D., Stahly, T.S. and Cromwell, G.L. (1989a) The effects of dietary fat and fibre addition on energy and nitrogen digestibility in lactating sows housed in a warm or hot environment. *Journal of Animal Science* 67, 473–481.

Schoenherr, W.D., Stahly, T.S. and Cromwell, G.L. (1989b) The effects of dietary fat or fibre addition on yield and composition of milk from sows housed in a warm or hot environment. *Journal of Animal Science* 67, 482–495.

Scott, M.L., Nesheim, M.C. and Young, R.J. (1982) *Nutrition of the Chicken*. M.L. Scott & Associates, Ithaca, New York.

Stern, S. and Andresen, N. (2003) Performance, site preferences, foraging and excretory behaviour in relation to feed allowance of growing pigs on pasture. *Livestock Production Sciences* 79, 257–265.

Stoll, P. and Hilfiker, J. (1995) Finishing pigs on pasture carries a price. *Agrarforschung* 2, 449–452.

Tadelle, D. and Ogle, B. (2000) Nutritional status of village poultry in the central highlands of Ethiopia as
assessed by analyses of crop contents. *Ethiopian Journal of Agricultural Science* 17, 47–57.

Tadelle, D. and Ogle, B. (2002) The feed resource base and its potential for increased poultry production
in Ethiopia. *World's Poultry Science Journal* 58, 77–87.

Tadelle, D. and Ogle, B. (2003) *Studies on Village Poultry Production in the Central Highlands of Ethiopia.*
Department of Animal Nutrition and Management, Uppsala, Sweden.

Thear, K. (1997) *Free-Range Poultry*, 2nd edn. Farming Press, Ipswich, UK.

Thomas, O.A. and Scott, M.L. (1962) Coconut oil meal as a protein supplement in practical poultry diets.
Poultry Science 41, 477–485.

Thompson, A. (1952) *The Complete Poultryman*, Faber and Faber, London.

Thorne, P.J., Wiseman, J., Cole, D.J.A. and Machin, D.H. (1992) Effect of level of inclusion of copra meal
in balanced diets supplemented with synthetic amino acids on growth and fat deposition and
composition in growing pigs fed *ad libitum* at a constant temperature of 25°C. *Animal Feed Science
and Technology* 40, 31–40.

Thornton, K. (1988) *Outdoor Pig Production.* Farming Press, Ipswich, UK.

Van der Poel, A.F.B., Den Hartog, L.A., Van den Abeele, T.H., Boer, H. and Van Zuilichem, D.J. (1989)
Effect of infrared irradiation or extrusion processing of maize on its digestibility in piglets. *Animal Feed
Science and Technology* 26, 29–43.

Verstegen, M.W.A., Brandsma, H.A. and Mateman, G. (1982) Feed requirements of growing pigs at low
environmental temperatures. *Journal of Animal Science* 55(1), 88–94.

Waldroup, P.W., Mitchell, R.J., Payne, J.R. and Hazen, K.R. (1976) Performance of chicks fed diets
formulated to minimize excess levels of essential amino acids. *Poultry Science* 55, 243.

Walker, A. and Gordon, S. (2003) Symposium on 'Nutrition of farm animals outdoors'. Intake of nutrients
from pasture by poultry. *Proceedings of the Nutrition Society* 62, 253–256.

Worthington, T.R. and Danks, P.W. (1992) Nitrate leaching and intensive outdoor pig production. *Soil Use
and Management* 8, 56–60.

Wu, J.F. (1980) Energy value of sweet potato chips for young swine. *Journal of Animal Science* 51,
1261–1265.

Zombade, S.S. and Ichhponani, J.S. (1983) Nutritive value of raw, parboiled, stabilised and deoiled rice
bran for growing chicks. *Journal of the Science of Food Agriculture* 34, 783–788.

19 Conclusion and Perspectives

J.M. Forbes

School of Biology, University of Leeds, Leeds LS2 9JT, UK

The scope of this book is very wide. Some chapters are specific, others quite broad in their coverage. It is part of the remit of this chapter to draw together the disparate approaches to the subject of feeding covered in this book, but I will not be paying equal attention to each.

In the introductory chapter the editor provides a synopsis of the coverage of the book and fills in some details that help to set the scene and prepare the reader for what follows.

Feeding Structures and Mechanisms

'[A]nimals aren't machines' (Provenza and Villalba, Chapter 13, this volume), meaning that in the context of feeding behaviour we cannot predict events by inspection, however close, of the detailed workings of the system. When it comes to the mechanics of feeding, however, we do have a good understanding of the physical and chemical processes of prehension, mastication, swallowing and digestion in all parts of the gastrointestinal (GI) tract. Gussekloo (Chapter 2, this volume) gives a detailed description of the anatomy of the bill and mouth of birds, of the prehension of food and of the digestive system and its processes in chickens.

A more functional approach is taken by Bels and Baussart (Chapter 3, this volume), who highlight the integration of movements of many components of the musculoskeletal system, from mouth and head to axial and appendicular systems not just in birds but in all vertebrates also. They discuss the grasping of food, its manipulation in the mouth and swallowing, and emphasize the importance of particle size in these activities. Both galliform (chicken) and anseriform (duck) birds are used as examples of the complex nature of the interactions between the various parts of the food-processing system.

Popowics and Herring (Chapter 5, this volume) describe the masticatory structures of mammals in general terms and then deal with several example species, including the major species of agricultural importance, in particular the ungulate herbivore, with its adaptations for crushing fibrous plant material. These authors mention that all species of domestic ungulates can suffer from periodontal disease but do not discuss this any further. Loss of teeth by old ewes is a well-known reason for culling and is of special concern where the sheep are fed root crops, which they have to gnaw directly from the soil. It is this grit that wears down the teeth so rapidly. We also know that silage can contain high levels of soil, especially when the grass from which it is made is cut shortly after heavy rain, when splashing occurs, or when the harvester is set to cut too close to the ground. An extreme example of intake of soil is ostriches grazing on very short pasture that repeatedly peck at plants until they are uprooted, in which case sand may comprise 31–46% of their stomach contents (Brand and Gous, Chapter 9, this volume).

Are there other causes of tooth problems? Do animals get toothache that might inhibit them chewing and thus limit their intake of food? If such problems do occur, they are more likely in older animals – breeding stock – than in younger ones.

The mechanics of suckling and the transition at weaning, sudden in intensive farming, are covered in considerable detail by German et al. (Chapter 4, this volume) and considered below in relation to welfare. A sufficiently detailed description of the anatomy of the mouth and digestive tract of the horse is given by Houpt (Chapter 12, this volume), who then proceeds to describe biting and chewing, including examples of bite size and frequency. Mastication and particle breakdown in the ruminant is covered by Baumont et al. (Chapter 14, this volume), while Gidenne and Lebas (Chapter 11, this volume) deal similarly with the rabbit.

In each of these cases we can see how the various hard and soft tissues have evolved together to enable a particular animal to obtain and process efficiently the food(s) available in its niche. The development of these tissues and organs is largely following the plan set out in its genes, and once developed to a functional state, it is possible to see how nervous and chemical signals cause the contraction of muscles, and the secretion of digestive juices.

The brevity with which I mention these structures and mechanisms is not to deny their importance, or the excellence of the scientific endeavour involved in unravelling their intricacies. It is simply that I believe there are much greater challenges currently in understanding the extent to which particle breakdown in the mouth and rumen are under the control of the animal, in order to try to achieve a particular goal in terms of maximizing or optimizing nutrient availability and how animals control their food intake and choice between foods.

Feeding Behaviour

Given that animals take their food in discrete meals, and that these meals include pauses, the definition of the intermeal interval is criti-

cal if feeding behaviour is to be described properly. In the past, an arbitrary criterion has often been used (10 min in the case of some studies with horses; Houpt, Chapter 12, this volume). Others (e.g. Metz, 1975) have used a break in the slope of the cumulative frequency of intermeal intervals. Baumont et al. (Chapter 14, this volume) mention without further comment that recently Tolkamp et al. (1998) proposed using log-normal models rather than log (cumulative) frequency models. To my mind this deserves much more than just a mention as it is a breakthrough in our understanding of what is meant by a meal, being much sounder biologically than previously used methods. The proposition is that if meals are satiating, the probability of an animal starting to eat again after a meal should initially be low but then increase as the satiating properties of the meal decline. Thus, there would be a large number of short (intrameal) intervals, and a large number of long (intermeal) intervals with relatively few of intermediate length. The low point at which the frequency distribution of intrameal intervals crosses that of the intermeal intervals is used as the critical criterion for separating meals and gives a much longer interval, i.e. fewer, larger meals, than the interval used by previous workers.

Using the wrong intermeal criterion, invariably one that is shorter than the one derived by the Tolkamp method, leads to false conclusions about the importance of meals. Those intervals classified as intermeal, which should in reality be intrameal pauses, will be followed by apparently small meals that are in fact just part of the bigger, true meal. Thus, false correlations between meal size and the interval before (hunger ratio) or after (satiety ratio) will be generated. This in turn gives more emphasis to the meal as being controlled than is justified. Meals seem to occur at random and to be of random size – how this short-term lack of control is integrated into a long-term control is a mystery yet to be unravelled.

Physiological Controls

The 'mechanisms' referred to in the title of the book are not specifically intended to include coverage of physiological mechanisms of the control of feeding. In one of this book's few references to physiological controls (Houpt, Chapter 12, this volume), the role of blood glucose is emphasized for horses: 'Physiologically, glucose levels in the blood appear to be more important in predicting meals, i.e. horses eat when their plasma glucose levels fall below 90 mg/dl'. More is required than simply observing that feeding is correlated with certain changes in blood metabolites to provide critical evidence of a control mechanism; such critical evidence shows that it is the fall in blood glucose that is responsible for the start of feeding, both in rats (Campfield et al., 1985) and in humans (Campfield and Smith, 1986).

Neural mechanisms are discussed by Pittroff and Soca (Chapter 16, this volume). They offer the opinion that the study of single neural agents in isolation 'will not specifically aid in the systematic, holistic understanding of feed intake'. I am in complete agreement but it is clearly necessary to study single factors first in order to design the appropriate experiments to study interactions and synergies.

Leptin was hailed a decade ago as the answer to our questions about feedback from adipose tissue to the central nervous system (CNS). However, in addition to its direct effects on feeding, it is now clear that leptin influences energy metabolism and reproduction, so many of its effects might be indirect. What started as a simple story becomes more complex as the years go by.

As the situation stands at present, the study of neurohormonal regulation pathways alone cannot provide functional insight. The link between the short-term fluctuations in hormones, metabolites and neurotransmitters, on the one hand, and the medium and long-term control of food intake and choice, on the other, is still far from clear (but see, e.g. Rhind et al., 2002). It is essential that research into these links continues, as it is through such endeavours that a more complete understanding of the physiology of intake control will develop, with attendant methods for manipulation, whether in animals or humans.

Welfare

There are several welfare paradoxes in relation to feeding, not the least of which is how cruel it is to purposely underfeed animals, i.e. offer them less than they would eat voluntarily – a common practice with broiler breeders and pregnant sows. Koene (Chapter 6, this volume) quotes McFarland (1989) as suggesting that it is not such a great insult to animals to treat them in ways that might be expected to occur in nature. As food shortage would be common in the wild ancestors and cousins of our domesticated animals, purposely underfeeding them might not be a problem for their welfare. On the other hand, underfeeding has psychological effects that we do not fully understand – the overriding necessity of eating regularly for survival means that hunger and thirst are the most pressing of motivations and something to which we should not subject animals under our dominion, at least not for prolonged periods. However, without hunger they would not be inclined to eat, or to drink without experiencing thirst, so hunger and thirst are not in themselves bad things. If we are concerned about underfeeding of intensively farmed animals, what about extensive farming in highly seasonal environments in which man has little control over food supply?

Not only does shortage of food threaten animal welfare but the composition of that food is also important. Animals do not simply eat for calories. Given a diet deficient in one or more essential nutrients, their behaviour can become distorted, with abnormalities such as biting or pulling feathers from conspecifics, gnawing at bones (even in herbivores), chewing wood and repetitively mouthing objects in their environment (stereotypic behaviour). In a given situation do animals not showing stereotypies have worse welfare than those that do show them because the former are 'bottling them up'?

Dietary dilution with indigestible or poorly digested materials has been investigated as a means of alleviating the severe hunger associated with feed restriction. Variable results have been obtained, but none of the materials tested could be fed ad libitum without increasing the body weight. Sugarbeet pulp, while promising in some studies (pregnant sows, Brouns et al., 1995); (broiler breeders, de Jong and Jones,

Chapter 8, this volume), has not given universally useful results.

As a footnote to the use of high-fibre diets, if the food is too low in fibre, stomach pH can fall too low and lead to erosion of the stomach wall and gastric bleeding in pigs, sometimes causing death. In horses, wood-chewing and cribbing has been ascribed to lack of roughage in the diet, while in ruminants very low-fibre diets can cause acidosis and a strong desire for fibrous food.

Various intake-suppressing compounds, such as propionic acid, have been tried but are not very effective. Scattering the restricted amount of food pellets on the floor to keep the birds busy for longer parts of the day, searching for food did not significantly improve broiler breeder welfare during rearing because there was no evidence that feelings of hunger were reduced nor did dividing the daily ration into two portions fed at different times of the day.

Genetic solutions have been investigated but, given that intake is primarily driven by nutrient requirements, the only real way to breed for low intake is to breed for low performance.

A major problem of low food intake in intensive livestock is after abrupt weaning in pigs. To be separated from the sow, transported to another building (or even another farm) and possibly mixed with unknown piglets is thought to be extremely stressful even when piglets have experienced solid food (creep food) and separate drinking water whilst still suckling. It takes several days for a level of food intake to be attained that will support the rate of growth achieved just before weaning. During this time the animal is undernourished and unprofitable, and its welfare is certainly not ideal. Understanding of what happens at weaning, such as provided by German *et al.* (Chapter 4, this volume), will surely help to improve the feeding and management of the young piglet. Creep food intakes are very variable between piglets, even in the same litter, and the evidence that creep feeding increases intake after weaning is uncertain – in some studies lighter piglets eat more creep, in other studies, heavier ones do (Nielsen *et al.*, Chapter 10, this volume). Again, there is considerable variation in the speed with which piglets start to eat after weaning and considerable

research effort is being expended to resolve the factors affecting the acceptance of solid food by young pigs.

Choice Feeding

There has been considerable misunderstanding about the ability of animals to select in a nutritionally wise manner from a choice of foods on offer. Forbes (Chapter 7, this volume) discusses the conditions necessary for animals to select a balanced diet; many 'experiments' on diet selection have not observed these guidelines and have ended up not demonstrating anything in particular, never mind helping us to understand how such selection is controlled.

Houpt (Chapter 12, this volume) includes a great deal on horses' choice between grasses, herbs and shrubs, but there is no information on which to base an explanation of the underlying drives. However, an example is given of horses being taught to avoid locoweed by giving them lithium chloride by gavage soon after they ingested this plant, thereby generating a learned aversion. As locoweed is itself toxic (otherwise why would it be necessary to teach horses to avoid it?), it is not clear why the animals did not simply learn to avoid this plant after feeling ill on first sampling it. As locoweed contains a neurotoxin and subclinical signs of poisoning take many days to become apparent, it may well be that the necessary learned association between eating the plant and feeling discomfort is not generated. The use of medicinal plants to control helminth and other infections is of great current interest, given the progressive ban in Europe of antibiotics and other drugs in animal production.

Sheep and goats on varied grazing (including trees and shrubs) can show very high intakes, which seem to be driven by the dietary diversity (Agreil *et al.*, Chapter 17, this volume). Is offering a choice itself a stimulant to intake or does choice offer the opportunity to balance the diet and thus maximize intake? How would we design experiments to answer this question?

As far as practical application of choice feeding is concerned, Gidenne and Lebas

(Chapter 11, this volume) conclude that regulation of intake in a free choice situation is difficult to predict and therefore in 'most practical situations of rabbit production the utilization of a complete balanced diet is advisable'. A more positive but still cautionary note is adopted with regard to ostriches, which are able to select higher-quality foods when given a choice: 'However, choice feeding on a commercial scale is not a trivial endeavour' (Brand and Gous, Chapter 9, this volume).

Given the practical difficulties in choice feeding, but recognizing the possibility that animals can tell us what is from their point of view an optimum diet, might there be a place for trials with a small proportion of the animals or birds penned in one corner of the shed being choice-fed? Their preferences could then be applied to the formulation of the single food given to the rest of the animals.

Philosophy and Modelling

Prevalent in the literature is the idea that food intake by ruminants is controlled to provide the animal's energy 'requirement' unless physical limitation of rumen capacity intervenes; this is likely to apply to other animals, including the rabbit (Gidenne and Lebas, Chapter 11, this volume). This two-phase hypothesis (TPH) is criticized by Pittroff and Soca (Chapter 16, this volume) and its shortcomings are highlighted. They champion Weston's concept (1996) of ruminants being able to regulate rumen load according to energy deficit and deduce that 'prediction of feed intake is neither possible from knowledge of animal properties nor from knowledge of feed properties *alone*'. I agree with the latter statement, but do not think it is necessary to postulate that energy deficit directly controls rumen load (see below).

Optimal foraging theory suggests that animals seek to maximize intake of dietary energy. However, Agreil *et al.* (Chapter 17, this volume) argue that sheep 'adopt a stabilizing feeding strategy, modelled by stabilization functions rather than maximization functions'. How this long-term stability is achieved, when there are large, apparently random fluctua-

tions in meal size and interval, and in daily intakes, is still a mystery.

Readers may have found difficulty in the early parts of the chapter of Provenza and Villalba (Chapter 13, this volume) when they deal with the philosophical background to food choice. It is worth persevering to understand the whole chapter, as later on there are numerous examples of research supporting their general lines of argument. Consider, as an example of the type of evidence put forward, that sheep choose to eat more polyethylene glycol (PEG), a substance that attenuates the aversive effects of tannins, the higher the concentration of tannins in the diet; they choose PEG over other similar substances that do not neutralize tannins, following a meal high in tannins; they forage in locations where they have learned PEG to be present, rather than where it is absent, when offered nutritious foods high in tannins in different locations.

'Creatures seek positive reinforcers' and 'avoid negative reinforcers' (Provenza and Villalba, Chapter 13, this volume). For some substances in the diet small doses can be beneficial and therefore act as positive reinforcers, while larger doses can be toxic and act as negative reinforcers. Such a continuum of effect is at the core of the debate on the control of food intake and choice. Pittroff and Soca (Chapter 16, this volume) question the generality of such a continuum because major support comes from the use of graded doses of lithium chloride, which they claim is 'unnatural'. But what is natural? If an animal comes across a toxin (or nutrient) that it or members of its family or even of its species have not come across before, surely it will be able to learn the appropriate response, depending on its physiological requirements.

There is clear evidence of a continuum of effects of infusion of casein (a 'natural' protein) into the rumen (Arsenos and Kyriazakis, 1999), in which small doses given to sheep on a protein-deficient diet induce preferences for the food flavour paired with the casein (because it alleviates protein deficiency), whereas large doses induce aversion (because an excess of protein intake is toxic).

The quotation I used at the beginning of this chapter continues: '[A]nimals aren't machines, set points never are, and feedback changes animal behaviour from conception to the death of the individual and among individuals across generations' (Provenza and Villalba, Chapter 13, this volume). The way in which animals organize their feeding, and their choice between foods, is not according to fixed set points (homeostasis), but according to the balance between a multitude of factors, many of which change with time, for which the term homeodynamics has been coined.

Palatability is a word that frequently crops up in discussions of food intake and choice. But what is palatability? Provenza and Villalba (Chapter 13, this volume) discuss the uselessness of the word, and I prefer not to use it unless I have a clear concept of all the factors that determine how much, and what mixture, of food a particular animal eats. Those whose primary interest is in animals are likely to consider palatability to be a function of the current state of the animal, while plant scientists are certain that it depends on the chemical and physical properties of the food(s). A true appreciation is not even obtained by a combination of these two approaches, but depends also on acknowledging that the history of interactions between this animal and this food(s) is necessary for the full understanding of why an animal chooses to eat more or less of a particular food.

Adding a flavour to a food and expecting this to lead to persistent changes in the weight of food eaten is not realistic. One example is that addition of an appetizer or a repellent to a normal food has a short-term effect on food intake by rabbits – even though when given a choice of foods with and without these substances, the preference is strongly for the one with appetizer or against the one with repellent (Gidenne and Lebas, Chapter 11, this volume). There must be some 'post-ingestive consequences associated with a particular flavour' (Houpt, Chapter 12, this volume) if horses (and other animals) are to learn to prefer or avoid particular foods. Learning to associate the sensory properties of a particular food with its post-ingestive consequences is key to successful food choice and, I would argue, is also important in determining what quantity of food is eaten.

Integration of Factors Affecting Intake and Welfare

Provenza and Villalba (Chapter 13, this volume) conclude that 'the conventional univariate focus mainly on energy must be replaced with multivariate approaches that recognize multiple biochemical interactions'. One way of combining more than one factor affecting intake is multiplication. However, this introduces interaction between independent variables, presents 'non-trivial requirements (usually not met) for model testing' (Pittroff and Soca, Chapter 16, this volume) and reduces the usefulness of extrapolation beyond observed data. Additivity of effects is a simpler way of combining many factors, supported by research results, and Forbes (Chapter 7, this volume) has outlined the Minimal Total Discomfort (MTD) principle, whereby more than one nutrient deficiency or excess can be combined additively to provide an overall picture of the imbalance between the animal and its food. I now propose that the same principle can be used to include discomforts additional to those that are nutrient-related. Note that it has been shown that multiple stressors increase heterophil/lymphocyte ratio and plasma corticosterone in an approximately additive manner (McFarlane and Curtis, 1989).

We will take an example in which three variables are causing discomfort and contributing to poor welfare: protein intake, environmental temperature and space allowance. Each of these has an optimum, the combination of which can be shown in 3D space (Fig. 19.1) as the 'desired' point. The current situation is shown as 'current'; the animal is overeating protein (perhaps because the dietary protein content is high in relation to energy concentration), is cold-stressed and is allowed less space than it requires for comfort. The most direct route from 'current' to 'desired' is along the straight line joining the two, and the length of this can be calculated from the square root of the sum of squares of the three individual deviations: dP, dE and dS (Pythagoras). However, in order to estimate the total discomfort the animal is suffering from these three factors, we need to express each in a common currency. As a simple expedient

I chose to express the deviation between current and desired as a proportion of the desired value.

Thus:

$$\text{Total discomfort} = \sqrt{\sum_{j=1}^{i} \left[\left(d_j - c_j \right) / d_j \right]^2} \quad (19.1)$$

where d is the desired value and c is the current value of each of j variables affecting welfare. This Pythagorean approach gives discomforts that are always positive, whether more or less than the optimal quantity of a resource is currently available; further, by squaring, large deviations have a disproportionately greater effect than small ones, in line with physiological expectations.

For example, for a current protein intake of 12 g/day and a desired intake of 10 g/day the deviation is 2 g/day and the proportional deviation is $2/12 = 0.17$. In order to quantify the effects of environmental temperature and space allowance we have to provide optimal values and ranges of possible values for each factor. It would be useless, for example, to take absolute zero as the starting point of the temperature scale and to say that an animal in a temperature of 293K (20°C), whose optimal environmental temperature was 303K (30°C), had a proportional deviation of $(303-293)/303 = 0.033$. However, if a lethally cold temperature was 0°C, we could use the range 0–30°C and calculate proportional deviation to be $(30-20)/30 = 0.33$, which seems much more likely than 0.033. Similarly, for space

allowance, a minimum realistic space might be the area that the animal itself occupies, say 0.1 m², while the optimum might be that space which it is prepared to work to attain, say 1 m². The range from minimal to optimal is thus 0.9 m², and an animal with a space allowance of 0.2 m² has a proportional deviation of $(0.9-0.2)/0.9 = 0.77$. The squared deviations for protein, temperature and space are, respectively, 0.029 (0.17^2), 0.11 (0.33^2) and 0.59 (0.77^2), the sum of which is 0.729. The square root of this, total discomfort, is 0.854. The extent to which the animal can minimize this is limited to eating more or less food, unless it has some freedom to seek a different environment, either in terms of temperature or space, or both. Of course this example is too simple – a fuller treatment would take into account the greater susceptibility of underfed animals to cold environments, interactions between dietary energy and protein (Forbes, 2005), increased space requirements and lower critical temperatures of fatter animals, and so on.

The calculations above suggest that space allowance has the largest effect on discomfort. Increasing this to 0.55 m² greatly reduces total discomfort to 0.14. This compares with an MTD of 0.22 for the dietary imbalance situation described earlier in this book (Forbes, Chapter 7, this volume).

Previous examples of the theory of MTD have only included food-related factors. Here I have attempted to extend it to include any

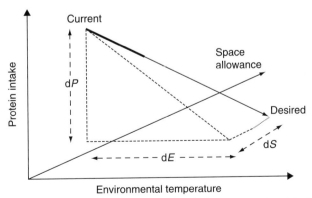

Fig. 19.1. Diagram of Pythagorean calculation of distance between current and desired position of an animal in relation to protein intake, environmental temperature and space allowance.

factor affecting an animal's welfare. Of course, all of this is highly speculative and time will tell whether it provides any advantage in solving 'the problem . . . to find the right parameters, variables and indicators and the right solutions using firm biological, psychological and statistical methods' (Koene, Chapter 6, this volume). Could such methodology be used to provide indices of welfare to be used in comparing different systems of production, e.g. battery vs. deep litter for poultry?

Grazing

Agreil *et al.* (Chapter 17, this volume) emphasize the need for animals used in studies to be fully familiar with the grazing situation and to be shepherded by their usual shepherds, not by research staff. Their method of direct observation by the human eye picks up subtleties of behaviour missed by automated recording methods but the great difficulty of estimating bite size still remains. They also emphasize the need to study the 'fourth dimension of the grazing process', i.e. time, as the system is dynamic, with interactions between animals and herbage. In addition, the social environment must be taken into account as lessons learned by one animal can be passed to others in a group, especially from mother to offspring (Provenza and Villalba, Chapter 13, this volume). Horses eat more readily when they can see another horse eating (Houpt, Chapter 12, this volume).

Despite the very detailed studies that have been carried out on variation in bite size and frequency, and therefore on intake rate (IR), in temperate grazing situations, there is still no comprehensive model available to guide grazers (Gordon and Benvenutti, Chapter 15, this volume).

A common method of increasing the protein content of grazed herbage is to sow mixtures of grass and clover. Small ruminants are capable of fine differentiation and selection between different materials in the pasture, whereas cattle have broad mouths and cannot make such selection. However, sowing grass and clover in alternating strips leads to higher selection for clover than where it is sown in an intimate mixture with ryegrass, with the implication that the animals were able to exercise selection in order to meet better their nutrient requirements. Further empirical research and dynamic modelling is required to allow such systems to be optimized for commercial use (Gordon and Benvenutti, Chapter 15, this volume).

With the increased interest in free-range production of pigs and poultry, questions such as 'How much nutrition do poultry and pigs obtain from grazing?' become important. Clearly it depends on herbage quantity and quality, and on stocking density. In UK poultry systems there should be 'provision of adequate suitable, properly managed vegetation, outdoor scratch whole grain feeding, a fresh supply of water and overhead cover, all sufficiently far from the house to encourage the birds to range' (Defra, 2002) – as usual with these codes of practice, it is not possible to give absolute levels. Do nutrients obtained from herbage even compensate for the extra demands of living outdoors, including exercise and additional heat production in cold weather? Intakes of pasture are likely to be much lower in intensive free-range systems than in village hens in developing countries, both due to the free availability of balanced concentrate foods in the former and to the greater freedom to roam and greater quantity and variety of browse on offer in the latter.

There is much uncertainty about the intake from pasture and Glatz *et al.* (Chapter 18, this volume) state that '[i]t is difficult to measure the intake of foraging pigs and poultry due to the lack of an appropriate method'. However, N-alkanes have been widely used as indigestible markers in ruminants (Mayes and Dove, 2000), and it is difficult to understand why my comprehensive search for publications on their use in pigs and poultry yielded nothing although there is published work with rabbits, pigeons and Galapagos tortoises.

Measurement of herbage intake by individual chickens would also provide information on between-animal variation. Various experimental results are quoted by Glatz *et al.* (Chapter 18, this volume), some estimating up to half the nutrient requirements taken as pasture intake. But this is very variable and unreliable as a guide to commercial practice, as in intensive free-range systems some birds rarely or never leave the house (personal observation).

Ostriches graze veld, even to destruction in extreme cases, but 'little research has been conducted on the utilization of natural veld as a feed source' (Brand and Gous, Chapter 9, this volume).

Conclusion

Although much of the research covered in this book is not designed to have immediate impact on the commercial use of animals, there are many lessons to be learned, and even more pointers to where further research is needed.

A case in point is plant–herbivore interactions in tropical pasture (Gordon and Benvenutti, Chapter 15, this volume), where little is known about the influence of plant structural characteristics. The effect is likely to be large 'and management interventions could have dramatic effects on improving sward structure and longevity, and ultimately animal performance. Given the expanding demand for ruminant livestock products in the tropics, because we would urge researchers to fill the gap in knowledge urgently!'.

Mathematical modelling is likely to continue to play a key role in our developing ability to predict the outcomes of management decisions and to improve the efficiency of animal production. Questions such as 'Is it better to increase or decrease protein content of feed in hot climate?' have no simple answer, but the combination of circumstances favouring either option can be pinpointed by consideration of many of the disparate factors involved. Models must be suited to their purpose: too detailed, and they will be impossible to parameterize; too simple, and they will not represent sufficient variables to be meaningful. Biologists and animal scientists, working closely with mathematicians where necessary, are likely to have the breadth of knowledge necessary to model effectively. If, for no other reason, attempts to model animal feeding systems highlight those areas where we still have insufficient information, qualitative and/or quantitative, to describe adequately the system under consideration.

The very first sentence of this book reads: 'The primary concern of all animals is to find and ingest food to cover their dietary needs.' The primary concern of scientists involved in this field is to use multiple approaches to enable a synthesis of sufficient detail to understand how animals control their feeding behaviour, food intake and dietary choice. This book is an excellent step along the road to achieving that goal.

References

Arsenos, G. and Kyriazakis, I. (1999) The continuum between preferences and aversions for flavoured foods in sheep conditioned by administration of casein doses. *Animal Science* 68, 605–616.

Brouns, F., Edwards, S.A. and English, P.R. (1995) Influence of fibrous feed ingredients on voluntary intake of dry sows. *Animal Feed Science and Technology* 54, 301–313.

Campfield, L.A. and Smith, F.J. (1986) Functional coupling between transient declines in blood glucose and feeding behavior: temporal relationships. *Brain Research Bulletin* 174, 427–433.

Campfield, L.A., Brandon, P. and Smith, F.J. (1985) On-line continuous measurement of blood glucose and meal pattern in free-feeding rats: the role of glucose in meal initiation. *Brain Research Bulletin* 14, 605–616.

Defra (2002) *Code of Recommendations for the Welfare of Livestock: Laying Hens.* Defra, London.

Forbes, J.M. (2005) Voluntary feed intake and diet selection. In: Dijkstra, J., Forbes, J.M. and France, J. (eds) *Quantitative Aspects of Ruminant Digestion and Metabolism.* CAB International, Wallingford, UK.

Mayes, R.W. and Dove, H. (2000) Measurement of dietary nutrient intake in free-ranging mammalian herbivores. *Nutrition Research Reviews* 13, 107–138.

McFarland, D. (1989) Suffering in animals. In: McFarland, D. (ed.) *Problems of Animal Behaviour.* Longman, Essex, UK, pp. 34–58.

McFarlane, J.M. and Curtis, S.E. (1989) Multiple concurrent stressors in chicks. 3. Effects on plasma–corticosterone and the heterophil–lymphocyte ratio. *Poultry Science* 68, 522–527.

Metz, J.H.M. (1975) Time patterns of feeding and rumination in domestic cattle. *Meded Landbouwhogeschool Wageningen* 75, 1–66.

Rhind, S.M., Archer, Z.A. and Adam, C.L. (2002) Seasonality of food intake in ruminants: recent developments in understanding. *Nutrition Research Reviews* 15, 43–65.

Tolkamp, B.J., Allcroft, D.J., Austin, E.J., Nielsen, B.L. and Kyriazakis, I. (1998) Satiety splits feeding behaviour into bouts. *Journal of Theoretical Biology* 194, 235–250.

Weston, R.H. (1996) Some aspects of constraint to forage consumption by ruminants. *Australian Journal of Agricultural Research* 47, 175–197.

Index